Advances in Delivery Science and Technology

Series Editor
Michael J. Rathbone

More information about this series at http://www.springer.com/series/8875

Camilla Foged • Thomas Rades
Yvonne Perrie • Sarah Hook
Editors

Subunit Vaccine Delivery

Springer

Editors
Camilla Foged
Department of Pharmacy
Faculty of Health and Medical Sciences
University of Copenhagen
Copenhagen, Denmark

Thomas Rades
Department of Pharmacy
Faculty of Health and Medical Sciences
University of Copenhagen
Copenhagen, Denmark

Yvonne Perrie
Pharmacy School
School of Life and Health Sciences
Aston University
Birmingham, UK

Sarah Hook
Division of Health Sciences
School of Pharmacy
University of Otago
Dunedin, New Zealand

ISSN 2192-6204 ISSN 2192-6212 (electronic)
ISBN 978-1-4939-1416-6 ISBN 978-1-4939-1417-3 (eBook)
DOI 10.1007/978-1-4939-1417-3
Springer New York Heidelberg Dordrecht London

Library of Congress Control Number: 2014954730

Printed on acid-free paper

Springer is part of Springer Science+Business Media (www.springer.com)

Preface

The objective of this book is to compile the concepts essential for the understanding of the pharmaceutical science and technology associated with the delivery of subunit vaccines. The book's goal is to provide a comprehensive overview of the scientific and regulatory challenges facing scientists who research and develop subunit vaccines. The scope of the book is wide. It is written in a manner that will enlighten newcomers to the field (e.g., Ph.D. students or experienced scientist switching fields) yet provides an in-depth knowledge that would benefit a skilled worker in the field.

A significant improvement in the safety of modern vaccines has been the development of subunit vaccines, as these are composed of very well-defined and highly pure components, often recombinant proteins. However, since protein-based antigens in general are weakly immunogenic by themselves, co-administration of adjuvants is required to induce potent and persistent specific immune responses. In recent years, there has been substantial progress in the discovery of new efficient adjuvants for subunit vaccines that are often classified into delivery systems (e.g., liposomes, emulsions, and polymeric nanoparticles) and immunopotentiating compounds that constitute pathogen-associated molecular patterns, such as the toll-like receptor ligands. The combination of delivery systems and immunopotentiators has created highly efficacious adjuvants due to concomitant enhanced antigen delivery and potent stimulation of immunity. Many of these adjuvants are of a particulate nature and mimic the structure and/or composition of microbes in a reductionist fashion. Examples are liposomes, polymeric nanoparticles, emulsions, and virus-like particles. However, there are a substantial number of pharmaceutical challenges associated with the subunit vaccine development process due to the complex nature of the antigen–adjuvant combinations. These challenges will be presented and discussed in this book.

Copenhagen, Denmark — Camilla Foged
Copenhagen, Denmark — Thomas Rades
Birmingham, UK — Yvonne Perrie
Dunedin, New Zealand — Sarah Hook

Contents

Contributors

Farah Al-Barwani Department of Microbiology and Immunology, School of Medical Sciences, University of Otago, Dunedin, New Zealand

Department of Pathology, School of Medicine, University of Otago, Dunedin, New Zealand

Mansoor M. Amiji Department of Pharmaceutical Sciences, School of Pharmacy, Northeastern University, Boston, MA, USA

Patric Baumann Department of Chemistry, University of Basel, Basel, Switzerland

Ben J. Boyd Faculty of Pharmacy and Pharmaceutical Sciences, Monash University, Melbourne, VIC, Australia

Luis A. Brito Novartis Vaccines & Diagnostics, Cambridge, MA, USA

Nico Bruns Department of Chemistry, University of Basel, Basel, Switzerland

Adolphe Merkle Institute, University of Fribourg, Marly, Switzerland

Anja Car Department of Chemistry, University of Basel, Basel, Switzerland

Yuan Cheng Department of Discovery Pharmaceutics, Pharmaceutical Candidate Optimization, Bristol-Myers Squibb, Princeton, NJ, USA

Dennis Christensen Department of Infectious Disease Immunology, Vaccine Adjuvant Research, Statens Serum Institut, Copenhagen, Denmark

Braeden Donaldson Department of Microbiology and Immunology, School of Medical Sciences, University of Otago, Dunedin, New Zealand

Department of Pathology, School of Medicine, University of Otago, Dunedin, New Zealand

Julia Engert Department of Pharmacy, Pharmaceutical Technology and Biopharmaceutics, Ludwig-Maximilians-University Munich, Munich, Germany

Camilla Foged Department of Pharmacy, Faculty of Health and Medical Sciences, University of Copenhagen, Copenhagen, Denmark

Henderik W. Frijlink Department of Pharmaceutical Technology and Biopharmacy, University of Groningen, Groningen, The Netherlands

Olivier Gasser Malaghan Institute of Medical Research, Wellington, New Zealand

Sarah Gordon Helmholtz Institute for Pharmaceutical Research Saarland (HIPS), Saarland University, Saarbrücken, Germany

Mette Hamborg Department of International Health, Microbiology and Immunology, Centre for Medical Parasitology, Faculty of Health and Medical Sciences, University of Copenhagen, Copenhagen, Denmark

Ian F. Hermans School of Biological Sciences, Victoria University Wellington, Wellington, New Zealand

Andrew J. Highton Department of Microbiology and Immunology, University of Otago, Dunedin, New Zealand

Wouter L.J. Hinrichs Department of Pharmaceutical Technology and Biopharmacy, University of Groningen, Groningen, The Netherlands

Sarah Hook School of Pharmacy, University of Otago, Dunedin, New Zealand

Anke L.W. Huckriede Molecular Virology Section, Department of Medical Microbiology, University of Groningen, Groningen, The Netherlands
University Medical Centre Groningen, Groningen, The Netherlands

Henriette B. Hübschmann Department of Pharmacy, Faculty of Health and Medical Sciences, University of Copenhagen, Copenhagen, Denmark

Pål Johansen Department of Dermatology, University Hospital Zurich, Zurich, Switzerland

Sangeeta B. Joshi Department of Pharmaceutical Chemistry, Macromolecule and Vaccine Stabilization Center, University of Kansas, Lawrence, KS, USA

Rie S. Kallerup Department of Pharmacy, Faculty of Health and Medical Sciences, University of Copenhagen, Copenhagen, Denmark

Elisabeth Kastner School of Life and Health Sciences, Aston University, Birmingham, UK

Roslyn A. Kemp Department of Microbiology and Immunology, University of Otago, Dunedin, New Zealand

Thomas M. Kündig Department of Dermatology, University Hospital Zurich, Zurich, Switzerland

Erik B. Lindblad Brenntag Biosector, Frederikssund, Denmark

Deborah Lowry Aston Pharmacy School, Aston University, Birmingham, UK

Wolfgang Meier Department of Chemistry, University of Basel, Basel, Switzerland

C. Russell Middaugh Department of Pharmaceutical Chemistry, Macromolecule and Vaccine Stabilization Center, University of Kansas, Lawrence, KS, USA

Hanne M. Nielsen Department of Pharmacy, Faculty of Health and Medical Sciences, University of Copenhagen, Copenhagen, Denmark

Martin Nussbaumer Department of Chemistry, University of Basel, Basel, Switzerland

Derek T. O'Hagan Novartis Vaccines & Diagnostics, Cambridge, MA, USA

Michele Pallaoro Novartis Vaccines, Siena, Italy

Yvonne Perrie School of Life and Health Sciences, Aston University, Birmingham, UK

Thomas Rades Department of Pharmacy, Faculty of Health and Medical Sciences, University of Copenhagen, Copenhagen, Denmark

Teerawan Rattanapak School of Pharmacy, University of Otago, Dunedin, New Zealand

Shakila B. Rizwan New Zealand's National School of Pharmacy, University of Otago, Dunedin, New Zealand

Regina Scherließ Department of Pharmaceutics and Biopharmaceutics, Kiel University, Kiel, Germany

Signe T. Schmidt Department of Pharmacy, Faculty of Health and Medical Sciences, University of Copenhagen, Copenhagen, Denmark

Thomas Schuster Department of Chemistry, University of Basel, Basel, Switzerland

Sarah Scullion Department of Microbiology and Immunology, School of Medical Sciences, University of Otago, Dunedin, New Zealand

Ruchi R. Shah Department of Pharmaceutical Sciences, School of Pharmacy, Northeastern University, Boston, MA, USA

Justin C. Thomas Eli Lilly and Company, Indianapolis, IN, USA

Wouter F. Tonnis Department of Pharmaceutical Technology and Biopharmacy, University of Groningen, Groningen, The Netherlands

David B. Volkin Department of Pharmaceutical Chemistry, Macromolecule and Vaccine Stabilization Center, University of Kansas, Lawrence, KS, USA

Vernon Ward Department of Microbiology and Immunology, School of Medical Sciences, University of Otago, Dunedin, New Zealand

Jitinder S. Wilkhu School of Life and Health Sciences, Aston University, Birmingham, UK

Alexander Wilkinson School of Life and Health Sciences, Aston University, Birmingham, UK

Sarah Young Department of Pathology, Dunedin School of Medicine, University of Otago, Dunedin, New Zealand

Vivienne Young Department of Microbiology and Immunology, School of Medical Sciences, University of Otago, Dunedin, New Zealand

About the Editors

Associate Professor Camilla Foged is Head of the Vaccine Design and Delivery Group in Section for Biologics, Department of Pharmacy, University of Copenhagen, Copenhagen, Denmark. She has an M.Sc. in Biochemistry from University of Copenhagen. Dr. Foged attained her Ph.D. from The Danish University of Pharmaceutical Sciences. Her main research interest is advanced drug delivery aiming at designing new vaccine and nucleic acid delivery systems to improve therapy. Her group efficiently has addressed drug delivery challenges with cutting-edge technologies, which has enabled high impact publications and innovative solutions in drug delivery. The group's research goal is to improve disease prevention and treatment by designing nanoparticulate vaccine and nucleic acid formulations through in-depth mechanistic knowledge of how the physicochemical properties of the particles affect their interaction with the environment, e.g., in formulation, in vitro, and in vivo. Dr. Foged is interested in lipid- and polymer-based nanoparticulate delivery systems using various approaches for targeting and membrane destabilization.

Professor Thomas Rades is the Research Chair in Pharmaceutical Design and Drug Delivery at the Department of Pharmacy, University of Copenhagen, Copenhagen, Denmark. Dr. Rades obtained his Ph.D. from the University of Braunschweig, Germany, and he has developed an international reputation for his research on solid dosage forms as well as in vaccine delivery using nanoparticulate systems. Research in both areas aims to improve drug therapy through appropriate formulation and physicochemical characterization of medicines and vaccines. It combines physical, chemical, and biological sciences and technology to optimally formulate drugs and vaccines for human and veterinary uses. Professor Rades has published more than 270 papers in international peer reviewed journals.

Professor Yvonne Perrie is Head of Pharmacy and Chair in Drug Delivery within Aston Pharmacy School, Aston University, Birmingham, UK. Dr. Perrie attained her Ph.D. from the University of London under the supervision of Professor Gregoriadis. Her research is multidisciplinary and is focused on the development of drug carrier systems to facilitate the delivery of drugs and vaccines, providing practical solutions for current healthcare problems. Dr. Perrie is a Fellow of the Society of Biology and an Eminent Fellow of the Academy of Pharmaceutical Sciences. Externally, she is a Director-at-large for the Controlled Release Society and Editor of the Controlled Release Society Newsletter. She is Editor-in-Chief of the *Journal of Liposome Research* and *Pharmaceutics*, and she is associate editor for the *Journal of Drug Targeting* and the *Journal of Pharmacy and Pharmacology*.

Professor Sarah Hook is the Professor of Biopharmaceutics of the School of Pharmacy at the University of Otago, Dunedin, New Zealand, where she obtained her Ph.D. in immunology. Her area of interest is in the development and testing (physical, chemical, and immunological) of novel formulations for the delivery of small and large molecule drugs for the prevention and treatment of both communicable and non-communicable diseases.

Part I
Background

Chapter 1
Immunological Background

Andrew J. Highton and Roslyn A. Kemp

1.1 Introduction

In the eighteenth century, Lady Mary Wortley Montagu, while living in Turkey, observed the local practice of variolation to protect against smallpox (Dinc and Ulman 2007). This involved transferring live smallpox virus harvested from smallpox blisters of a patient to another person. Lady Montagu brought this principle home to Britain to mixed success—3 % of patients contracted the virus and died, although this was still lower than the 20–40 % mortality caused by natural infection.

By the end of the eighteenth century, several researchers had tested the idea that a related virus could induce protection against smallpox—this was primarily noted in the low infection rates of those women who were in close contact with cowpox-infected cows. In 1775, Benjamin Jesty used cowpox virus on his family during a smallpox outbreak (Pead 2003), but it was not until Edward Jenner tested the same principle in a series of patients in 1796 (Jenner 1798) that the idea of vaccination was widely understood and accepted.

Over the next 200 years, vaccination was used against many diseases and was ultimately successful in eradicating smallpox entirely (Breman and Arita 1980). It is arguably the most clinically and cost-effective public health measure ever introduced. Live virus vaccines, such as that for smallpox, can be very effective but because of the associated dangers, many researchers now concentrate on using only parts of infectious organisms in vaccines, which can induce a similar immune response as to the original pathogen.

A.J. Highton • R.A. Kemp (✉)
Department of Microbiology and Immunology, University of Otago,
PO Box 56, Dunedin, New Zealand
e-mail: andrew.highton@gmail.com; roslyn.kemp@otago.ac.nz

C. Foged et al. (eds.), *Subunit Vaccine Delivery*, Advances in Delivery Science and Technology, DOI 10.1007/978-1-4939-1417-3_1

Despite the success of vaccination against many pathogens in preventing infection or reducing disease course, effective vaccines generating long-lived immunity to many other pathogens have yet to be developed. The problems faced by vaccinologists include (1) complexity of the pathogens (e.g., plasmodium causing malaria), (2) high mutation rates of pathogens (e.g., HIV and influenza), and (3) the ability to generate the right type of immune response (e.g., a cell-mediated response rather than an antibody-mediated response). The development of therapeutic vaccines, i.e., those that induce an immune response to a noninfectious disease such as cancer is another area of intense interest.

The immune response to pathogens has three characteristics – inducibility, specificity, and memory – these characteristics are also important for vaccine development. Inducibility refers to the fact that an immune response is quiescent but can be invoked quickly and effectively upon infection or vaccination. Specificity refers to the ability of the immune system to respond to specific pathogens (or vaccines) – and thus ensures that an immune response is ideal for fighting off a particular infection. The memory component of an immune response is its ability to respond more quickly and effectively upon re-exposure to an infection. Vaccination exploits all three characteristics – it induces an immune response; this response is specific to the antigen(s) administered in the vaccine; and the vaccine serves to prime the immune response and generate memory to allow for quicker response upon infection with the real pathogen.

1.2 The Innate Immune System

The immune system is broadly divided into the innate and adaptive immune systems. In evolution, the innate immune system developed early and is shared by many vertebrates. Innate responses are generally non-antigen specific but are generated early after exposure to infection and are therefore the first line of defense. In contrast, the adaptive immune system evolved much later and is a targeted response, i.e., antigen specific, and takes a much longer time to become activated (see Sect. 1.5).

The cells of the innate immune system recognize pathogen-associated molecular patterns (PAMPs), rather than specific antigens. They induce inflammation and recruit the adaptive immune system in order to eradicate infection. The innate immune system consists of three key cell types:

Macrophages. These cells circulate through the blood and tissue and remove particulate matter of foreign origin as well as removing damaged or aged host cells by the process of phagocytosis (active uptake of pathogen; Ovchinnikov 2008).

Neutrophils. Neutrophils are polymorphonuclear granulocytes that are recruited to sites of infection by inflammation-induced chemokines (e.g., those produced by macrophages). Neutrophils contain granules that destroy microorganisms and also form neutrophil extracellular traps (NETs) around an infection site to trap pathogens (Brinkmann et al. 2004).

Dendritic cells (DCs). Similarly to macrophages, DCs can phagocytose microorganisms but their main role is to link the innate and adaptive immune systems. Following phagocytosis of a microorganism, DCs can process their components into antigens for presentation to lymphocytes (see Sect. 1.3).

The innate immune system controls whether an adaptive immune response develops and thus must determine if the stimulus is foreign or self, and dangerous or benign. Pathogens are recognized as foreign because they express PAMPs. These are generally a unique structure for a pathogen, e.g., lipopolysaccharide (LPS) in Gram-negative bacteria, flagellin in flagellated bacteria or single-stranded RNA in some viruses (Kawai and Akira 2010). Cells of the innate immune system (and also other host cells such as epithelial cells) express a range of pattern-recognition receptors (PRRs) to detect these PAMPs.

Toll-like receptors (TLRs) were the first PRRs to be discovered and were named based on their structural homology to the Toll receptors used by Drosophila for detecting infection (Janeway and Medzhitov 2002). TLRs form a family of at least ten different receptors, each recognizing a distinct PAMP (Kawai and Akira 2010), although there is considerable redundancy between receptors. TLRs signal via an adaptor molecule such as MyD88, which results in activation of transcription factors and transcription of inflammatory mediators (Akira and Takeda 2004). Other PRRs include NOD-like receptors, RIG-I receptors (intracellular PRRs), and C-type lectin receptors specialized for detection of fungal infections (Kawai and Akira 2010).

The inflammatory response is a key requirement for both the initiation and quality of the subsequent adaptive immune response. Inflammatory cytokines and chemokines produced during the innate response recruit and activate other cells, and therefore inflammation must be carefully controlled during vaccination.

Cells of the innate immune system can discriminate between foreign and self by using PRRs and can also respond to the so-called danger signals. NOD-like receptors (among others) are intracellular PRRs that respond to stimuli indicative of danger—including crystals, microbial peptides, cell death, stress, and some drugs (Franchi et al. 2009). The signaling pathway downstream of these stimuli is a cytoplasm complex of scaffolding and sensing proteins called the inflammasome (Franchi et al. 2009). The inflammasomes (of which there are several types) act to cleave inflammatory pro-cytokines into active cytokines and therefore enhance the inflammatory response (Martinon et al. 2002).

Cytokines are proteins that control the type, duration, and amplitude of an immune response, and form a communication network between cells of the immune system and also between the immune system and other body systems. Cytokines bind to specific cytokine receptors expressed on the surface of cells and therefore the effects of cytokines are dependent on the expression of the receptors. While a balance of cytokines is necessary to maintain immune homeostasis, during infection, an increase in inflammatory cytokines is necessary for induction of an effective immune response. Pro-inflammatory cytokines, such as tumor necrosis factor α (TNF-α), interleukin 6 (IL-6), and IL-1, mediate inflammation by (a) increasing thermoregulation set points (fever), (b) upregulating synthesis of other pro-inflammatory cytokines (positive feedback), (c) stimulating the production of acute phase proteins, and

(d) attracting inflammatory cells (Dinarello 2007). Unregulated inflammation can result in a "cytokine storm," which can be dangerous (Osterholm 2005).

Chemokines are a family of chemoattractant cytokines that, like other cytokines, bind to specific receptors to mediate their effects (Murphy and Baggiolini 2000). Unlike cytokines, chemokines are much less pleiotropic and do not induce the production of other cytokines or chemokines. Chemokines form a concentration gradient for cell migration in an immune response (Moser and Loetscher 2001), and thus chemokines produced by the initially infected cells will recruit other immune cells, such as neutrophils, to the site of infection.

1.3 Antigen Uptake and Presentation

The role of the innate immune system is to recognize an infection, destroy the pathogenic microorganisms, and to activate a specific and more effective adaptive immune response. The adaptive immune system is specific for a defined pathogen, is based on antigen recognition (rather than PAMP recognition) and involves the expansion of lymphocytes expressing receptors specific for antigens.

Pathogens are taken up at the site of infection by macrophages, DCs, and other cells of the innate immune system. They do this via phagocytosis, pinocytosis (passive osmosis of pathogen), or receptor-mediated uptake (Aderem and Underhill 1999).

During phagocytosis, pathogens are destroyed by a decrease in pH and the action of cytotoxic molecules such as reactive oxygen species within phagolysosomes. This process can also result in the degradation of whole proteins into peptides that can be presented to T-cells of the adaptive immune system. DCs are present in small numbers in tissues in contact with the environment, and process and present antigen on their cell surface on major histocompatibility complex (MHC) molecules.

Phagocytosed bacteria are degraded in the acidic environment of the phagolysosomes into short peptides (~20 amino acids; Turley 2000). Vesicles from the Golgi contain MHC class II molecules, with an invariant chain blocking the antigen-binding site. These vesicles fuse with the phagolysosomes and the short peptide antigen replaces the invariant chain. The entire complex then migrates to the cell surface for recognition by $CD4^{+}$ T-cells (Wolf and Ploegh 1995). In contrast, endogenous antigens (e.g., in virus-infected cells) are created by degradation of the pathogen by the proteasome complex into short peptides (~8–10 amino acids). Antigen is transported by transporters associated with antigen processing (TAP) to the Golgi where it is loaded onto assembling MHC class I molecules. These peptide–MHC complexes are then transported to the surface of the cells for recognition by $CD8^{+}$ T-cells. In reality, there is significant variation in the MHC pathways for different pathogens; for example, exogenously acquired proteins can be loaded onto MHC class I molecules in a process called cross-presentation (Amigorena and Savina 2010). Most pathogens will initiate both a $CD4^{+}$ and $CD8^{+}$ T-cell response.

1.4 Generation of T- and B-Cells

In order to activate an adaptive immune response, lymphocytes must be able to recognize antigen, either in the context of MHC (T-cells) or as a native antigen (B-cells). An antigen can be defined as a processed macromolecule and is recognized by both T- and B-cells. The antigenic determinant (epitope) is the part of the antigen that binds to a specific receptor on T-cells or B-cells. To induce an immune response, antigenicity is determined by a combination of size, complexity, physical form (e.g., aggregates or soluble), dose, route, and degree of difference to self (Wang et al. 2011).

Because of the breadth of pathogens that must be recognized by the immune system, it would be impossible to have effector T- and B-cells present in sufficient numbers in the body to eliminate all possible pathogens at any one time. The human genome contains around 25,000 genes, but humans have around 50 million different receptors on T-cells or B-cells. T- and B-cells use a system of genetic recombination and clonal expansion so that the specific immune response can be induced only when needed.

The B-cell receptor (BCR) is a surface protein of the immunoglobulin superfamily. Following activation, this receptor can be produced in a secreted form as antibody, thus B-cells produce a soluble version of the specific antigen receptor. The receptor consists of light and heavy chains (bound to the membrane of the cell as BCR). Both chains contribute to the antigen specificity of the receptor (variable region) while only the heavy chain contributes to the core structure (constant region). Each heavy chain and light chain (of which there are two types, κ and λ) is made from a combination of V (variable), C (constant), and J (joining) genes (Market and Papavasiliou 2003). There are a variable number of each of these genes present in the genome and recombination of these genes for one heavy and one light chain creates one of approximately 50 million possible combinations for the structure of the BCR. Thus, a population exists with the ability to respond to several million possible antigens.

B-cells are also controlled by allelic exclusion, whereby only one allele of each of the heavy and light chains is arranged—if this is successful, the second allele is turned off (Pernis et al. 1965). If unsuccessful, the second allele will recombine, and if both are unsuccessful the cell dies. This prevents expression of multiple antigen receptors on one cell, and reduces the likelihood of an autoimmune response.

The T-cell receptor is of a similar structure to the B-cell, and consists of an α chain and a β chain, with a variable region (for antigen recognition) and a constant region. Both α and β chains are coded for by a combination of V, C, D, and J genes and this is handled in a similar way to the recombination of the genes involved in the BCR.

T-cells undergo further development in the thymus, where potentially self-reactive cells are eliminated. In the thymus, new T-cells are exposed to self-peptides presented by specialized antigen-presenting cells (APCs) (Klein et al. 2009). If the TCR binds strongly to the host MHC, and will therefore have the ability to be activated in the periphery by APCs presenting foreign antigen, the cell will survive. Cells binding weakly to host MHC will die “by neglect,” i.e., will not receive a survival signal. Cells binding strongly to self-antigen receive a death signal and are destroyed, to prevent autoimmune responses in the periphery.

1.5 Activation of T- and B-Cells

Following infection, DCs traffic to the local lymph nodes to present antigens to the pool of circulating T-cells. T-cells with a receptor specific for antigen will bind to the APC presenting antigen in the context of MHC. Binding of CD4 and CD8 molecules on T-cells to MHC class II and class I, respectively, on APCs stabilizes this interaction (and provides a signaling component; Artyomov et al. 2010; Wooldridge et al. 2005). A co-stimulatory signal, via CD28 on T-cells and CD80/86 on APCs, provides a second signal to activate T-cells (Nurieva et al. 2009). Finally, the inflammatory milieu of cytokines produced by APCs activates T-cells. The end result is antigen-specific T-cells stimulated to divide and acquire effector functions, including the ability to traffic to the site of infection.

B-cells can respond to native (unprocessed) antigen and can also act as APCs. B-cell activation requires help from T-cells via both cytokine production and ligation of CD40. B-cells binding antigens upregulate expression of CD40 on the surface and this binds to CD40L expressed on T-cells activated by the same antigen. Cytokines produced by activated T-cells help to activate B-cells and trigger antibody production (see Sect. 1.6). In addition, B-cells acquiring antigen can present it via MHC to T-cells and be directly activated by cytokines from the responding T-cells. B-cells recognizing native antigen can also be activated without these extra T-cell signals (T-cell independent activation).

1.6 Effector Functions of B-Cells

The primary effector function of B-cells is the production of antibodies. After activation, B-cells clonally proliferate and produce antigen-specific antibodies. Antibodies must be able to clear pathogens that invade different sites in the body; therefore, there are different classes of antibodies. These classes retain the antigen-specificity of the original B-cells and are thus specific to the pathogen.

Following activation of B-cells, a process called class switching can occur. In this process, the constant region of the BCR is changed (Malisan and Brière 1996). During B-cell proliferation, recombination of the C genes occurs and the constant chain can become a different class—this class switching can result in five different types of antibodies (Schroeder and Cavacini 2010). The class switch decision is influenced by cytokines produced by T-cells. Antibodies of the IgG class enter tissues readily and are therefore the most abundant class, IgA is present in serum and mucosal tissues, IgM is optimized for complement activation and the first class induced upon activation. IgE degranulates mast cells and is therefore responsible for allergic symptoms as well as clearing parasitic infections, and IgD is of function unknown. Each class of antibody is specific for the same antigen as the activated B-cell, but now is able to perform different functions in different sites.

A second improvement in antibody is affinity maturation that occurs through a process called somatic recombination (Chaudhuri et al. 2007). During B-cell proliferation, errors occur in DNA replication, some of which result in small changes in the variable region of the BCR/antibody. These mutations can alter the binding affinity of the receptor to the antigen, and those cells with increased affinity receive a stronger survival signal and are selected to continue dividing. Thus, the B-cell response evolves continuously throughout the immune response to a pathogen.

Antibodies have multiple roles in an immune response—they can bind directly to the pathogen and destroy it, or label it as more accessible for phagocytosis (opsonization). They can neutralize receptors for toxins on microorganisms or bind to pathogens to initiate the complement cascade, an innate immune mechanism that destroys pathogens quickly and efficiently (Schroeder and Cavacini 2010). Long-lived antibody continues to circulate through the body after the immune response clears the pathogen, and is therefore readily available to counter a re-infection with the same pathogen. This forms the basis of the bulk of successful vaccines. However, to resolve many infections, a T-cell response is required.

1.7 Effector Functions of T-Cells

T-cells have many roles in an immune response. They produce several types of cytokines that influence antibody class switching, activate, and differentiate other T-cells and cells of the innate immune system, and can lyse infected cells. $CD4^+$ T-cells produce a variety of cytokines that can improve the function of other $CD4^+$ T-cells, as well as $CD8^+$ T-cells and APCs. $CD8^+$ T-cells also produce cytokines, but their primary role is their cytolytic activity. However, both $CD4^+$ and $CD8^+$ T-cells are capable of cytotoxic function and enhanced cytokine production, depending on the pathogen and the inflammatory milieu.

Cytokines produced by T-cells can activate cells of the innate immune system. In particular, interferon γ (IFN-γ) will increase phagocytosis by macrophages and increase expression of MHC molecules on target cells (Schroder et al. 2004). As discussed in Sect. 1.5, T-cell derived cytokines activate B-cells to become antibody-producing cells and also influence antibody class switching. T-cells also influence the development of other T-cells—for example, IL-4 produced by T-cells will bind to IL-4 receptors on other T-cells and induce more IL-4 along with production of IL-5 and IL-13.

$CD8^+$ and, to a lesser extent, $CD4^+$ T-cells are able to kill infected cells and tumors. Because all cells can be infected by a pathogen, all cells express MHC class I and are therefore able to be recognized by antigen-specific $CD8^+$ T-cells at the site of infection. Activated antigen-specific $CD8^+$ T-cells traffic from the lymph node to the site of infection where they recognize antigen in the context of MHC class I presented on infected cells. The three most studied mechanisms of killing include perforin/granzyme, Fas–Fas ligand (FasL) and cytokine-mediated cytotoxicity. When $CD8^+$ T-cells become activated they produce granules containing toxic granzymes and perforin, a protein that forms pores in cell and vesicle walls (Smyth and Trapani 1995). These pores provide entry for granzymes that destroy the cell via a caspase apoptosis pathway (Trapani et al. 1998). Alternatively, FasL is upregulated on the surface of activated $CD8^+$ T-cells and its ligation with Fas on the target cell also results in apoptosis (Yonehara et al. 1989). Finally, cytokines such as IFN-γ can bind to their cognate receptors on the target cells and also induce a caspase pathway leading to apoptosis (Schroder et al. 2004).

At the end of the infection, T-cells produce suppressive cytokines and regulatory T-cells are induced. Regulatory T-cells suppress T-cell responses, and therefore vaccine strategies must take into account any effect on this population (Sakaguchi et al. 1995).

1.8 Immune Memory

At the end of an immune response, the majority of effector cells die, but T and B lymphocytes form memory populations, whereby a small number of antigen-specific cells are retained at a quiescent but easily activated state (Wakim and Bevan 2010). Upon re-infection with the same pathogen, these memory cells respond more quickly and efficiently than during the primary response. Immune memory prevents re-infection or reduces disease severity in a second infection. A memory response is more effective than the primary naïve response in a number of ways. It results in a higher number of cells recognizing the antigen than naïve cells, a faster activation in response to specific antigens than naïve cells, a more effective response than naïve cells (e.g., targeted cytokine production), and the presence of memory cells at the site of infection (resident memory). All of these factors mean that a secondary memory response is significantly more effective than the primary naïve response.

There are two types of memory B-cells—long-lived plasma cells that reside in the bone marrow, receiving survival signals from stromal cells, and producing long lasting high affinity class switched antibodies; and memory B-cells which express surface immunoglobulin but do not secrete antibodies (Klein et al. 1998; Slifka et al. 1998). Upon re-exposure to antigen, memory B-cells quickly proliferate and generate antibody-producing plasma B-cells.

Memory T-cells exist in at least three different activation states—effector memory T-cells circulate through the lymph and tissue and are quickly activated by antigen presented on APCs. Central memory T-cells are much more effective at producing cytokines and granzyme but reside in the lymphoid organs. Finally, resident memory T-cells do not circulate through the body; instead they are present permanently in different sites of infection in the body such as the gut, the lungs, and the skin (Masopust et al. 2001; Sallusto et al. 1999). These cells recruit other effector cells quickly to the site of infection as well as acting against the pathogen themselves (Schenkel et al. 2013).

1.9 Challenges for Vaccination

Pathogen. Microorganisms constantly evolve—the difficulties of finding effective and long-lasting vaccines to viruses such as influenza and HIV is partly due to the mutation rate of these organisms (Barouch et al. 2002). Identifying a specific and constant antigen is often not possible. In rapidly dividing infections, the immune response may drive these mutations.

Many microorganisms, particularly viruses, have evolved immune evasion strategies. These can be homologs of chemokines or proteins that interfere with antigen presentation or lymphocyte activation (Alcami and Koszinowski 2000). Handling these evasion systems must also be factored into vaccine design.

Simple bacteria maintain core structures vital for life and growth. More complex pathogens such as parasites have more complex structures, making it more difficult to identify a simple antigen to use in a vaccine. In addition, many pathogens have life cycles passing through different hosts (e.g., influenza in ducks and pigs; malaria in mosquitoes and humans). This means that antigens expressed in one host may not be the dominant antigen in another host.

Host Immune Response. Antibodies circulate in the blood, but T and B lymphocytes reside in different parts of the body. For a local immune response, it may be necessary to have a resident population of memory cells (Woodland and Kohlmeier 2009). In this situation, vaccination needs to induce both an effector response at the site of infection, and a memory response. How to deliver vaccines to different sites in the body is another challenge for vaccine development.

To date, few adjuvants have been approved for use in the clinic (Mbow et al. 2010). In order to generate the inflammatory response required for an effective immune response, the incorporation of appropriate adjuvants has become increasingly important. Development of new adjuvants for vaccines would involve inducing inflammation (via the inflammasome), differentiation of the immune response towards the right cell type and the right cytokine profile, all of which can be retained in the memory population.

Successful vaccines either prevent infection or inhibit the replication of the infectious organisms and reduce the disease burden. In many situations, the immune pathology resulting from infection can lead to death (e.g., influenza)—the goal of a vaccine can therefore, in these cases, be shifted to reduce pathology or disease progression rather than prevent infection entirely.

Host Populations. The developing world has a higher proportion of young children than other demographics whereas the developed world has an ageing population. Both groups have impaired immune systems that may not respond to vaccines as well as young adults (Effros 2007). Vaccine development must take into account the underdeveloped immune system of young children and the less functional immune response of the elderly.

Another peculiar dichotomy is that part of the world is malnourished and part of the world is obese. The immune system is affected greatly by nutrition intake and both depleted and excess nutrition can influence the type and efficacy of an immune response to a pathogen. To generate a vaccine that works with such an impaired immune system but also on a healthy population will be difficult.

Finally, one third of the world's population is infected with tuberculosis (TB), and many of these people are co-infected with HIV (Sharma 2005). Moreover, other populations are co-infected with parasites or chronic viral infections that may influence the type and duration of immune response to vaccination.

References

Aderem A, Underhill DM (1999) Mechanisms of phagocytosis in macrophages. Annu Rev Immunol 17:593–623. doi:10.1146/annurev.immunol.17.1.593

Akira S, Takeda K (2004) Toll-like receptor signalling. Nat Rev Immunol 4(7):499–511. doi:10.1038/nri1391

Alcami A, Koszinowski UH (2000) Viral mechanisms of immune evasion. Trends Microbiol 8(9):410–418, http://www.ncbi.nlm.nih.gov/pubmed/10989308

Amigorena S, Savina A (2010) Intracellular mechanisms of antigen cross presentation in dendritic cells. Curr Opin Immunol 22(1):109–117. doi:10.1016/j.coi.2010.01.022

Artyomov M, Lis M, Devadas S, Davis M, Chakraborty A (2010) CD4 and CD8 binding to MHC molecules primarily acts to enhance Lck delivery. Proc Natl Acad Sci U S A 107(39):16916–16921. doi:10.1073/pnas.1010568107/-/DCSupplemental.www.pnas.org/cgi/doi/10.1073/pnas.1010568107

Barouch DH, Kunstman J, Kuroda MJ, Schmitz JE, Santra S, Peyerl FW, Krivulka GR, Beaudry K, Lifton MA, Gorgone DA, Montefiori DC, Lewis MG, Wolinsky SM, Letvin NL (2002) Eventual AIDS vaccine failure in a rhesus monkey by viral escape from cytotoxic T lymphocytes. Nature 415(6869):335–339. doi:10.1038/415335a

Breman J, Arita I (1980) The confirmation and maintenance of smallpox eradication. N Engl J Med 303(22):1263–1273, http://www.ncbi.nlm.nih.gov/pubmed/6252467

Brinkmann V, Reichard U, Goosmann C, Fauler B, Uhlemann Y, Weiss DS, Weinrauch Y, Zychlinsky A (2004) Neutrophil extracellular traps kill bacteria. Science 303(5663):1532–1555. doi:10.1126/science.1092385

Chaudhuri J, Basu U, Zarrin A, Yan C (2007) Evolution of the immunoglobulin heavy chain class switch recombination mechanism. Adv Immunol 94(06):157–214. doi:10.1016/S0065-2776(06)94006-1

Dinarello C (2007) Historical review of cytokines. Eur J Immunol 37(Suppl 1):1–19. doi:10.1002/eji.200737772.Historical

Dinc G, Ulman YI (2007) The introduction of variolation "A La Turca" to the West by Lady Mary Montagu and Turkey's contribution to this. Vaccine 25(21):4261–4265. doi:10.1016/j.vaccine.2007.02.076

Effros RB (2007) Role of T lymphocyte replicative senescence in vaccine efficacy. Vaccine 25(4):599–604. doi:10.1016/j.vaccine.2006.08.032

Franchi L, Warner N, Viani K, Nunez G (2009) Function of Nod-like receptors in microbial recognition and host defense. Immunol Rev 227(1):106–128. doi:10.1111/j.1600-065X.2008.00734.x.Function

Janeway CA, Medzhitov R (2002) Innate immune recognition. Annu Rev Immunol 20(2): 197–216. doi:10.1146/annurev.immunol.20.083001.084359
Jenner, E. (1798). An inquiry into the causes and effects of the variolae vaccinae: a disease discovered in some of the western counties of England, particularly Gloucestershire, and known by the name of the cow pox. Printed, for the author, by Sampson Low … ; and sold by Law … and Murray and Hihghley. http://books.google.co.nz/books?id=tj3h3ZA0ppAC
Kawai T, Akira S (2010) The role of pattern-recognition receptors in innate immunity: update on Toll-like receptors. Nat Immunol 11(5):373–384. doi:10.1038/ni.1863
Klein U, Rajewsky K, Küppers R (1998) Human immunoglobulin (Ig)M+IgD+peripheral blood B cells expressing the CD27 cell surface antigen carry somatically mutated variable region genes: CD27 as a general marker for somatically mutated (memory) B cells. J Exp Med 188(9):1679–1689, http://www.pubmedcentral.nih.gov/articlerender.fcgi?artid=2212515&tool=pmcentrez&rendertype=abstract
Klein L, Hinterberger M, Wirnsberger G, Kyewski B (2009) Antigen presentation in the thymus for positive selection and central tolerance induction. Nat Rev Immunol 9(12):833–844. doi:10.1038/nri2669
Malisan F, Brière F (1996) Interleukin-10 induces immunoglobulin G isotype switch recombination in human CD40-activated naive B lymphocytes. J Exp Med 183(3):937–947, http://www.pubmedcentral.nih.gov/articlerender.fcgi?artid=2192363&tool=pmcentrez&rendertype=abstract
Market E, Papavasiliou FN (2003) V(D)J recombination and the evolution of the adaptive immune system. PLoS Biol 1(1):E16. doi:10.1371/journal.pbio.0000016
Martinon F, Burns K, Tschopp J (2002) The inflammasome: a molecular platform triggering activation of inflammatory caspases and processing of proIL-beta. Mol Cell 10(2):417–426, http://www.ncbi.nlm.nih.gov/pubmed/12191486
Masopust D, Vezys V, Marzo AL, Lefrançois L (2001) Preferential localization of effector memory cells in nonlymphoid tissue. Science 291(5512):2413–2417. doi:10.1126/science.1058867
Mbow ML, De Gregorio E, Valiante NM, Rappuoli R (2010) New adjuvants for human vaccines. Curr Opin Immunol 22(3):411–416. doi:10.1016/j.coi.2010.04.004
Moser B, Loetscher P (2001) Lymphocyte traffic control by chemokines. Nat Immunol 2(2):123–128. doi:10.1038/84219
Murphy P, Baggiolini M (2000) International union of pharmacology. XXII. Nomenclature for chemokine receptors. Pharmacol Rev 52(1):145–176, http://www.ncbi.nlm.nih.gov/pubmed/10699158
Nurieva R, Liu X, Dong C (2009) Yin–Yang of costimulation: crucial controls of immune tolerance and function. Immunol Rev 229(1):88–100. doi:10.1111/j.1600-065X.2009.00769.x
Osterholm M (2005) Preparing for the next pandemic. N Engl J Med 346:364, http://www.ncbi.nlm.nih.gov/pubmed/23344310
Ovchinnikov DA (2008) Macrophages in the embryo and beyond: much more than just giant phagocytes. Genesis 46(9):447–462. doi:10.1002/dvg.20417
Pead P (2003) Benjamin Jesty: new light in the dawn of vaccination. The Lancet 362:2104–2109, http://www.sciencedirect.com/science/article/pii/S0140673603151112
Pernis B, Chiappino G, Kelus A, Gell P (1965) Cellular localization of immunoglobulins with different allotypic specificities in rabbit lymphoid tissues. J Exp Med 122(5):853–876, http://jem.rupress.org/content/122/5/853.abstract
Sakaguchi S, Sakaguchi N, Asano M, Itoh M, Toda M (1995) Immunologic self-tolerance maintained by activated T cells expressing IL-2 receptor α-chains (CD25). J Immunol 155(3): 1151–1164
Sallusto F, Lenig D, Förster R, Lipp M, Lanzavecchia A (1999) Two subsets of memory T lymphocytes with distinct homing potentials and effector functions. Nature 401(6754):708–712. doi:10.1038/44385
Schenkel JM, Fraser KA, Vezys V, Masopust D (2013) Sensing and alarm function of resident memory CD8(+) T cells. Nat Immunol 14(5):509–513. doi:10.1038/ni.2568
Schroder K, Hertzog PJ, Ravasi T, Hume DA (2004) Interferon-γ: an overview of signals, mechanisms and functions. J Leukoc Biol 75(2):163–189. doi:10.1189/jlb.0603252.Journal

Schroeder H, Cavacini L (2010) Structure and function of immunoglobulins. J Allergy Clin Immunol 125(2):S41–S52. doi:10.1016/j.jaci.2009.09.046
Sharma S (2005) HIV-TB co-infection: epidemiology, diagnosis & management. Indian J Med Res 121(4):550–567, http://www.ncbi.nlm.nih.gov/pubmed/15817963
Slifka M, Antia R, Whitmire J, Ahmed R (1998) Humoral immunity due to long-lived plasma cells. Immunity 8(3):363–372, http://www.sciencedirect.com/science/article/pii/S1074761300805415
Smyth MJ, Trapani JA (1995) Granzymes: exogenous proteinases that induce target cell apoptosis. Immunol Today 16(4):202–206, http://www.ncbi.nlm.nih.gov/pubmed/7734049
Trapani JA, Jans DA, Jans PJ, Smyth MJ, Browne KA, Sutton VR (1998) Efficient nuclear targeting of granzyme B and the nuclear consequences of apoptosis induced by granzyme B and perforin are caspase-dependent, but cell death is caspase-independent. J Biol Chem 273(43):27934–27938. doi:10.1074/jbc.273.43.27934
Turley SJ (2000) Transport of peptide-MHC class II complexes in developing dendritic cells. Science 288(5465):522–527. doi:10.1126/science.288.5465.522
Wakim LM, Bevan MJ (2010) From the thymus to longevity in the periphery. Curr Opin Immunol 22(3):274–278. doi:10.1016/j.coi.2010.03.003
Wang Y, Wu W, Negre NN, White KP, Li C, Shah PK (2011) Determinants of antigenicity and specificity in immune response for protein sequences. BMC Bioinformatics 12(1):251. doi:10.1186/1471-2105-12-251
Wolf PR, Ploegh HL (1995) How MHC class II molecules acquire peptide cargo: biosynthesis and trafficking through the endocytic pathway. Annu Rev Cell Dev Biol 11:267–306. doi:10.1146/annurev.cb.11.110195.001411
Woodland DL, Kohlmeier JE (2009) Migration, maintenance and recall of memory T cells in peripheral tissues. Nat Rev Immunol 9(3):153–161. doi:10.1038/nri2496
Wooldridge L, van den Berg HA, Glick M, Gostick E, Laugel B, Hutchinson SL, Milicic A, Brenchley JM, Douek DC, Price DA, Sewell AK (2005) Interaction between the CD8 coreceptor and major histocompatibility complex class I stabilizes T cell receptor-antigen complexes at the cell surface. J Biol Chem 280(30):27491–27501. doi:10.1074/jbc.M500555200
Yonehara S, Ishii A, Yonehara M (1989) A cell-killing monoclonal antibody (anti-Fas) to a cell surface antigen co-downregulated with the receptor of tumor necrosis factor. J Exp Med 169(5):1747–1756, http://jem.rupress.org/content/169/5/1747.abstract

Chapter 2
Classification of Vaccines

Rie S. Kallerup and Camilla Foged

2.1 Introduction

The introduction of human vaccines has had a tremendous impact on global health by dramatically reducing the mortality and morbidity caused by infectious diseases, and next to the wider availability of potable water, it is considered the most cost-effective and successful medical intervention ever introduced. Vaccines have inevitably prevented disease, complications, and the death of millions of infants and children by protecting against many deadly infectious diseases (Bloom et al. 2005; Ehreth 2003).

Although vaccines have mainly demonstrated their value to human society during the past century, the principle of vaccination has been used in China and India for more than a thousand years as the practice of variolation, where individuals were inoculated with live and virulent smallpox virus to achieve protection against a later encounter. Although the procedure did lead to protection, it was not without the risk of death or causing an epidemic. However, Edward Jenner is generally honored for the pioneering development of the first vaccine more than 200 years ago by demonstrating that exposure of humans to cowpox virus induced cross-protective immunity towards smallpox (Riedel 2005). The word vaccine was in fact coined by Jenner, and is derived from the Latin word *vacca*, which means cow.

R.S. Kallerup (✉) • C. Foged
Department of Pharmacy, Faculty of Health and Medical Sciences,
University of Copenhagen, Universitetsparken 2, 2100 Copenhagen Ø, Denmark
e-mail: rie.kallerup@sund.ku.dk; camilla.foged@sund.ku.dk

C. Foged et al. (eds.), *Subunit Vaccine Delivery*, Advances in Delivery Science and Technology, DOI 10.1007/978-1-4939-1417-3_2

Subsequently, the development of vaccines have for more than a century been based on Louis Pasteur's principle of isolating, purifying, and injecting the causative microorganisms in order to induce protective immunity (Rappuoli 2007). After World War II more systematic childhood vaccination programs became a widespread tool for improving public health (Bloom et al. 2005). The mortality caused by serious and life-threatening diseases has been dramatically reduced as a result of these successful global childhood vaccination programs, and the introduction of vaccines has led to the eradication of smallpox and near eradication of infectious diseases such as polio (Ehreth 2003; Rappuoli 2007). The World Health Organization (WHO) currently recommends routine immunization against 12 different diseases (Table 2.1). Furthermore, additional vaccines are recommended for populations at high risk or regions with special needs.

Despite this true medical success story, current vaccination efforts do face a number of obstacles. Three million people are estimated to die annually from vaccine-preventable illnesses, and infectious diseases still remain the leading cause of death worldwide for several reasons. The rapid progress towards universal vaccination coverage in the 1970s and 1980s has slowed in the past decade, and several childhood illnesses have started to re-emerge as a result of inefficient vaccine coverage. This may be due to public perception of vaccination, where an individual may find it rational to refuse vaccination in order to avoid the possible side effects, or due to political reasons. The consequence has been the reemergence of diseases such as measles and pertussis in certain industrialized countries and of polio in certain developing countries (Bloom et al. 2005).

Infectious disease-caused mortality can also be explained by lack of efficacious vaccines where conventional vaccinology has failed due to factors such as antigenic drift, and by the existence of more difficult target diseases, for example, tuberculosis (TB), human deficiency virus-acquired immune deficiency syndrome (HIV-AIDS), and malaria. Antigenic drifts represent a challenge for vaccine development,

Table 2.1 WHO recommendations for routine immunization (WHO 2012)

Disease/antigen	Age group
Bacillus Calmette-Guérin (BCG)	Children
Hepatitis B	Children (+adolescents/adults in high risk groups)
Polio	Children
Diphtheria	Children, adolescents, and adults
Tetanus	Children, adolescents, and adults
Pertussis	Children
Haemophilus influenzae type B	Children
Pneumococcal	Children
Rotavirus	Children
Measles	Children
Rubella	Children
Human papilloma virus (HPV)	Adolescent girls

and the success stories in vaccinology arise to a large extent from development of vaccines against pathogens with no or little antigenic drift, for example, vaccines against diphtheria and tetanus, where there is no antigenic drift in the target toxin antigen. Antigenic shift can result in changes in surface antigens and the influenza virus is an example of a pathogen where such changes occur annually. This antigenic variability is overcome by altering the vaccine on a yearly basis. However, pathogens where antigens change faster, e.g., human immunodeficiency virus (HIV), are more difficult to approach by conventional vaccinology. To date, conventional vaccinology has been most successful in vaccines against pathogens for which protection is antibody mediated. The difficult vaccine targets represent to a large extent pathogens for which antibodies cannot provide sufficient protection (Rappuoli 2007). An example is the intracellular pathogen *Mycobacterium tuberculosis*. In 2012, 8.6 million people were infected with *M. tuberculosis* and approximately 1.3 million people died from TB (WHO, Fact sheet 104, 2012). Numbers like these put great emphasis on the acute need for new prophylactic as well as therapeutic vaccines against global killers like TB, malaria, HIV-AIDS, and cancer. However not only new vaccines are needed since improvements to conventional vaccines could have a tremendous impact on vulnerable population groups such as the elderly, since this population is immunologically hyporesponsive. Several vaccines approved for human use are listed in Table 2.2.

2.2 Classification of Vaccines

Traditionally vaccines have been based on live attenuated pathogens, whole inactivated organisms or inactivated bacterial toxins and are most often sufficiently immunogenic. Traditional vaccines based on the whole-cell concept possess intrinsic immune stimulatory capacity, which is adequate for the induction of long-lived protective immunity. However, a great disadvantage related to this approach is that these live systems have associated adverse effects that in some cases are mild but can be severe or even fatal in others (Huang et al. 2004). Safety is of major concern in vaccine development and limits the use of the traditional approach in the development of new vaccines as traditional vaccines may cause disease in immune-compromised hosts or revert back to virulence (Robinson and Amara 2005). With these issues, new parenteral vaccines are unlikely to be live attenuated vaccines.

In light of these limitations, new strategies for vaccine development are emerging, and vaccine development is moving away from the whole-cell based approach of live attenuated or inactivated vaccines and towards the safer spilt and subunit vaccine technology. The field of vaccinology has undergone tremendous breakthroughs over the past 30–40 years. An important contribution to these breakthroughs is provided by the introduction of recombinant DNA technology, which solved the problem of antigen manufacturing. Also the development of conjugate vaccines, subunit vaccines, and the non-replicating recombinant virus-like particles (VLPs) has had an enormous impact on vaccine development and success (Rappuoli 2007).

Table 2.2 Vaccines licensed for human use (non-exhaustive list)

Trade name	Disease target	Causative agent	Vaccine type	Vaccine components	Adjuvant/ comment
AGRIFLU®—Novartis Vaccines and Diagnostics, Inc. (various others)	Influenza	Influenza virus	Inactivated	Trivalent for influenza type A and B (whole virus based)	None
BCG vaccine "SSI"	Tuberculosis	*Myco Bacterium tuberculosis*	Live attenuated	Bacillus Calmette Guérins (BCG), Danish strain 1331	None
BioThrax® (Emergent Biodefense Operations Lansing, Inc.)	Anthrax	*Bacillus anthracis*	Subunit vaccine	Cell-free filtrates of microaerophilic cultures of an avirulent, nonencapsulated strain of *Bacillus anthracis* and proteins, including the 83 kDa protective antigen (PA) protein, released during the growth period	Alum
Cervarix® (GlaxoSmithKline)	Cervical cancer	Human papillomavirus (HPV)	Subunit—virus like particles	Recombinant L1 protein, the major antigenic protein of the capsid, of oncogenic HPV types 16 and 18	AS04
COMVAX® (Merck & Co.)	Hib-induced diseases (pneumonia, meningitis)	*Haemophilus influenzae* type B	Conjugate vaccine	Haemophilus b conjugate (meningococcal protein conjugate) and Hepatitis B (recombinant) vaccine	Alum
	Hepatitis B	Hepatitis B virus			
FluMist® Quadrivalent (MedImmune, LLC)	Influenza	Influenza virus	Live inactivated	Quadrivalent vaccine for administration by intranasal spray. FluMist Quadrivalent contains four vaccine virus strains: an A/H1N1 strain, an A/H3N2 strain, and two B strains	None
				Cold adapted, temperature sensitive, and attenuated	
Gardasil® (Merck & Co.)	Cervical cancer	Human papillomavirus (HPV)	Subunit—virus like particles	Recombinant quadrivalent vaccine prepared from the purified virus-like particles (VLPs) of the major capsid (L1) protein of HPV Types 6, 11, 16, and 18	Alum

Havrix® (GlaxoSmithKline)	Hepatitis A	Hepatitis A virus	Inactivated virus	The virus (strain HM175) is propagated in MRC-5 human diploid cells. Treatment with formalin ensures viral inactivation	Alum
Influenza A (H1N1) 2009 Monovalent Vaccine Sanofi Pasteur, Inc.	Influenza	Influenza virus	Inactivated (against influenza disease caused by pandemic (H1N1) 2009 virus)	Monovalent split vaccine whole cell approach	None
IPOL® Sanofi Pasteur	Polio	Poliovirus (type 1, 2 and 3)	Live inactivated	Inactivated at +37 °C for at least 12 days with 1:4,000 formalin	None
Infanrix® (GlaxoSmithKline)	Diphtheria	*Corynebacterium diphtheriae*	Subunit	Toxoid of diphtheria and tetanus and the acellular pertussis antigens (inactivated pertussis toxin (PT), filamentous hemagglutinin (FHA), and pertactin)	Alum
	Tetanus	*Clostridium tetani*			
	Pertussis	*Bordetella pertussis*			
Pediarix® (GlaxoSmithKline)	Diphtheria	*Corynebacterium diphtheria*	Subunit	As Infarix in combination with HBsAg and type 1, 2, and 3 polio viruses	Alum
	Tetanus	*Clostridium tetani*		Inactivated with formaldehyde	
	Pertussis	*Bordetella pertussis*			
	Polio	Poliovirus			
	Hepatitis B	Hepatitis B virus			
Recombivax HB® (Merck & Co.)	Hepatitis B	Hepatitis B virus	Subunit viral vaccine	Derived from hepatitis B surface antigen (HBsAg)	Alum
Zostavax® (Merck & Co.)	Herpes Zoster	Herpes Zoster virus	Live attenuated	Oka/Merck strain of VZV	

2.2.1 Live Attenuated Vaccines

Conventional vaccines have been based on live attenuated pathogens, and contain laboratory-weakened versions of the original pathogen. The rationale for using live attenuated vaccines is that they mimic the natural infection, which results in an effective vaccination strategy. The advantage of this type of vaccine is that both a strong cellular and an antibody response are produced. Usually, long-term protection is also achieved, and a single inoculation is often sufficient. The attenuation of the microorganism results in a non-pathogenic microorganism, which still possesses all the pathogenic features as the original microorganism (Clem 2011).

Attenuation can be achieved via different approaches. Edward Jenner's approach was to use a virus pathogenic in a different host but not pathogenic to humans, as he isolated pus from cows with cowpox, and this provided the basis for his smallpox vaccine (Riedel 2005). Naturally occurring attenuated strains can also be used, exemplified by the use of type 2 poliovirus. Attenuation is also possible by applying harsh conditions on a virulent virus strain (e.g., cold adaption of influenza virus).

The Bacillus Calmette Guérin (BCG) vaccine against TB is an example of an attenuated live vaccine. The currently used vaccine strains are all descendants of the original *M. bovis* isolate that Calmette and Guérin passaged through many cycles. Further passages, under different laboratory conditions, have resulted in a variety of new BCG strains with phenotypic and genotypic difference.

One such strain is the 1331 strain produced at the Danish Serum Institute (WHO 2004). As adults with lung TB are the major source of disease transmission, BCG vaccination of children has had very limited influence on the global epidemic. Another very important limitation of BCG is the lack of effect in the two billion individuals already infected with TB, which underlines the need for the development of new TB vaccines (WHO 2004).

Another example of an attenuated live viral vaccine is the measles, mumps, and rubella vaccine (MMR). This vaccine has been available in the United States since 1971 (Ravanfar et al. 2009). Priorix® is a marketed MMR vaccine produced by GlaxoSmithKline. The vaccine contains attenuated MMR viruses. Each of these attenuated virus strains, measles (the Schwarz strain), mumps (the RIT 4385 strain), and rubella (the Wistar RA 27/3 strain) is obtained separately by propagation in chick embryo tissue cultures (mumps and measles) or MRC5 human diploid cells (rubella) (Wellington and Goa 2003).

2.2.2 Inactivated Vaccines

The main advantage of killed or inactivated vaccines over attenuated vaccines is safety. Since these vaccines are based on killed/inactivated pathogens, the concerns regarding reverting back to virulence are obviated. However, this also constitutes a huge disadvantage since the lack of replication results in a fast clearance from the body leading to a decreased efficacy, as compared to the live vaccines. Killed/inactivated vaccines do, however, give rise to a more complex or greater inflammatory immune response in comparison to the newer subunit vaccines due to the fact that most of the pathogenic components are preserved.

Inactivated vaccines are used widely. An example of such a vaccine is the Hepatitis A vaccine Epaxal® from Crucell. This vaccine is based on a hepatitis A virus (strain RG-SB) which is inactivated by formalin treatment. The inactivated vaccine is adsorbed onto a virosome formulation, which constitutes the adjuvant system (Bovier 2008).

2.2.3 Subunit Vaccines

Subunit vaccines are, by definition, vaccine agents that comprise one or more components of a pathogen rather than the entire pathogen. Subunit vaccines are composed of one or several recombinant peptides/proteins or polysaccharides normally present in the structure of the target pathogen (Dudek et al. 2010). In terms of safety and cost of production, these vaccines offer considerable advantages over the traditional vaccines, as these are composed of very well-defined and highly pure components. This approach results in a more appealing safety profile due to the lack of replication and the removal of material that may initiate unwanted host responses (Robinson and Amara 2005).

For bacterial subunit vaccines, two main types exist. The first type is the toxoid vaccines which are generated against bacteria where toxins are the main disease-causeing agents. The toxins are inactivated by converting the toxins into detoxified versions (toxoids), for instance by treatment with formaldehyde. These toxoids can then safely be used for vaccination purposes. The close resemblance of the toxoid to the toxin enables the immune system to neutralize and fight the natural toxins via generation of anti-toxoid antibodies. Examples of toxoid vaccines are the different vaccines against diphtheria, tetanus, and pertussis. The second major group of bacterial subunit vaccines as based on the capsular polysaccharides of encapsulated bacteria. There are several examples of vaccines of this type, including vaccines against *Streptococcus pneumoniae*, *Neisseria meningitidis*, and *Haemophilus influenzae* type b (Hib). A variation of this is the conjugate vaccine, which is created by covalently attaching an antigen (often the bacterial polysaccharides) to a carrier protein, e.g., tetanus toxoid, resulting in the generation of more efficacious vaccines. Common virus subunit vaccines are the split virus vaccines where the structure of the viruses has been disrupted, resulting in a mixture of the various viral components.

Alternatively, subunit vaccines may consist of one or more viral or bacterial proteins, or peptide fragments of these. In some cases, such antigens might be sufficiently immunogenic by themselves. This is the case for the subunit vaccine for influenza comprising the two purified surface antigens hemagglutinin (HA) and neuraminidase (NA). These two proteins are isolated for the seasonal flu vaccine from three selected virus strains and combined in a trivalent vaccine, with or without an adjuvant. Also for the hepatitis B vaccine, the surface antigen, HBsAg, is sufficiently immunogenic, and a vaccine based on recombinant HBsAg was the first genetically engineered vaccine product produced commercially and used worldwide.

However, in many cases the highly purified subunit antigens lack many of the intrinsic pathogenic features which render these protein-based antigens weakly immunogenic by themselves and co-administration of adjuvants is often required. The addition of adjuvants not only enables the induction of an effective immune response, but also provides the potential to modulate the immune response (Reed et al. 2009; O'Hagan 2001). The use of adjuvants can also allow for a dose-sparring effect or can reduce the number of required administrations.

2.2.3.1 Adjuvants

A vaccine adjuvant is defined as a component that potentiates the immune response to an antigen and/or modulates it towards a desired immune response. The term adjuvant is derived from the Latin word *adjuvare*, which means to help. The most commonly used adjuvants are the aluminum salts commonly, although incorrectly, referred to as alum (Chap. 3). The adjuvant effect of alum was discovered by Glenny in 1926, and alum has now been utilized for more than 70 years in vaccines (Glenny et al. 1926). For many years alum was the only adjuvant approved worldwide and it has been used in large numbers of vaccines for human use (Clements and Griffiths 2002). Formulation is achieved by adsorption of antigen onto highly charged aluminum particles (Reed et al. 2009).

In recent years, there has been substantial progress in the discovery of new efficient adjuvants for subunit vaccines [reviewed by (Foged 2011)], and a handful of these have been marketed as components of approved licensed vaccines. Examples of adjuvants are emulsions, liposomes, polymeric nanoparticles, immune-stimulating complexes (ISCOMs), and VLPs, which are described in the following chapters.

Adjuvants can broadly be classified into delivery systems and immunopotentiating compounds, generally pathogen-associated molecular patterns (PAMPs) such as the toll-like receptor (TLR) ligands. The function of delivery systems is to effectively deliver the vaccine components to the target antigen-presenting cells (APCs) and thereby enhance the amount of antigen reaching the cells or tissue responsible for induction of immune responses. Delivery systems are often particulate in nature and mimic nature in terms of size and shape resulting in a delivery system with similar dimensions as a given pathogen, which is a natural target for APCs. The combination of delivery systems and immunopotentiators has great potential due to concomitant enhanced antigen delivery and potent stimulation of innate immunity [reviewed by (Reed et al. 2009, 2013)].

Thus adjuvants are a heterogenous group of compounds that can have many different functions, i.e., depot or targeting functions and immunostimulatory or immunomodulatory functions (Guy 2007). Adjuvants utilize very different mechanisms in order to potentiate an immune response: (a) depot effect; (b) up-regulation of cytokines and chemokines; (c) cellular recruitment at the site of injection; (d) increased antigen uptake and presentation to APCs; (e) activation and maturation of

APCs and migration to the draining lymph nodes; and (f) activation of the inflammasome [reviewed by (Awate et al. 2013)]. Understanding of the adjuvant mechanism of action can be utilized to develop vaccines with a very specific and tailored effect. The mechanism behind adjuvanticity is however in many cases poorly understood since immune responses to vaccines involve a very complex cascade of events and the isolated effect of an adjuvant can be very difficult to dissect.

The antigen can be associated to a delivery system by surface adsorption or encapsulation, depending on the mode of preparation. In this sense, delivery systems provide the potential to control antigen kinetics and dynamics. This is done (a) by stabilizing as well as protecting the antigen from degradation; (b) by inhibiting/delaying clearance of the antigen from the injection site; (c) targeting and also carrying the antigen to the APCs; (d) prolonging the time of exposure of antigen to the immune cells; (e) enhancing the antigen uptake in the APCs; and (f) controlling the antigen release and intracellular trafficking (reviewed by Foged 2011; O'Hagan and De Gregorio 2009).

Immunopotentiators function via direct activation of the innate immune system by interacting with the APCs through pattern recognition receptors (PRRs) (O'Hagan and Valiante 2003). Examples of such immunopotentiators are ligands of innate immune receptors, the TLRs, NOD-like receptors (NLRs), C-type lectin receptors (CLRs), and RIGI-like receptors (RLRs) [reviewed by (Reed et al. 2013; Foged 2011; Guy 2007)]. A wide variety of PAMPs are recognized through TLRs, examples thereof are lipopolysaccharide (LPS) and its derivatives which are recognized through TLR4, peptidoglycans from Gram-positive bacteria and lipopeptides are recognized through TLR2, RNA is recognized through TLR3, bacterial flagellin through TLR5, single-stranded RNA and imidazoquinolines signal through TLR7 and TLR8, and unmethylated CpG motifs in bacterial DNA are recognized through TLR9 (Gay and Gangloff 2007; Medzhitov 2001).

A growing body of preclinical and clinical data demonstrates that TLR agonists are potent vaccine adjuvants and provide the opportunity for tailoring and modulating the immune response against a vaccine by inducing distinct cytokine profiles (Duthie et al. 2011). Monophosphoryl lipid A (MPL) is the most studied TLR agonist for vaccination purposes. MPL is derived from LPS which is found in the cell wall of Gram-negative bacteria (Casella and Mitchell 2008). The adjuvant formulation AS04 from GlaxoSmithKline is based on MPL adsorbed to alum (Garcon 2010) and is approved for the hepatitis B vaccine Fendrix™ (Garcon et al. 2007) and the HPV vaccine Cervarix™ in combination with VLPs (Schwarz 2009; Romanowski et al. 2009). In addition new and synthetic TLR agonists are being developed and the availability of such immunopotentiators has expanded.

Hence rational development and formulation of adjuvant systems can result in a wide variety of ways to modulate the immune response in a desired direction.

The non-TLRs are not as well described as the TLRs and include intracellular innate receptors such as the RLRs, the soluble NLRs, and CLRs. The surface-expressed CLRs include the mannose receptor and DC-SIGN that are able to bind a wide range of viruses, bacteria, and fungi through recognition of sugar moieties (Guy 2007).

Adjuvant systems are defined as functional excipients and are in that sense components of a specific vaccine. Table 2.3 lists adjuvant delivery systems used in vaccines approved for human use. The aluminum salts are described further in Chap. 3 of this book, the oil-in-water emulsions MF59 and AS03 are described in Chap. 4, and VLPs are discussed in Chap. 9.

In order to achieve the optimal immunological effect, an adjuvant appropriate for the formulation must be considered. The choice of formulation is in turn dependent upon the choice of antigenic components, the type of immune response that is needed, the optimal/desired route of administration, any potential adverse effects, and the stability of the vaccine. These factors must be considered in the early phases of development. Also the adjuvant must be chemically as well as physically stable in order to face the quality control criteria (see Chap. 19) which ensures reproducible manufacturing as well as activity (Reed et al. 2009).

The inclusion of adjuvants in vaccine formulations should be justified. Efficacy, safety, and tolerability are the most important factors for vaccine development. The use of adjuvants should therefore be considered in relation to the target population and should be selected based on a risk/benefit ratio. For example, a higher risk is more acceptable for cancer patients than for healthy children.

2.2.4 DNA Vaccines

DNA vaccines represent a new generation of vaccines that are attractive due to their simplicity in addition to several other advantages they have over conventional vaccines. The principle underlying DNA vaccination is to induce immunity by transiently transfecting host cells with plasmid DNA (pDNA) encoding antigen, as opposed to injecting antigen in the form of a peptide or protein. Upon DNA vaccination, host cells produce the protein (antigen) encoded by the DNA and immunity against this particular protein is subsequently induced (Bins et al. 2013; Senovilla et al. 2013). The great advantages associated with DNA vaccines are that they can be manufactured relatively easily at low costs, and both humoral and cellular immune responses can be elicited. In addition, pDNA is fairly stable at room temperature (Bins et al. 2013), which renders the normally required cold chain redundant for DNA vaccine storage. This is certainly of high importance for the effectiveness of vaccine programs in developing countries.

As yet no DNA vaccines have been approved for human use. Several clinical trials are being conducted at this point in time for different cancers and HIV-AIDS. Some DNA vaccines are approved/registered for veterinary use (Bins et al. 2013; Senovilla et al. 2013).

2.2.5 Dendritic Cell-Based Vaccines

Another type of vaccination strategy is based on dendritic cells (DCs). The function of these cells is to acquire, process and present antigens to T-cells, and provide the stimulatory signals and cytokines required to induce T-cell proliferation and differentiation into effector cells (Chap. 1). Therefore, a much-studied vaccination

Table 2.3 Adjuvant delivery systems used in vaccines approved for human use

Delivery system	Adjuvant name	Vaccine	Disease target	Company	Reference
Mineral salt	Aluminum hydroxide/phosphate	Various	Various, e.g., Diphtheria, tetanus, hepatitis		Clements and Griffiths (2002)
Oil-in-water emulsion	MF59	Fluad®	Influenza	Novartis	Schultze et al. (2008) and Podda (2001)
		Focetria®			Banzhoff et al. (2009) and Gasparini et al. (2010)
		Aflunov®			Galli et al. (2009a) and Galli et al. (2009b)
Oil-in-water emulsion	AS03	Arepanrix® Prepandrix® Pandemrix®	Influenza	GlaxoSmith Kline	Roman et al. (2010) Walker and Faust (2010)
Water-in-oil emulsion	Montanide	CimaVax EGT™	Cancer	Bioven	Rodriguez et al. (2010)
Virosomes		Exapal®	Hepatitis A	Crucell	Bovier (2008)
		Inflexal V®	Influenza	Crucell	Gluck and Metcalfe (2002)
		Invivac™	Influenza	Solvay	de Bruijn et al. (2006)
VLPs adsorbed onto alum		Gardasil®	HPV	Merck	Schiller et al. (2008)
MPL adsorbed onto alum	AS04	Fendrix®	Hepatitis B	GlaxoSmith Kline	Garcon et al. (2007)
VPL+MPL adsorbed onto alum		Cervarix®	HPV	GlaxoSmith Kline	Schwarz (2009)

MPL monophosphoryl lipid A, *VLP* virus-like particle

strategy is to load in vitro-generated DCs with antigens and infuse them into a patient so as to elicit T-cell-mediated responses, particularly in the context of cancer where DC function in vivo is often blunted or subverted by factors released by the tumor (Chap. 13). While preclinical studies have repeatedly shown that DC-based vaccines can delay or prevent tumor progression, human clinical trials have been disappointing in comparison, offering only marginal benefit for patients. There is therefore still a need to improve the stimulatory capacity of the injected cells, and strategies for how to achieve this are discussed further in Chap. 13.

2.3 Pharmaceutical and Delivery Challenges for the Development of Subunit Vaccines

Research in the field of modern vaccinology is to a large extent conducted in the absence of knowledge of how the physicochemical properties of the subunit formulations impact the efficacy, safety, and mechanism of action (Mortellaro and Ricciardi-Castagnoli 2011). In order to move towards a more rational process regarding vaccine development it is of crucial importance to increase understanding of vaccine formulation, which is a great challenge since vaccines are often very complex systems (Reed et al. 2009). An in-depth understanding of the physicochemical properties and what effect production and biological processes impose on safety and efficacy is desirable during development of subunit vaccines, also from a stability and quality control point of view. Therefore, there are a substantial number of pharmaceutical challenges associated with the subunit vaccine development process. With these complex systems a tremendous amount of work on development, formulation, and characterization is needed. Also the regulatory challenges facing scientists who research and develop subunit vaccines are of great importance for the successful development of subunit vaccines. The pharmaceutical analysis and quality control of vaccines are described further in Chaps. 19–21 of this book.

A crucial aspect in addressing the challenges in vaccine development is vaccine delivery, which encompasses (a) administration of the vaccine formulation to specific sites of the body and (b) delivery of the antigen to, and activation of, relevant cells of the immune system. Administration of vaccine formulations to specific sites of the body can be achieved by various routes, and the most commonly used routes have been intramuscular (i.m.) and subcutaneous (s.c.) injection. During the past decades, much effort has been devoted to exploring the use of minimally invasive or noninvasive administration routes, such as nasal delivery, pulmonary delivery, transcutaneous delivery, oral delivery, and sublingual/buccal delivery. Such alternative routes of administration might allow for easier and more convenient administration, e.g., needle-free approaches, and might eventually result in increased vaccine coverage by increasing the willingness of the public to be vaccinated. In addition, the use of alternative administration routes might affect the quality of the immune response. One example is mucosal vaccination. Most pathogens access the body through the mucosal membranes. Therefore, effective vaccines that protect at these sites are much needed. However, despite early success with the live attenuated oral polio vaccine, only a few new mucosal vaccines have been approved for human use. This is partly due to problems with developing safe and effective mucosal adjuvants.

Each of these immunization routes requires specially designed formulations (e.g., suspensions, emulsions, powders, tablets) and specially designed delivery devices (such as microneedles, nasal sprayers, and pulmonary inhalers). To license a product for vaccination applying alternative administration routes, the combination of formulation and device should be licensed as a whole. For this reason, formulation development and development of a suitable device should go hand in hand. In Chaps. 14–18, different administration routes are discussed together with formulations and devices used specifically for these routes.

Finally, the development of stable vaccine formulations is important to consider, in particular the development of thermostable vaccines that can be distributed independently of the expensive cold chain are highly in demand for the developing countries. Processes for drying of vaccines such as spray drying, spray freeze drying, and supercritical fluid technology are further described for pulmonary formulations in Chap. 16.

2.4 Conclusions

Prophylactic vaccination is the medical intervention with by far the largest impact on public health and has greatly reduced the incidences of bacterial and viral infections. Despite this the field of vaccinology faces a number of challenges, and there is still an unmet medical need for new vaccines due to the existence of a number of infectious diseases for which no effective vaccine is available (e.g., HIV-AIDS, malaria), or for which existing vaccines provide insufficient immunity (e.g., TB) or are unaffordable for those most in need (e.g., Pneumococcal disease). Conventional vaccines include the live, attenuated, or inactivated whole organism vaccines. Novel vaccine development strategies aim towards more safe, efficient, and stable vaccines in the future. New generation vaccines are usually of the subunit vaccine type, which are based on highly purified recombinant or synthetic antigens. A number of adjuvant technologies are used to enhance efficacy and there are efforts ongoing to explore the usage of noninvasive administration routes. This poses special demands in terms of formulation development and device technology for optimizing the delivery of antigens and immunopotentiators to the immune system.

References

Awate S, Babiuk LA, Mutwiri G (2013) Mechanisms of action of adjuvants. Front Immunol 4:114

Banzhoff A, Gasparini R, Laghi-Pasini F, Staniscia T, Durando P, Montomoli E, Capecchi PL, Di Giovanni P, Sticchi L, Gentile C, Hilbert A, Brauer V, Tilman S, Podda A (2009) MF59-adjuvanted H5N1 vaccine induces immunologic memory and heterotypic antibody responses in non-elderly and elderly adults. PLoS One 4:e4384

Bins AD, Van Den Berg JH, Oosterhuis K, Haanen JB (2013) Recent advances towards the clinical application of DNA vaccines. Neth J Med 71:109–117

Bloom DE, Canning D, Weston M (2005) The value of vaccination. World Economics 6:15
Bovier PA (2008) Recent advances with a virosomal hepatitis A vaccine. Expert Opin Biol Ther 8:1177–1185
Casella CR, Mitchell TC (2008) Putting endotoxin to work for us: monophosphoryl lipid A as a safe and effective vaccine adjuvant. Cell Mol Life Sci 65:3231–3240
Clem AS (2011) Fundamentals of vaccine immunology. J Glob Infect Dis 3:73–78
Clements CJ, Griffiths E (2002) The global impact of vaccines containing aluminium adjuvants. Vaccine 20(Suppl 3):S24–S33
De Bruijn IA, Nauta J, Gerez L, Palache AM (2006) The virosomal influenza vaccine Invivac: immunogenicity and tolerability compared to an adjuvanted influenza vaccine (Fluad in elderly subjects). Vaccine 24:6629–6631
Dudek NL, Perlmutter P, Aguilar MI, Croft NP, Purcell AW (2010) Epitope discovery and their use in peptide based vaccines. Curr Pharm Des 16:3149–3157
Duthie MS, Windish HP, Fox CB, Reed SG (2011) Use of defined TLR ligands as adjuvants within human vaccines. Immunol Rev 239:178–196
Ehreth J (2003) The global value of vaccination. Vaccine 21:596–600
Foged C (2011) Subunit vaccines of the future: the need for safe, customized and optimized particulate delivery systems. Ther Deliv 2:1057–1077
Galli G, Hancock K, Hoschler K, Devos J, Praus M, Bardelli M, Malzone C, Castellino F, Gentile C, McNally T, Del Giudice G, Banzhoff A, Brauer V, Montomoli E, Zambon M, Katz J, Nicholson K, Stephenson I (2009a) Fast rise of broadly cross-reactive antibodies after boosting long-lived human memory B cells primed by an MF59 adjuvanted prepandemic vaccine. Proc Natl Acad Sci U S A 106:7962–7967
Galli G, Medini D, Borgogni E, Zedda L, Bardelli M, Malzone C, Nuti S, Tavarini S, Sammicheli C, Hilbert AK, Brauer V, Banzhoff A, Rappuoli R, Del Giudice G, Castellino F (2009b) Adjuvanted H5N1 vaccine induces early CD4+ T cell response that predicts long-term persistence of protective antibody levels. Proc Natl Acad Sci U S A 106:3877–3882
Garcon N (2010) Preclinical development of AS04. Methods Mol Biol 626:15–27
Garcon N, Chomez P, Van Mechelen M (2007) GlaxoSmithKline Adjuvant Systems in vaccines: concepts, achievements and perspectives. Expert Rev Vaccines 6:723–739
Gasparini R, Schioppa F, Lattanzi M, Barone M, Casula D, Pellegrini M, Veitch K, Gaitatzis N (2010) Impact of prior or concomitant seasonal influenza vaccination on MF59-adjuvanted H1N1v vaccine (Focetria) in adult and elderly subjects. Int J Clin Pract 64:432–438
Gay NJ, Gangloff M (2007) Structure and function of Toll receptors and their ligands. Annu Rev Biochem 76:141–165
Glenny AT, Pope CG, Waddington H, Wallace U (1926) Immunological notes. XVII–XXIV. J Pathol Bacteriol 29:31–40
Gluck R, Metcalfe IC (2002) New technology platforms in the development of vaccines for the future. Vaccine 20(Suppl 5):B10–B16
Guy B (2007) The perfect mix: recent progress in adjuvant research. Nat Rev Microbiol 5:505–517
Huang DB, Wu JJ, Tyring SK (2004) A review of licensed viral vaccines, some of their safety concerns, and the advances in the development of investigational viral vaccines. J Infect 49:179–209
Medzhitov R (2001) Toll-like receptors and innate immunity. Nat Rev Immunol 1:135–145
Mortellaro A, Ricciardi-Castagnoli P (2011) From vaccine practice to vaccine science: the contribution of human immunology to the prevention of infectious disease. Immunol Cell Biol 89:332–339
O'Hagan DT, De Gregorio E (2009) The path to a successful vaccine adjuvant—'the long and winding road'. Drug Discov Today 14:541–551
O'Hagan DT, Valiante NM (2003) Recent advances in the discovery and delivery of vaccine adjuvants. Nat Rev Drug Discov 2:727–735
O'Hagan DT (2001) Recent developments in vaccine delivery systems. Curr Drug Targets Infect Disord 1:273–286

Podda A (2001) The adjuvanted influenza vaccines with novel adjuvants: experience with the MF59-adjuvanted vaccine. Vaccine 19:2673–2680

Rappuoli R (2007) Bridging the knowledge gaps in vaccine design. Nat Biotechnol 25:1361–1366

Ravanfar P, Satyaprakash A, Creed R, Mendoza N (2009) Existing antiviral vaccines. Dermatol Ther 22:110–128

Reed SG, Bertholet S, Coler RN, Friede M (2009) New horizons in adjuvants for vaccine development. Trends Immunol 30:23–32

Reed SG, Orr MT, Fox CB (2013) Key roles of adjuvants in modern vaccines. Nat Med 19:1597–1608

Riedel S (2005) Edward Jenner and the history of smallpox and vaccination. Proc (Bayl Univ Med Cent) 18:21–25

Robinson HL, Amara RR (2005) T cell vaccines for microbial infections. Nat Med 11:S25–S32

Rodriguez PC, Rodriguez G, Gonzalez G, Lage A (2010) Clinical development and perspectives of CIMAvax EGF, Cuban vaccine for non-small-cell lung cancer therapy. MEDICC Rev 12:17–23

Roman F, Vaman T, Gerlach B, Markendorf A, Gillard P, Devaster JM (2010) Immunogenicity and safety in adults of one dose of influenza A H1N1v 2009 vaccine formulated with and without AS03A-adjuvant: preliminary report of an observer-blind, randomised trial. Vaccine 28:1740–1745

Romanowski B, De Borba PC, Naud PS, Roteli-Martins CM, De Carvalho NS, Teixeira JC, Aoki F, Ramjattan B, Shier RM, Somani R, Barbier S, Blatter MM, Chambers C, Ferris D, Gall SA, Guerra FA, Harper DM, Hedrick JA, Henry DC, Korn AP, Kroll R, Moscicki AB, Rosenfeld WD, Sullivan BJ, Thoming CS, Tyring SK, Wheeler CM, Dubin G, Schuind A, Zahaf T, Greenacre M, Sgriobhadair A (2009) Sustained efficacy and immunogenicity of the human papillomavirus (HPV)-16/18 AS04-adjuvanted vaccine: analysis of a randomised placebo-controlled trial up to 6.4 years. Lancet 374:1975–1985

Schiller JT, Castellsague X, Villa LL, Hildesheim A (2008) An update of prophylactic human papillomavirus L1 virus-like particle vaccine clinical trial results. Vaccine 26(Suppl 10): K53–K61

Schultze V, D'agosto V, Wack A, Novicki D, Zorn J, Hennig R (2008) Safety of MF59 adjuvant. Vaccine 26:3209–3222

Schwarz TF (2009) Clinical update of the AS04-adjuvanted human papillomavirus-16/18 cervical cancer vaccine, Cervarix. Adv Ther 26:983–998

Senovilla L, Vacchelli E, Garcia P, Eggermont A, Fridman WH, Galon J, Zitvogel L, Kroemer G, Galluzzi L (2013) Trial watch: DNA vaccines for cancer therapy. Oncoimmunology 2:e23803

Walker WT, Faust SN (2010) Monovalent inactivated split-virion AS03-adjuvanted pandemic influenza A (H1N1) vaccine. Expert Rev Vaccines 9:1385–1398

Wellington K, Goa KL (2003) Measles, mumps, rubella vaccine (Priorix; GSK-MMR): a review of its use in the prevention of measles, mumps and rubella. Drugs 63:2107–2126

WHO (2004) BCG vaccine. WHO position paper. Wkly Epidemiol Rec 79:27–38

WHO (2012) Summary of WHO position papers—recommendation for routine immunization in: immunization, vaccines and biologicals. World Health Organization, Geneva

WHO Fact Sheet 104. (2012). http://www.who.int/mediacentre/factsheets/fs104/en/

Part II
Delivery Systems for Subunit Vaccines

Chapter 3
Aluminum Adjuvants: Basic Concepts and Progress in Understanding

Erik B. Lindblad

3.1 Introduction

In 1926 A.T. Glenny and co-workers demonstrated the adjuvant effect of aluminum compounds (Glenny et al. 1926). He was inspired by the work of Hartley who showed that antigen–antibody complexes induced higher titers than injection of the antigen alone. Glenny prepared a variety of diphtheria toxoid precipitates and investigated their immunogenicity (Glenny et al. 1926, 1931). Among these were toxoids precipitated by addition of potassium alum ($KAl(SO_4)_2 \cdot 12H_2O$). He observed that injecting the diphtheria toxoid as an alum precipitate led to a significant increase in the immune response against the toxoid. Vaccines prepared in accordance with this principle have been used for vaccination and are generally referred to as ***alum-precipitated vaccines***.

Such preparations are usually highly heterogeneous (Holt 1950), depending on which anions (e.g., bicarbonate, sulfate, or phosphate) are present at the time of precipitation, e.g., as buffer constituents or growth media residuals in the antigen solution. It soon became clear that a number of disadvantages are associated with using this approach. As the adjuvant itself is formed upon co-precipitation with the antigen, it is not possible to characterize the adjuvant before mixing. In addition, the precipitation with antigen requires alkaline conditions, which are detrimental to some antigens. Therefore, the attention was shifted towards preformed aluminum hydroxide hydrated gels that have the ability to adsorb protein antigens from an aqueous solution, and such gels can be preformed in a well-defined and standardized

E.B. Lindblad (✉)
Brenntag Biosector, Elsenbakken 23, 3600 Frederikssund, Denmark
e-mail: erik.lindblad@brenntag-nordic.com

C. Foged et al. (eds.), *Subunit Vaccine Delivery*, Advances in Delivery Science and Technology, DOI 10.1007/978-1-4939-1417-3_3

way (Maschmann et al. 1931). Vaccine preparations based on this approach are generally referred to as ***aluminum adsorbed vaccines***. Aluminum phosphate was introduced as an adjuvant somewhat later. In 1946 Ericsson devised a method where diphtheria toxoid was co-precipitated into a matrix of aluminum phosphate (Ericsson 1946). Holt shortly after demonstrated that preformed aluminum phosphate prepared using equimolar amounts of aluminum chloride and trisodium phosphate, acted as an adsorbant and was adjuvant active with diphtheria (Holt 1947).

In addition to the alum precipitation and the preformed aluminum hydroxide and aluminum phosphate gels, a number of other aluminum compounds have been investigated as adjuvants, e.g., aluminum hydrochloride, which was used to raise anti-snake venom (Kawamura and Sawai 1989a, b), and aluminum silicate, which was applied for studying the immunoglobulin E (IgE) synthesis in experimental animal models (Fujimaki et al. 1984).

Finally, aluminum compounds are a constituent of some patented composite adjuvant formulations, such as AS04 (a formulation of aluminum hydroxide with monophosphoryl lipid A) (Garcon et al. 2006; Giannini et al. 2006) and of Algammulin™, a composite adjuvant formulation consisting of aluminum hydroxide and gamma-inulin (Cooper et al. 1991).

In the literature the word alum is often, but incorrectly, used to describe both aluminum hydroxide and aluminum phosphate gels. The aluminum salt used for co-precipitation with antigen, potassium alum, ($KAl(SO_4)_2 \cdot 12H_2O$), is in accordance with the chemical definition of alum, whereas neither aluminum hydroxide nor aluminum phosphate is. The general term aluminum adjuvants is therefore preferred in this chapter.

Data on the use of alum-precipitated vaccines can be found in older literature (Volk and Bunney 1942), but in modern vaccination the adsorption onto preformed aluminum hydroxide and aluminum phosphate gels is now preferred over the alum precipitation in vaccine preparations.

3.2 Application in Vaccines

Aluminum adjuvants are by far the most commonly used adjuvants in vaccine formulations for both human and veterinary use. For human prophylactic vaccination, aluminum adjuvants have primarily been used in vaccines as part of standard childhood vaccination programs in tetanus, diphtheria, pertussis and poliomyelitis vaccines and various polyvalent combination vaccines thereof. Later aluminum adjuvants were also introduced in vaccines against hepatitis A virus, hepatitis B virus, *Haemophilus influenzae b* (Kanra et al. 1999, 2003) and Japanese encephalitis virus (Dubischar-Kastner et al. 2010), in vaccines (LYMErix™) against Lyme's Disease (tick-borne *Borrelia burgdorferi* infection) (Poland and Jacobson 2001) and in vaccines against meningococcal disease. Among the more recent successful vaccines containing aluminum adjuvants are the vaccines against human papilloma virus (HPV) to protect against cervical cancer and genital warts, e.g., Cervarix™

and Gardasil™, (Giannini et al. 2006; Han et al. 2010). Aluminum-adsorbed vaccines against, e.g., anthrax, although not part of standard vaccination programs, are available for special risk groups, including military servicemen at risk of being exposed to weapons of biological warfare. Use of aluminum adjuvants in therapeutic vaccines is known from hyposensitizing treatment of allergic patients using aluminum-adsorbed allergens.

In veterinary medicine aluminum adjuvants have been used in a large number of vaccine formulations against viral (Pini et al. 1965; Hyslop and Morrow 1969; McDougall 1969; Sellers and Herniman 1974; Wilson et al. 1977) and bacterial diseases (Nagy and Penn 1974; McCandlish et al. 1978; Ris and Hamel 1979; Thorley and Egerton 1981), as well as in attempts to make anti-parasite vaccines (Gamble et al. 1986; Carlow and Bianco 1987; Leland et al. 1988; Monroy et al. 1989) (Table 3.1).

Hundreds of millions of doses of aluminum-adjuvanted vaccines have been administered over a time span comparable to the average life time of a human being. Hence, the long-term safety profile is very well documented and these adjuvants are generally regarded as safe when used in accordance with current vaccination schedules and recommended dose limitations (WHO Technical Report Series vol 595 1976; Edelman 1980; Goldenthal et al. 1993). For an extensive review, see Lindblad 2007.

For vaccines intended for use in humans there are well-defined limitations for the content of aluminum allowed. These limits are 1.25 mg aluminum per dose in Europe (Ph.Eur. 3 ed. 1997). In USA 0.85 mg aluminum per dose is accepted if determined by assay, 1.14 mg if determined by calculation and 1.25 mg if safety and efficacy data justifies it (Code of Federal Regulations 21 2003).

In veterinary vaccines there is no defined maximum limit for the allowed content of aluminum. Here the dose is normally set from a balance between efficacy and local reactogenicity, which may differ considerably in different animal species. The optimal dose of adjuvant is usually determined empirically in a pilot trial.

Table 3.1 Examples of application of aluminum adjuvants in veterinary vaccinology

Viral vaccines	Bacterial vaccines	Experimental anti-parasite vaccines
Avian infectious bronchitis virus	*Bacteroides nodosus*	*Cooperia punctata*
Canine hepatitis virus	*Bordetella bronchiseptica*	*Nematospiroides dubius*
Foot-and-mouth disease virus	*Clostridium botulinum*	*Onchocerca lienalis*
Newcastle disease virus	*Clostridium chauvoei*	*Trichinella spiralis*
	Clostridium novyi	
	Clostridium perfringens	
	Clostridium septicum	
	Clostridium sordellii	
	Haemophilus somnus	
	Leptospira interrogans	
	Pasteurella multocida	

3.3 Limitations to Use

The choice of adjuvant for a particular application should obviously be based on the idea that it should primarily stimulate those reaction patterns of the immune response that are responsible for protection against the target disease. For some diseases, protection is provided by an adequate antibody response. Other diseases require cellular immunity for protection. Aluminum adjuvants are efficiently generating an early, high-titered and lasting antibody response, biased towards the Th2 immune response profile (Grun and Maurer 1989). This introduces an obvious limitation to the use of these adjuvants. A Th2-biased immune response is not likely to protect efficiently against diseases for which Th1 immunity and major histocompatibility complex (MHC) class I-restricted cytotoxic T-lymphocytes (CTLs) are essential for protection, such as intracellular parasites or tuberculosis (Lindblad et al. 1997).

For vaccines against typhoid fever and seasonal influenza, aluminum adjuvants have failed to provide satisfactory augmentation of the immune response in a number of cases (Cvjetanovic and Uemura 1965; Davenport et al. 1968).

Both aluminum hydroxide and aluminum phosphate adjuvants have been tested in a few DNA vaccine formulations (Ulmer et al. 2000; Kwissa et al. 2003) including experimental vaccines against hepatitis B virus and in vaccine formulations where the DNA oligonucleotides were combined with aluminum-adsorbed protein antigens (Kwissa et al. 2003). Here it was shown that aluminum hydroxide had an inhibiting effect, whereas aluminum phosphate adjuvant augmented the immune response against the antigen encoded by the DNA. A very plausible explanation is that the content of phosphate in the DNA molecule apparently gives a high binding affinity of the nucleotides to the aluminum hydroxide, which in turn prevents the transcription of RNA and subsequent translation of the nucleotides into protein (Kwissa et al. 2003).

There is some discussion about the ability of aluminum adjuvants to potentiate the immune response against peptide-based vaccines (Francis et al. 1987). In some cases, e.g., with foot-and-mouth disease (FMD) virus peptides, the problem could be overcome by conjugating the small peptides to a larger carrier molecule (Francis et al. 1985), whereas in other cases this is not a viable strategy (Lew et al. 1988; Geerligs et al. 1989).

The overall picture that aluminum adjuvants give a fairly clear Th2 immune response profile in mice (Grun and Maurer 1989), however, dates back from a time when Th1 and Th2 were the only effector T-cell subsets known. Since then, additional effector T-cell subsets have been identified, e.g., Th17 T-cells, regulatory T-cells (T_{regs}), and follicular T-helper cells (T_{FH}). Further research is required to elucidate to what extent aluminum adjuvants may encompass the stimulation of these additional T-cell subsets.

3.3.1 Thermostability of Aluminum Adjuvants

The EP monograph 1664 for aluminum hydroxide prescribes that the aluminum hydroxide adjuvant should be stored at temperatures between 4 and 30 °C. However, the characteristics in terms of thermostability introduce some limitations to the use of aluminum adjuvants.

Traditional aluminum-adsorbed vaccines are frost sensitive, and attempts to lyophilize adsorbed vaccines in the absence of lyoprotectants have therefore failed. Freezing of aluminum adjuvants leads to the destruction of the hydrated colloid structure and induces the formation of larger aggregates. For samples that have been frozen completely this is easily recognized by a collapse of the gel, leaving behind a significantly increased fraction of clear supernatant and quick sedimentation of the precipitate (Fig. 3.1). Partial freeze damages may be less easily detected visually, but as the freeze destruction is accompanied by a loss of protein adsorption capacity, partial destruction of the adjuvant will be accompanied by partial loss of protein adsorption capacity (Lindblad and Schønberg 2010).

Freezing of adsorbed vaccines, i.e., the situation where commercial vaccines are exposed to freezing, has also been studied by a group under the World Health Organization (WHO) (Kurzatkowski et al. 2013). Also here a faster sedimentation (2–15 times as fast), as compared to non-freeze damaged vials, was seen.

Fig. 3.1 Two vials of aluminum hydroxide gel adjuvant. The vial to the *left* has not been exposed to freezing. The vial to the *right* has been frozen and subsequently thawed: The gel structure in the vial is collapsed and the colloid bound water released into the supernatant

Electron microscopy and X-ray analysis demonstrated the formation of larger aggregates, primarily consisting of aluminum salt in the precipitate.

Similar findings have led to the development of the so-called *shake test* (Kartoglu et al. 2010), which was validated against phase contrast microscopy analysis. Non-frozen samples showed a fine-grain structure under phase contrast microscopy, but freeze-damaged samples showed large agglomerates of massed precipitates with amorphous, crystalline, solid, and needle-like structures. Particles in the non-frozen samples had an average diameter above 1 μm (vaccines against diphtheria-tetanus-pertussis; Haemophilus influenzae type b; hepatitis B; diphtheria-tetanus-pertussis-hepatitis B) to 20 μm (diphtheria and tetanus vaccines, alone or in combination). By contrast, aggregates in the freeze-damaged samples measured up to 700 μm (max value) for diphtheria-tetanus-pertussis and 350 μm on average (Kartoglu et al. 2010).

Attempts are made to develop procedures that allow freezing of vaccines without the described damages by adding lyoprotectants such as trehalose to the adjuvant/vaccine formulations (Clausi et al. 2008; Hassett et al. 2013).

Both aluminum hydroxide and aluminum phosphate adjuvants can be autoclaved to obtain sterility, not only as part of their production process, but can also be re-autoclaved later if sterility is questioned.

What is referred to as aluminum hydroxide adjuvant is poorly crystalline aluminum oxyhydroxide corresponding to the mineral form Boehmite (Shirodkar et al. 1990) whereas aluminum phosphate adjuvant is better described as aluminum hydroxy phosphate (Shirodkar et al. 1990). Accordingly X-ray diffraction crystallography can be used to study possible changes induced by autoclaving. This is typically done by studying the width of the diffraction bands at half their height. When aluminum hydroxide adjuvant is autoclaved the main diffraction band decreased in width, indicating an increase in crystallinity. Also the pH value is slightly decreased following autoclaving indicating deprotonization (Burrell et al. 1999). Aluminum phosphate adjuvant is amorphous when studied by X-ray diffraction and retains the amorphous pattern after autoclaving (Burrell et al. 1999). However, autoclaving, as well as re-autoclaving, leads to a slight reduction of the protein adsorption capacity for both aluminum hydroxide and aluminum phosphate adjuvants (Burrell et al. 1999).

3.4 Mechanisms of Action

In the older literature, the function of a depot-forming or repository adjuvant was originally described as to delay the clearance of the antigen from the injection site and to sustain the release of adsorbed antigen from the inoculated depot (Glenny et al. 1931), thereby ensuring a prolonged exposure of the antigen to the cells of the immune system. Although it quickly became obvious that the sustained release was inadequate for explaining the adjuvant mechanism of aluminum adjuvants, the physical adsorption of antigen onto the adjuvant is still considered a very important parameter for the effect of these adjuvants in the early phases of the immune response.

3.4.1 Mechanisms of Antigen Adsorption

The physicochemical mechanisms behind the antigen adsorption process are complex and depend on the nature and composition of the antigen itself, as well as on the chemical composition of the buffer applied during the adsorption process. The two main mechanisms for adsorption are electrostatic attraction and ligand exchange.

It is well-established that charged particles in aqueous suspension attract ions of opposite charges (counter ions), as illustrated by the Gouy-Chapman double layer model (Fig. 3.2). This model describes the existence of an inner layer, the *Stern layer*, and an outer, more diffuse layer surrounding the particles. The electrical potential decreases with the distance from the particle surface. The electrical potential at the outer rim of the diffuse layer is known as the zeta potential.

Electrostatic attraction between the aluminum adjuvant and a protein antigen is possible when the adjuvant and the antigen have opposite electrical charges. Aluminum hydroxide has an alkaline *point of zero charge* (PZC). At pH values

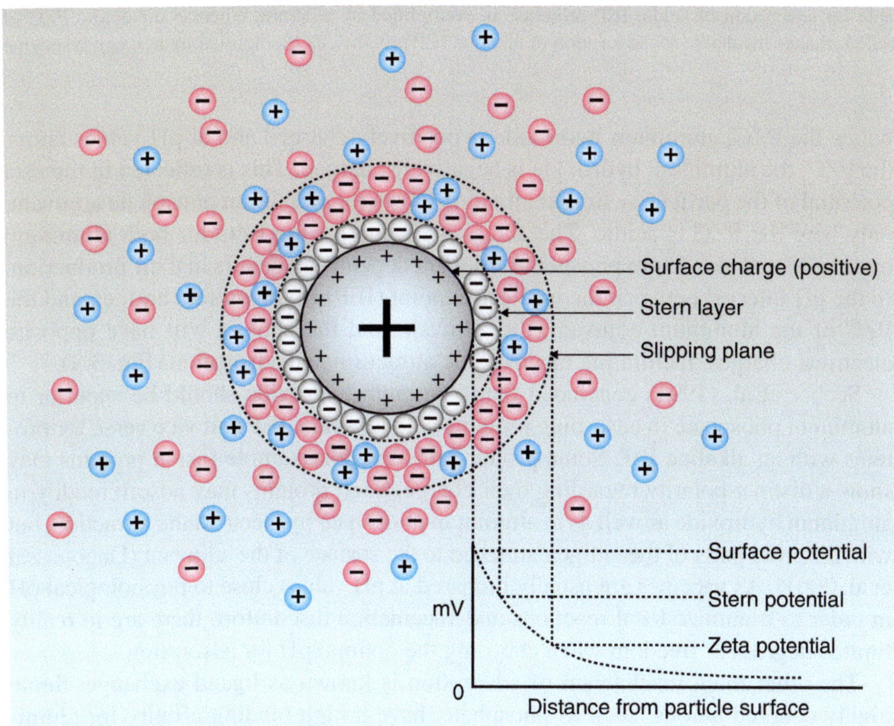

Fig. 3.2 The Gouy-Chapman double layer model

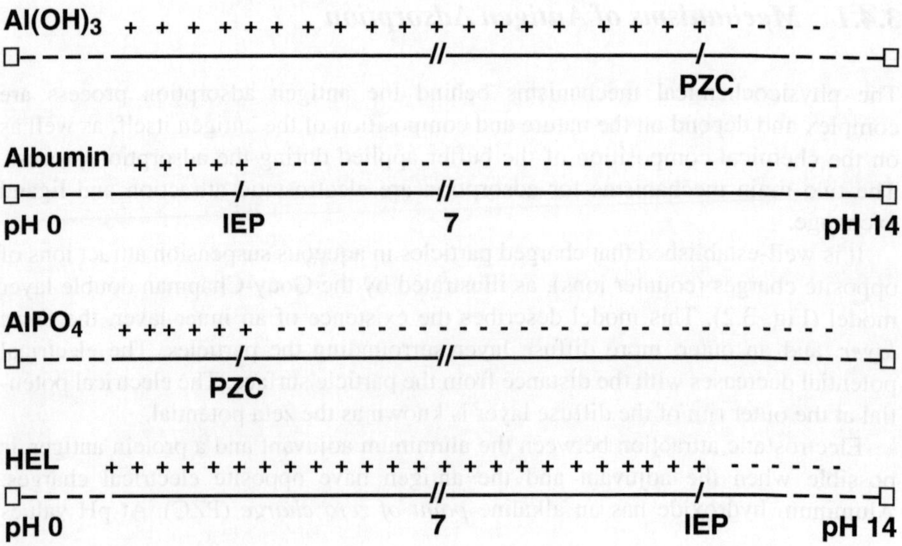

Fig. 3.3 In the pH range between the IEP of the antigen and the PZC of the mineral adjuvant there is basis for electrostatic attraction, due to opposite charge. The alkaline PZC for $Al(OH)_3$ makes it suitable for adsorption of acidic IEP proteins, as exemplified by albumin, whereas the acidic PZC of $AlPO_4$ makes it suitable for adsorption of alkaline IEP proteins, as exemplified by hen egg lysozyme

below the PZC, aluminum hydroxide is positively charged and at pH values above the PZC, the aluminum hydroxide is negatively charged. This is reflected in the zeta potential of the particle. A similar situation exists for aluminum phosphate adjuvant, only here the PZC is acidic. The actual pH value for the PZC for both aluminum hydroxide and aluminum phosphate adjuvant depends on details in their production. In the pH interval between the isoelectric point (IEP) of the protein antigen and the PZC of the aluminum adjuvant, the adjuvant and the antigen will have opposite electrical charges, facilitating electrostatic attraction and adsorption (Fig. 3.3).

Seeber et al. (1991) concluded that aluminum hydroxide should be superior to aluminum phosphate in adsorbing proteins with an acidic IEP and vice versa for proteins with an alkaline IEP. Some protein antigens, for example fusion proteins may show a distinct polarity regarding their charge. Such proteins may adsorb readily to aluminum hydroxide as well as to aluminum phosphate by electrostatic attraction, but with different parts of the antigen attached to the surface of the adjuvant (Dagouassat et al. 2001). As vaccines are usually prepared at pH values close to physiological pH in order to minimize local reactions and vaccination discomfort, there are in reality limited degrees of freedom when choosing the optimal pH for adsorption.

The other main mechanism of adsorption is known as ligand exchange. Some highly charged anions, such as phosphate, have a high binding affinity for aluminum adjuvants, in particular aluminum hydroxide, and are able to displace surface-bound hydroxyl groups. If the antigen contains phosphorylated groups, for example if there is phosphoserine in the amino acid sequence, this mechanism may account for a high-affinity binding to the adjuvant. Ligand exchange is also important for the adsorption of, e.g., HBsAg particles, which contain fragments of phospholipid bilayer (Iyer et al. 2004), and has been demonstrated in experiments using phosphorylated alpha-casein as model antigen (Iyer et al. 2003). Ligand exchange provides a stronger binding than electrostatic attraction, since it might account for adsorption of proteins even when repulsive electrostatic forces between antigen and adjuvant exist (Iyer et al. 2003).

In addition, other intermolecular binding forces, like hydrophobic interactions, may play a role in protein adsorption (Al-Shakhshir et al. 1995) and each type of binding force may play distinct roles for a given antigen-adjuvant combination, depending on the nature of the antigen and the chemical environment (pH, ionic strength, presence of surfactants, etc.) (Rinella et al. 1995, 1998a, b).

3.4.2 Antigen Targeting

Adsorption of soluble antigen to aluminum adjuvant particles leads to presentation of the antigenic epitopes as particulate structures. This is of importance because antigen-presenting cells (APCs) are believed to take up particles by phagocytosis more efficiently than they take up soluble protein by pinocytosis. This line of research was pioneered by Mannhalter and co-workers (1985), who investigated the uptake of soluble tetanus toxoid (TT) versus Al-adsorbed TT in human macrophages. When these were subsequently incubated with autologous T-cells in the absence of antigen, a highly increased T-cell proliferation was seen in the cultures of macrophages that had been exposed to Al-adsorbed antigen, as compared to cultures exposed to soluble antigen. Also, cultures of macrophages pulsed with Al-adsorbed ^{125}I radiolabelled TT expressed higher radioactivity than cultures, which had been pulsed with soluble ^{125}I-TT (Mannhalter et al. 1985). This led to the conclusion that adsorbing a soluble antigen onto aluminum adjuvants favors antigen uptake by APCs (antigen targeting).

3.4.3 The Inflammatory Focus

Upon injection of the adsorbed vaccine a mild inflammatory reaction is established at the injection site. Neutrophils, eosinophils, lymphocytes, macrophages, and immature dendritic cells (DCs) are attracted to the injection site (White et al. 1955; Walls 1977). At this point, weakly bound antigens may be partly dissociated from the aluminum adjuvant under influence of interstitial fluid, whereas tighter bound antigens may remain adsorbed (Chang et al. 2001; Morefield et al. 2005; Jiang et al. 2006). Antigen thus released can reach the lymph nodes through the afferent lymphatics and react directly with surface-bound immunoglobulin on B-cells within minutes (Hem and HogenEsch 2007). Alternatively, the antigen-adjuvant complex can be taken up by immature DCs and transported to the lymph nodes for T-cell presentation (Guermonprez et al. 2002). This involves cellular migration and may take 8–12 h (Itano et al. 2003; Sixt et al. 2005; Hem and HogenEsch 2007). Further, an encapsulation process may take place at the injection site resulting in granuloma formation as part of the inflammatory reaction *sensu latu*. The relative significance of these reaction patterns may vary from case to case, depending on the specific antigen and the host response.

The formation of an inflammatory focus seems to contribute to stimulating the immune response. This can be seen as a consequence of disturbing the tissue integrity, release of uric acid crystals from dying cells and heat-shock proteins (HSP, e.g., HSP-70) as danger signals (Goto et al. 1997; Asea et al. 2000, 2002; Lindblad 2006). John Naim and co-workers found a direct correlation between the magnitude of the antibody response and the local inflammatory response elicited after injection of different plurivalent metal oxides, tested as adjuvants, including aluminum (Naim et al. 1997).

3.4.4 The NALP3 Inflammasome

Until approximately 10 years ago vaccine adjuvants were traditionally subdivided into immunomodulators and delivery systems or vehicles. Immunomodulators are defined as compounds, which interact with pattern-recognition receptors (PRRs) such as the Toll-like receptors (TLRs), whereas delivery systems do not. Little was known about the mechanisms of action of the adjuvants, which had a documented adjuvant effect, but did not function as TLR agonists. Brewer and co-workers investigated cultures of DCs in the presence of aluminum adjuvants and did not observe increased expression of MHC class II and co-stimulatory molecules (Sun et al. 2003). This demonstration in vitro led to the conclusion that aluminum adjuvants did not act through TLRs (Sun et al. 2003), which is consistent with the fact that mineral adjuvants are devoid of pathogen-associated molecular patterns (PAMPs).

In 2002 Jürg Tschopp's group at University of Lausanne defined the inflammasome as a molecular platform triggering activation of inflammatory caspases and processing of pro-IL-1β (Martinon et al. 2002). Further work lead to a possible explanation for the mechanisms of action of aluminum adjuvants. According to this, uptake of Al-adjuvanted vaccines by DCs is accompanied by K^+ efflux and the three intracellular proteins NALP3, CARDINAL, and ASC then join to form the NALP3 inflammasome (Fig. 3.4).

NALP3 is belonging to the Nucleotide-binding oligomerization domain (NOD)-like receptors (NLRs). CARDINAL is also known as CARD-8, or Caspase recruitment domain-8 and ASC is an Apoptosis-associated Speck-like protein including a CARD. Once the NALP3 inflammasome is assembled it induces cleavage of the 45 kDa pro-caspase-1 to the active caspase-1 enzyme, which is a cysteine-rich asparaginase able to cleave pro-IL-1β and pro-IL-18 into their active counterparts IL-1β and IL-18, which are secreted by DCs as active, pro-inflammatory cytokines

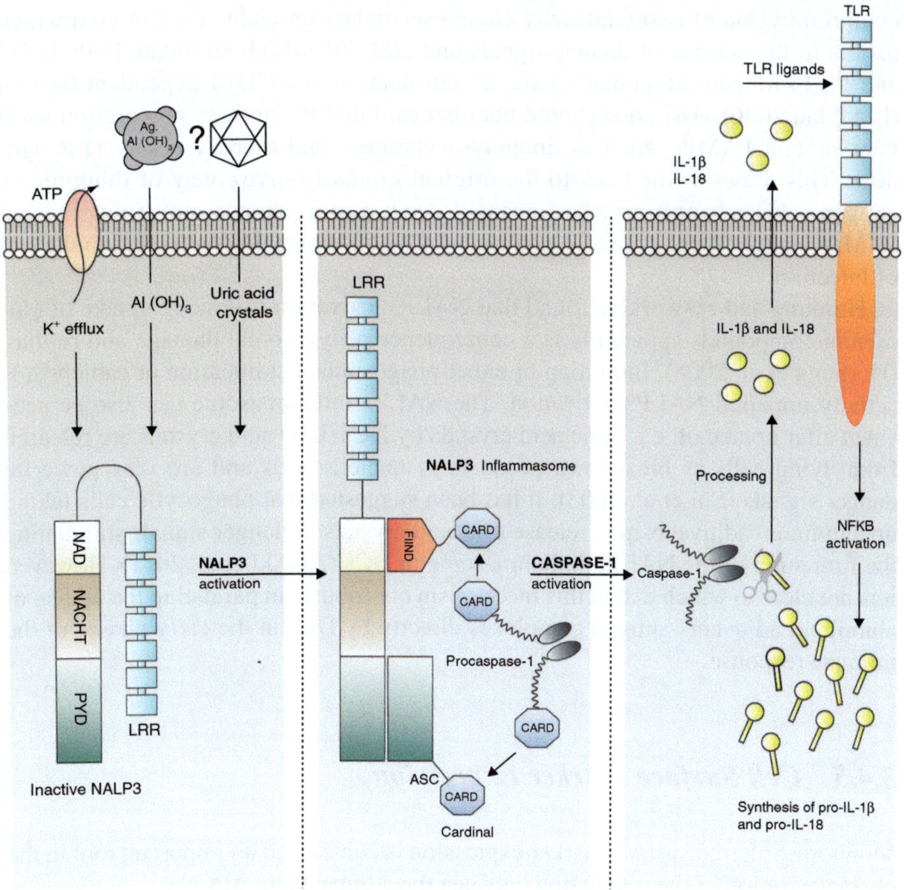

Fig. 3.4 Assembly of the NALP3 inflammasome in DCs after uptake of aluminum adjuvant (or uric acid crystals) followed by activation of caspase-1 and cleavage of pro-IL-1β and pro-IL-18 into active pro-inflammatory cytokines (*graphics inspired by Leigh Church et al. 2008, ref. 111*)

(Kool et al. 2008; Li et al. 2007, 2008; Eisenbarth et al. 2008). In NALP3 knockout mice, vaccination with aluminum-adjuvanted antigen resulted in IL-1β levels comparable to the levels seen in mice receiving saline or antigen alone (Kool et al. 2008). Apparently, this pathway does not require MyD88, as the process could take place in MyD88-deficient mice (Li et al. 2007). The synthesis of pro-IL-1β and pro-IL-18 is affected by TLR agonists reacting with TLRs on the surface of DCs thus resulting in activation of the nuclear factor-κB (NF-κB) pathway leading to the transcriptional activation of the genes for pro-IL-1β and pro-IL-18 in the nucleus of DCs.

Additional mechanisms may hypothetically contribute to the upregulation of the precursor molecules pro-IL-1β and pro-IL-18 in vivo. One such possible mechanism draws a line back to my previous hypothesis (Lindblad 2006), suggesting that the inflammatory focus induced by injection of Al-adjuvanted vaccines and the compromisation of tissue integrity elicit a secondary cascade effect in vivo which results in the release of danger signals and HSP-70, which stimulate both TLR2 and TLR4 for pro-inflammatory signal transduction in a CD14-dependent fashion (Lindblad 2006). Also, one should not disregard that the antigen preparation itself may contain PAMPs, such as lipopolysaccharides, and thereby act as TLR agonists. This draws a line back to the original *gradual-release* way of thinking, as antigen released from an Al-adjuvanted depot would expose antigen-associated PAMPs making them available to surface TLRs on nearby DCs attracted to the site of injection.

Hornung and co-workers found that NALP3 activation following uptake of aluminum compounds apparently is a consequence of lysosomal damage and rupture (Hornung et al. 2008). Inhibition of either phagosomal acidification or cathepsin B activity impaired NALP3 activation. The NALP3 inflammasome can also be activated after uptake of, e.g., uric acid crystals by DCs. Uric acid crystals are released from dying cells as breakdown products of nucleic acids and are very powerful danger signals (Shi et al. 2003). It has been suggested that phagocytic cells taking up aluminum adjuvants may release uric acid crystals as danger signals stimulating the formation of the NALP3 inflammasome in DCs (Al-Akl et al. 2011). However, it is not clear to which extent this mechanism contributes in parallel to the uptake of aluminum adjuvants antigen complexes directly by DCs in the early phases of the immune response.

3.4.5 Cell Surface Marker Expression

Monitoring altered surface marker expression is considered an important tool in the characterization of the interaction between the adjuvant and APCs.

Studies have comprised both bone marrow-derived DCs and peripheral blood mononuclear cells (PBMCs).

Brewer and co-workers (Sun et al. 2003) monitored the expression of MHC Class II, as well as CD40, CD80 and CD86 in cultures of BALB/c bone marrow DCs after incubation with various adjuvants and using ovalbumin (OVA) as model antigen. They found no significant upregulation of mentioned surface markers after incubation with aluminum hydroxide and OVA when compared to incubation with lipopolysaccharide (LPS, a well-known TLR 4 agonist) and OVA. They concluded that aluminum adjuvant failed to demonstrate the ability to activate DCs in vitro (Sun et al. 2003).

In a later study Sokolovska and co-workers looked at surface marker expression on DCs from BALB/c bone marrow after incubating with aluminum hydroxide or aluminum phosphate respectively (Sokolovska et al. 2007) and with OVA as antigen. They found that both adjuvants increased the expression of CD86, but only aluminum hydroxide also increased the expression of CD80, although only modestly (Sokolovska et al. 2007). Apparently they found no significant stimulation of the expression of neither CD40 nor CD275 after treatment with aluminum adjuvants.

Ulanova et al. investigated the direct effect of aluminum hydroxide in cultures of PBMCs (Ulanova et al. 2001). They found an increase in the expression of co-stimulatory and adhesion molecules: MHC Class II, CD40, CD54 (formerly known as ICAM-1), CD58 (formerly known as LFA-3), CD83 (maturation marker) and CD86 (formerly known as B7-2) on monocytes as well as an increase of mRNA for IL-4. However, in the presence of anti-IL4 antibody or in highly purified monocyte cultures (i.e., depleted for $CD4^+$ T-cells) there was no increase in MHC Class II expression. So apparently aluminum adjuvant-induced monocyte-derived cytokines stimulate $CD4^+$ T-cells to secrete IL-4 which in turn stimulate MHC class II expression on the monocyte surface.

Rimaniol et al. also investigated surface marker expression in cultures of PBMCs from the blood of healthy donors (Rimaniol et al. 2004). They found that aluminum-treated macrophages expressed surface markers similar to those described for cultured DCs ($HLA\text{-}DR^{high}$, $CD86^{high}$ and $CD14^-$). They also screened for CD1a and CD83. Incubation with aluminum hydroxide induced a readily detectable CD83 expression, but no CD1a expression (Rimaniol 2004). The conclusion was that the macrophages, thus treated, develop into a cell type which, with respect to surface marker expression, resembles DCs, although with a phenotype distinctly different from these.

3.4.6 Cytokine Profiles

Studies of cytokines in model systems were initiated almost 35 years ago and were subsequently intensified thanks to the later introduction of gene-disrupted mice in research. Among the early observations in classical animal models was the demonstration that aluminum-adsorbed TT led to an increase in antigen-induced T-cell proliferation, apparently due to increased release of IL-1 (Mannhalter et al. 1985). On the other hand, there was a lack of importance of IL-1 in the augmentation of the primary antibody response in rabbits immunized with aluminum adjuvant (Sagara et al. 1990).

The first to conclude that aluminum adjuvants stimulated a Th2-biased immune response in animal models were Grun and Maurer (1989). They demonstrated that anti-IL-1α or anti-IL-4 antibody was able to inhibit an antigen-specific T-cell proliferative response after immunization with aluminum adjuvant, and the proliferative response was inhibited by an anti-CD4 antibody. This indicated that the proliferating $CD4^+$ T-cells from mice immunized using aluminum adjuvant were of the Th2 subset. Later studies, including studies in knock-out mice, have also confirmed a clear Th2 profile in mice (Brewer et al. 1999). Lindblad and co-workers found a stimulatory effect looking at IL-4 and IL-10 specific mRNA in the regional draining lymph nodes at day 7 following vaccination of C57BL/6J mice with aluminum-adjuvanted vaccine (Lindblad et al. 1997).

Important studies by Brewer et al. demonstrated the importance of IL-4 in the function of aluminum adjuvants and the impact on Th subset determination. In IL-4 gene-disrupted mice, immunization with OVA and $Al(OH)_3$ gave IgG2a titers of a similar magnitude as when OVA was injected together with Freund's Adjuvant (Brewer et al. 1996). Interestingly, the group immunized with OVA and $Al(OH)_3$ continued to produce IL-5 (a cytokine normally associated with the Th2 profile). This was in support of the idea that the major role of aluminum-induced IL-4 in Th-subset stimulation is to down-regulate the Th1 response. In a later study, Brewer's group showed that aluminum adjuvants could induce Th2-associated IL-4 and IL-5 production in the absence of IL-4 signaling in mice deficient in either IL-4Rα or STAT6. It was concluded that the Th2 responses could not be due to IL-13 as the IL-13 response is also impaired in IL-4Rα- or STAT6-deficient mice (Brewer et al. 1999).

As discussed above, uptake of aluminum adjuvants by DCs followed by inflammasome-induced activation of caspase-1 leads to cleavage of pro-IL-1β and pro-IL18 into the active pro-inflammatory cytokines (Kool et al. 2008; Li et al. 2007, 2008; Eisenbarth et al. 2008). It has been shown that IL-18 deficient mice immunized with OVA and $Al(OH)_3$ have reduced IL-4 production in lymph node cells, as compared to wild-type mice. However, if exogenous IL-18 was added it did not further enhance the aluminum-induced Th2 response (Pollock et al. 2003). Although the aluminum adjuvant led to reduced IL-4 production in IL-18–/– mice, this was not accompanied by a reduced level of serum IgG1. Apparently, there is poor correlation between this particular antibody subclass and IL-4 production (Pollock et al. 2003).

The overall Th2-biased reaction profile of aluminum adjuvants may be modified by adding IL-12 to the formulation. An ($Al(OH)_3$/IL-12) complex induced a Th1-like response, rather than a Th2 response, when used as an adjuvant (Jankovic et al. 1999) and the Th1 promoting effect of the $Al(OH)_3$/IL-12 complex was greatly augmented by the co-administration of exogenous IL-18 (Pollock et al. 2003).

3.5 In Vivo Clearance of Aluminum Adjuvants

In contrast to a number of other metal ions, no physiological function in the mammalian organism, e.g., as essential trace element, as coenzyme or otherwise, has been established for aluminum. However, due to the abundance of aluminum in the environment, both humans and animals are fairly constantly being exposed to aluminum in various forms (Martyn et al. 1989; Tomljenovic 2011; HogenEsch 2013). As a consequence, aluminum is normally found in the blood and serum of humans and animals, whether or not they have been vaccinated using aluminum adjuvants (Martyn et al. 1989; Flarend et al. 1997; Tomljenovic 2011). The major source of this aluminum is apparently oral intake of food and drinking water, which was reported to be in the magnitude of 5–10 mg daily for humans in Britain (Martyn et al. 1989). This aluminum uptake is excreted with the urine by individuals with normal renal functions.

Previous claims that aluminum adjuvants are not broken down in vivo and subjected to excretion, however, have been challenged. The clearance in vivo of aluminum adjuvants was investigated in rabbits using adjuvants prepared from the isotope

^{26}Al (Flarend et al. 1997). Excretion of ^{26}Al was followed in the urine and in the blood using accelerator mass spectroscopy for a period of 28 days. As early as 1 h following i.m. injection radioactively labeled Al could be detected in the blood. Significant pharmacokinetic differences were found in the excretion rates of aluminum hydroxide and aluminum phosphate adjuvant. It was found that approx. three times more ^{26}Al was excreted from animals vaccinated with aluminum phosphate than from those vaccinated with aluminum hydroxide. Assumingly, interstitial fluid containing organic acids with an α-hydroxy carboxylic acid, able to chelate Al, was more capable in dissolving aluminum phosphate than aluminum hydroxide (Flarend et al. 1997).

According to the calculation of Flarend it seems that the amount of aluminum administered via vaccination does not contribute significantly to the general exposure to aluminum in humans and serum levels of aluminum.

It is likely that the excretion through blood and urine described above primarily involves Al dissolved by interstitial fluid, whereas the radioactivity detected in lymph nodes and spleen might involve also Al-adjuvant taken up by APCs. In addition to this, one might expect to find a residual level of radioactivity encapsulated at the injection site.

3.6 Side Effects

Any visible or palpable reaction at the injection site after vaccination is in principle *non grata*. Aluminum hydroxide and aluminum phosphate adjuvants have been used for more than half a century now and are generally regarded as safe when used according to current immunization schedules (WHO Technical Report Series vol. 595, 1976; Edelman 1980). In 1993 the U.S. NCVDG Working Group on Safety Evaluation of Vaccine Adjuvants with the participation of the Food and Drug Administration (FDA) representatives concluded that "the extensive experience with this class of adjuvant for vaccine use has indicated that it is safe" (Goldenthal et al. 1993). Contact hypersensitivity to aluminum is not commonly seen, and aluminum adjuvants are not considered pyrogenic, carcinogenic, or teratogenic in themselves (Edelman 1980; Böhler-Sommeregger and Lindemayr 1986). However, as discussed above, injection of a vaccine adjuvanted with a repository adjuvant will normally be accompanied by the formation of an inflammatory focus at the injection site (WHO Technical Report Series vol. 595 1976).

3.6.1 Local Reactions

Several cases of local reactions after administration of aluminum-adjuvanted vaccines have been reported (White et al. 1955; Frost et al. 1985; Böhler-Sommeregger and Lindemayr 1986; Vogelbruch et al. 2000; Bergfors et al. 2003). These can be seen *sensu latu* as a consequence of the inflammatory focus at the injection site.

They comprise swellings, indurations, erythemas, and cutaneous nodules that can persist for up to 8 weeks or sometimes longer (Frost et al. 1985). The reports often describe cases of hyposensitization of allergic patients who receive a large number of injections of adsorbed allergenic extracts over a limited period of time, or cases where injections were made incorrectly (Vogelbruch et al. 2000). In a vaccination program in Sweden, itching local reactions were found in 0.8 % out of 76,000 vaccinees (Bergfors et al. 2003). A follow-up study was carried out 5–9 years later comprising 241 children, who all reacted positively in the previous study in a test for delayed (type IV) hypersensitivity. Of the 241 individuals tested, 186, corresponding to 77.2 %, were now negative, 37 had a weak positive reaction whereas 15 had a positive reaction and 3 had a strongly positive reaction (Lidholm et al. 2013).

There are inconsistent observations whether adsorption onto aluminum adjuvants leads to increased or decreased vaccine reactogenicity (Butler et al. 1969; Collier et al. 1979).

On one hand Collier and co-workers (1979) found that booster vaccination with Al-adsorbed tetanus-induced more frequent local reactions than vaccination with plain toxoid. This could in part be explained by the plain toxoid vaccine being dispersed from the injection site before a local inflammatory reaction was established.

On the other hand, Butler's group found that adsorption onto aluminum hydroxide significantly reduced the side-effects with combined DTP vaccines (Butler et al. 1969).

It is conceivable that in the presence of reactogenic or toxic vaccine constituents, like pertussis toxin, peptidoglycans from Gram-negative cell walls or LPS the acute toxicity is reduced in adsorbed vaccines simply by blocking or delaying their release from the injection site.

The binding of LPS to aluminum hydroxide is well established and is much higher than to aluminum phosphate (283 μg/mg Al vs. 3 μg/mg Al, respectively) (Shi et al. 2001). This is ascribed to the phosphate content of LPS enabling a strong binding by ligand exchange onto the surface of aluminum hydroxide. Norimatsu found that adsorption of LPS onto aluminum hydroxide prior to injection inhibited or mitigated systemic effects like the trembling, transient leucopenia and elevated serum tumor necrosis factor (TNF)-α otherwise observed following i.m. injection of LPS in saline (Norimatsu et al. 1995). Also the level of IL-6 after administration of LPS was reduced when the LPS was adsorbed to aluminum hydroxide prior to injection (Shi et al. 2001).

3.6.2 *Impact of Administration Route*

Vaccinations may be given subcutaneously (s.c.) or intramuscularly (i.m.) and the injection modus is not without importance in relation to the perception of local reactogenicity. When immunizing by the s.c. route, the vaccine inoculum is introduced into a compartment with numerous sensory neurons (in contrast to the intramuscular compartment). The introduction of a local inflammatory response here may more easily give lead to irritation and itching reactions. Besides, a transient swelling, as a consequence of the inflammatory focus formed, may more easily be palpable through the skin. When immunizing by the intramuscular route even a similar size swelling may be less easily visible and palpable as it is located in deeper lying tissue.

3.6.3 *Macrophagic Myofasciitis*

Local granuloma formation in inflammatory foci after injection of aluminum adjuvants were described already back in the mid-1950s (White et al. 1955). Attempts were later made to link the presence of a local inflammatory focus in the myofascii (the condition is referred to as macrophagic myofasciitis or MMF) after i.m. injections of Al-adjuvanted vaccines to conditions like myalgia and muscle fatigue. Such manifestations can be partly explained by the formation of adjuvant granulomas in the muscle. However, MMF has also been claimed to be statistically correlated to neurological disorders with no obvious etiological relation to the vaccination (Authier et al. 2001). Such correlations are, however, associated with statistical problems. The vaccination coverage in the western countries is very high. Hence, it is expected statistically that patients suffering from a wide range of etiologically unrelated diseases would all have been vaccinated with Al-containing vaccines at some point in their medical history. Another problem is that adequate statistical control groups of non-vaccinated individuals may be hard to find in the same population (Batista-Duharte et al. 2011). In a controlled study in primates by Verdier and co-workers in France it was not possible to detect any histological changes after injection of an aluminum adjuvanted vaccine besides the local inflammatory focus itself and they found no abnormal clinical signs associated to it (Verdier et al. 2005).

3.6.4 *The IgE Controversy*

It is a well-established fact that aluminum adjuvants in addition to stimulating IgG1 also stimulate the production of IgE as part of the overall Th2 profile in mice (Hamaoka et al. 1973; Kenney et al. 1989). This has often been mentioned as a disadvantage, as it may hypothetically introduce a risk of inducing allergic conditions. However, it has been difficult to demonstrate cases where vaccination with aluminum adjuvants has led to IgE-mediated allergy towards the vaccine antigen. In contrast, aluminum adjuvants have been used to hyposensitize allergic patients for many years with good results.

Much of the work on the IgE/Th2 stimulation by aluminum adjuvants in rodent models has been carried out in a dual setup model where groups of animals have been immunized with either aluminum hydroxide adjuvant or Freund's complete adjuvant (FCA) using the same antigen, (FCA is known as having a suppressive effect on IgE), subsequently assessing antibody and cytokine profiles in comparison.

Early studies using keyhole limpet hemocyanin as antigen showed that glycosylation-enhancing factors and $Fc\gamma R^+$ T-cells were involved in a regulatory pathway whereby aluminum adjuvant stimulated the synthesis of IgE (Uede et al. 1982; Uede and Ishizaka 1982). Later it was found (Brewer et al. 1996) that IgE production was abrogated in IL-4 gene disrupted mice (IL-4−/−) regardless whether aluminum adjuvant or FCA was used as adjuvant. This suggests that IL-4 is an essential prerequisite for the induction of IgE by aluminum adjuvants.

3.6.5 Evaluation of Reactogenicity

It is important to realize that the inflammatory response described is part of a normally functioning immune system. Hence, it may not be achievable to use repository adjuvants without temporarily also inducing an inflammatory focus around the inoculums (Technical Report Series vol 595, WHO 1976).

Realistically, evaluating reactogenicity is a question of when to judge that a mild and normal local reaction moves from being a temporary and minor cosmetic problem and develops into a toxic adverse reaction (*A. Batista Duharte*, *personal communication*).

3.6.6 Other Safety Considerations

Significant resources have been spent on throwing light on a possible link between aluminum exposure and the prevalence of Alzheimers disease (AD) (Tomljenovic 2011).

Some researchers have found aluminum deposits in AD brain tissue biopsies (Perl and Brody 1980; Andrasi et al. 2005) whereas others have not (Chafi et al. 1991; Landsberg et al. 1992). In a later report it was suggested that the aluminum detection was an artifact caused by the staining reagents used in the preparation of the specimen (Landsberg et al. 1993).

Persons with normal kidney function are known to excrete aluminum with the urine whereas persons with impaired renal function may to some extent accumulate it and may over a life-long exposure reach Al-levels associated with adverse reactions.

What is important in the present context, but often overlooked, are the proportions. The exposure to aluminum from vaccination, seen over a life time, is minimal compared to the daily intake of aluminum by drinking water, antiperspirants, and food additives in convenience food. For example, bread made with aluminum-based baking powder may contain up to 15 mg aluminum per slice and processed American cheese as much as 50 mg aluminum per slice (Mitkus et al. 2011). Even if it is taken into consideration that only as little as 0.25 % of the ingested aluminum may be taken up from the GI tract (Tomljenovic 2011) exposure to aluminum from the use of adsorbed vaccines in normal vaccination schedules will still be minimal.

The Canadian Alzheimers Society (http://www.alzheimer.ca/en/Research/Alzheimer-s-disease-research/Aluminum) concluded on their webpage (accessed February 2014): "Most researchers no longer regard aluminum as a risk factor for Alzheimer's disease. However, some researchers are still examining whether some people are at risk because their bodies have difficulties in handling foods containing the metals copper, iron, and aluminum".

3.7 Conclusive Remarks

Slowly we are beginning to unveil the mechanisms of how aluminum compounds exert their function as adjuvants.

Although there is still a lot of research to be done, altered surface marker expression on monocytes after uptake of aluminum adjuvants, as well as the discovery of the inflammasome has provided important information about the mechanisms by which the aluminum adjuvants lead to expression of co-stimulatory molecules on APCs and elicit the release of pro-inflammatory cytokines, such as IL-1β and IL-18 from DCs.

The inflammatory reaction at the injection site may give lead to *danger signals*, such as uric acid crystals as breakdown products of nucleic acids released from phagocytic cells at the injection site. These may act as co-activators of the inflammasome and HSPs may interact with TLRs stimulating the NF-κB pathway. However, cellular interaction in the early phases of the immune response is highly complex and T-cell derived IL-4 seems important for the stimulation of increased expression of MHC class II molecules on APCs.

When evaluating an adjuvant for possible new applications very few adjuvants can match the extremely comprehensive cohorts that are available for aluminum adjuvants in terms of records of efficacy and safety profiles. After almost 70 years of application with very few problems, the use of aluminum adjuvants may reach practically over a life-long time span in humans. It is interesting that evaluation of the relative contribution of aluminum from vaccination and from the diet in infants did not give lead to concerns when held up against the so-called *Minimal Risk Level* (MRL) as set up by the Agency for Toxic Substances and Disease Registry (Keith et al. 2002; Mitkus et al. 2011).

The aluminum adjuvants have their limitations, especially due to their Th2-biased reactivity, which means that there are vaccines in which they will have little or no effect. However, in the future modified formulations with a more balanced Th1–Th2 profile may find their way into practical vaccinology. AS04, a composite formulation in which aluminum hydroxide is combined with MPL, is one such example that is now used in practical vaccination. The potential application of the $Al(OH)_3$/IL-12 complex, as well as other approaches to expand the use of aluminum adjuvants into new vaccine applications where a more balanced immune response is desired, are yet to be explored in detail.

Acknowledgements I would like to dedicate this chapter to the memory of professor Stanley L. Hem (1940–2011), whose work was a source of inspiration to me for more than 25 years. I also thank Mrs. Jane Momsen for the artwork of the illustrations.

References

Al-Akl NS, Chaktoura M, Kazzi NF, Usta J, Chamoun CA, Abdelnoor AM (2011) Uric acid: a possible mediator of the adjuvant effect of alum in mice immunized with ovalbumin. World J Vaccines 1:148–155

Al-Shakhshir RH, Regnier FE, White JL, Hem SL (1995) Contribution of electrostatic and hydrophobic interactions to the adsorption of proteins by aluminium-containing adjuvants. Vaccine 13(1):41–44

Andrasi E, Pali N, Molnar Z, Kosel S (2005) Brain aluminum, magnesium and phosphorus contents of control and Alzheimer-diseased patients. J Alzheimers Dis 7:273–284

Asea A, Kraeft S-K, Kurt-Jones EA, Stevenson MA, Chen LB, Finberg RW, Koo GC, Calderwood SK (2000) HSP70 stimulates cytokine production through a CD14-dependant pathway, demonstrating its dual role as a chaperone and a cytokine. Nat Med 6(4):435–442

Asea A, Rehli M, Kabingu E, Boch JA, Bare O, Auron PE, Stevenson MA, Calderwood SK (2002) Novel signal transduction pathway utilized by extracellular HSP70. J Biol Chem 277(17): 15028–15034

Authier FJ, Cherin P, Creange A, Bonnotte B, Ferrer X, Abdelmoumni A, Ranoux D, Pelletier J, Figarella-Branger D, Granel B, Maisonobe T, Coquet M, Degos JD, Gherardi RK (2001) Central nervous system disease in patients with macrophagic myofasciitis. Brain 124(Pt 5): 974–983

Batista-Duharte A, Lindblad EB, Oviedo-Orta E (2011) Progress in understanding adjuvant immunotoxicity mechanisms. Toxicol Lett 203(2):97–105

Bergfors E, Trollfors B, Inerot A (2003) Unexpectedly high incidence of persistent itching nodules and delayed hypersensitivity to aluminium in children after use of adsorbed vaccines from a single manufacturer. Vaccine 22:64–69

Böhler-Sommeregger K, Lindemayr H (1986) Contact sensitivity to aluminium. Contact Dermatitis 15:278–281

Brewer JM, Conacher M, Satoskar A, Bluethmann H, Alexander J (1996) In interleukin-4-deficient mice, alum not only generates T helper 1 responses equivalent to Freund's complete adjuvant, but continues to induce T helper 2 cytokine production. Eur J Immunol 26:2062–2066

Brewer JM, Conacher M, Hunter CA, Mohrs M, Brombacher F, Alexander J (1999) Aluminium hydroxide adjuvant initiates strong antigen-specific Th2 responses in the absence of IL-4- or IL-13-mediated signaling. J Immunol 163(12):6448–6454

Burrell LS, Lindblad EB, White JL, Hem SL (1999) Stability of aluminium-containing adjuvants to autoclaving. Vaccine 17:2599–2603

Butler NR, Voyce MA, Burland WL, Hilton ML (1969) Advantages of aluminium hydroxide adsorbed combined diphtheria, tetanus and pertussis vaccines for the immunization of infants. Br Med J 1:663–666

Carlow CKS, Bianco AE (1987) Resistance of *Onchocerca lienalis* microfilariae in mice conferred by egg antigens of homologous and heterologous onchocerca species. Parasitology 94(3):485–496

Chafi AH, Hauw JJ, Rancurel G, Berry JP, Galle C (1991) Absence of aluminium in Alzheimer's disease brain tissue: electron microprobe and ion microprobe studies. Neurosci Lett 123:61–64

Chang M, Shi Y, Nail SL, HogenEsch H, Adams SB, White JL, Hem SL (2001) Degree of antigen adsorption in the vaccine or interstitial fluid and its effect on the antibody response in rabbits. Vaccine 19(20–22):2884–2889

Church LD, Cook GP, McDermott MF (2008) Primer: inflammasomes and interleukin 1β in inflammatory disorders. Nat Clin Pract Rheumatol 4(1):34–42

Clausi AL, Merkley SA, Carpenter JF, Randolph TW (2008) Inhibition of aggregation of aluminum hydroxide adjuvant during freezing and drying. J Pharm Sci 97(6):2049–2061

Code of Federal Regulations 21, vol. 7: sec. 610.15 (Constituent Materials), revised April 1. 2003

Collier LH, Polakoff S, Mortimer J (1979) Reactions and antibody responses to reinforcing doses of adsorbed and plain tetanus vaccines. Lancet 1(8131):1364–1368

Cooper PD, McComb C, Steele EJ (1991) The adjuvanticity of Algammulin, a new vaccine adjuvant. Vaccine 9(6):408–415

Cvjetanovic B, Uemura K (1965) The present status of field and laboratory studies of typhoid and paratyphoid vaccines. Bull WHO 32:29–36

Dagouassat N, Robillard V, Haeuw JF, Plotnicky-Gilquin H, Power UF, Corvaïa N, Nguyen T, Bonnefoy JY, Beck A (2001) A novel bipolar mode of attachment to aluminium-containing adjuvants by BBG2Na, a recombinant subunit hRSV vaccine. Vaccine 19:4143–4152

Davenport FM, Hennessy AV, Askin FB (1968) Lack of adjuvant effect of $AlPO_4$ on purified influenza virus hemagglutinins in man. J Immunol 100(5):1139–1140

Dubischar-Kastner K, Kaltenboeck A, Klingler A, Jilma B, Schuller E (2010) Safety analysis of a Vero-cell culture derived Japanese encephalitis vaccine, IXIARO (IC51), in 6 months of follow-up. Vaccine 28(39):6463–6469

Edelman R (1980) Vaccine adjuvants. Rev Infect Dis 2(3):370–383

Eisenbarth SC, Colegio OR, O'Connor W, Sutterwala FS, Flavell RA (2008) Crucial role for the Nalp3 inflammasome in the immunostimulatory properties of aluminium adjuvants. Nature 453(7198):1122–1126

Ericsson H (1946) Purification and adsorption of diphtheria toxoid. Nature 158:350–351

Flarend RE, Hem SL, White JL, Elmore D, Suckow MA, Rudy AC, Dandashli EA (1997) In vivo absorption of aluminium containing vaccines using ^{26}Al. Vaccine 15:1314–1318

Francis MJ, Fry CM, Rowlands DJ, Brown F, Bittle JL, Houghten RA, Lerner RA (1985) Immunological priming with synthetic peptides of foot-and-mouth disease virus. J Gen Vir 66:2347–2354

Francis MJ, Fry CM, Rowlands DJ, Bittle JL, Houghten RA, Lerner RA, Brown F (1987) Immune response to uncoupled peptides of foot-and-mouth disease virus. Immunology 61:1–6

Frost L, Johansen P, Pedersen S, Veien N, Østergaard P, Nielsen M (1985) Persistent subcutaneous nodules in children hyposensitized with aluminium-containing allergen extracts. Allergy 40:368–372

Fujimaki H, Ozawa M, Imai T, Kubota K, Watanabe N (1984) Adjuvant effects of aluminium silicate on IgE and Ig G1 antibody production in mice. Int Arch Allergy Appl Immunol 75(4):351–356

Gamble HR, Murrell KD, Marti HP (1986) Inoculation of pigs against trichinella spiralis using larval excretory-secretory antigens. Am J Vet Res 47(11):2396–2399

Garcon N, van Mechelen M, Wettendorff M (2006) Development and evaluation of AS04, a novel and improved immunological adjuvant system containing MPL and aluminium salt. In: Schijns V, O'Hagan D (eds) Immunopotentiators in modern vaccines. Elsevier Academic Press, London, pp 161–177

Geerligs HJ, Weijer WJ, Welling GW, Welling-Wester S (1989) The influence of different adjuvants on the immune response to a synthetic peptide comprising amino acid residues 9–21 of herpes simplex virus type 1. J Immunol Methods 124(1):95–102

Giannini SL, Hanon E, Moris P, Van Mechelen M, Morel S, Dessy F, Fourneau MA, Colau B, Suzich J, Losonksy G, Martin MT, Dubin G, Wettendorff MA (2006) Enhanced humoral and memory B cellular immunity using HPV16/18 L1 VLP vaccine formulated with the MPL/aluminium salt combination (AS04) compared to aluminium salt only. Vaccine 24(33–34):5937–5949

Glenny AT, Pope CG, Waddington H, Wallace U (1926) Immunological notes XVII to XXIV. J Pathol 29:31–40

Glenny AT, Buttle GAH, Stevens MF (1931) Rate of disappearance of diphtheria toxoid injected into rabbits and guinea-pigs: toxoid precipitated with alum. J Pathol 34:267–275

Goldenthal KL, Cavagnaro JA, Alving C, Vogel FR (1993) Safety evaluation of vaccine adjuvants. AIDS Res Hum Retroviruses 9(suppl 1):s47–s51

Goto N, Kato H, Maeyama J-I, Shibano M, Saito T, Yamaguchi J, Yoshihara S (1997) Local tissue irritating effects and adjuvant activities of calcium phosphate and aluminium hydroxide with different physical properties. Vaccine 15:1364–1371

Grun JL, Maurer PH (1989) Different T helper cell subsets elicited in mice utilizing two different adjuvant vehicles. The role of endogenous interleukin-l in proliferative responses. Cell Immunol 121(1):134–145

Guermonprez P, Valladeau J, Zitvogel L, Thery C, Amigorena S (2002) Antigen presentation and T cell stimulation by dendritic cells. Annu Rev Immunol 20:621–667

Hamaoka T, Katz DH, Bloch KJ, Benacerraf B (1973) Hapten-specific IgE antibody responses in mice. I. Secondary IgE response in irradiated recipients of syngeneic primed spleen cells. J Exp Med 138:306–311

Han JE, Kim HK, Park SA, Lee SJ, Kim HJ, Son GH, Kim YT, Cho YJ, Kim HJ, Lee NG (2010) A nontoxic derivative of lipopolysaccharide increases immune responses to Gardasil HPV vaccine in mice. Int Immunopharmacol 10(2):169–176

Hassett KJ, Cousins MC, Rabia LA, Chadwick CM, O'Hara JM, Nandi P, Brey RN, Mantis NJ, Carpenter JF, Randolph TW (2013) Stabilization of a recombinant ricin toxin A subunit vaccine through lyophilization. Eur J Pharm Biopharm 85(2):279–286, pii: S0939-6411(13)00118-5

Hem SL, HogenEsch H (2007) Aluminum-containing adjuvants: properties, formulation, and use. In: Singh M (ed) Vaccine adjuvants and delivery systems. Wiley, Hoboken, pp 81–114

HogenEsch H (2013) Mechanism of immunopotentiation and safety of aluminum adjuvants. Front Immunol 3:1–13

Holt LB (1947) Purified precipitated diphtheria toxoid of constant composition. Lancet 1:282–285

Holt LB (1950) Developments in diphtheria prophylaxis. Wm. Heinemann, London

Hornung V, Bauernfeind F, Halle A, Samstad EO, Kono H, Rock KL, Fitzgerald KA, Latz E (2008) Silica crystals and aluminum salts activate the NALP3 inflammasome via phagosomal destabilization. Nat Immunol 9(8):847–856

Hyslop NS, Morrow AW (1969) The influence of aluminium hydroxide content, dose volume and the inclusion of saponin on the efficacy of inactivated foot-and-mouth disease vaccines. Res Vet Sci 10(2):109–120

Itano AA, McSorley SJ, Reinhardt RL, Ehst BD, Ingulli E, Rudensky AY, Jenkins MK (2003) Distinct dendritic cell populations sequentially present antigen to CD4 T cells and stimulate different aspects of cell-mediated immunity. Immunity 19:47–57

Iyer S, HogenEsch H, Hem SL (2003) Effect of the degree of phosphate substitution in aluminium hydroxide adjuvant on the adsorption of phosphorylated proteins. Pharm Dev Technol 8:81–86

Iyer S, Robinett RSR, HogenEsch H, Hem SL (2004) Mechanism of adsorption of hepatitis B surface antigen by aluminium hydroxide adjuvant. Vaccine 22:1475–1479

Jankovic D, Caspar P, Zweig M, Garcia-Moll M, Showalter SD, Vogel FR, Sher A (1999) Adsorption to aluminium hydroxide promotes the activity of IL-12 as an adjuvant for antibody as well as type 1 cytokine responses to HIV-1 gp 120. J Immunol 163:4481–4488

Jiang D, Morefield GL, HogenEsch H, Hem SL (2006) Relationship of adsorption mechanism of antigens by aluminum-containing adjuvants to in vitro elution in interstitial fluid. Vaccine 24(10):1665–1669

Kanra G, Viviani S, Yurdakök K, Ozmert E, Yalçin S, Baldini A, Mutlu B, Kara A, Ceyhan M, Podda A (1999) Safety, tolerability and immunogenicity of a Haemophilus influenzae type b vaccine containing aluminum phosphate adjuvant administered at 2, 3 and 4 months of age. Turk J Pediatr 41(4):421–427

Kanra G, Viviani S, Yurdakök K, Ozmert E, Anemona A, Yalçin S, Demiralp O, Bilgili N, Kara A, Cengiz AB, Mutlu B, Baldini A, Marchetti E, Podda A (2003) Effect of aluminum adjuvants on safety and immunogenicity of Haemophilus influenzae type b-CRM197 conjugate vaccine. Pediatr Int 45(3):314–318

Kartoglu U, Ozgüler NK, Wolfson LJ, Kurzatkowski W (2010) Validation of the shake test for detecting freeze damage to adsorbed vaccines. Bull WHO 88(8):624–631

Kawamura Y, Sawai Y (1989a) Study on Indian cobra venom toxoid. The Snake 21:6–8
Kawamura Y, Sawai Y (1989b) Study on the immunogenicity of purified toxoid of Siamese cobra (*Naja Naja kaouthia*) venom. Snake 21:81–84
Keith LS, Jones DE, Chou CH (2002) Aluminum toxicokinetics regarding infant diet and vaccinations. Vaccine 20(Suppl 3):S13–S17
Kenney JS, Hughes BW, Masada MP, Allison AC (1989) Influence of adjuvants on the quantity, affinity, isotype and epitope specificity of murine antibodies. J Immunol Methods 121:157–166
Kool M, Pétrilli V, de Smedt T, Rolaz A, Hammad H, van Nimwegen M, Bergen IM, Castillo R, Lambrecht BN, Tschopp J (2008) Cutting edge: alum adjuvant stimulates inflammatory dendritic cells through activation of the NALP3 inflammasome. J Immunol 181(6):3755–3759
Kurzatkowski W, Kartoglu U, Staniszewska M, Gorska P, Krause A, Wysocki MJ (2013) Structural damages in adsorbed vaccines affected by freezing. Biologicals 41:71–76
Kwissa M, Lindblad EB, Schirmbeck R, Reimann J (2003) Co-delivery of a DNA vaccine and a protein vaccine with aluminium phosphate stimulates a potent and multivalent immune response. J Mol Med 81(8):502–510
Landsberg JP, McDonald B, Watt F (1992) Absence of aluminium in neuritic plaque cores in Alzheimer's disease. Nature 360:65–68
Landsberg J, McDonald B, Grime G, Watt F (1993) Microanalysis of senile plaques using nuclear microscopy. J Geriatr Psychiatry Neurol 6:97–104
Leland SE, Sofield WL, Minocha HC (1988) Immunogenic effects of culture-derived exoantigens of *Cooperia punctata* on calves before and after challenge exposure with infective larvae. Am J Vet Res 49(3):366–379
Lew AM, Anders RF, Edwards SJ, Langford CJ (1988) Comparison of antibody avidity and titer elicited by peptide as a protein conjugate or as expressed in vaccinia. Immunology 65(2):311–314
Li H, Nookala S, Re F (2007) Aluminum hydroxide adjuvants activate caspase-1 and induce IL-1beta and IL-18 release. J Immunol 178(8):5271–5726
Li H, Willingham SB, Ting JP, Re F (2008) Cutting edge: inflammasome activation by alum and alum's adjuvant effect are mediated by NLRP3. J Immunol 181(1):17–21
Lidholm AG, Bergfors E, Inerot A, Blomgren U, Gillstedt M, Trollfors B (2013) Unexpected loss of contact allergy to aluminium by vaccine. Contact Dermatitis 68(5):286–292
Lindblad EB (2006) Are mineral adjuvants triggering TLR2/TLR4 on dendritic cells by a secondary cascade reaction in vivo through the action of heat shock proteins and danger signals? (Editorial Letter). Vaccine 24:697–698
Lindblad EB (2007) Safety evaluation of vaccine adjuvants (Chapter 18). In: Singh M (ed) Vaccine adjuvants and delivery systems. Wiley, Hoboken, pp 421–444
Lindblad EB, Schønberg NE (2010) Aluminum adjuvants—preparation, application, dosage and formulation with antigen. In: Davies G (ed) Vaccine adjuvants. Methods in molecular biology. Humana, Totowa, pp 41–58
Lindblad EB, Elhay MJ, Silva R, Appelberg R, Andersen P (1997) Adjuvant modulation of immune responses to tuberculosis subunit vaccines. Infect Immun 65(2):623–629
Mannhalter JW, Neychev HO, Zlabinger GJ, Ahmad R, Eibl MM (1985) Modulation of the human immune response by the non-toxic and non-pyrogenic adjuvant aluminium hydroxide: effect on antigen uptake and antigen presentation. Clin Exp Immunol 61(1):143–151
Martinon F, Burns K, Tschopp J (2002) The inflammasome: a molecular platform triggering activation of inflammatory caspases and processing of pro-IL-β. Mol Cell 10(2):417–426
Martyn CN, Barker DJP, Osmond C, Harris EC, Edwardson JA, Lacey RF (1989) Geographical relation between Alzheimers disease and aluminium in drinking water. Lancet 1(8629):59–62
Maschmann E, Küster E, Fischer W (1931) Über die Fähigkeit des Tonerde-Präparates B, Diphtherie-Toxin zu adsorbieren. Ber Dtsch Chem Ges 64:2174–2178
McCandlish IAP, Thompson H, Wright NG (1978) Vaccination against canine bordetellosis using an aluminium hydroxide adjuvant vaccine. Res Vet Sci 25:51–57
McDougall JS (1969) Avian infectious bronchitis: the protection afforded by an inactivated virus vaccine. Vet Rec 85:378–380

Mitkus RJ, King DB, Hess MA, Forshee RA, Walderhaug MO (2011) Updated aluminum pharmacokinetics following infant exposures through diet and vaccination. Vaccine 29(51): 9538–9543

Monroy FG, Adams JH, Dobson C, Bast IJ (1989) *Nematospiroides dubius*: influence of adjuvants on immunity in mice vaccinated with antigens isolated by affinity chromatography from adult worms. Exp Parasitol 68(1):67–73

Morefield GL, Jiang D, Romero-Mendez IZ, Geahlen RL, Hogenesch H, Hem SL (2005) Effect of phosphorylation of ovalbumin on adsorption by aluminum-containing adjuvants and elution upon exposure to interstitial fluid. Vaccine 23(12):1502–1506

Nagy LK, Penn CW (1974) Protective antigens in bovine pasteurellosis. Dev Biol Stand 26:65–76

Naim JO, van Oss CJ, Wu W, Giese RF, Nickerson PA (1997) Mechanisms of adjuvancy: I—metal oxides as adjuvants. Vaccine 15:1183–1193

Norimatsu M, Ogikubo Y, Aoki A, Takahashi T, Watanabe G, Taya K, Sasamoto S, Tsuchiya M, Tamura Y (1995) Effects of aluminium adjuvant on systemic reactions of lipopolysaccharides in swine. Vaccine 13(14):1325–1329

Perl DP, Brody AR (1980) Alzheimer's disease: X-ray spectrometric evidence of aluminum accumulation in neurofibrillary tangle-bearing neurons. Science 208:297–299

Ph. Eur. (1997) Vaccines for human use, 3rd edn. p 1697

Pini A, Danskin D, Coackley W (1965) Comparative evaluation of the potency of beta-propiolactone inactivated Newcastle disease vaccines prepared from a lentogenic and a velogenic strain. Vet Rec 77(5):127–129

Poland GA, Jacobson RM (2001) The prevention of Lyme disease with vaccine. Vaccine 19:2303–2308

Pollock KGJ, Conacher M, Wei X-Q, Alexander J, Brewer JM (2003) Interleukin-18 plays a role in both the alum-induced T helper 2 response and the T helper 1 response induced by alum-adsorbed interleukin-12. Immunology 108:137–143

Rimaniol AC, Gras G, Verdier F, Capel F, Grigoriev VB, Porcheray F, Sauzeat E, Fournier JG, Clayette P, Siegrist CA, Dormont D (2004) Aluminum hydroxide adjuvant induces macrophage differentiation towards a specialized antigen-presenting cell type. Vaccine 22(23–24):3127–3135

Rinella JV, White JL, Hem SL (1995) Effect of anions on model aluminium-adjuvant-containing vaccines. J Colloid Interface Sci 172:121–130

Rinella JV, White JL, Hem SL (1998a) Effect of pH on the elution of model antigens from aluminium-containing adjuvants. J Colloid Interface Sci 205:161–165

Rinella JV, Workman RF, Hermodson MA, White JL, Hem SL (1998b) Elutability of proteins from aluminium-containing vaccine adjuvants by treatment with surfactants. J Colloid Interface Sci 197:48–56

Ris DR, Hamel KL (1979) Leptospira interrogans *serovar*. pomona vaccines with different adjuvants in cattle. New Zealand Vet J 27:169–171

Sagara T, Mori S, Ohkawara S, Goto F, Takagi K, Yoshinaga M (1990) A limited role of IL-1 in immune-enhancement by adjuvants. Immunology 71(2):251–257

Seeber SJ, White JL, Hem SL (1991) Predicting the adsorption of proteins by aluminium-containing adjuvants. Vaccine 9:201–203

Sellers RF, Herniman KAJ (1974) Early protection of pigs against foot-and-mouth disease. Br Vet J 130:440–445

Shi Y, HogenEsch H, Regnier FE, Hem SL (2001) Detoxification of endotoxin by aluminium hydroxide adjuvants. Vaccine 19:1747–1752

Shi Y, Evans JE, Rock KL (2003) Molecular identification of a danger signal that alerts the immune system to dying cells. Nature 425(6957):516–521

Shirodkar S, Hutchinson RL, Perry DL, White JL, Hem SL (1990) Aluminium compounds used as adjuvants in vaccines. Pharm Res 7(12):1282–1288

Sixt M, Kanazawa N, Selg M, Samson T, Roos G, Reinhardt DP, Pabst R, Lutz MB, Sorokin L (2005) The conduit system transports soluble antigens from the afferent lymph to resident dendritic cells in the T cell area of the lymph node. Immunity 22:19–29

Sokolovska A, Hem SL, HogenEsch H (2007) Activation of dendritic cells and induction of CD4(+) T cell differentiation by aluminum-containing adjuvants. Vaccine 25(23):4575–4585
Sun H, Pollock KGJ, Brewer JM (2003) Analysis of the role of vaccine adjuvants in modulating dendritic cell activation and antigen presentation in vitro. Vaccine 21:849–855
Technical Report Series 595 (1976) Immunological adjuvants. World Health Organization, Geneva
Thorley CM, Egerton JR (1981) Comparison of alum-adsorbed or non-alum-adsorbed oil emulsion vaccines containing, either pilate or non-pilate bacteroides nodosus cells in inducing and maintaining resistance of sheep to experimental foot rot. Res Vet Sci 30:32–37
Tomljenovic L (2011) Aluminum and Alzheimer's disease: after a century of controversy, is there a plausible link? J Alzheimers Dis 23:567–598
Uede T, Ishizaka K (1982) Formation of IgE binding factors by rat T lymphocytes. VI. Cellular mechanisms for the formation of IgE-potentiating factor and IgE-suppressive factor by antigenic stimulation of antigen primed spleen cells. J Immunol 129(4):1391–1397
Uede T, Huff TF, Ishizaka K (1982) Formation of IgE binding factors by Rat T lymphocytes. V. Effect of adjuvant for the priming immunization on the nature of IgE binding factors formed by antigenic stimulation. J Immunol 129(4):1384–1390
Ulanova M, Tarkowski A, Hahn-Zoric M, Hanson LA (2001) The common vaccine adjuvant aluminum hydroxide up-regulates accessory properties of human monocytes via an interleukin-4-dependent mechanism. Infect Immun 69(2):1151–1159
Ulmer JB, DeWitt CM, Chastain M, Friedman A, Donelly JJ, McClements WL, Caulfield MJ, Bohannon KE, Volkin DB, Evans RK (2000) Enhancement of DNA vaccine potency using conventional aluminum adjuvants. Vaccine 18:18–28
Verdier F, Burnett R, Michelet-Habchi C, Moretto P, Fievet-Groyne F, Sauzeat E (2005) Aluminium assay and evaluation of the local reaction at several time points after intramuscular administration of aluminium containing vaccines in the Cynomolgus monkey. Vaccine 23(11):1359–1367
Vogelbruch M, Nuss B, Korner M, Kapp A, Kiehl P, Bohm W (2000) Aluminium-induced granulomas after inaccurate intradermal hyposensitization injections of aluminium-adsorbed depot preparations. Allergy 55(9):883–887
Volk VK, Bunney WE (1942) Diphtheria immunization with fluid toxoid and alum-precipitated toxoid. Am J Public Health 32:690–699
Walls RS (1977) Eosinophil response to alum adjuvants: involvement of T-cells in non-antigen-dependent mechanisms. Proc Soc Exp Biol Med 156(3):431–435
White RG, Coons AH, Conolly JM (1955) Studies on antibody production: III. The alum granuloma. J Exp Med 102(1):73–82
Wilson JHG, Hermann-Dekkers WM, Leemans-Dessy S, de Meijer JW (1977) Experiments with an inactivated hepatitis leptospirosis vaccine in vaccination programmes for dogs. Vet Rec 100:552–554

Chapter 4
Emulsions as Vaccine Adjuvants

Ruchi R. Shah, Luis A. Brito, Derek T. O'Hagan, and Mansoor M. Amiji

4.1 Introduction

Early vaccines were based either on whole inactivated pathogens or attenuated pathogens; the advent of recombinant DNA technology facilitated expression of recombinant antigens from cells, thus reducing complexity, improving purity, and addressing unmet medical needs that live attenuated pathogens or bacterial toxins could not meet (new targets and limitation in method of production of non-cultivable pathogens), *e.g.*, meningitis type B vaccine Bexsero (Hansson et al. 2000; Rappuoli 2000; Ott and Nest 2006). However, due to the high purity of these recombinant proteins, their immunogenicity is low compared to whole pathogens, due to lack of pathogen-associated molecular patterns (PAMP), structures normally found on and within whole organism's surface (Tritto et al. 2009).

Adjuvants improve the effectiveness of vaccines by enhancing and sustaining immune responses, reducing the dose of antigen, enhancing the breadth of the immune response, increasing immunological memory associated with the adaptive immune system, and reducing the frequency of vaccination (Foged et al. 2012). Emulsion adjuvants have been shown to promote or enhance T-cell responses that are typically absent after recombinant protein immunization and have been used successfully within a diverse group of patients for flu, from pediatric to geriatric and from healthy individuals to immune-compromised patients (Lima et al. 2004). Vaccines are different than other biopharmaceuticals as they are administered to

R.R. Shah • M.M. Amiji (✉)
Department of Pharmaceutical Sciences, School of Pharmacy,
Northeastern University, Boston, MA, USA
e-mail: shah.ru@husky.neu.edu; m.amiji@neu.edu

L.A. Brito • D.T. O'Hagan
Novartis Vaccines & Diagnostics, 350 Massachusetts Avenue, Cambridge, MA 02139, USA
e-mail: luis.brito@novartis.com; derek.ohagan@novartis.com

C. Foged et al. (eds.), *Subunit Vaccine Delivery*, Advances in Delivery Science and Technology, DOI 10.1007/978-1-4939-1417-3_4

healthy individuals some as young as a few hours old. Potent and well-tolerated adjuvants will be needed for treating diseases using the immune system *e.g.*, therapeutic vaccines.

4.2 Role of Vaccine Adjuvants

An immune response is generated as a reaction to foreign substances termed antigens, which induce *anti*body *gen*eration (Murphy et al. 2008). The immune response against an antigen is collectively produced by innate immunity working to clear a pathogen and adaptive immunity providing immune memory. It is a consequence of a complex cascade of events involving immune cells, cytokines, and chemokines. Adjuvants can be used to modulate the immune response, *e.g.*, the squalene water-in-oil emulsion MF59 can enhance both antibody titers and T-cell-mediated immune responses (O'Hagan et al. 2012). Adjuvants are known to work through a variety of different mechanisms including (1) improved antigen presentation to the immune system, (2) facilitating transport of antigens, (3) activation of immune cells, and (4) induction of production of cytokines (Schijns and Lavelle 2011). A single adjuvant such as MF59 can engage in all of the above mentioned functions wherein, the adjuvant provides "danger signals" to the surrounding tissue and generates a controlled induction of proinflammatory cytokines, subsequently recruiting immune cells to the site of injection and promoting antigen uptake and trafficking to the lymph node, thus improving the immune response when compared to an unadjuvanted vaccine (Batista-Duharte et al. 2011).

4.3 History of Vaccine Adjuvants

The use of adjuvants has spanned almost a century with alum being the most commonly used adjuvant for over 80 years. Aluminum compounds which have been used as adjuvants (Chap. 3), are incorrectly, but collectively known as alum and include aluminum phosphate, aluminum hydroxide, and other aluminum adsorbed vaccine preparations (Gupta and Rost 2000). Despite 70 years of clinical experience, the exact mechanism of action of alum is still an active area of research. Recently alum was found to activate the NALP3 inflammasome which induces the production of IL1β, innate responses and increase in the number of antigen-specific T-cells (Kool et al. 2008; Eisenbarth et al. 2008). These act as the signals for the recruitment of inflammatory monocytes and immature dendritic cells (DCs) to the site of injection (Kool et al. 2008; Eisenbarth et al. 2008). Antigen adsorption on alum and antigen presentation in particulate form enhances phagocytosis by macrophages, DCs, and B-cells (Lambrecht et al. 2009). It has been found that macrophages and monocytes are the primary targets of alum and induce secretion of chemokines, which leads to recruitment of immune cells, enhanced endocytosis by monocytes, and the differentiation to DCs which in turn prime T-cells in a TLR-independent mechanism (Seubert et al. 2008).

Emulsion adjuvants also have an extensive history of clinical use dating back to the 1930s. Emulsions are biphasic systems comprising of a hydrophilic phase and hydrophobic phase which require surfactants to stabilize the oil–water interface and are classified as oil-in-water, water-in-oil, or multiple emulsions like water-in-oil-in-water or oil-in-water-in-oil emulsions (O'Hagan 2007). Freund's Complete Adjuvant (FCA) and Freund's Incomplete Adjuvant (FIA) are water-in-oil emulsion-based adjuvants that were first described in 1937 and 1950s, respectively (Vogel et al. 2009). FCA is composed of 85–90 % (v/v) oil with 15–10 % mannide monooleate emulsifier (Arlacel A) and 500 μg of heat-killed *Mycobacteria tuberculosis* (FIA has the same composition but does not contain any mycobacteria). The mixture is homogenized with an equal volume of antigen dissolved in the aqueous phase and then emulsified prior to use (Lindblad 2000). FCA was found to be highly reactogenic due partly to the quality of mineral oil used for formulation (Whitehouse et al. 1974; Stills 2005; Stuewart-Tull et al. 1976). Some components of Freund's adjuvant were also found to be oncogenic in preclinical studies excluding FCA and FIA from widespread clinical use (Murray et al. 1972; Hilleman 1966). The reactogenicity of FIA was found to be directly proportional to the immunogenicity further limiting its use clinically. Taken together this adjuvant illustrated that emulsions are able to produce potent immune responses, but required further refinement before being used extensively in the clinic.

Several alternative formulations have been identified; the most prominent being Montanide ISA 51 VG, Montanide ISA 720 VG, and Adjuvant 65. Montanides ISA 51 and 720 manufactured by Seppic are water-in-oil emulsion adjuvants that are similar to IFA and have been evaluated for therapeutic applications in cancer, malaria, AIDS, and other autoimmune diseases. Montanide ISA 51 is composed of white medicinal oil of mineral origin, while Montanide ISA 720 is composed of squalene oil. The surfactants used in Montanides are from the mannide monooleate family (esters of oleic acid and mannitol). In case of Montanide ISA 51 the oil-to-water ratio is 50:50, while in case of Montanide ISA 720 it is 70:30 (Aucouturier et al. 2002). An alternative emulsion adjuvant commonly described is Adjuvant 65 which is composed of 45 % peanut oil, 3 % synthetic emulsifier isomannide monooleate, 2 % aluminum monostearate, mixed 1:1 (v/v) with various strains of influenza vaccine. Clinical data have shown that by using Adjuvant 65 it is possible to reduce the antigen dose by fourfold, but as peanut oil is the main component, it has never been broadly adopted due to concerns with peanut allergies (Smith et al. 1975; Hilleman 1969).

Biodegradable oil-in-water emulsions have been explored in efforts to improve tolerability by decreasing the amount of oil in the adjuvant and improving the biodegradability of the oil. Initial formulations included immune stimulators such as threonyl muramyl dipeptide (t-MDP), bacterial cell wall (BCW), and monophosphoryl lipid A (MPL). Syntex Adjuvant Formulation (SAF) consists of 5 % squalane,

Pluronic L121, Tween 80, and t-MDP (Hjorth et al. 1997). A squalene oil emulsion developed at Ribi called DETOX™, included in the vaccine Melacine® for the treatment of melanomas, is composed of 1 % squalene and 0.2 % Tween 80 with immune stimulators like bacterial cell wall and MPL (Fox 2009; Ribi et al. 1984). Ribi's Adjuvant System (RAS) has the composition of 2 % squalene, 0.2 % Tween 80, and synthetic trehalose dicorynomycolate, bacterial cell wall, and MPL (Fox 2009; Ribi et al. 1975). Using metabolizable oils in oil-in-water emulsion adjuvants improves the tolerability, but it was not until the late 1990s when it was discovered that MF59 (an oil-in-water squalene emulsion) improved immune responses without the need of muramyl tripeptide phosphatidylethanolamine (MTP-PE), an additional immunostimulator.

We will now describe MF59, Adjuvant System 03 (AS03) and AF03 with a focus on clinical experience, formulation, and mechanism of action. MF59 and AS03 are components of licensed vaccines, whereas AF03 was initially approved but subsequently (Withdrawal by European Medical Agency 2011) (EMA) since June 2011 (European Medicines Agency 2011). Table 4.1 summarizes all emulsion adjuvants included in licensed vaccines in addition to other emulsion adjuvants that are being evaluated in preclinical and clinical settings.

4.4 MF59

MF59, the emulsion adjuvant developed by Novartis Vaccines & Diagnostics, was originally developed by Ciba Geigy and Chiron Corporation in the 1990s (O'Hagan et al. 2013). With over 150 million administered doses in >35 countries, MF59 has been the most successful emulsion adjuvant (O'Hagan et al. 2013). Originally developed as a delivery vehicle for MTP-PE, MF59 was later found to be better tolerated and equally immunogenic without the inclusion of MTP-PE, and the removal of MTP-PE was a key milestone in the development of MF59 (O'Hagan and Singh 2007). It is an oil-in-water emulsion which has been licensed in Europe since 1997 as part of the influenza vaccine Fluad® (Table 4.1). It has also been included as an adjuvant in the pandemic influenza vaccine Aflunov® against H5N1 and in Focetria® and Celtura® against pandemic H1N1 influenza (Gasparini et al. 2012). MF59 has been administered to various populations from infants to elderly, provides improved protection, lasting immunity, is well tolerated and was found to improve immunogenicity in vaccines for influenza, human immunodeficiency virus (HIV), herpes simplex virus (HSV), hepatitis B/hepatitis C virus (HBV/HCV), parvovirus, human papilloma virus (HPV), and cytomegalovirus (CMV) in the clinic (O'Hagan et al. 2013). MF59 has been shown to provide cross clade immunity in flu vaccines, *e.g.*, broader responses for avian H5N1 vaccine (Khurana et al. 2010).

Table 4.1 Emulsion adjuvants in licensed vaccines, clinical and preclinical stage

Formulation	Type of emulsion	Particle size (nm)	Proposed mode of action	Significance	References
MF59	Oil-in-water emulsion	~160	Creates an immunocompetent environment and enhancing production of cytokines and chemokines	First approved squalene oil-based emulsions for flu vaccines Fluad®, Aflunov®, Focetria®, and Celtura®	(Tritto et al. 2009; Calabro et al. 2011; Seubert et al. 2008; O'Hagan et al. 2012; Mosca et al. 2008)
AS03	Oil-in-water emulsion	~155	By spatial and colocalization of antigen at the site of injection and creating an immunocompetent environment	Influenza vaccine Prepandrix® and Pandemrix®	(Morel et al. 2011; Garcon et al. 2012; Vogel et al. 2009; Leroux-Roels 2009)
AF03	Water-in-oil emulsion	~100	No data available	Influenza vaccine Humenza® (approved in Europe)	(Klucker et al. 2012; Caillet et al. 2010; Girard et al. 2011)
GLA-SE	Oil-in-water	~110	Activation of DCs and generation of Th1 response	Combined adjuvanticity by squalene oil and TLR4 agonist GLA	(Coler et al. 2010, 2011; Behzad et al. 2012; Windish et al. 2011; Anderson et al. 2010)
Montanides	Water-in-oil	~1 μm	Antigen depot formation, enhancing antibody and CTL levels	Adjuvant for malaria and cancer	(Aucouturier et al. 2002)
AS02	Oil-in-water		Antibody production and cell-mediated immunity	In Phase II Hepatitis B vaccine	(Surquin et al. 2011; Beran et al. 2010)
W805EC	Oil-in-water	~325	Systemic and mucosal immunity with a Th1-biased response	Hepatitis B, Influenza and Anthrax	(Bielinska et al. 2007, 2008a, b)

4.4.1 Formulation Composition

The composition of MF59 in a human dose is described in Table 4.2 (Tsai et al. 2010). Squalene is a biodegradable oil that can be obtained from shark liver, olive oil, yeast, carrot, germ oil, and other natural sources (Fox 2009). It is a precursor to cholesterol and is an intermediate in hormone synthesis. It is found in the human body, particularly in the skin where the level is the highest (478.1 μg/g dry tissue), but is also found in other tissues such as adipose tissue, the liver, and the small intestine (Liu et al. 1976). Span® 85 and Tween® 80 are nonionic surfactants that are added to stabilize the emulsion in a citrate buffer aqueous phase at pH 6.5 (Peek et al. 2008). MF59 is made by first preparing a primary emulsion consisting of an outer aqueous phase of polysorbate 80 (Tween® 80) dissolved in citrate buffer and an inner oil phase of sorbitan trioleate (Span® 85) dissolved in squalene oil. The micron-sized emulsion is then passed through a high-pressure homogenizer at 12,000 PSI (O'Hagan and Singh 2007). MF59 is subsequently sterile filtered using a 0.22 μm membrane filter to remove larger particles. The resultant emulsion has a mean droplet size of approximately 160 nm, contains less than 0.1 % particles greater than 1.2 μm in size and has 4.3 % (v/v) of shark liver-derived squalene oil (Seubert et al. 2008). MF59 has shown comparable activity with plant-derived squalene oil as well in mice, identifying a potential renewable source of oil over the animal-derived squalene oil (Brito et al. 2011), though extensive clinical data are needed to justify a switch-in-oil source.

4.4.2 Mechanism of Action

The mechanism of action of MF59 is still an area of active research, though progress has been made (Tritto et al. 2009; Seubert et al. 2008; Calabro et al. 2011; O'Hagan et al. 2012). Unlike alum, MF59 does not create a depot of antigen at the injection site; MF59 and the antigen are independently cleared from the site of injection. MF59 creates an immunocompetent environment at the injection site by recruiting immune cells like neutrophils, macrophages, granulocytes, monocytes, and other immune cells which take up the antigen; monocytes are subsequently differentiated into DCs in a TLR-independent manner (Seubert et al. 2008). These DCs prime naïve T-cells and generate potent immune responses in the local draining

Table 4.2 Composition of MF59 and AS03 in a single dose

Component	Milligrams in MF59	Milligrams in AS03	Role
Squalene	9.75	10.69	Oil
Tween 80	1.17	4.86	Surfactant
Span 85	1.17		Surfactant
α-Tocopherol		11.86	Immune potentiator

lymph node (Seubert et al. 2008). The activated immune cells at the injection site also induce secretion of chemokines which attract additional immune cells initiating a self-limiting positive immune feedback loop. Administration of MF59 was found to upregulate genes that enhance the production of chemokines and cytokines in mice (Mosca et al. 2008). The individual components of MF59 do not act as adjuvants, only when formulated together as an emulsion do they improve the immune response (Calabro et al. 2013).

4.5 AS03

AS03 developed by GlaxoSmithKline is an oil-in-water emulsion adjuvant that contains squalene oil and alpha-tocopherol. It has been used as an adjuvant in Prepandrix® (pre-pandemic H5N1) and Pandemrix® (pandemic H1N1) influenza vaccines (Morel et al. 2011; Garcon et al. 2012). Both vaccines consist of a split inactivated virus mixed with the emulsion adjuvant. AS03 was selected as the final formulation from a pool of more than 70 formulations of the family of Adjuvant System on the basis of size, ability to be sterile filtered, and stability (Garcon et al. 2012). AS03 was shown to have dose sparing effect and cross-clade immunogenicity (Vogel et al. 2009). Recently there has been concern related to the safety and tolerability of vaccines containing AS03 due to an association of narcolepsy in children post vaccination with the adjuvanted vaccine in Europe (Vajdy 2011; Nohynek et al. 2012; Miller et al. 2013). However, immunization with the adjuvanted vaccine in Canada has shown no increased incidence of narcolepsy (Nohynek et al. 2012). This is an active area of research that is ongoing and it will likely take years to understand what factors led to an association of AS03 with narcolepsy incidences.

4.5.1 Formulation Composition

AS03 is comprised of two oils, squalene and α-tocopherol. Alpha-tocopherol is the most bioavailable form of vitamin E and has been shown to enhance both cell-mediated and humoral immunity by acting as an immune potentiator (Vajdy 2011). The adjuvant composition for humans consists of 4.86 mg of polysorbate 80, 10.69 mg of squalene and 11.86 mg of α-tocopherol in phosphate-buffered saline (PBS) as aqueous vehicle (Garcon et al. 2012). The dose of hemagglutinin (antigen) in Prepandrix®, an AS03-adjuvanted prepandemic A/Vietnam/1194/2004 NIBRG-14 influenza vaccine, is 3.8 μg for adults in the age group of 18–60 years and is a multi-vial vaccine preparation with the antigen and adjuvant mixed in a 1:1 ratio immediately prior to administration (Leroux-Roels 2009). The final composition of AS03 in a human dose is described in Table 4.2 (Morel et al. 2011).

4.5.2 *Mechanism of Action (Morel et al. 2011; Garcon et al. 2012)*

The mechanism of action of AS03 was found to be dependent upon the spatial and temporal colocalization of antigen after immunization (Garcon et al. 2012). It induces production of cytokines, chemokines, and regulation of certain genes by activating the transcription factor nuclear factor-κB (NF-κB). These upregulated cytokines, chemokines, and genes regulate differentiation and recruitment of immune cells which increase in number in muscles and draining lymph nodes after immunization. In contrast to MF59, AS03 directly activates innate immunity in the draining lymph node, due to the presence of α-tocopherol (Morel et al. 2011). AS03 enhances antigen uptake by targeting macrophages, monocytes, and DCs, which leads to an increase in $CD4^+$ T-cells, neutralizing antibodies, and antigen-specific memory B-cells in the draining lymph node. A mixed Th1/Th2 response has also been observed with AS03 (Coffman et al. 2010). Unlike MF59 where none of the components are immunogenic, AS03 has the immune potentiator α-tocopherol, which must be incorporated in the adjuvant to achieve full potency as an adjuvant.

4.6 AF03

AF03 is a squalene oil-based adjuvant that has been included in the pandemic influenza split virion vaccine Humenza® developed by Sanofi Pasteur. Although Humenza® was licensed but never commercialized, various clinical trials conducted in the age groups of 3–17 years indicated that it produces strong seroprotective titers against H1N1 pandemic strain with just one immunization (Vesikari et al. 2012). In a large study involving children and infants for pandemic influenza A H1N1, it was observed that the AF03-adjuvanted vaccine generated high titers and reduced the HA needed per dose (Vesikari et al. 2012).

4.6.1 *Formulation Composition*

AF03 is composed of 32.5 % (w/w) squalene, 6.2 % eumulgin™ B1-PH, 4.8 % montane™ 80 PH, 6 % mannitol, and 50.5 % PBS prior to administration (Klucker et al. 2012). AF03, unlike other emulsion adjuvants such as MF59 and AS03 is not microfluidized, it is prepared by a process called the phase inversion temperature (PIT) emulsification process. Briefly, an oil phase consisting of montane (surfactant) and squalene and an aqueous phase consisting of Eumulgin (surfactant), PBS, and mannitol are prepared. These two phases are mixed into a stainless-steel jacketed mixing vessel. The mixture is then emulsified at 1,000 rpm for 4 min under nitrogen and subsequently heated up to 57 °C (PIT) under constant stirring at 400 rpm. Once the conductivity becomes zero (inversion from o/w emulsion to w/o emulsion) the emulsion is cooled to room temperature. This process produces an initial bulk emulsion that is further diluted to a final squalene concentration of 5 % or 3.3 % depending on the target patient population (Klucker et al. 2012). The dose of AF03 in children in the ages of 3–17 years is 2.5 % (w/w), whereas in children in the ages of 6–35 months it is halved to 1.25 % (w/w) (Vesikari et al. 2012). The average particle size of the resulting emulsion is around 100 nm and is easily filterable through a 0.2 μm membrane filter.

4.7 Recent Developments in Emulsion Adjuvants

Addition of immune potentiators to emulsion adjuvants can improve the overall immune response and can also impact the T-cell bias and is an active area of emulsion adjuvant research. However one must first question whether it is necessary to incorporate an additional stimulator. If a vaccine provides protection while being safe and tolerable, there is little need to include an immune potentiator and risk raising the reactogenicity of the vaccine (Fox and Haensler 2013). There are many adjuvants available to vaccinologists; one should evaluate approved and well-accepted adjuvants (e.g., alum) before venturing into developing a novel adjuvant. MF59 has been tested for enhancement of immunogenicity by mixing the TLR9 agonist CpG and by incorporating the TLR4 agonist E6020 (Baudner et al. 2009). The results using the trivalent flu vaccine and Balb/c mice indicated that although MF59 did not increase the immune response with TLRs, it shifted the immune response to a more Th1-biased response.

Stable emulsions (SE) developed by Ribi consists of 10 % (v/v) squalene oil, 1.9 % (w/v) lecithin, 0.091 % (w/v) Pluronic F68, 0.05 % (w/v) α-tocopherol, and 1.8 % (v/v) glycerol in 25 mM ammonium phosphate buffer pH 5.1 and is prepared similarly to MF59 (Anderson et al. 2010). Recently SE containing the TLR4 agonist glucopyranosyl lipid A (GLA) has been tested as a potential new adjuvant and is in clinical trials for influenza vaccine (Coler et al. 2010). GLA-SE has been shown to be effective in combination with split virus vaccine by enhancing Th1 responses. GLA-SE has also shown promise in a tuberculosis vaccine where it exhibited a Th1-biased response, reduced the bacterial burden, and enhanced the cellular infiltration in the lung (Windish et al. 2011). Infectious Disease Research Institute (IDRI) launched a phase I clinical trial with GLA-SE by incorporating it in a vaccine against leishmaniasis with antigen LEISH-F3 (IDRI 2012). An alternative emulsion adjuvant containing immune potentiators is being tested by GlaxoSmithKline in a HBV vaccine that is in clinical trials. The vaccine is adjuvanted by an oil-in-water emulsion adjuvant, AS02, which contains MPL and QS21 (saponin). This vaccine called HB-AS02 was developed for patients with renal insufficiency and for individuals who show reduced or no response to recombinant

HBV vaccine (Surquin et al. 2010). Surquin et al. showed that HB-AS02 is successful in inducing antibody response in patients with renal insufficiency (Surquin et al. 2011). In Beran et al., the HB-AS02 vaccine showed lot-to-lot consistency in terms of immune responses and also seroprotection after two doses of the vaccine (Beran et al. 2010). AS02 is also included in a malaria vaccine RTS, S/AS02 (currently in phase 2 trials) where it has shown antibody and T-cell responses and proven to be safe and well tolerated (Bojang et al. 2001).

4.8 Future Trends in Emulsion Adjuvants

MF59 and AS03 have shown success as adjuvants and have encouraged researchers to explore and incorporate emulsions to improve immune responses for prophylactic and therapeutic vaccines. Complexity of the immune system, compatibility with the paired antigen, stability of the vaccine, tolerability of the adjuvant, and strict regulatory guidelines are some reasons for relatively few licensed emulsion-adjuvanted vaccines in the market (Mbow et al. 2010). This section looks forward to future trends to what we believe will be key areas of interest for emulsion adjuvants.

As noted throughout this chapter success of an adjuvant not only depends on its efficacy, but also on its safety and tolerability. Use of adjuvants should be considered in relation to the target patient population and should be judged as a ratio of risk to benefit, *e.g.*, prophylactic vaccines for healthy infants versus therapeutic vaccines for cancer patients. While vaccines for infants must be well tolerated and safe, the main concern in cancer vaccines is efficacy and survival of the patient. Initial studies of vaccines are performed on smaller animals, *e.g.*, mice, rats, rabbits, etc. where it is difficult to predict adverse reactions that can correlate with humans. This is compounded with the small size of early clinical trials, illustrating why safety issues are sometimes not known immediately. Applying the burgeoning field of systems biology to vaccine development can help to organize and analyze existing and new clinical data (Oberg et al. 2011). Exploratory clinical studies to understand the human response to adjuvanted vaccines will accelerate the development of the next generation vaccines (Rappuoli and Aderem 2011). Long-term stability for emulsion adjuvants must be examined on a case-by-case basis. Properly formulated emulsions on their own are stable formulations, *e.g.*, MF59 has a shelf life of 3 years at 2–8 °C (Ott et al. 2000). Mixing of antigens within this complex mixture can lead to destabilization and unfolding of the antigen due to the presence of hydrophobic interfaces and an aqueous environment. As the field of structural vaccinology delivers well-defined antigens, formulation scientists will need better analytical tools to understand the impact of changes in the antigen conformation (Dormitzer et al. 2012). Approaches such as lyophilization of antigen and emulsion in one vial can be explored to develop a single vial vaccine which can be reconstituted prior to administration. Alternatively, mixing of the antigen with the adjuvant can be done bedside for highly labile antigens.

Administration to alternative immune compartments is an area of potential interest for next generation emulsion adjuvants. A large number of pathogens invade the body through various mucosal routes. For example, in a study, chitosan-coated emulsion adjuvants were applied for intranasal delivery where they proved to retain the antigen long enough for mucosal uptake (Nagamoto et al. 2004). In contrast, MF59 tested mucosally showed no differences in responses when comparing adjuvanted and unadjuvanted vaccine (Boyce et al. 2000). Recently a novel mucosal adjuvant delivery system has been described, $W_{80}5EC$, composed of soybean oil emulsified in water to form nano-sized oil droplets and developed by NanoBio Corporation. This nano-sized emulsion (NE) is prepared by emulsification of an aqueous phase composed of 1 % cetylpryridium chloride, 5 % Tween® 20 and 8 % ethanol with soyabean oil (64 %) in a high-speed emulsifier. The NE obtained has an average particle size in the range of 300 nm. This NE has shown potential in intranasal vaccine delivery for smallpox, HBV, anthrax, and influenza (Makidon et al. 2008; Bielinska et al. 2007, 2008a; Hamouda et al. 2011) by enhancing both mucosal and systemic immunity with a Th-1 biased immune responses. Extensive toxicology studies concluded that this NE is well tolerated and is found to be safe in mice and larger animals like dogs (Makidon et al. 2008).

An alternate way to change the bio-distribution of an emulsion adjuvant is to make changes to biophysical properties of the emulsion adjuvant such as, size, charge, etc., to modulate the bio-distribution and immune responses by passively targeting them to alternate sites. The size of MF59 was shown to be one of the factors responsible for its potent immune response (Ott et al. 1995). Thus far the field has struggled to prepare adjuvants smaller than 100 nm in size. The trend is that smaller is better for polymeric particles, but is there a theoretical limit to this (Fifis et al. 2004) In Figs. 4.1 and 4.2 we have highlighted structures of some of the oils and immunostimulators commonly used or studied for adjuvant formulation.

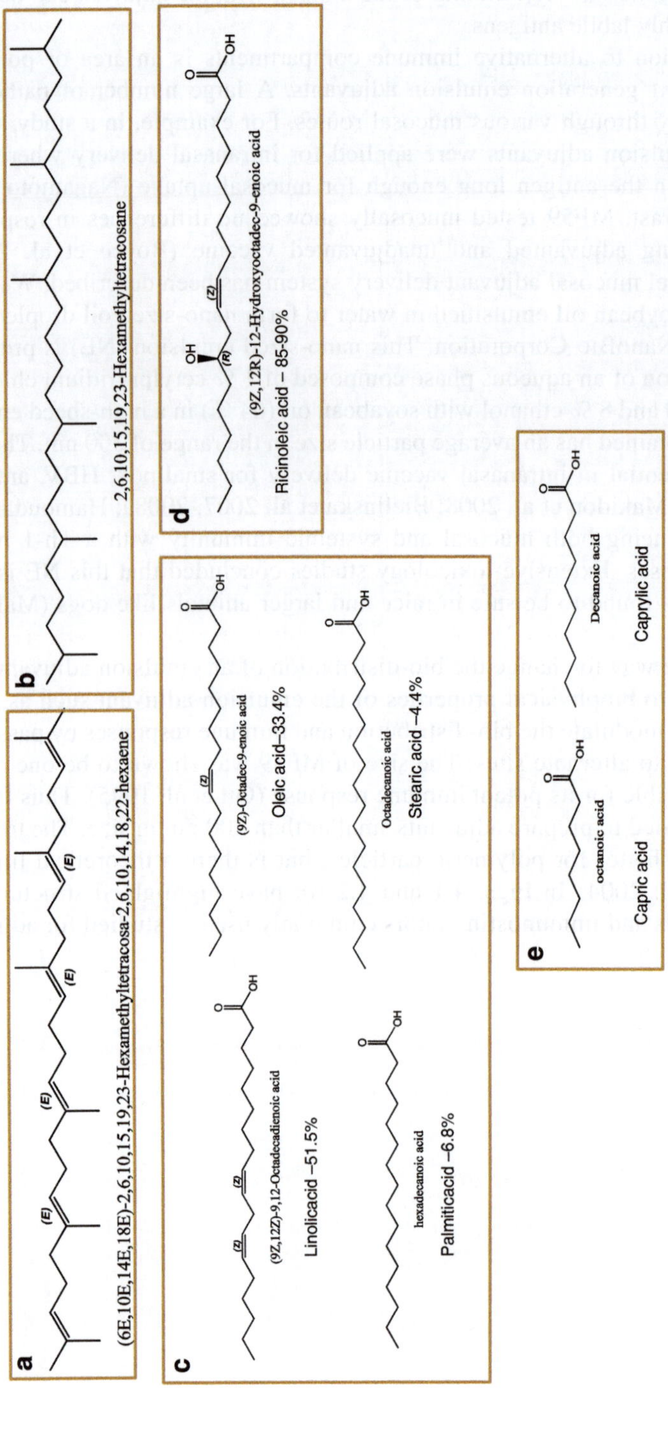

Fig. 4.1 Oils being explored in adjuvants **(a, b)** Squalene, **(c)** Soyabean oil—composed of triglycerides of majorly linoleic acid oleic acid, palmitic acid, and stearic acid (Baughman and Jamieson 1922), **(d)** Castor Oil—composed of triglycerides of majorly richoleic acid and **(e)** Miglyol 810—composed of triglycerides of caprylic acid and capric acid

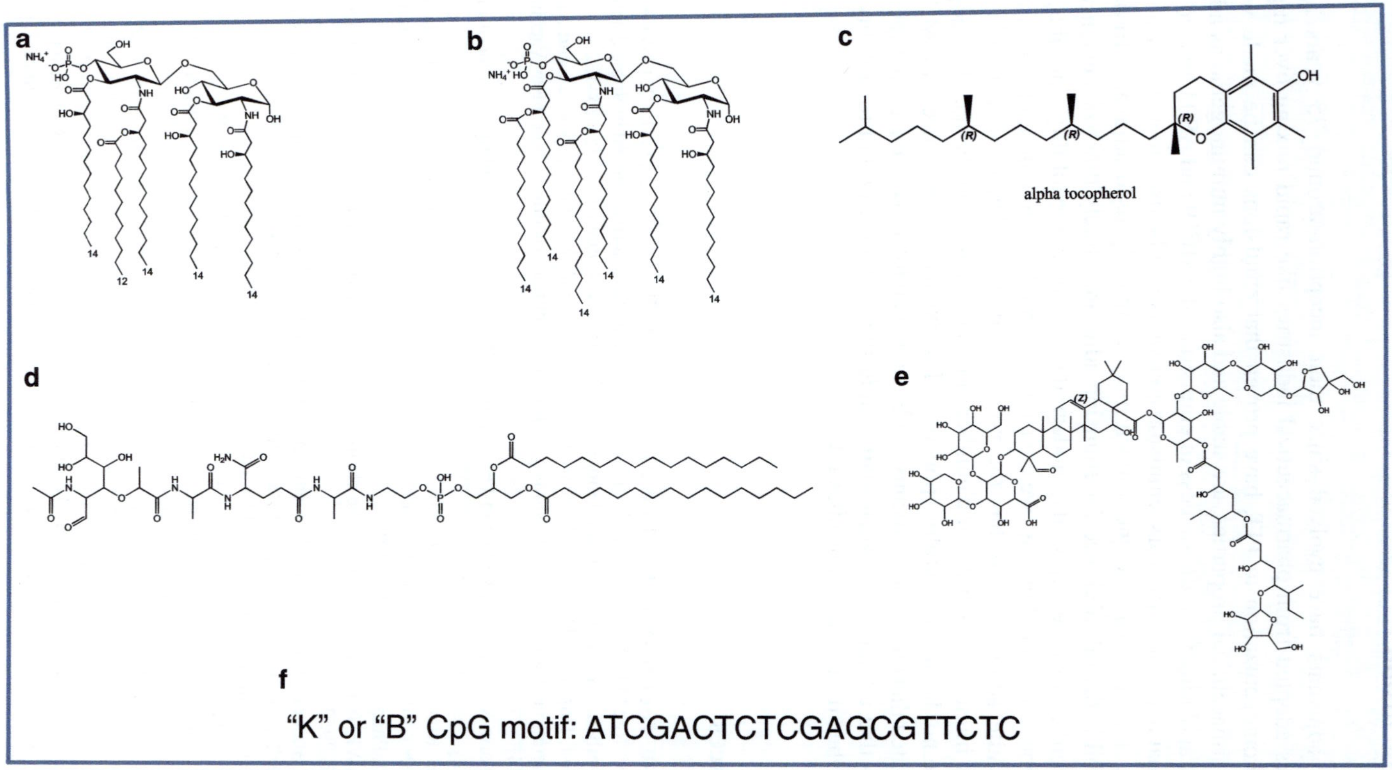

Fig. 4.2 Structures of commonly studied immune potentiators: (**a**) Lipid A, (**b**) Monophosphoryl Lipid A (MPL-A)—Glucopyranosyl Lipid Adjuvant (GLA) and E6020 are derivatives of MPLA, (**c**) Vitamin E, (**d**) Muramyl tripeptide phosphoethanolamine, (**e**) QS 21 and (**f**) CpG motif (Linman et al. 2004)

4.9 Conclusion

Emulsion adjuvants have evolved since their inception around 75 years ago. Techniques adopted from pharmaceutical industries like rapid mixing over high-pressure homogenization to PIT, have proven that emulsions can be made with proper pharmaceutical properties, are stable and also highly immunogenic as adjuvants. Extrapolating from the success of emulsions like MF59 and AS03, one could imagine that in the next 10 years, emulsion adjuvants will be used in a greater number of vaccines and will become a much more commonly used adjuvant. To date squalene oil is the oil of choice for emulsion adjuvants, likely due to its extensive safety record in vaccine adjuvants, biodegradability, biocompatibility, and ubiquitous distribution within humans; we envision that trend to continue going forward. This versatile adjuvant class has the potential to impact modern medicine by improving immune responses for a number of unmet medical needs that alum adjuvants have not been successfully applied to. The extensive experience with MF59 has illustrated that emulsion adjuvants can be safe, well tolerated, and immunogenic indicating that a balance between maintaining immunogenicity and tolerability is possible when developing new adjuvants.

References

Anderson RC, Fox CB, Dutill TS, Shaverdian N, Evers TL, Poshusta GR, Chesko J, Coler RN, Friede M, Reed SG, Vedvick TS (2010) Physicochemical characterization and biological activity of synthetic TLR4 agonist formulations. Colloids Surf B Biointerfaces 75:123–132

Aucouturier J, Dupuis L, Deville S, Ascarateil S, Ganne V (2002) Montanide ISA 720 and 51: a new generation of water in oil emulsions as adjuvants for human vaccines. Expert Rev Vaccines 1:111–118

Batista-Duharte A, Lindblad EB, Oviedo-Orta E (2011) Progress in understanding adjuvant immunotoxicity mechanisms. Toxicol Lett 203:97–105

Baudner BC, Ronconi V, Casini D, Tortoli M, Kazzaz J, Singh M, Hawkins LD, Wack A, O'Hagan DT (2009) MF59 emulsion is an effective delivery system for a synthetic TLR4 agonist (E6020). Pharm Res 26:1477–1485

Baughman WF, Jamieson GS (1922) The chemical composition of soya bean oil. J Am Chem Soc 44:2947–2952

Behzad H, Huckriede AL, Haynes L, Gentleman B, Coyle K, Wilschut JC, Kollmann TR, Reed SG, Mcelhaney JE (2012) GLA-SE, a synthetic toll-like receptor 4 agonist, enhances T-cell responses to influenza vaccine in older adults. J Infect Dis 205:466–473

Beran J, Hobzova L, Wertzova V, Kuriyakose S, Leyssen M, Surquin M, Houard S (2010) Safety and immunogenicity of an investigational adjuvanted hepatitis B vaccine (HB-AS02V) in healthy adults. Hum Vaccin 6:578–584

Bielinska AU, Janczak KW, Landers JJ, Makidon P, Sower LE, Peterson JW, Baker JR Jr (2007) Mucosal immunization with a novel nanoemulsion-based recombinant anthrax protective antigen vaccine protects against Bacillus anthracis spore challenge. Infect Immun 75(8):4020–4029

Bielinska AU, Chepurnov AA, Landers JJ, Janczak KW, Chepurnova TS, Luker GD, Baker JR Jr (2008a) A novel, killed-virus nasal vaccinia virus vaccine. Clin Vaccine Immunol 15(2): 348–358

Bielinska AU, Janczak KW, Landers JJ, Markovitz DM, Montefiori DC, Baker JR Jr (2008b) Nasal immunization with a recombinant HIV gp120 and nanoemulsion adjuvant produces Th1 polarized responses and neutralizing antibodies to primary HIV type 1 isolates. AIDS Res Hum Retroviruses 24:271–281

Bojang KA, Milligan PJ, Pinder M, Vigneron L, Alloueche A, Kester KE, Ballou WR, Conway DJ, Reece WH, Gothard P, Yamuah L, Delchambre M, Voss G, Greenwood BM, Hill A, Mcadam KP, Tornieporth N, Cohen JD, Doherty T (2001) Efficacy of RTS, S/AS02 malaria vaccine against Plasmodium falciparum infection in semi-immune adult men in The Gambia: a randomised trial. Lancet 358:1927–1934

Boyce TG, Hsu HH, Sannella EC, Coleman-Dockery SD, Baylis E, Zhu Y, Barchfeld G, Difrancesco A, Paranandi M, Culley B, Neuzil KM, Wright PF (2000) Safety and immunogenicity of adjuvanted and unadjuvanted subunit influenza vaccines administered intranasally to healthy adults. Vaccine 19:217–226

Brito LA, Chan M, Baudner B, Gallorini S, Santos G, O'Hagan DT, Singh M (2011) An alternative renewable source of squalene for use in emulsion adjuvants. Vaccine 29:6262–6268

Caillet C, Piras F, Bernard MC, de Montfort A, Boudet F, Vogel FR, Hoffenbach A, Moste C, Kusters I (2010) AF03-adjuvanted and non-adjuvanted pandemic influenza A (H1N1) 2009 vaccines induce strong antibody responses in seasonal influenza vaccine-primed and unprimed mice. Vaccine 28:3076–3079

Calabro S, Tortoli M, Baudner BC, Pacitto A, Cortese M, O'Hagan DT, de Gregorio E, Seubert A, Wack A (2011) Vaccine adjuvants alum and MF59 induce rapid recruitment of neutrophils and monocytes that participate in antigen transport to draining lymph nodes. Vaccine 29: 1812–1823

Calabro S, Tritto E, Pezzotti A, Taccone M, Muzzi A, Bertholet S, de Gregorio E, O'Hagan DT, Baudner B, Seubert A (2013) The adjuvant effect of MF59 is due to the oil-in-water emulsion formulation, none of the individual components induce a comparable adjuvant effect. Vaccine 31:3363–3369

Coffman RL, Sher A, Seder RA (2010) Vaccine adjuvants: putting innate immunity to work. Immunity 33:492–503

Coler RN, Baldwin SL, Shaverdian N, Bertholet S, Reed SJ, Raman VS, Lu X, Devos J, Hancock K, Katz JM, Vedvick TS, Duthie MS, Clegg CH, Van Hoeven N, Reed SG (2010) A synthetic adjuvant to enhance and expand immune responses to influenza vaccines. PLoS One 5:e13677

Coler RN, Bertholet S, Moutaftsi M, Guderian JA, Windish HP, Baldwin SL, Laughlin EM, Duthie MS, Fox CB, Carter D, Friede M, Vedvick TS, Reed SG (2011) Development and characterization of synthetic glucopyranosyl lipid adjuvant system as a vaccine adjuvant. PLoS One 6:e16333

Dormitzer PR, Grandi G, Rappuoli R (2012) Structural vaccinology starts to deliver. Nat Rev Microbiol 10:807–813

Eisenbarth SC, Colegio OR, O'Connor W, Sutterwala FS, Flavell RA (2008) Crucial role for the Nalp3 inflammasome in the immunostimulatory properties of aluminium adjuvants. Nature 453:1122–1126

European Medicines Agency (2011) Humenza (Pandemic influenza vaccine (H1N1) split virion, inactivated, adjuvanted)

Fifis T, Gamvrellis A, Crimeen-Irwin B, Pietersz GA, Li J, Mottram PL, Mckenzie IF, Plebanski M (2004) Size-dependent immunogenicity: therapeutic and protective properties of nanovaccines against tumors. J Immunol 173:3148–3154

Foged C, Hansen J, Agger EM (2012) License to kill: Formulation requirements for optimal priming of CD8(+) CTL responses with particulate vaccine delivery systems. Eur J Pharm Sci 45: 482–491

Fox CB (2009) Squalene emulsions for parenteral vaccine and drug delivery. Molecules 14: 3286–3312

Fox CB, Haensler J (2013) An update on safety and immunogenicity of vaccines containing emulsion-based adjuvants. Expert Rev Vaccines 12:747–758

Garcon N, Vaughn DW, Didierlaurent AM (2012) Development and evaluation of AS03, an adjuvant system containing alpha-tocopherol and squalene in an oil-in-water emulsion. Expert Rev Vaccines 11:349–366
Gasparini R, Amicizia D, Lai PL, Panatto D (2012) Aflunov(®): a prepandemic influenza vaccine. Expert Rev Vaccines 11:145–157
Girard MP, Katz JM, Pervikov Y, Hombach J, Tam JS (2011) Report of the 7th meeting on evaluation of pandemic influenza vaccines in clinical trials, World Health Organization, Geneva, 17–18 February 2011. Vaccine 29:7579–7586
Gupta RK, Rost BE (2000) Aluminium compounds as vaccine adjuvants. In: O'Hagan D (ed) Vaccine adjuvants: preparation methods and research protocols. Humana, Totowa
Hamouda T, Sutcliffe JA, Ciotti S, Baker JR Jr (2011) Intranasal immunization of ferrets with commercial trivalent influenza vaccines formulated in a nanoemulsion-based adjuvant. Clin Vaccine Immunol 18:1167–1175
Hansson M, Nygren PA, Stahl S (2000) Design and production of recombinant subunit vaccines. Biotechnol Appl Biochem 32(Pt 2):95–107
Hilleman MR (1966) Critical appraisal of emulsified oil adjuvants applied to viral vaccines. Prog Med Virol 8:131–182
Hilleman MR (1969) The roles of early alert and of adjuvant in the control of Hong Kong influenza by vaccines. Bull World Health Organ 41:623–628
Hjorth RN, Bonde GM, Piner ED, Goldberg KM, Levner MH (1997) The effect of Syntex adjuvant formulation (SAF-m) on humoral immunity to the influenza virus in the mouse. Vaccine 15:541–546
IDRI (2012) First vaccine against fatal visceral leishmaniasis enters clinical trial. February 22, 2012
Khurana S, Chearwae W, Castellino F, Manischewitz J, King LR, Honorkiewicz A, Rock MT, Edwards KM, Del Giudice G, Rappuoli R, Golding H (2010) Vaccines with MF59 adjuvant expand the antibody repertoire to target protective sites of pandemic avian H5N1 influenza virus. Sci Transl Med 2(15):15ra5
Klucker MF, Dalencon F, Probeck P, Haensler J (2012) AF03, an alternative squalene emulsion-based vaccine adjuvant prepared by a phase inversion temperature method. J Pharm Sci 101:4490–4500
Kool M, Petrilli V, de Smedt T, Rolaz A, Hammad H, Van Nimwegen M, Bergen IM, Castillo R, Lambrecht BN, Tschopp J (2008) Cutting edge: alum adjuvant stimulates inflammatory dendritic cells through activation of the NALP3 inflammasome. J Immunol 181:3755–3759
Lambrecht BN, Kool M, Willart MA, Hammad H (2009) Mechanism of action of clinically approved adjuvants. Curr Opin Immunol 21:23–29
Leroux-Roels G (2009) Prepandemic H5N1 influenza vaccine adjuvanted with AS03: a review of the pre-clinical and clinical data. Expert Opin Biol Ther 9:1057–1071
Lima KM, Dos Santos SA, Rodrigues JM Jr, Silva CL (2004) Vaccine adjuvant: it makes the difference. Vaccine 22:2374–2379
Lindblad EB (2000) Freund's adjuvant. In: O'Hagan D (ed) Vaccine adjuvants: preparation methods and research protocols. Humana, Totowa
Linman D et al (2004) Use of CpG oligodeoxynucleotides as immune adjuvants. Immunol Rev 199:201–216
Liu GC, Ahrens EH Jr, Schreibman PH, Crouse JR (1976) Measurement of squalene in human tissues and plasma: validation and application. J Lipid Res 17:38–45
Makidon PE, Bielinska AU, Nigavekar SS, Janczak KW, Knowlton J, Scott AJ, Mank N, Cao Z, Rathinavelu S, Beer MR, Wilkinson JE, Blanco LP, Landers JJ, Baker JR Jr (2008) Pre-clinical evaluation of a novel nanoemulsion-based hepatitis B mucosal vaccine. PLoS One 3(8):e2954
Mbow ML, de Gregorio E, Valiante NM, Rappuoli R (2010) New adjuvants for human vaccines. Curr Opin Immunol 22:411–416
Miller E, Andrews N, Stellitano L, Stowe J, Winstone AM, Shneerson J, Verity C (2013) Risk of narcolepsy in children and young people receiving AS03 adjuvanted pandemic A/H1N1 2009 influenza vaccine: retrospective analysis. BMJ 346:f794

Morel S, Didierlaurent A, Bourguignon P, Delhaye S, Baras B, Jacob V, Planty C, Elouahabi A, Harvengt P, Carlsen H, Kielland A, Chomez P, Garcon N, Van Mechelen M (2011) Adjuvant System AS03 containing alpha-tocopherol modulates innate immune response and leads to improved adaptive immunity. Vaccine 29:2461–2473

Mosca F, Tritto E, Muzzi A, Monaci E, Bagnoli F, Iavarone C, O'Hagan D, Rappuoli R, de Gregorio E (2008) Molecular and cellular signatures of human vaccine adjuvants. Proc Natl Acad Sci U S A 105:10501–10506

Murphy K, Travers P, Walport M (eds) (2008) Janeway's immunobiology. Garland Science, London

Murray R, Cohen P, Hardegree MC (1972) Mineral oil adjuvants: biological and chemical studies. Ann Allergy 30:146–151

Nagamoto T, Hattori Y, Takayama K, Maitani Y (2004) Novel chitosan particles and chitosan-coated emulsions inducing immune response via intranasal vaccine delivery. Pharm Res 21:671–674

Nohynek H, Jokinen J, Partinen M, Vaarala O, Kirjavainen T, Sundman J, Himanen SL, Hublin C, Julkunen I, Olsen P, Saarenpaa-Heikkila O, Kilpi T (2012) AS03 adjuvanted AH1N1 vaccine associated with an abrupt increase in the incidence of childhood narcolepsy in Finland. PLoS One 7:e33536

O'Hagan DT (2007) MF59 is a safe and potent vaccine adjuvant that enhances protection against influenza virus infection. Expert Rev Vaccines 6:699–710

O'Hagan DT, Singh M (2007) Vaccine adjuvants and delivery systems. Wiley, Hoboken

O'Hagan DT, Ott GS, de Gregorio E, Seubert A (2012) The mechanism of action of MF59 - an innately attractive adjuvant formulation. Vaccine 30:4341–4348

O'Hagan DT, Ott GS, Nest GV, Rappuoli R, Giudice GD (2013) The history of MF59(®) adjuvant: a phoenix that arose from the ashes. Expert Rev Vaccines 12:13–30

Oberg AL, Kennedy RB, Li P, Ovsyannikova IG, Poland GA (2011) Systems biology approaches to new vaccine development. Curr Opin Immunol 23:436–443

Ott G, Nest GV (2006) Development of vaccine adjuvants: a historical perspective. In: Singh M (ed) Vaccine adjuvants and delivery systems. John Wiley, Hoboken

Ott G, Barchfeld GL, Van Nest G (1995) Enhancement of humoral response against human influenza vaccine with the simple submicron oil/water emulsion adjuvant MF59. Vaccine 13:1557–1562

Ott G, Radhakrishnan R et al (2000) The adjuvant MF59: a 10 year perspective. In: O'Hagan D (ed) Vaccine adjuvants: preparation methods and research protocols. Humana, Totowa

Peek LJ, Middaugh CR, Berkland C (2008) Nanotechnology in vaccine delivery. Adv Drug Deliv Rev 60:915–928

Rappuoli R (2000) Reverse vaccinology. Curr Opin Microbiol 3:445–450

Rappuoli R, Aderem A (2011) A 2020 vision for vaccines against HIV, tuberculosis and malaria. Nature 473:463–469

Ribi E, Meyer TJ, Azuma I, Parker R, Brehmer W (1975) Biologically active components from mycobacterial cell walls. IV. Protection of mice against aerosol infection with virulent mycobacterium tuberculosis. Cell Immunol 16:1–10

Ribi E, Cantrell JL, Takayama K, Qureshi N, Peterson J, Ribi HO (1984) Lipid A and immunotherapy. Rev Infect Dis 6:567–572

Schijns VE, Lavelle EC (2011) Trends in vaccine adjuvants. Expert Rev Vaccines 10:539–550

Seubert A, Monaci E, Pizza M, O'Hagan DT, Wack A (2008) The adjuvants aluminum hydroxide and MF59 induce monocyte and granulocyte chemoattractants and enhance monocyte differentiation toward dendritic cells. J Immunol 180:5402–5412

Smith JW, Fletcher WB, Peters M, Westwood M, Perkins FJ (1975) Response to influenza vaccine in adjuvant 65-4. J Hyg (Lond) 74:251–259

Stills HF Jr (2005) Adjuvants and antibody production: dispelling the myths associated with Freund's complete and other adjuvants. ILAR J 46(3):280–293

Stuewart-Tull DE, Shimono T, Kotani S, Knights BA (1976) Immunosuppressive effect in mycobacterial adjuvant emulsions of mineral oils containing low molecular weight hydrocarbons. Int Arch Allergy Appl Immunol 52:118–128

Surquin M, Tielemans CL, Kulcsar I, Ryba M, Voros P, Mat O, Treille S, Dhaene M, Stolear JC, Kuriyakose SO, Leyssen MX, Houard SA (2010) Rapid, enhanced, and persistent protection of patients with renal insufficiency by AS02(V)-adjuvanted hepatitis B vaccine. Kidney Int 77:247–255

Surquin M, Tielemans C, Nortier J, Jadoul M, Peeters P, Ryba M, Roznovsky L, Doman J, Barthelemy X, Crasta PD, Messier M, Houard S (2011) Anti-HBs antibody persistence following primary vaccination with an investigational AS02(v)-adjuvanted hepatitis B vaccine in patients with renal insufficiency. Hum Vaccin 7:913–918

Tritto E, Mosca F, de Gregorio E (2009) Mechanism of action of licensed vaccine adjuvants. Vaccine 27:3331–3334

Tsai T, Kyaw MH, Novicki D, Nacci P, Rai S, Clemens R (2010) Exposure to MF59-adjuvanted influenza vaccines during pregnancy–a retrospective analysis. Vaccine 28:1877–1880

Vajdy M (2011) Immunomodulatory properties of vitamins, flavonoids and plant oils and their potential as vaccine adjuvants and delivery systems. Expert Opin Biol Ther 11:1501–1513

Vesikari T, Pepin S, Kusters I, Hoffenbach A, Denis M (2012) Assessment of squalene adjuvanted and non-adjuvanted vaccines against pandemic H1N1 influenza in children 6 months to 17 years of age. Hum Vaccin Immunother 8:1283–1292

Vogel FR, Caillet C, Kusters IC, Haensler J (2009) Emulsion-based adjuvants for influenza vaccines. Expert Rev Vaccines 8:483–492

Whitehouse MW, Orr KJ, Beck FW, Pearson CM (1974) Freund's adjuvants: relationship of arthritogenicity and adjuvanticity in rats to vehicle composition. Immunology 27:311–330

Windish HP, Duthie MS, Misquith A, Ireton G, Lucas E, Laurance JD, Bailor RH, Coler RN, Reed SG (2011) Protection of mice from Mycobacterium tuberculosis by ID87/GLA-SE, a novel tuberculosis subunit vaccine candidate. Vaccine 29:7842–7848

Withdrawal of the marketing authorisation in the European Union (2011). In: Evaluation, HMDA (ed). Based on a report issued by European Medical Agency. Report number is EMA/418809/2011 London.

Chapter 5
The Application of Liposomes as Vaccine Adjuvants

Elisabeth Kastner, Signe T. Schmidt, Alexander Wilkinson, Dennis Christensen, and Yvonne Perrie

5.1 Liposomes: A Brief Introduction

Liposomes, meaning lipid bodies, were first identified by Bangham in 1965 who used phospholipid and cholesterol-based liposomes as a model of cell walls to investigate receptors (Bangham et al. 1965). Building on this, it was Gregoriadis who first used liposomes as a vehicle for delivery of drugs, and suggested they were effective as vaccine delivery systems (Allison and Gregoriadis 1974), and extensive research and development of liposome technologies has brought several candidates into clinical trials and use in humans (Watson et al. 2012). When Gregoriadis first suggested the use of liposomes for vaccine delivery, very little was known of the intricacies of the immune system. As knowledge pertaining to the function of the immune system has expanded, so has the underpinning principles of liposome design and their formulations have been modified and adapted to the requirements of effective adjuvants.

In a subunit vaccine, the non-immunogenic antigen must be delivered in association with immunostimulatory compounds that drive the immune response. The antigens in a subunit vaccine are often proteins or peptides derived from the pathogen, while the immunostimulatory compounds can be lipids, proteins, peptides, DNA, or RNA-derived. All these different compounds require a carrier system that facilitates the

---

Author contributed equally with all other contributors

E. Kastner • A. Wilkinson • Y. Perrie (✉)
School of Life and Health Sciences, Aston University, Birmingham B4 7ET, UK
e-mail: kastnere@aston.ac.uk; y.perrie@aston.ac.uk

S.T. Schmidt
Department of Pharmacy, Faculty of Health and Medical Sciences,
University of Copenhagen, 2100 Copenhagen, Denmark

D. Christensen
Department of Infectious Disease Immunology, Vaccine Adjuvant Research,
Statens Serum Institut, Copenhagen, Denmark

C. Foged et al. (eds.), *Subunit Vaccine Delivery*, Advances in Delivery Science and Technology, DOI 10.1007/978-1-4939-1417-3_5

delivery to the antigen presenting cells (APCs). However, the carrier system must be readily modifiable to allow incorporation of different compound groups. Liposomes are good candidates for such a carrier system, as they offer numerous design possibilities for incorporation of a range of moieties including non-lipid compounds and can be tailored for various delivery routes and immune response required.

5.2 The Versatility of Liposomes: Formulation Considerations in the Preparation of Liposomal Adjuvants

Liposomes are by definition vesicles comprised of lipids, with polar head groups and cylindrical structure of the hydrophobic carbon tails, that cause spontaneous formation of lipid bilayers when suspended in an aqueous medium (Fig. 5.1), rather than for example micelles, where the lipids are oriented with their hydrophobic carbon chains in one lipid layer and the hydrophilic head group in contact with the aqueous medium (Lasic 1998). The formation of liposomes only occurs when lipids are above their transition temperature (T_m) and can be described as in a fluid state. The lipid bilayer fluidity of liposomes is mainly determined by the choice of lipids, and in particular their hydrocarbon tails. At temperatures below the phase transition temperature, the

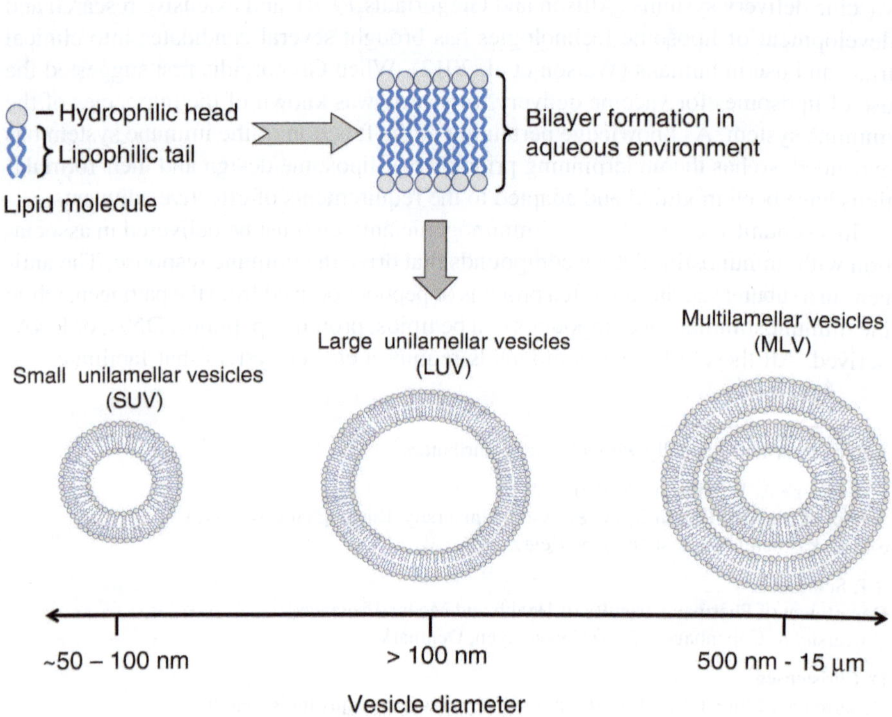

Fig. 5.1 The structural attributes of liposomes

lipid bilayers are in an ordered, solid gel-like phase which changes to a liquid crystalline phase upon heating above the $T_{m.}$ (Fig. 5.1). As would be expected, the choice of lipid in the formulation of liposomes, combined with their method of manufacture determine important physicochemical parameters such as vesicle size, surface charge, adjuvant loading, and membrane rigidity (Watson et al. 2012; Lasic 1998).

Liposomes can be classified by their physical sizes (diameters) and numbers of membrane bilayers (Mozafari 2005). Typical sizes range between 50 nm to several microns; and are furthermore categorized by the number of bilayer membranes incorporated (Fig. 5.1) (Taylor et al. 2005).

1. Small unilamellar vesicles (SUV); size ≤0.1 μm; single bilayer membrane
2. Large unilamellar vesicles (LUV); size >0.1 μm; single bilayer membrane
3. Multilamellar vesicles (MUV); size >0.1 μm; multi-bilayer membrane
4. Multivesicular vesicles (MVV); size >0.1 μm multi-bilayer membrane incorporation of several vesicles into a single bilayer

The orientation of the lipids in the bilayer gives a hydrophobic core and polar surfaces surrounding a closed-off aqueous volume. This unique feature enables incorporation of both hydrophilic and hydrophobic compounds, entrapped in the aqueous interior or embedded in the lipid bilayer, respectively. Furthermore, compounds can be attached to the surface of the liposomes, for example by electrostatic association or covalent bonding to lipophilic anchors which are then incorporated into the lipid bilayer (Watson et al. 2012). The extent of antigen incorporation in the aqueous interior or lipid bilayer depends on the liposome composition and manufacturing method.

5.3 Liposome Manufacturing Methods

Common to the variety of methods for the manufacturing of liposomes is the use of an aqueous buffer system (Taylor et al. 2005). Methods can be either categorized as "top down" method, referring the initial production of larger-sized liposomes, which are subsequently reduced in size and lamellarity by mechanical methods. "Bottom up" methods refer to the direct manufacturing of small vesicles, often accompanied by a kind of fluid flow or fluid channel system (Fig. 5.2).

5.3.1 Mechanical Methods

5.3.1.1 Lipid Hydration Method

The lipid film hydration method, or "Bangham-method" refers to Bangham, who first synthesized liposome in 1961 (Bangham et al 1965; Bangham 1961). Initially, lipids are dissolved in a solvent, often a mixture of chloroform and methanol. The organic solvent is removed by applying a vacuum to the lipids; this process is

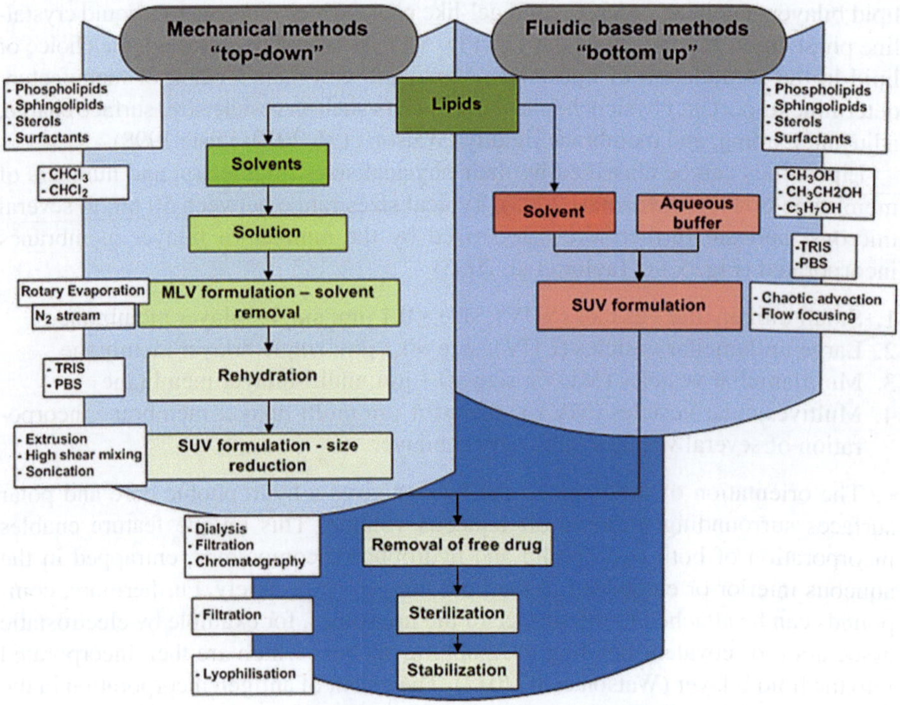

Fig. 5.2 Summary of the various methods used in the preparation of liposomes

usually performed by rotary evaporation and allows for the formation of a dried lipid film. Residues of solvent are removed by flushing the film under a stream of nitrogen, followed by the addition of an aqueous vehicle. This is referred to as the hydration step, which should be performed above the critical transition temperature of the lipids (Bangham et al. 1965; Szoka and Papahadjopoulos 1980). Due to addition of the polar phase, accompanied by mechanical stresses like agitation, the lipids arrange themselves into MLV by swelling lamella from the flask bottom. This method leads to the formation of MLV, ranging up to several microns in size and polydisperse. Controlling the size of the vesicles is not easily achieved with this method and generally dictated by the choice of lipids, the hydration media, and temperatures in the process (Bangham et al. 1965; Gregoriadis et al. 2002). Using this method, drug loading is achieved by adding the compound of interest into the aqueous phase during the hydration stage (aqueous solvent compounds) or addition into the initial solvent phase (lipophilic compounds). Unfortunately, despite the ease of this method, encapsulation efficiencies are usually quite low with this method (Riaz 1996).

5.3.1.2 Detergent Depletion Method

The detergent depletion method is based on the initial formation of micelles. Following, the detergent is removed in order to form liposomes (Brunner et al. 1976). This relatively mild process is compromised by a high dilution of resulting liposomes as well as low entrapment of nonpolar compounds and the rate of detergent removal can affect the size and heterogeneity of resulting liposome formulation (Brunner et al. 1976).

5.3.1.3 Reverse-Phase Evaporation Method: Emulsification Methods

As first described by Szoka and Papahadjopoulos in 1978, lipids are initially dissolved in an organic phase followed by addition of a small volume of aqueous vehicle in order to form inverted micelles (Szoka and Papahadjopoulos 1980). The aqueous vehicle may contain the compound to be encapsulated. After sonication, the two-phase system forms a one-phase dispersion. The organic solvent is removed by rotary evaporation step, which results in the formation of LUV that are often referred to as reverse-phase evaporation vesicles (REV). Even though high encapsulation efficiencies are reported, issues with this process arise as the compounds entrapped are in contact with the organic phase, which may compromise their stability and efficacy (Meure et al. 2008).

5.3.1.4 Homogenization and Extrusion

As vesicles can easily be disrupted by shear or pressure forces, those methods rely on size reduction of preformed MLV. The liposomes are introduced into a microfluidization, high-pressure homogenization, or other shear force-induced homogenizer (Szoka and Papahadjopoulos 1980; Wagner and Vorauer-Uhl 2010), which allow for a high and constant pressure and can be used for continuous and scalable downsizing method (Wagner and Vorauer-Uhl 2010). At pressures as high as 20,000 psi, the MLV dispersions are forced through a small gap and collide with a stainless steel wall, where turbulences or high shear forces result in the break-up of larger vesicles into smaller ones (Barnadas-Rodriguez and Sabés 2001; Bergstrand et al. 2003). Many systems are available in continuous processing mode and pump the liposome solution through a system for continuous reduction of liposome sizes, which makes the resulting liposome sizes dependent on pressure and number of recirculation cycles. The same principle is applied by extrusion methods, where MLV are forced though a polycarbonate membrane or other filter mesh at low or medium pressures in order to produce smaller-size liposomes. The resulting liposome diameter is dictated by the pore size of the filter and recirculation cycles (Meure et al. 2008).

5.3.1.5 Sonication

Probe, cup, or bath sonication is used for breaking down larger vesicles into smaller unilamellar vesicles at sonication temperatures above the lipid transition temperature (Wagner and Vorauer-Uhl 2010). Internal volume/encapsulation efficiencies and industrial scalability are often low and a main drawback of this method (Riaz 1996). Additionally, degradation of lipids or encapsulated material might occur. At very high energy input, the MLV are broken down into smaller SUV. The energy input releases heat, which necessitates a water/ice bath. In a bath sonication process, the liposome dispersion does not come directly into contact with a probe, which is an advantage as the probe can provide a source of contamination (Kataria et al. 2011).

5.3.2 *Methods Based on Fluidic Control*

These methods involve the control of fluids by means of pumps, channels, or micro-channels by either stepwise addition or continuous flow.

5.3.2.1 Supercritical Fluid Method

The supercritical fluid method was described by Frederiksen et al., in 1997 which results in the formation of SUV (Frederiksen et al. 1997). Here, the lipids are dissolved in supercritical carbon dioxide. Liposomes are formed upon expansion into the aqueous vehicle that may also contain the hydrophilic drug to be encapsulated. Nevertheless, encapsulation efficiencies reported here were generally low (Karn et al. 2013).

5.3.2.2 Ethanol Injection

The ethanol injection method was first described in the 1970s by Batzri and Korn (1973). Initially, the lipids are dissolved in a solvent system that is injected into an aqueous buffer system. The rapid injection into an aqueous buffer system leads to the precipitation of lipids and the subsequent formation of SUV. Despite the simplicity of the method, the formation of the liposomes and resulting heterogeneity strongly depends on the solubility of the lipids in ethanol. Remaining solvent is removed by heating up the solution as in the related ether-injection method. High encapsulation efficiencies were reported for the use of a hydrophobic drug, whereas the encapsulation of a hydrophilic drug was relatively poor (Jaafar-Maalej et al. 2010). Injection methods are generally relatively scalable, rapid, simple, and easy to use for standard lipid solution and low transition temperature lipids. Hauschild et al. recently developed a variation of the ethanol injection method, called inkjet method, where high control of liposome sizes are reported (Hauschild et al. 2005).

5.3.2.3 Microfluidics Methods

The usage of microfluidics for the size-controlled preparation of liposomes is based on controlled mixing of streams in a micro-sized channel. The process of liposome formation is dependent on the increase in polarity and liposome formation is dictated by resulting nanoprecipitation reaction. Using microfluidic channels, lipids in a solvent are mixed with an aqueous buffer system and pumped through the mixing chamber, often aided by syringe pumps. Flow in all microchannels is generally categorized as laminar, which may lead to tight control of liposome sizes. Advantages in all microfluidics methods are the tight control of flow rates as well as the flow rate ratios (solvent to aqueous buffer stream) which can provide highly reproducible vesicle formulations. Furthermore, much smaller liquid volumes, along with reduced time for sample handling, mixing, and detection are required (Weigl et al. 2003). Systems are designed for high-throughput with the option of continuous manufacturing by parallelization. The two most common methods are described below.

There are a range of options that can be considered in microfluidics. For example, Zhigaltsev et al. demonstrated the application of a staggered herringbone micromixer (SHM), as a passive micromixer based on chaotic advection, for limit-size synthesis of liposomes. The main advantage here is the control of resulting liposome size by alteration in flow rate and flow rate ratios (Zhigaltsev et al. 2012; Belliveau et al. 2012). The fluid streams are passed through the series of herringbone structures and the chaotic flow profile and increasing advection and diffusion leads to the increase in polarity (Zhigaltsev et al. 2012; Belliveau et al. 2012). An alternative option considered the use of flow-focusing techniques for size-controlled preparation of liposomes as described by Jahn et al. and Valencia et al. with liposome size ranges from 35 to 180 nm (Jahn et al., 2007; Valencia et al. 2010). Here, the central lipid-solvent stream is passed between two streams of aqueous buffer, where mixing occurs at the interfaces and liposome sizes can be tightly controlled.

5.4 Industrial Manufacturing of Liposomes: Requirements

As outlined, there are a range of methods that can be adopted for the preparation of liposomal adjuvants and the protocol adopted will be dependent on a range of factors including industrial applicability and the resultant liposomal characteristics can be dictated by the method employed. For example, comparing four of the main methods for laboratory manufacturing of liposomes outlined above (rotary evaporation, sonication, high shear mixing, microfluidics) by dynamic light scattering (DLS) the difference in intensity-based size distributions of MLV (500 nm), SUV produced by sonication (120 nm), and homogenization (230 nm) as well as SUV produced with microfluidics-chaotic advection (160 nm) is depicted and the polydispersities being relatively unaffected by the respective methods, (Fig. 5.3). However for the industrial manufacturing of liposomes several key characteristics are crucial, based on the FDA draft-guide "Liposomes Drug Products" (www.fda.gov).

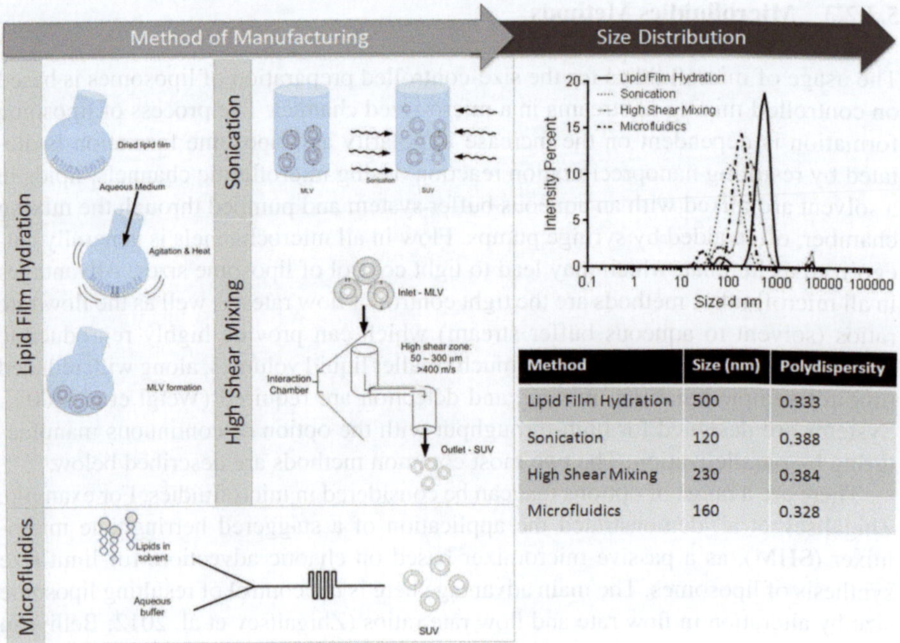

Method	Size (nm)	Polydispersity
Lipid Film Hydration	500	0.333
Sonication	120	0.388
High Shear Mixing	230	0.384
Microfluidics	160	0.328

Fig. 5.3 The effects of four commonly used manufacturing method on liposomal size

- *Scalability*: The process, as developed and optimized at bench-scale, should be easily scalable without alterations in key liposomal parameters.
- *Organic solvent residuals*: Any in vivo application should be free of Class I and II organic solvent residues (classification according to US Pharmacopeia).
- *Sterility*: A Large-scale manufacturing process should be sterile, easy to sterilize or with implementation of disposable unit operations in order to avoid cross contamination between batches. Liposomes should be sterile and pyrogen free.
- *Consistency*: The process should provide minimal batch-to-batch variability and consistent results, including physicochemical properties, morphology, lamellarity, net charge, entrapment volume, liposome size (mean and distribution profile), phase transition temperature, spectroscopic data, in vitro release of the drug substance encapsulated or attached to the liposomal vesicles, osmotic properties, and light scattering index.
- *Throughput*: The ideal process should have a high throughput and the option for continuous manufacturing in order to increase throughput.

Table 5.1 summarizes the key advantages and disadvantages of several methods used in liposome preparation; as can be seen there are now laboratory protocols, such as microfluidics and homogenization that fit much of the above requirements.

Table 5.1 Comparison of the advantages and disadvantages of various methods used to prepare liposomes

Method			Advantages	Disadvantages	Resulting vesicles	Sizes reported	Reference
Mechanical methods	Rotary evaporation		• Easy • Simple	• Organic solvent residue • No control of vesicle size • Additional method for size reduction required • Heterogeneity • Low encapsulation	MLV	Up to several microns	Bangham (1961)
	"Top-down" methods	Homogenization, extrusion, high shear mixing	• Easy • Simple design • Bulk production • Continuous	• High pressures • Heat generation • Filter clogging (extrusion method)	SUV	Dependent on no of cycles and pressure	Bally et al., (1991)
		Sonication	• Easy • Relatively quick	• Sterility • Degradation • Encapsulation low • Limited scalability	SUV	> 90 nm	Wagner and Vorauer-Uhl (2010)
	Reverse phase evaporation		• High encapsulation efficiencies	• Contact with organic phase • Time intensive • Limited scalability	LUV	Intermediate nm to um	Meure et al., (2008)

(continued)

Table 5.1 (continued)

Method			Advantages	Disadvantages	Resulting vesicles	Sizes reported	Reference
Fluid control "bottom up" methods	Injection	Ethanol injection	• Rapid • Easy • No special equipment required • High encapsulation of hydrophobic drugs	• Solvent residue • Dilution	SUV	Dependent on needle diameter pressure and concentration <100 nm	Batzri and Korn, (1973); Jaafar-Maalej et al. (2010)
		Ether injection	• Size controlled SUV • Higher encapsulation efficiencies		SUV	100–300 nm	(Deamer and Bangham (1976); Deamer (1978)
		Inkjet	• Small unilamellar SUV • Loading in combination with liposome formation • High reproducibility		SUV	50–200 nm	Hauschild et al., (2005)
		Supercritical fluid	• Very small particles reported	• High costs, • High pressures • Encapsulation Lower than with conventional methods	SUV	>20 nm	Frederiksen et al., (1997); Karn et al., (2013)
	Micro-fluidic	Chaotic advection	• Size controlled SUV • Bulk production through parallelization of chips	• Solvent residue • Solvent compatibility with chip	SUV	20–200 nm	Zhigaltsev et al. (2012); Belliveau et al., (2012)
		Flow focusing	• Flexible • High reproducibility	• Dilution effect		50–150 nm	Jahn et al. (2007, 2004)

5.5 Liposomal Characteristics That Influence the Immune Response

The above impact of the liposome preparation method on vesicle size is an important factor when considering adjuvant efficacy (Fig. 5.4). Particles less than 200 nm, with an optimum at 40 nm, are generally believed to be able to drain from the site of injection at the classical parenteral administration routes to the draining lymph nodes, and thereby promote cross-presentation and a CTL-driven immune response (Bachmann and Jennings 2010; Manolova et al. 2008). Larger particles are retained at the site of injection as they cannot cross the loose epithelium of the lymphatic ducts (Swartz 2001). In a DNA-vaccine, the reduction of the particle size of the cationic liposomal adjuvant improved the humoral and CTL responses (Carstens et al. 2011). Moreover, SUV less than 155 nm have also been shown to induce primarily a Th2 response identified by increased IL-5 production, while lipid vesicles larger than 225 nm induced a Th1 response based on the presence of IgG2a in plasma and IFN-γ production (Brewer et al. 1998). In line with this, vaccination with a cationic liposomal adjuvant resulted in induction of an increased Th1 response at particle sizes of 685 nm compared to smaller and larger liposomes (Henriksen-Lacey et al. 2011). Thus, the adjuvant particle size can determine the localization of the vaccine particles by controlling the draining kinetics, and thereby determine the obtained immune response.

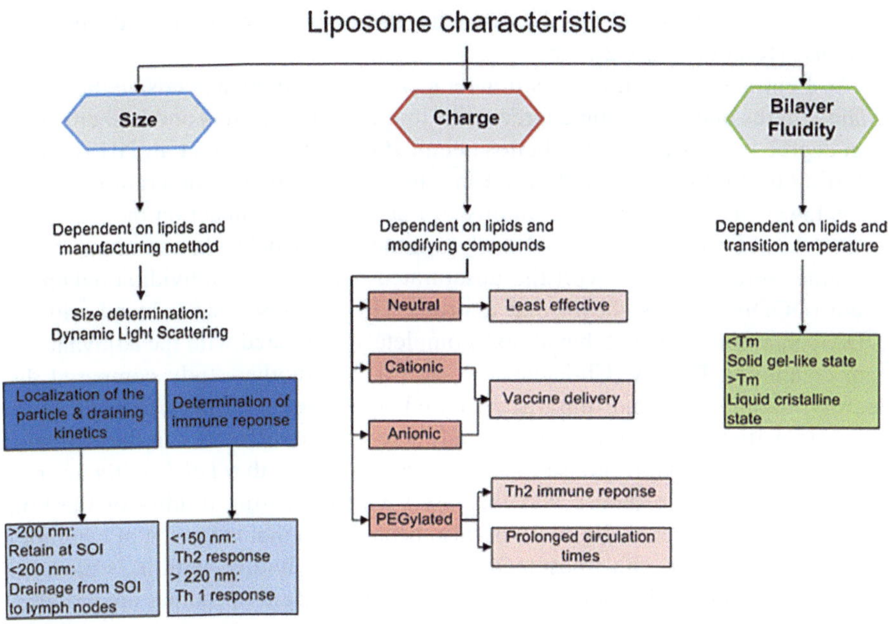

Fig. 5.4 Summary of the impact of vesicle characteristics on liposomal adjuvant action

However in addition to the particle size, the vesicle charge is an important consideration. The charge of the liposomal delivery system is controlled via the choice of lipids and added modifying compounds. Neutral liposomes are the least effective inducers of the immune response, whereas both cationic and anionic liposomes have been used as vaccine delivery systems with success (Watson et al. 2012). Especially, cationic liposomes have been extensively investigated as vaccine delivery systems (Christensen et al. 2007, 2011), and have been shown to elicit increased immune responses compared to a neutral analogue (Carstens et al. 2011). The adjuvant effect of charged liposomes may be caused by their ability to interact with cell membranes; several studies have shown enhanced cell association in vitro of positively charged adjuvants compared to the neutral or anionic counterparts (Foged et al. 2004; Miller et al. 1998; Li et al. 2011). However it is unlikely that such cationic vesicles will only interact with cell membranes, and aggregation with any anionic moieties found at the injection site is most likely (Henriksen-Lacey et al. 2010a, b, c). The charge of liposomes can be further modified by grafting the lipid bilayer with hydrophilic polymer-linked lipid molecules, of which the most used is poly(ethylene glycol) (PEG), thereby creating the so called "stealth" liposomes (Romberg et al. 2008). The grafting of liposomes with PEG-lipids shields the surface charge which hampers the association with immune cells (Foged et al. 2004; Miller et al. 1998). This feature has been applied in liposomes intended for drug delivery, such as Doxil, because it prolongs circulation times in the body, due to decreased uptake by macrophages (Barenholz 2012). However, it has been shown that even extensively grafted cationic liposomes can induce an immune response. The PEGylated liposomes skewed the immune response towards a Th2 response, compared to the non-PEGylated cationic liposomes, which primarily induced a Th1 response (Kaur et al. 2012).

As mentioned, the phase transition temperature of a lipid depends on the carbon chain lengths and saturation degree, with longer carbon chains and higher saturation degree increasing the T_m (Feitosa et al. 2006; Christensen et al. 2012). When referring to fluid liposomes, they are in a fluid state at body temperature, i.e., the T_m is lower than 37 °C and the fluidity of liposomes can impact on their adjuvant action. In one study, liposomes based on dimethyldioctadecylammonium (DDA) bromide were compared with the unsaturated analogue dimethyldioleoylammonium (DODA); it was shown that a strong Th1 response was induced with the DDA-containing adjuvant, but almost completely abrogated with the adjuvant system containing DODA (Christensen et al. 2012). Another study compared distearoyl, dipalmitoyl, and dimyristoyl-based liposomes with progressively lower phase transition temperatures, and found there was no difference in the induction of a humoral immune response following vaccination with a *Leishmania* antigen. However, the CMI response was increased with increasing rigidity of the lipid bilayer (Mazumdar et al. 2005). It has also been shown that induction of a humoral response is optimal with liposomes of intermediate fluidity, with T_m's close-to-body temperature (van Houte et al. 1981). Generally, rigid liposomes are preferred as vaccine delivery systems, as they overall are more potent inducers of both humoral and CMI responses (Christensen et al. 2012).

5.6 The Effect of Liposomal Composition on the Targeting of Specific APCs

The choice of lipid composition of liposomes and thereby the obtained physical characteristics, such as particle size, membrane fluidity and charge, are important when formulating delivery systems for targeting of specific APCs. In addition to this, the formation of a depot of vaccine at the site of injection is controlled by the delivery system. Positively charged liposomes are retained at the site of injection to a much higher degree than neutral or PEGylated liposomes because of interaction with negatively charged interstitial proteins (Kaur et al. 2012; Henriksen-Lacey et al. 2010b; Khan et al. 2013), while decreasing the particle size enhances free drainage away from the site of injection (Carstens et al. 2011). Furthermore, increased membrane fluidity decreased the depot formation but also increased the amount of antigen positive and adjuvant negative APCs in the draining lymph nodes (Christensen et al. 2012). Studies compared DDA-based, depot forming cationic liposomes with fluid, uncharged or PEGylated analogues with less depot formation at the injection site. In both cases the Th1 response induced by the depot forming delivery system was skewed towards a Th2 response or completely abrogated by the analogues (Kaur et al. 2012; Christensen et al. 2012; Henriksen-Lacey et al. 2010b). Supporting this, retention at the injection site and ability to induce a CMI response was positively correlated comparing 3β [N-(N′, N′-dimethylaminoethane)-carbamoyl] cholesterol (DC-Chol) -, DDA- and 1,2-dioleoyl-3-dimethylammonium propane (DOTAP)-based liposomes with the latter performing poorly compared to the other formulations (Henriksen-Lacey et al. 2010a. Seemingly, there is a connection between the ability of a given adjuvant to form a depot at the site of injection and the induction of a Th1-driven immune response, possibly because of the recruitment of monocytes at the site of injection (Henriksen-Lacey et al. 2010c).

Targeting of the lymphatics for the induction of a CTL response is the goal of many studies, as there is a lack of CTL-inducing vaccines against intracellular pathogens such as viruses (Rappuoli 2007). As mentioned above, a small particle size is an important factor for the targeting of the draining lymph nodes, while negatively charged liposomes are reported to drain faster than positively charged liposomes to the lymph nodes due to electrostatic repulsion at the site of injection (Khan et al. 2013). In one study, negatively charged liposomes were recovered in the lymph nodes at a higher percentage than positively charged liposomes, but since the study focuses on drug delivery and not vaccines, it does not investigate the charge-determined uptake by APCs (Kaur et al. 2008). It is likely though that the negatively charged liposomes would not associate with the APCs to the same degree as positively charged liposomes cf. the studies mentioned above (Miller et al. 1998).

The presented variety of lipids and physicochemical characteristics illustrates the versatility of liposomes, and how they can be tailored to induce certain immune responses. Thus, it is possible to design a liposomal delivery system to suit the vaccine by knowing the pathogen in question, effective antigens and the immune responses that are most effective to prevent infection. However, sometimes the choice to design one parameter may at once enhance and work against the desired immunostimulatory ability of the adjuvant. For example, cationic liposomes interact to a high degree with APCs in the lymph nodes, but their positive charge cause them to form a depot at the site of injection, so they never reach the lymph nodes. In those cases it may be necessary to apply more sophisticated designs such as surface coatings (Romberg et al. 2008). In other cases tailoring of different parameters may counteract each other, such as when small, cationic liposomes are used as adjuvants. They should be able to drain to the lymph nodes, but the charge may cause retention at the site of injection. It is difficult to predict which immune response will be predominant by administration of such an adjuvant.

5.7 Incorporation of Immunomodulating Compounds

Choice of lipids and manufacturing method can control the action of the liposomes by determining depot formation at the site of injection and interaction with APCs. However, to specify the immune response towards a certain pathogen, immunomodulators and pathogen-specific antigen must be incorporated into the liposomes. Because liposomes resemble cell walls with their bilayer structure, it is often relatively easy to attach immunomodifying molecules and the antigen. This is because these molecules are often derived from pathogen cell-wall-associated complexes such as LPS, mycobacterial cord factor, and flagella (Mifsud et al. 2014; Rosenkrands et al. 2005).

The association between the liposomal adjuvant and the antigen is important for the induction of a Th1 and Th17 response, as vaccination with a Tuberculosis antigen, Ag85B-ESAT-6, electrostatically associated with a DDA-based cationic liposomal delivery system showed significantly increased responses compared to administration with the same antigen and adjuvant administered unassociated. The same study also showed that prior administration of the free antigen abrogated the Th1/Th17 response (Kamath et al. 2012). Supporting this, lysozyme, unable to electrically associate with cationic liposomes, was found as free antigen in draining lymph nodes, while Ag85B-ESAT-6 to a much higher degree localized at the site of injection with the liposomal delivery system (Henriksen-Lacey et al. 2010b). Similar results were obtained in a study with human MUC1 peptides either encapsulated inside or associated to the surface of phospholipid-based liposomes containing MPL. Only peptides associated with the liposomes induced a Th1 immune response, whereas free peptides were non-immunogenic. Moreover, peptides associated to the surface of the liposomes were able to induce a humoral immune response, whereas encapsulated peptide did not (Guan et al. 1998). Surface-associated antigens are probably better inducers for a humoral immune response, as the B cell has a direct access to the antigen with no internalization required.

When liposomes are used for vaccine delivery, it is in most cases necessary to modify the liposomes by incorporation of immunostimulatory or –potentiating compounds to obtain acceptable levels of immune responses. Especially (synthetic)

PAMPs which are ligands for the PRRs have been used to modulate the immune response, as they are often specific to natural ligands from certain types of pathogens. When using PAMPs, dendritic cells (DCs) are targeted and activated, forming a link between the innate and adaptive immune responses (Hafner et al. 2013). The PRRs include TLRs, NLRs, RLRs, and CLRs (Hafner et al. 2013; Kumar et al. 2011).

The synthetic double-stranded RNA, polyI:C has been investigated for the induction of a CTL response (Hafner et al. 2013; Nordly et al. 2011a). PolyI:C is a ligand of TLR3, which has the endogenous double-stranded RNA produced solely by viruses as a ligand. TLR3 is located in endosomes of lymph node resident DCs, and the polyI:C-containing vaccine particles must reach the lymph nodes unchanged to be taken up by the TLR3 expressing DCs and induce a CTL response (Jelinek et al. 2011). Incorporation of polyI:C into liposomes can be done both by encapsulation or electrostatic association if the liposomes are cationic. Besides acting as a delivery vehicle for the polyI:C, the liposomes abrogate the inflammatory responses polyI:C cause when injected in a non-complexed form (Nordly et al. 2011a).

The TLR9-ligand, the bacterially derived DNA-sequences CpG have also been used as immunostimulators after incorporation into liposomal adjuvants. TLR9 is located in the endosomes, just as TLR3, and it is probably therefore CpG has the same CTL inducing effect as polyI:C (Kumar et al. 2011). CpG-associated lecithin and cholesterol-based liposomes increased antigen uptake by and maturation of APCs, causing increased CD8+ T-cell IFN-γ production and induced higher antigen-specific antibody titers (Neeland et al. 2014).

Immunostimulators can also be of a lipophilic origin. TDB and MMG are synthetic analogues of the immunostimulatory cord factor derived from *M. Tuberculosis* cell walls and have been incorporated into DDA-based liposomes. Without either TDB or MMG, the liposomes were not able to induce an immune response, while incorporation of TDB- or MMG-induced strong Th1/Th17 responses (Rosenkrands et al. 2005; Nordly et al. 2011b).

By targeting PRRs with the adjuvant system, the antigen is delivered directly to the main link of the innate and adaptive immune system, the DCs. The immunopotentiating compounds used e.g. as ligands for TLRs, cause the DCs to recognize the vaccine particles as a certain class of pathogen be it a virus, a bacterium, or another type. The DCs will react accordingly, promoting an immune response directed towards this class of pathogens, e.g., a Th1-, Th2-, humoral, or CTL-response or a combination of these (Mifsud et al. 2014).

5.8 Summary

To develop a vaccine against a given pathogen it is desirable to know which immune response is optimal to achieve the best protection against the pathogen. Knowledge of the required immune response enables rational development of vaccines targeting certain cells in the body. Cellular targeting is controlled by the adjuvant used in subunit vaccines combined with the administration route. By exploiting the versatility of liposomes in terms of lipid compounds, manufacturing methods and further incorporation of immunostimulatory compounds it is possible to tailor the vaccine to induce the correct immune response.

References

Allison AC, Gregoriadis G (1974) Liposomes as immunological adjuvants. Nature 252(5480): 252–252

Bachmann MF, Jennings GT (2010) Vaccine delivery: a matter of size, geometry, kinetics and molecular patterns. Nat Rev Immunol 10(11):787–796

Bally MB, Cullis PR, Hope MJ, (1991). Extrusion technique for producing unilamellar vesicles. Google Patents

Bangham A (1961) A correlation beween surface charge and coagulant action of phospholipids. Nature 192:1197–1198

Bangham A, Standish M, Watkins J (1965) Diffusion of univalent ions across the lamellae of swollen phospholipids. J Mol Biol 13(1):238–IN27

Barenholz Y (2012) Doxil®—The first FDA-approved nano-drug: Lessons learned. J Control Release 160(2):117–134

Barnadas-Rodriguez R, Sabés M (2001) Factors involved in the production of liposomes with a high-pressure homogenizer. Int J Pharm 213(1):175–186

Batzri S, Korn ED (1973) Single bilayer liposomes prepared without sonication. Biochim Biophys Acta 298(4):1015–1019

Belliveau NM et al (2012) Microfluidic synthesis of highly potent limit-size lipid nanoparticles for in vivo delivery of siRNA. Mol Ther Nucleic Acids 1(8):e37

Bergstrand N et al (2003) Interactions between pH-sensitive liposomes and model membranes. Biophys Chem 104(1):361–379

Brewer JM et al (1998) Lipid vesicle size determines the Th1 or Th2 response to entrapped antigen. J Immunol 161(8):4000–4007

Brunner J, Skrabal P, Hausser H (1976) Single bilayer vesicles prepared without sonication physico-chemical properties. Biochim Biophys Acta 455(2):322–331

Carstens MG et al (2011) Effect of vesicle size on tissue localization and immunogenicity of liposomal DNA vaccines. Vaccine 29(29–30):4761–4770

Christensen D et al (2007) Cationic liposomes as vaccine adjuvants. Expert Rev Vaccines 6(5):785–796

Christensen D et al (2011) Cationic liposomes as vaccine adjuvants. Expert Rev Vaccines 10(4): 513–521

Christensen D et al (2012) A cationic vaccine adjuvant based on a saturated quaternary ammonium lipid have different in vivo distribution kinetics and display a distinct CD4 T cell-inducing capacity compared to its unsaturated analog. J Control Release 160(3):468–476

Deamer D, Bangham A (1976) Large volume liposomes by an ether vaporization method. Biochimica et Biophysica Acta (BBA)-Biomembranes 443, 629–634

Deamer DW., 1978. Preparation and Properties of Ether-Injection Liposomes. Annals of the New York Academy of Sciences 308, 250–258

FDA. Guidance for Industry Liposome Drug Products, Chemistry, Manufacturing, and Controls; Human Pharmacokinetics and Bioavailability; and Labeling Documentation. 2002 Apri 2013; Available from: http://www.fda.gov/drugs/guidancecomplianceregulatoryinformation/guidances/ucm064979.htm.

Feitosa E, Jansson J, Lindman B (2006) The effect of chain length on the melting temperature and size of dialkyldimethylammonium bromide vesicles. Chem Phys Lipids 142(1–2):128–132

Foged C et al (2004) Interaction of dendritic cells with antigen-containing liposomes: effect of bilayer composition. Vaccine 22(15–16):1903–1913

Frederiksen L et al (1997) Preparation of liposomes encapsulating water-soluble compounds using supercritical carbon dioxide. J Pharm Sci 86(8):921–928

Gregoriadis G et al (2002) A role for liposomes in genetic vaccination. Vaccine 20:B1–B9

Guan HH et al (1998) Liposomal formulations of synthetic MUC1 peptides: effects of encapsulation versus surface display of peptides on immune responses. Bioconjug Chem 9(4):451–458

Hafner AM, Corthésy B, Merkle HP (2013) Particulate formulations for the delivery of poly(I:C) as vaccine adjuvant. Adv Drug Deliv Rev 65(10):1386–1399

Hauschild S et al (2005) Direct preparation and loading of lipid and polymer vesicles using inkjets. Small 1(12):1177–1180

Henriksen-Lacey M et al (2010a) Comparison of the depot effect and immunogenicity of liposomes based on dimethyldioctadecylammonium (DDA), 3β-[N-(N′, N′-dimethylaminoethane) carbomyl] cholesterol (DC-chol), and 1,2-dioleoyl-3-trimethylammonium propane (DOTAP): prolonged liposome retention mediates stronger Th1 responses. Mol Pharm 8(1):153–161

Henriksen-Lacey M et al (2010b) Liposomal cationic charge and antigen adsorption are important properties for the efficient deposition of antigen at the injection site and ability of the vaccine to induce a CMI response. J Control Release 145(2):102–108

Henriksen-Lacey M et al (2010c) Liposomes based on dimethyldioctadecylammonium promote a depot effect and enhance immunogenicity of soluble antigen. J Control Release 142(2): 180–186

Henriksen-Lacey M, Devitt A, Perrie Y (2011) The vesicle size of DDA:TDB liposomal adjuvants plays a role in the cell-mediated immune response but has no significant effect on antibody production. J Control Release 154(2):131–137

Jaafar-Maalej C et al (2010) Ethanol injection method for hydrophilic and lipophilic drug-loaded liposome preparation. J Liposome Res 20(3):228–243

Jahn A, Vreeland WN, Gaitan M, Locascio LE, (2004). Controlled vesicle self-assembly in microfluidic channels with hydrodynamic focusing. Journal of the American Chemical Society 126, 2674–2675

Jahn A, Vreeland WN, DeVoe DL, Locascio LE, Gaitan M, (2007). Microfluidic directed formation of liposomes of controlled size. Langmuir 23, 6289–6293

Jelinek I et al (2011) TLR3-specific double-stranded RNA oligonucleotide adjuvants induce dendritic cell cross-presentation, CTL responses, and antiviral protection. J Immunol 186(4): 2422–2429

Kamath AT et al (2012) Synchronization of dendritic cell activation and antigen exposure is required for the induction of Th1/Th17 responses. J Immunol 188(10):4828–4837

Karn PR, Cho W, Hwang S-J (2013) Liposomal drug products and recent advances in the synthesis of supercritical fluid-mediated liposomes. Nanomedicine 8(9):1529–1548

Kataria S et al (2011) Stealth liposomes: a review. Int J Res Ayur Pharm 2(5):1534–1538

Kaur CD, Nahar M, Jain NK (2008) Lymphatic targeting of zidovudine using surface-engineered liposomes. J Drug Target 16(10):798–805

Kaur R et al (2012) Pegylation of DDA:TDB liposomal adjuvants reduces the vaccine depot effect and alters the Th1/Th2 immune responses. J Control Release 158(1):72–77

Khan A et al (2013) Advanced drug delivery to the lymphatic system: lipid-based nanoformulations. Int J Nanomedicine 8:2733–2744

Kumar H, Kawai T, Akira S (2011) Pathogen recognition by the innate immune system. Int Rev Immunol 30(1):16–34

Lasic DD (1998) Novel applications of liposomes. Trends Biotechnol 16(7):307–321

Li Y et al (2011) Relationships between liposome properties, cell membrane binding, intracellular processing, and intracellular bioavailability. AAPS J 13(4):585–597

Manolova V et al (2008) Nanoparticles target distinct dendritic cell populations according to their size. Eur J Immunol 38(5):1404–1413

Mazumdar T, Anam K, Ali N (2005) Influence of phospholipid composition on the adjuvanticity and protective efficacy of liposome-encapsulated Leishmania Donovani antigens. J Parasitol 91(2):269–274

Meure LA, Foster NR, Dehghani F (2008) Conventional and dense gas techniques for the production of liposomes: a review. AAPS PharmSciTech 9(3):798–809

Mifsud EJ, Tan AC-L, Jackson DC (2014) TLR agonists on the modulation of innate immune response and their potential as agents against infectious diseases. Front Immunol 5

Miller CR et al (1998) Liposome–cell interactions in vitro: effect of liposome surface charge on the binding and endocytosis of conventional and sterically stabilized liposomes. Biochemistry 37(37):12875–12883

Mozafari MR (2005) Liposomes: an overview of manufacturing techniques. Cell Mol Biol Lett 10(4):711–719

Neeland MR et al (2014) Incorporation of CpG into a liposomal vaccine formulation increases the maturation of antigen-loaded dendritic cells and monocytes to improve local and systemic immunity. J Immunol 192(8):3666–3675

Nordly P et al (2011a) Immunity by formulation design: Induction of high CD8+ T-cell responses by poly(I:C) incorporated into the CAF01 adjuvant via a double emulsion method. J Control Release 150(3):307–317

Nordly P et al (2011b) Incorporation of a synthetic mycobacterial monomycoloyl glycerol analogue stabilizes dimethyldioctadecylammonium liposomes and potentiates their adjuvant effect in vivo. Eur J Pharm Biopharm 77(1):89–98

Rappuoli R (2007) Bridging the knowledge gaps in vaccine design. Nat Biotechnol 25(12): 1361–1366

Riaz M (1996) Liposomes preparation methods. Pak J Pharm Sci 9(1):65–77

Romberg B, Hennink W, Storm G (2008) Sheddable coatings for long-circulating nanoparticles. Pharm Res 25(1):55–71

Rosenkrands I et al (2005) Cationic liposomes containing mycobacterial lipids: a New powerful Th1 adjuvant system. Infect Immun 73(9):5817–5826

Swartz MA (2001) The physiology of the lymphatic system. Adv Drug Deliv Rev 50(1–2):3–20

Szoka F Jr, Papahadjopoulos D (1980) Comparative properties and methods of preparation of lipid vesicles (liposomes). Annu Rev Biophys Bioeng 9(1):467–508

Taylor TM et al (2005) Liposomal nanocapsules in food science and agriculture. Crit Rev Food Sci Nutr 45(7–8):587–605

Valencia PM et al (2010) Single-step assembly of homogenous lipid–polymeric and lipid–quantum Dot nanoparticles enabled by microfluidic rapid mixing. ACS Nano 4(3):1671–1679

van Houte A et al (1981) Characterization of immunogenic properties of haptenated liposomal model membranes in mice. V effect of membrane composition on humoral and cellular immunogenicity. Immunology 44(3):561–568

Wagner A, Vorauer-Uhl K (2010) Liposome technology for industrial purposes. J Drug Deliv 2011

Watson DS, Endsley AN, Huang L (2012) Design considerations for liposomal vaccines: Influence of formulation parameters on antibody and cell-mediated immune responses to liposome associated antigens. Vaccine 30(13):2256–2272

Weigl BH, Bardell RL, Cabrera CR, Catherine R (2003) Lab-on-a-chip for drug development. Adv Drug Deliv Rev 55:349–377

Zhigaltsev IV et al (2012) Bottom-up design and synthesis of limit size lipid nanoparticle systems with aqueous and triglyceride cores using millisecond microfluidic mixing. Langmuir 28(7): 3633–3640

Chapter 6
Developing Bilayer-Based Delivery Systems for Oral Delivery of Subunit Vaccines

Jitinder S. Wilkhu and Yvonne Perrie

6.1 The Advantages Offered by Oral Vaccination

Whilst advanced public health strategies are embedded in most developed countries, many areas of the world still lack the required infrastructure to support mass-scale public health vaccination programmes (Di Fabio and de Quadros 2001; Graham et al. 2012). Currently, most vaccines available are injectables; these have several disadvantages including the need for administration by trained personnel, irritation at the site of injection causing a lack of patient compliance, and needle phobia in adults and children which can make immunisation a stressful procedure (Breau et al. 2001; Wilkhu et al. 2011). In addition, injuries associated with needles and the reuse of needles and syringes is leading to unsafe injection practices. For example, the number of HIV infections resulting from the reuse of needles by healthcare providers has been reported as 80,000–160,000 annually (Kane et al. 1999). Therefore it is essential to find a new delivery route for vaccines, with needle-free immunisations being an important goal.

The oral route for vaccine delivery offers a range of advantages compared to the parenteral route due to its non-invasive nature and low infection risk and, most importantly, mucosal immunity can be promoted by oral vaccines by offering strong resistance against many pathogens that infect via the mucosal lining (Clark et al. 2001). The safety profile of oral vaccines is also more acceptable as side effects, such as flu-like symptoms, fever and diarrhoea associated with parenteral formulations, are reduced or absent (Santiago 1995). The use of oral vaccines may also improve vaccine efficacy as we age given that the mucosa-associated lymphoid tissue (MALT) function does not diminish through the ageing process, unlike the lymphoid tissue exploited by intramuscularly administered vaccines (Santiago 1995). However, there

J.S. Wilkhu • Y. Perrie (✉)
School of Life and Health Sciences, Aston University, Birmingham B4 7ET, UK
e-mail: jitinder@gmail.com; y.perrie@aston.ac.uk

C. Foged et al. (eds.), *Subunit Vaccine Delivery*, Advances in Delivery Science and Technology, DOI 10.1007/978-1-4939-1417-3_6

is conflicting data that the mucosal immune response in the GI tract is compromised by ageing: In a review, Ogra suggests that ageing reduces the number of Peyer's patches, which are required for the uptake of antigens and carrier systems (Ogra 2010).

However, despite the oral route being the most accessible route with good patient compliance, only a small number of vaccines is administered via this route due to problems with vaccine stability within the GI tract. Within the GI tract vaccines are subjected to metabolism and degradation, where low pH and proteolytic enzymes promote degradation (Russell-Jones 2000; Wilkhu et al. 2011). In addition, the residence time of the vaccines at the immune induction sites within the GI tract is relatively short due to normal GI transit. As a result, higher doses, or an increased frequency of dosing can be required to supply a sufficient amount of antigen to elicit an immune response (Webster et al. 2003). However, it is also essential to avoid promoting systemic tolerance and low secretion of antigen-specific IgA through excessive dosing (Mowat 2003; Russell-Jones 2000).

6.2 Currently Licensed Oral Vaccines

An effective vaccine can be described as one which prevents infection and disease with the capability of eliciting specific immune responses with the aim of providing protective immunity. A range of vaccines have been developed and these may be divided into three categories; live attenuated, inactivated and subunit vaccines depending upon their properties. Currently, commercially available oral vaccines are mostly based on the live attenuated or inactivated vaccines (Table 6.1).

This is mainly due to the efficacy of live attenuated vaccines, which promote strong immune protection; they are very effective in initiating mucosal and humoral immune responses as they are able to mimic a natural infection (Webster et al. 2003). The ability of some live attenuated vaccines to replicate intracellularly allows sufficient quantities of the antigenic peptides to be produced which are presented by major histocompatibility complex (MHC) class I molecules, in turn activating cytotoxic T lymphocyte (CTL) responses (Webster et al. 2003; Burgdorf et al. 2007). However, in general live vaccines tend to have higher associated risks compared to other types of vaccines, including the risk of live attenuated vaccines reverting back to the wild form, and inducing a diseased state, especially in immunocompromised

Table 6.1 Current licensed commercially available oral vaccines

Vaccine	Brand name	Type
Adenovirus Type 4 and 7[a]		Live
Cholera	Dukoral	Inactivated
Polio		Live attenuated
Rotavirus[a]	Rotarix	Live attenuated
Rotavirus Pentavalent[a]	RotaTeq	Live
Typhoid[a]	Vivotif	Live attenuated

[a]FDA approved updated on 11th August 2013

individuals (Chadwick et al. 2010). Indeed, the oral polio vaccine (OPV) developed in 1958 by Dr. Albert Sabin, which is an attenuated wild-type polio virus, has been replaced by the inactivated polio vaccine (IPV) in many public healthcare policies due to considerations of risk vs. benefit. The most common form of the OPV is the trivalent system, which contains live attenuated strains of the three serotypes of poliovirus (Fine and Carneiro 1999). As attenuated vaccines show transient growth, an advantage of OPV is that it allows prolonged exposure of the immune system to the attenuated organism. As a result, the OPV confers long-lasting immunity. However, as OPV is a live attenuated vaccine, there is a risk of the attenuated form of the polio virus reverting back to the virulent form. This is a major problem in developing countries as other GI tract viruses can interfere with the replication of the attenuated polio vaccine virus in the intestine (Fine and Carneiro 1999). Indeed, the use of OPV can cause vaccine-associated paralytic poliomyelitis (VAPP), either in the vaccinees or people who are in close proximity to them (Troy et al. 2011). The OPV mutations are associated with VAPP and more recently, prolonged replication of the OPV can cause up to 15 % mutation to a vaccine-derived poliovirus (VDPV). This, in turn, can cause outbreaks of poliomyelitis (Minor 2009). The VDPV revertant strains are reported to be transmissible and hence pose significant population risks (Hull and Minor 2005). Thus in countries where the risk of polio is low, IPV which offers lower immune protection but also less associated side effects, is now more widely used to prevent risk of outbreaks from the regular use of wild-type OPV and its mutant forms (Troy et al. 2011; Heinsbroek and Ruitenberg 2010). Whilst the use of inactivated vaccines offer advantages in terms of reduced side-effect profiles compared to live vaccines, inactivated vaccines tend to offer reduced efficacy as they are unable to multiply in the host to give a strong signal to the adaptive immune system hence, further doses are required when using inactivated vaccines (Baxter 2007).

A third group of vaccines are the subunit vaccines which are based on recombinant protein or synthetic peptide-based antigens and therefore offer advantages in terms of improved safety profiles compared to other types of vaccines. However, subunit vaccines are generally limited by their poor immunogenicity and are unable to elicit strong CTL responses when administered without adjuvants. Adjuvants augment the effects of vaccines due to the increased stimulation of the immune system in its response to the vaccine, allowing increased immunity to a specific disease (Schijns 2000). Therefore, whilst live vaccines such as OPV have been shown to be effective when administered via the oral route, the move to subunit vaccines offers reduced side effects but these vaccines present problems in terms of delivery and stability within the GI tract.

6.3 Developing New Oral Vaccines

Given the safe profiles of the various vaccine systems outlined above, research has focused on the development of effective ways to deliver subunit vaccines orally, and a range of systems are being investigated including edible vaccines and particulate-based systems. Oral delivery of plant-based vaccines has been

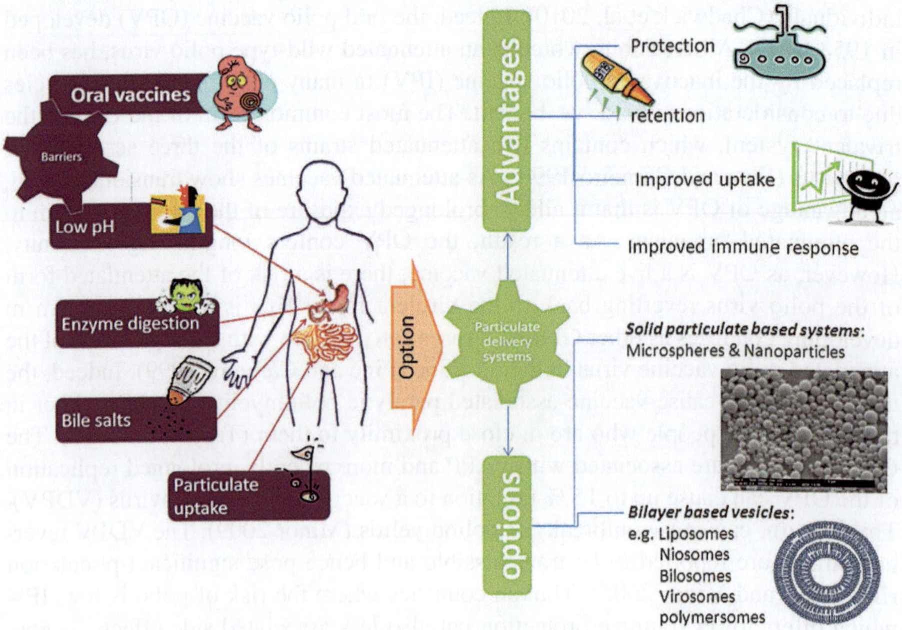

Fig. 6.1 Overview of the gastrointestinal (GI) tract and oral delivery systems highlighting barriers and advantages of particulate delivery systems

investigated as they offer a range of advantages including (1) relatively easy scale up of production and (2) plant components such as unmethylated CpGs have adjuvant properties (Jacob et al. 2013). Such edible vaccines are engineered to contain antigens without genes that would enable pathogens to form, thus eliminating the risk of reversion of the vaccine to virulence. Studies have been undertaken in a variety of plant systems including bananas, potatoes and tomatoes as alternatives to injectable vaccines. However, there are several challenges in growing edible vaccines; for example potatoes generally require cooking which can denature the antigens, bananas take years to mature and the fruit spoils fairly rapidly, and tomatoes can vary from batch to batch as they are cultivated broadly (Langridge 2000). Transgenic plants also present issues in ensuring uniformity of dosage, purity, potency and safety of antigens that would be viable against human or other infectious diseases (Langridge 2000; Jacob et al. 2013). In addition, the complexity of producing a highly expressive plant variety containing stable antigen production is currently a long and expensive task (Jacob et al. 2013).

Particulate delivery systems have also been considered for the oral delivery of subunit vaccines. They offer a range of advantages given that their size, surface characteristics and immunogenicity can be easily manipulated to improve antigen retention and appropriate uptake thereby improving immune responses (Fig. 6.1). Fundamentally, for the delivery of soluble antigens, carrier systems should be designed with the target site in mind. Therefore, for oral vaccines these carriers must be sufficiently acid/enzyme resistant and be formulated to promote uptake by the gut-associated lymphoid tissue (GALT). There are a range of particulate constructs that are being considered for drug and vaccine delivery including solid particulate systems (which can be built from polymers, lipids, proteins, etc.) and bilayer type vesicles (which are built from molecules with surfactant type properties) (Fig. 6.1).

6.4 Barriers for Oral Vaccines: The Journey Down the Gastrointestinal Tract

The GI tract presents a range of barriers to an oral vaccine including the acid environment, enzyme digestion and solubilisation by bile salts (Fig. 6.1). The mucosal surface area is extremely large (in the range of 400 m^2) and it offers several functions (Baudner and O'Hagan 2010): the main functions of the mucosal immune system are to protect the mucous membrane from invasion and colonisation of potentially harmful microbes, whilst preventing the uptake of foreign proteins derived from food sources or airborne matter. This hinders the development of inappropriate immune responses against ingested foreign proteins (Holmgren and Czerkinsky 2005).

The MALT acts as a physical barrier between the external environment and the internal cavities of the body, protecting and preventing invasion by bacteria, foreign pathogens and micro-organisms. However, due to the large surface area of the overall mucosal surface, it is highly susceptible to many pathogenic microorganisms, and therefore has a highly specialised immune system in order to confer immune protection. The immune system is adapted to establish a difference between pathogens that enter via mucosal surfaces to those entering via the blood supply or through tissues (Neutra and Kozlowski 2006). It is essential to develop vaccines which target the effective cells required to achieve immunity. For example, injected vaccines are poor inducers of mucosal immunity hence, are less effective against infection at mucosal surfaces (Levine 2000). Mucosal immune responses are initiated via the MALT which is comprised of several sites including Peyer's patches, mesenteric lymph nodes (MLN), appendix, tonsils and adenoids in the respiratory tract (Kiyono and Fukuyama 2004). In terms of oral vaccination, the key aspects are to deliver vaccines to the Peyer's patches which contain specialised epithelial cells known as the Microfold 'M' cells which capture the antigens (Fig. 6.2). These specialised epithelial cells deliver antigens from the lumen of the gut to the underlying epithelial lymphocytes and subsequently to the subepithelial lymphoid tissues (Mitragotri 2005; Mahato et al. 2003; des Rieux et al. 2006) and process and deliver the antigen to local antigen-presenting cells (APCs) such as dendritic cells (DCs), macrophages and B lymphocytes (Fig. 6.2) (Bilsborough and Viney 2004; des Rieux et al. 2006).

Coombes and Powrie (2008) discuss the immunological properties of DCs; findings indicate that DCs within the Peyer's patches secrete higher levels of

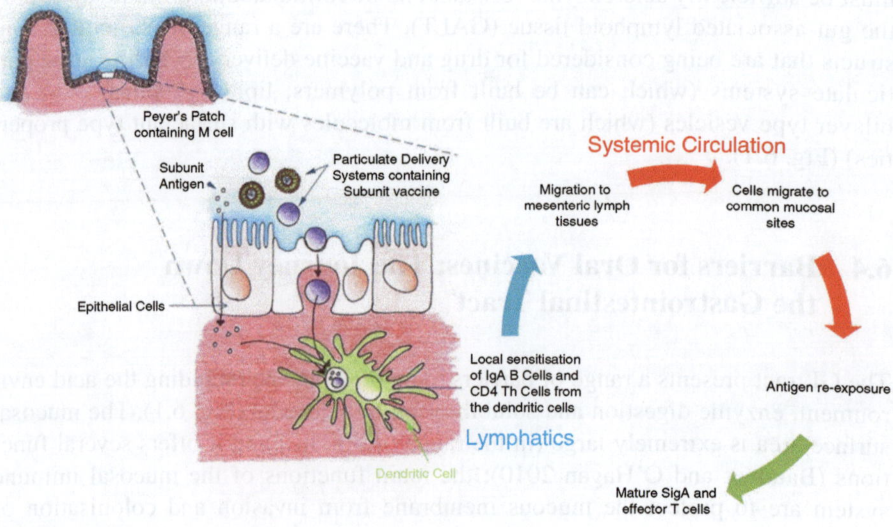

Fig. 6.2 Stimulation of an immune response upon antigen and vesicle exposure to Peyer's patches and M cells

Interleukin-10 (IL-10) than those DCs present within the spleen (Coombes and Powrie 2008). In addition, the CD4$^+$ T-cells (which are activated by the Peyer's patches) produce higher levels of IL-4 and IL-10 (which are represented by a T helper 2 phenotype) than those of the spleen DCs (Iwasaki and Kelsall 1999).

Depending on their size, particulates can enter the intestinal mucosa via a series of routes that include the tips of villi where enterocytes are extruded, across or between enterocytes, or by M cells (Norris et al. 1998). Figure 6.2 outlines how particulates and subunit antigens can be taken up by M cells or epithelial cells which predominantly consist of enterocytes and mucus-secreting goblet cells. M cells have sparse, irregular microvilli on their apical surface in addition to the presence of a basolateral cytoplasmic invagination which creates a pocket containing lymphocytes and macrophages (Clark et al. 2001; Norris et al. 1998). These features of the M cells are believed to be involved in establishment of mucosal immunity by allowing a route for antigens to be delivered to underlying lymphoid tissues where a secretory immune response is initiated. The epithelial cells play a major role in mucosal defence as they are able to detect foreign bodies through pattern-recognition receptors (PRRs) such as the Toll-like receptors (TLRs). The release of cytokine and chemokine signals to the DCs and macrophages which underline the mucosal cells, as shown in Fig. 6.2, allow an innate and adaptive immune response to be triggered (Kagnoff and Eckmann 1997; Izadpanah et al. 2001; Kraehenbuhl and Neutra 2000; Neutra and Kozlowski 2006). Although the mechanism of action upon absorption by the Peyer's patches remains unknown, the M cells remain good targets for subunit particulate and other oral vaccine delivery systems.

6.5 Strategies for Effective Delivery to M-Cells

As Fig. 6.1 demonstrates, to reach the target site, an orally delivered vaccine must firstly overcome the low pH conditions within the stomach and avoid enzymatic degradation to then face the bile acids/salts prior to reaching the small intestine where particulate uptake occurs. The small intestine is the site of absorption for electrolytes, nutrients and fluids. However, it also acts as a barrier which prevents the absorption of potentially toxic or harmful substances. In order for vaccine to be absorbed in the small intestine, they must penetrate through two barriers, the mucus gel layer and the mucosa. The intestinal mucosa comprises of epithelial cells and their associated glands and a mucus layer which is a secretion of high molecular weight glycoproteins and other non-specific defences including mucins and antimicrobial peptides (Kraehenbuhl and Neutra 2000). These structures of the mucosal lining trap enzymes and create an environment which is highly degradative to substances (Clark et al. 2001). This layer plays a major role in protecting the mucosa from harmful bacteria, pathogens and chemicals. In addition to protecting the mucosa, the mucus layer maintains the pH difference that is observed between the GI lumen and the mucosa. Therefore, the mucus layer will hinder the diffusion of certain compounds and as a result, size, pH and electric charge will be major design parameters when targeting the delivery of oral vaccines to these areas. As well as the mucus layer acting as a diffusion barrier, it also acts as a physical barrier by preventing the absorption of particles. The mucus entraps the particles, causing agglomeration in the mucus layer which then causes an increase in net size and a decrease in the diffusion coefficient through the mucus layer.

The surface charge also has a significant impact upon the uptake of particulate delivery systems, e.g. studies by Tabata and Ikada have shown that the peak uptake of particles occurs with a zeta potential of −70 mV (Tabata and Ikada 1988). This zeta potential also shows excellent stability and avoids coagulation of the particles due to the high surface charge (Freitas and Müller 1998; González-Rodríguez and Rabasco 2011). Based on fluorescence microscopy studies, anionic liposomes formulated using phosphatidyl serine (PS) were also shown to have improved uptake by Peyer's Patches compared to liposomes formulated without the inclusion of PS (Tomizawa et al. 1993; Aramaki et al. 1993). Norris et al. (1998) and Eldridge et al. (1990) also found that the uptake of negatively charged particulates was favoured (Norris et al. 1998; Eldridge et al. 1990). The work carried out by Shakweh et al. (2005) also compared the binding and uptake of differently charged PLGA microsphere formulations and noted that whilst the binding to Peyer's Patch free tissue was similar for all microspheres over 48 h, the uptake by Peyer's patches was notably higher for negatively charged or neutral particles in mice compared to positively charged particles (Shakweh et al. 2005).

However cationic formulations (e.g. employing cationic lipids or chitosan), often used for DNA delivery, have also been shown to be effective oral vaccine adjuvants. For example, a cationic liposomal oral mycobacterial DNA vaccine formulated using Lipofectamine™ 2000 (Invitrogen Corporation, USA) was shown to be effective in protecting and delivering plasmid DNA via the oral route to the epithelium,

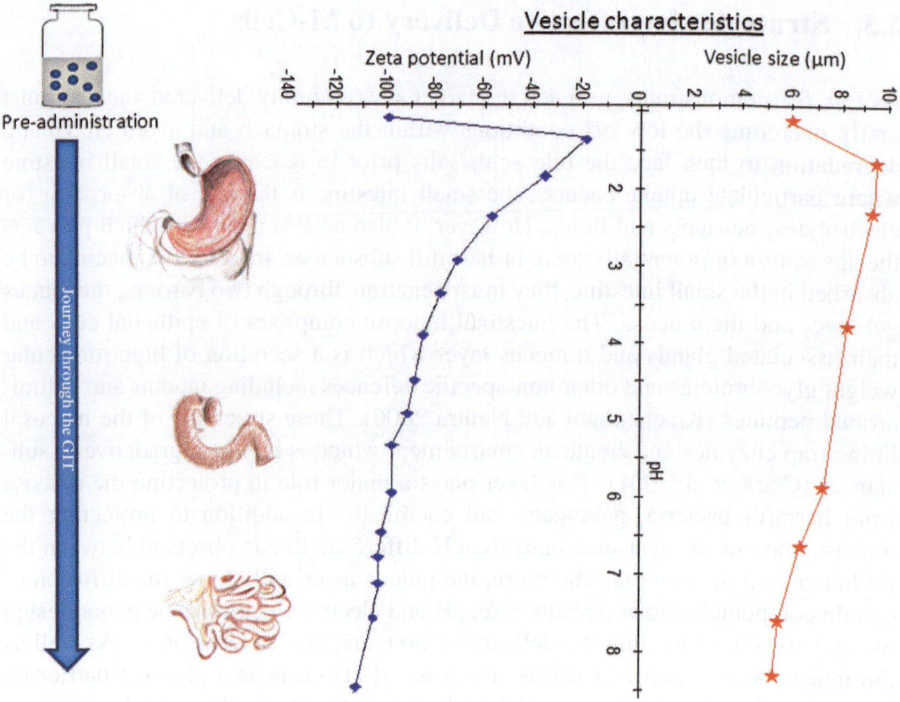

Fig. 6.3 The effects of pH changes during GI transit upon vesicle characteristics such as vesicle size and zeta potential. Adapted from Wilkhu et al. 2013a, b

M cells, DCs and Peyer's patches of the small intestine of mice, and subsequently they were able to promote expression of the encoded Ag85A antigen at these sites. By doing so, the cationic liposome formulation was able to induce higher specific mucosal cellular and humoural immune responses compared to DNA alone (Wang et al. 2010). The use of oral DNA vaccines using cationic particulates is also being extensively investigated for vaccination of fish (Rajesh Kumar et al. 2008; Ning et al. 2009; de las Heras et al. 2010).

Overall, it is important to remember that the zeta potential measured in an electrolyte double layer surrounding particulates is linked to the electrolytes present (both their concentration and valency) within the suspending aqueous media. Indeed the zeta potential of cationic liposomes was shown to be dramatically different (~40 mV vs. ~15 mV) when the liposomes were in simulated gastric or intestinal media respectively (Perrie et al. 2002). In recent studies, we have also investigated the impact of GI tract pH on the zeta potential of anionic lipid vesicles and have shown that whilst these vesicles were prepared with a zeta potential of −100 mV (sodium bicarbonate buffer pH 7.6), as they move through the various regions of the GI tract the zeta potential varies from −20 mV in acidic conditions (pH 1.2) back to −100 mV in intestinal pH conditions (pH 8.4, Fig. 6.3). Whilst changes in the zeta potential of the vesicles, due to changes in pH, are often reversible, such changes may impact on antigen loading which relies on electrostatic interactions.

6.6 Solid Particulate Delivery Systems for Oral Vaccines

Solid particulate delivery systems have also been extensively investigated for their potential as vaccines adjuvants. They can be prepared in a range of sizes (from nanoparticles to microspheres), with a choice of surface characteristics and include a selection of immunomodulators. Polymeric particles are generally formulated from natural or synthetic polymers with the most commonly studied polymers being those which are biodegradable such as poly(lactide-co-glycolide) (PLGA), polylactic acid (PLA), polycaprolactone (PCL) and polysaccharides (particularly chitosan). These polymers offer the advantage that they are well-characterised and used in a range of clinical products, particularly PLGA. Alternatively, solid particulate delivery systems can be prepared from solid (high melting point) lipids dispersed in an aqueous phase. Examples of lipids used include solid triglycerides, saturated phospholipids and fatty acids which are well tolerated by the body. Due to their composition, they are sometimes described as 'solidified' o/w emulsions in which the oil globule is replaced by solidified lipids. In general, in these solid particles the antigen is incorporated within the solid polymer or lipid matrix of the particle or by attaching the antigen to the surface of the particles.

Microspheres for vaccine delivery have been useful in the delivery of antigens which are ingested by the immunocompetent cells, in turn providing controlled antigen release and lasting immunity. Microspheres have been reported to protect the antigen from intestinal bile salts, the variation in pH along the GI tract and from the degradative enzymes (Hanes et al. 1995; O'Hagan and Illum 1990). Several studies with microspheres have shown that antigens encapsulated in microspheres can elicit secretory IgA and circulating antibody IgG responses in comparison to unencapsulated antigen (Langer et al. 1997). Studies by Eldridge et al. (1989) have suggested that the Peyer's patches readily take up particles less than 10 μm in size and prefer hydrophobic surfaces (Florence 2005). In addition to vesicle size, surface charge and hydrophobicity are also crucial factors in the uptake of microparticles via the Peyer's patches. Both are essential as particles have to translocate between a hydrophobic mucus layer and then across a hydrophilic interior of the cells (Hussain et al. 2001).

Given the biodegradable nature and sustained release properties that PLGA can offer, microspheres formulated from these polymers appear to be an ideal candidate for the delivery of subunit vaccines, due to their relative biocompatibility and controllable drug release profile. Given their particulate nature, such delivery systems can promote also uptake, transport and presentation of the antigen to APCs. For example, it has been reported that PLGA microparticles exhibit an adjuvant effect for both humoral (Eldridge et al. 1990; O'Hagan et al. 1991) and cell-mediated immunity (Audran et al. 2003). Indeed, sub-10 μm PLGA microspheres are readily recognised and ingested by macrophages and DCs, an important property for stimulating an immune response (Storni et al. 2005). After oral administration, PLGA microspheres have been shown to elicit immune responses, in some cases comparable to immune responses achieved via parenteral routes; for example, Igartua et al. (1998) compared subcutaneous and oral immunisation using PLGA microspheres. They showed that, whilst by a single oral administration of microspheres it was not possible to induce appropriate responses, dosing on three consecutive days enabled the induction of IgG responses similar to those induced by subcutaneous immunisation, and a later booster dose provided no further advantage (Igartua et al. 1998). The success of the PLGA systems as oral delivery systems was attributed to their ability to promote an initial burst release (~20 %) followed by a sustained release (~50 % by day 35; Igartua et al. 1998). Sarti et al. (2011) were also able to stimulate immune responses to ovalbumin (OVA) delivered orally in PLGA-based particulates, with both IgG and IgA responses being measured and shown to be higher than those of OVA given orally in PBS. However, there is the possibility of PLGA microspheres inducing oral tolerance to their entrapped antigen (Fattal et al. 2002) and whilst PLGA based systems have demonstrated their ability give protection to animals in challenge studies (e.g. Conway et al. 2001), studies have not yet reported a protective immunity induced in humans.

Whilst the use of polymer-based particulates as vaccine adjuvants has been strongly investigated, more recent work that has refocused investigations into the potential advantage of using these systems in the nano range. Alongside stability and protection, the polymers used to formulate nanoparticles can be modulated to control physicochemical characteristics such as zeta potential and hydrophobicity. In a review by des Rieux et al. (2006) it is suggested that in addition to the above advantages, the drug/antigen release properties and biological behaviour of the nanoparticles can also be modified (des Rieux et al. 2006). The resulting nanoparticles are subsequently taken up by the epithelial cells within the mucosa, M cells and Peyer's patches (Gelperina et al. 2005).

When considering the formulation of such nanoparticles, the main consideration for oral delivery involves finding the balance between the desired surface properties and the matrix used in the formulation. These factors control nanoparticle uptake and stability; hence it is vital to establish a relationship between these two factors. Nanoparticles work on a mechanism of releasing their loaded antigen either by particle degradation, erosion, diffusion out of the matrix or swelling (Gelperina et al. 2005). Due to these release mechanisms, the particulates must be sensitive to the local milieu of the environment such as pH, temperature or enzymes including the presence of other particulates such as food. The size of the nanoparticles also plays an important role as colloidal instability leads to aggregation and flocculation. Similarly the chemical stability of a nanoparticle matrix is crucial for its biodegradability and release of the encapsulated antigen (Florence 2005). A review by Chadwick et al (2010) summarises, in depth, the different polymers available for producing nanoparticles, and how the size of the particulates vary depending on the polymer used and its properties.

6.7 Bilayer Vesicle-Based Delivery Systems for Oral Vaccines

Bilayer vesicles systems, which can be built from a range of surfactant molecules have also been widely investigated as oral vaccine delivery systems. The most commonly studied delivery systems are the liposomes. These systems were first recognised by Bangham (Bangham and Horne 1964) and then used as delivery systems by Gregoriadis and Ryman (1971). Liposomes were identified as being effective immunological adjuvants by Allison and Gregoriadis (1974) where the ability of negatively charged liposomes (prepared with the inclusion of dicetyl phosphate (DCP)) to deliver and potentiate immune responses against diphtheria toxoid (DT) was demonstrated; from this work a plethora of investigations have subsequently steamed. There are a wide number of variations, based on these initial liposome systems, that have been developed including stealth liposomes to improve circulation profiles (Lasic et al. 1991), vesicles built from non-ionic surfactants also known as non-ionic surfactant vesicles (NISVs) or niosomes, e.g. for cosmetics (Handjani-Vila et al. 1979) or as drug/antigen carriers (Azmin et al. 1985; Baillie et al. 1985), surfactant polymers (e.g. polymersomes (Okada et al. 1995), cationic systems which can electrostatically bind DNA (e.g. lipoplexes (Felgner et al. 1987), vesicles incorporating bile salts to improve stability (e.g. bilosomes; (Conacher et al. 2001; Wilkhu et al. 2013b), or virus components (e.g. virosomes, Almeida et al. 1975) to name but a few. Many of these systems use alternatives to phospholipids to circumvent potential issues related to storage instabilities and cost (e.g. synthetic-based systems), to improve stability within harsh biological environments (e.g. bilosomes and polymersomes), or alternatively to modulate the properties of the vesicles in terms of immunological efficacy (e.g. virosomes).

Liposomes have been used in vaccine technology as carriers of antigens, which are either encapsulated into the aqueous space, incorporated into the bilayer of the liposomes or associated with the surface of the vesicles. Liposomes as carriers of antigens allow a reduction in biodistribution but enhance targeting which are vital for immunotherapy (Gregoriadis et al. 1999). In general, lipid-based vesicles provide several advantages of antigen delivery where they can be tailored to have desired effects in vivo. A major issue in the use of lipid-based vesicles for oral delivery is their stability when exposed to intestinal bile salts, which can cause the membrane of the lipid vesicles to deform and lyse resulting in the release of macromolecules from the vesicle prior to it reaching its intended site of action, thereby resulting in poor vaccine efficacy (Chen et al. 1996). The potential of vesicle lysis and drug leakage of lipid-based delivery systems when administered orally has been the key consideration in the development of lipid-based vesicles which have increased stability including liposomes (e.g. (Lasic et al. 1991), polymerised liposomes (e.g. (Okada et al. 1995; Chen et al. 1996; Gaucher et al. 2010), niosomes and bilosomes (Wilkhu et al. 2013b; Azmin et al. 1985; Baillie et al. 1985; Conacher et al. 2001).

In the development of oral liposomal vaccines the use of high-transition temperature lipids has been shown to be advantageous. For example, we have studied cationic liposomes as oral vaccine delivery systems for DNA vaccines where DNA encoded HBsAg was entrapped within multilamellar vesicles (MLV) (Perrie et al. 2002). Studies in mice orally dosed with liposomal-DNA vaccines, revealed that secretory IgA responses against the encoded antigen were substantially higher after dosing with 100 mg liposome-entrapped DNA compared to naked DNA and IgA responses in mice were consistently higher with cationic formulations containing the high-transition temperature lipid, distearoylphosphatidylcholine (DSPC), which correlated well with their ability to protect the DNA loading in simulated GI tract conditions. To investigate gene expression sites, mice were also orally dosed with liposome-entrapped plasmid DNA expressing the enhanced green fluorescent protein; fluorescence intensity in the draining MLN was much greater in mice dosed with liposomal DNA than in animals dosed with the naked DNA (Perrie et al. 2002). These results demonstrate the potential of these systems for oral delivery of DNA vaccines.

The potential of vesicle lysis when administered orally has also promoted the development of bilayer vesicles using non-ionic surfactants or niosomes. It has been suggested that these allow a greater chance for the antigen/drug to reach the site of action and to have increased absorption through the GI tract compared to free antigen/drug (Azmin et al. 1985; Yoshida et al. 1992). Hence, further modifications to the niosome vesicles by incorporation of bile salts to form bilosomes have also been investigated. Bilosomes protect antigens from the enzymes present in the GI tract, are resistant to bile acids, and can act as potent immunological adjuvants. Alexander and Brewer first developed and formulated bilosomes by exploiting the NISV technology (Conacher et al. 2001). They achieved this by incorporating bile salts such as sodium deoxycholate into the formulation thereby increasing the stability of the carrier thus, preventing premature release of the protein/antigen via the oral route. Schubert et al. (1983) suggested that by incorporating bile salts into the vesicles they resist degradation and disruption from the digestive enzymes (Fig. 6.4) therefore making the formulation more stable and giving the potential rise for an oral route of delivery for vaccines (Schubert et al. 1983). By preventing premature release, bilosomes deliver the vaccines to the mucosal tissue and thus smaller concentrations of antigen are required to elicit an effective immune response. The beneficial effect of bile salts may not be limited to non-ionic systems; Hu et al. (2013) also note that liposomes containing bile salts (sodium glycocholate) retained significantly higher levels of insulin than conventional liposomes prepared with phosphatidylcholine and cholesterol (Hu et al. 2013). It was concluded that liposomes containing bile salts protected the insulin from degradation and release through the GI tract and contributed to enhanced oral absorption of insulin. Furthermore, studies by Niu et al. (2012) confirm that the presence of bile salts (sodium glycocholate) within conventional liposomes resulted in increased hypoglycaemic effects and oral bioavailability which was attributed to better protection of the encapsulated insulin with the highest oral bioavailability of 11 % in diabetic rats (Niu et al. 2012). In addition, the pharmacological actions of the insulin-loaded vesicles were deemed to be dose- and size-dependent.

As vaccine delivery systems, various studies using bilosomes have proven to be successful in animal models with a range of antigens, e.g. the A/panama (Mann et al. 2004), tetanus toxoid (Mann et al. 2006), and hepatitis B (Shukla et al. 2008). These studies on the various antigens have shown increased antibody production, lower temperatures and reduced side effects (in relation to IM injections) when dosing with influenza haemagglutinin entrapped bilosomes (Bennett et al. 2009). Bile salt incorporation within the vesicles stabilises and protects the vesicles from the GI tract and other enzymes present in the body allowing bilosomes to be used orally as vaccine drug delivery system. In several bilosome studies (e.g. Bennett et al. 2009; Mann

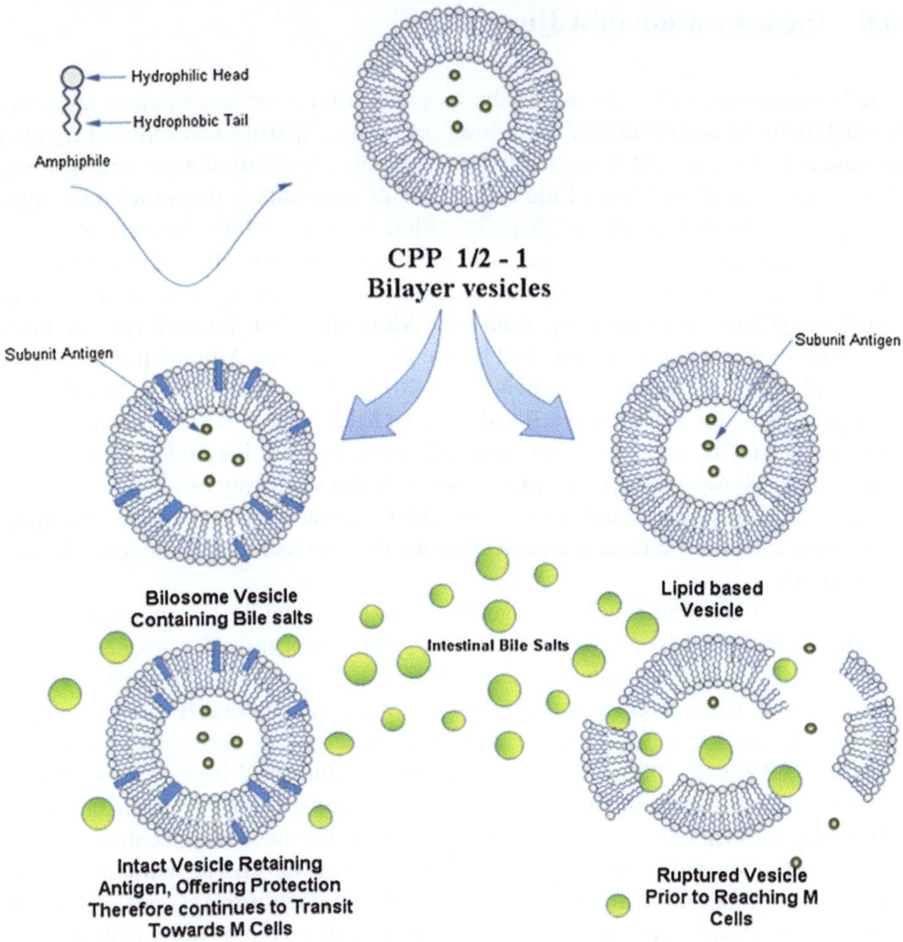

Fig. 6.4 Formation of bilayer vesicles and modifications with the addition of bile salts to improve GI stability

et al. 2004, 2006) the anionic surfactant, DCP is used and these studies all show that, by orally administering the bilosome vesicles with the antigen entrapped, a mucosal immune response is elicited with specific IgA production increased from the mucosal cells in the small intestine. The incorporation of DCP within the bilosome formulation, giving the vesicles a highly negative surface charge, could potentially be the reason for the increased immunity achieved given that previous studies showed that the peak uptake in the epithelial cells of particles occurs with a zeta potential of −70 mV (Tabata and Ikada 1988). This is also in line with studies by Eldridge et al. (1990) and Norris et al. (1998) where they also found that the uptake of negatively charged particulates is favoured. In addition, work by Shakweh and co-workers also suggest that negatively charged or neutral particles in mice have a greater affinity for the Peyer's patches than positively charged particles (Shakweh et al. 2005).

6.8 Incorporation of Adjuvants

The inclusion of an adjuvant within the vaccine formulation can enhance immunological memory and coverage, and allows for antigen sparing and reduced number of doses (Tritto et al. 2009). Currently there are only a handful of adjuvants licensed for human clinical use. One of the oldest type of adjuvants is the aluminium adjuvants, also referred to as alum (Chap. 3), which, since the 1920s, has been incorporated into various human vaccines as an aluminium salt such as aluminium hydroxide ($Al(OH)_3$) and aluminium phosphate ($AlPO_4$). More recently ASO_4 [oil-in-water emulsion (Chap. 4) combining alum and Monophosphoryl Lipid A] has been approved by European regulators and by the Food and Drug Administration (FDA) as an adjuvant for vaccines (Baldwin et al. 2012; Mbow et al. 2010). In addition, European regulators have approved and licensed MF59 (squalene oil-in-water emulsion) as an adjuvant, e.g. in the flu vaccine known as Fluad. Preclinical and clinical studies have shown that alum is often less potent than adjuvants such as the oil-in-water emulsion ones and has shown to be a poor inducer of protective Th-1 immune responses which are crucial for protection against intracellular pathogens (Mbow et al. 2010).

A range of immunostimulatory agents are available and have been used in conjunction with lipid-based systems such as monophosphoryl lipid A, *Quillaja saponaria* (Quil-A), CpG oligodeoxynucleotides, and trehalose 6,6-dibhenate (TDB). Currently, the most potent mucosal adjuvants which are studied for immunisation are the cholera toxin (CT) and the heat labile enterotoxin *Escherichia coli* (LT) (Tamura and Kurata 2000, Holmgren et al. 2003; Plant and Williams 2004). These mucosal adjuvants provide long-lasting immune responses either systemic or mucosal in turn, allowing the body to induce immunity at low antigen concentrations and doses (Holmgren and Czerkinsky 1992; Rappuoli et al. 1999). The major drawback of these two mucosal adjuvants is that they are extremely toxic, hence are not in human clinical trials. As a result, research is being carried out on acceptable derivatives of these toxins with reduced toxicity but retaining the adjuvant activity for use in humans. One of these products includes a non-toxic recombinant produced Cholera toxin B (CTB) subunit which promotes mucosal immunity sIgA which is important in the elucidation of an oral immune response. In addition to mucosal immunity, the CTB also provides anti-inflammatory tolerance to self-antigens (Holmgren and Czerkinsky 2005; Stanford et al. 2004). Other examples of current mucosal adjuvant derivatives include detoxified mutants of LT where the active toxic subunit has been modified to remove the toxic components. This has resulted in the loss of adjuvanticity; however, studies show that there are a few proteins with significant adjuvanticity, even without the presence of the detectable toxic component when administered intranasally (Pizza et al. 2001). A review by Baudner and O'Hagan (2010) and studies by Holmgren and Czerkinsky (2005) show different mucosal adjuvants and their effectiveness; however, the majority of studies focus on nasal administration (Holmgren and Czerkinsky 2005). A promising approach has been identified where hybrid molecules of the cholera toxin subunit has been linked to specific APC (CTA1-DD) from a bacteria *Staphylococcus aureus* (Lycke 2004). The incorporation of this CTA1-DD and antigen into immune-stimulating complexes (ISCOMs, Chap. 8) may provide oral use, possibilities by maintaining adjuvant effectiveness (Holmgren and Czerkinsky 2005).

Novasome™ adjuvants, also referred to as non-phospholipid liposomes (Chambers et al. 2004) are multicomponent adjuvant systems made up of dioxyethylene cetyl ether, cholesterol, and oleic acid have been licensed for veterinary application (for the immunisation of fowl against Newcastle virus disease and avian rheovirus). The Novasome™ technology, made with glycerol monostearate and butyl alcohol and the potent adjuvant monophosphoryl lipid A (Chambers et al. 2004), was shown to offer protection of guinea pigs against an aerosol challenge with virulent *Mycobacterium bovis*.

Overall, mucosal adjuvants show promising data in terms of eliciting an immune response by increasing the immunity to low concentration antigens; however, due to the toxicity of the adjuvants they have been rendered unsafe to use for human clinical trials. Mucosal adjuvants essentially if formulated to show minimal toxicity can be administered and used in combination with other particulate delivery systems to enhance mucosal immune responses therefore, reducing the requirement for multiple dosing or high concentrations of antigen use. Despite these systems, further adjuvants and vaccine delivery systems are needed.

6.9 Case Study: Development of Non-ionic-Based Vesicles for an Oral Influenza Vaccine

Through a series of studies within our laboratories we have developed a delivery system for an influenza subunit antigen. Based on the above outlined attributes, we selected non-ionic-based vesicles as our delivery platform and through a series of investigations we built a carrier system that can be used to facilitate the design of these systems for oral vaccines.

6.9.1 Designing the Vesicles: Surfactant Selection

In terms of design parameters for bilayer vesicles for oral vaccines, first is the selection of the correct lipids/surfactants to enable antigen entrapment, retention and protection during transit down the GI tract. There are a wide range of surfactants that can be used to prepare bilayer vesicles; non-ionic surfactants offer advantages in terms of low cost, good chemical stability and low toxicity profile. In general, the formation of non-ionic surfactants into bilayer vesicles is dependent on many factors including temperature, surfactant concentration, electrostatic and electrodynamic interactions of the surfactants within the aqueous phase. The molecular geometry of the surfactant is also an important parameter, as this influences their packing arrangement and the subsequent structure formed. The shape of a surfactant may be expressed as its critical packing parameter (CPP) which is defined as:

$$CPP = \frac{v}{a_{o} l_{c}} \tag{6.1}$$

where CPP is the critical packing parameter, v is the molecular volume of the hydrophobic part of the lipid, a_o is the surface area of the hydrophilic head group, and l_c is the length of the hydrocarbon chain.

CPP can be used to predict the geometry of surfactants and the resultant structures they can form: a CPP value <0.5 (which will arise from a large hydrophilic head group in the molecule) suggests the surfactant forms spherical micelles, a value between $0.5 < CPP < 1$ forms bilayer vesicles and a $CPP > 1$ (due to large contribution from the hydrophobic group volume) results in inverted micelles (Israelachvili and Mitchell 1975; Israelachvili et al. 1977; Uchegbu and Vyas 1998). Therefore, for bilayer vesicles a CPP of between 0.5 and 1 is required. However, many non-ionic surfactants are single chain with a $CPP < 0.5$. These will not form niosomes without additional components. For example, soluble surfactants such as solulan C-24 and polysorbate 20 readily form micelles. To use these surfactants to form bilayer vesicles, the addition of cholesterol is required (Uchegbu and Florence 1995). This is due to cholesterol increasing the overall CPP of the surfactant mixture; in cases where a mixture of surfactants and additives such as cholesterol is used to prepare vesicles, the operational CPP values will consider the average of the overall components (Kumar 1991) and Manosroi et al. (2003) have confirmed that as cholesterol is incorporated within a Tween 61 surfactant mixture, an average CPP value between 0.5 and 1 was obtained and hence bilayer vesicles rather than micelles are formed (Manosroi et al. 2003). Similarly, Tween 20 has a CPP below 0.5 but can form niosomes when mixed with cholesterol (Carafa et al. 1998) as does with Span 60 (e.g. Ning et al. 2005; Uchegbu et al. 1995). The incorporation of cholesterol has also been shown to stabilise the vesicles by enhancing the hydrophobic bonding within the bilayer (Srinivas et al. 2010; Uchegbu and Vyas 1998).

In the development of a particulate-based oral influenza vaccine, we used monopalmitoyl-glycerol (MPG) as our main surfactant. This has been commonly used to formulate niosomes, e.g. (Brewer and Alexander 1994), and due to its low CPP requires the inclusion of additional surfactants to support vesicle construction. To improve stability we incorporated cholesterol in the system. In terms of surface charge, for most oral vaccine particulate systems, anionic charges have been used. Therefore we combined MPG with cholesterol and DCP to form niosomes. Vesicles were prepared by high shear homogenisation using different blends of MPG, CHO and DCP. In the case of bilosomes, sodium deoxycholate (bile salt) was also incorporated. Briefly, the lipids were heated to 120 °C for 10 min in an oil bath and whilst maintaining the melted lipid solution an emulsion was created by the addition of 5.2 mL of 25 mM sodium bicarbonate buffer pH 7.6 (50 °C) and homogenised for 2 min. Whilst homogenising, various concentrations of bile salt in 25 mM sodium bicarbonate buffer (pH 9.7) was added and homogenised for a further 3 min. Upon cooling, the vesicle formulations were incubated for 2 h with gentle shaking at 220 rpm. The size of the bilosomes was determined using a sympatec 2005 (Helos/BF) analyser and zeta potential was measured using a zeta plus Brookhaven instrument analyser.

When considering the impact of surfactant blend ratios, our results (Wilkhu et al. 2013c) showed that vesicle size was not notably influenced by the ratio of cholesterol to DCP included in the vesicles. However, size was influenced by bile salt content, with increasing bile salt concentrations reducing vesicle size: vesicles reduced by approximately 1 μm in size (from 5.38 to 4.26 μm) with the addition of 800 mM bile salt. A previous study incorporating bile salt within lipid vesicles by Chen et al. (2009) showed a similar decrease in vesicle size with increasing bile salt concentration and the authors attributed this to enhanced bilayer flexibility and a lowering of surface tension between the vesicles (Chen et al. 2009). In terms of zeta potential, the bile salt concentration made no significant difference; however, the cholesterol and DCP content made a significant impact on the zeta potential of the vesicles, as might be expected given the anionic nature of DCP. However, results were not as expected: a higher DCP content resulted in a reduced negativity of the vesicles (Fig. 6.5). We subsequently demonstrated this was a result of the DCP influencing the pH of the vesicle suspension. As increasing concentrations of

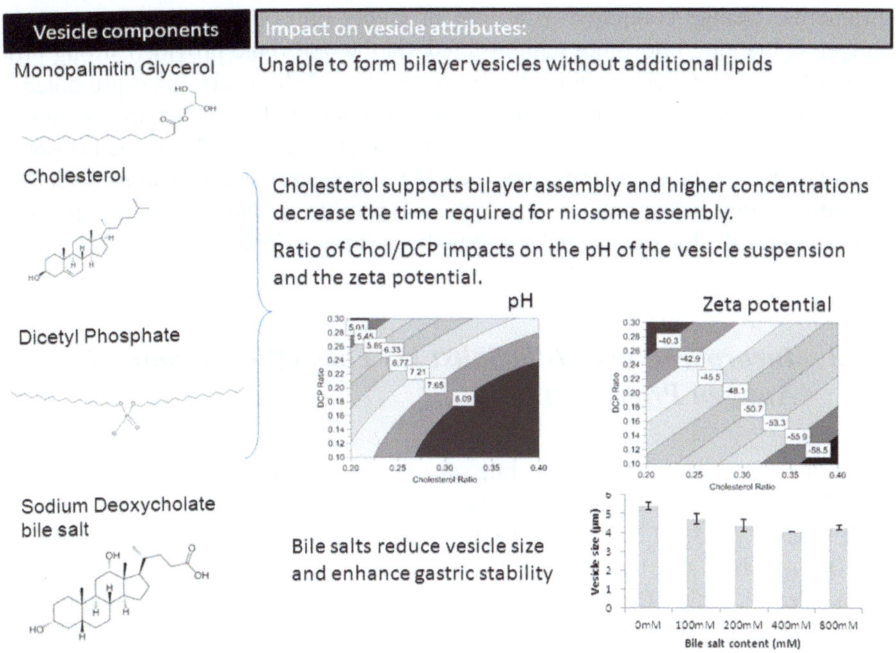

Fig. 6.5 The impact of vesicle components on the vesicle attributes. Adapted from Wilkhu et al. 2013a, b

DCP was added to the formulations, the pH of the vesicles suspension became more acidic. Through dissociation of the H^+ ion from DCP, the pH drops, and reduces the zeta potential. This was confirmed by measuring the zeta potential of a fixed (5:4:1 weight ratio MPG:Chol:DCP with 100 mM bile salt) bilosome formulation over a pH range from 1 to 10. Our results (Wilkhu et al. 2013c showed that as we increase the pH from 1 to 10, the zeta potential of the vesicles changes from around −15 to −120 mV confirming the controlling role of pH on the anionic nature of the vesicles. Therefore, by varying the DCP content in the bilosome formulations, this modulates the pH of the suspension system which in turn impacts on the zeta potential of the vesicles (Wilkhu et al. 2013c).

6.9.2 Vaccine Efficacy of Optimised Vesicle Formulation

From the above studies, we then progressed to investigate the efficacy of the optimised bilosome formulations containing the rHA directed against influenza viruses. Ferrets were orally on days 0, 3, 14, and 17, and 14 days later challenged with a clinical rHA isolate of influenza (Wilkhu et al. 2013c). The median temperature differential and inflammatory cell counts in nasal washes were measured to follow protection against fever. Ferrets immunised with the bilosome vaccine incorporating the recombinant haemagglutinin (rHA) shows a reduced median temperature differential change compared to a dose of empty bilosomes suggesting that the antigen-containing bilosomes administered via the oral route provided strong protection from fever. Furthermore, the bilosome plus rHA vaccine also promoted a reduction in viral cell load counts in ferrets compared to those that received bilosomes administered without antigen (Wilkhu et al. 2013c). Thus, the bilosome formulation containing the influenza vaccine promoted protection against fever and suppressed lung inflammation to extents comparable to empty vesicles showing promising results for vaccine delivery via the oral route (Wilkhu et al. 2013c).

6.9.3 Incorporation of Bile Salts: Do They Offer Enhanced Antigen Protection?

Given that the primary role of the bile salts in the formulation was to improve the stability of the vesicles and protect the vaccine antigen in transit through the GI tract, the NISVs (5:4:1 MPG:Chol:DCP, weight ratio) with and without the addition of 100 mM bile salt were tested in simulated gastric and intestinal conditions (Wilkhu et al. 2013a). Our results showed that the niosome preparation significantly decreased ($p<0.05$) in volume mean diameter from 6.54 ± 0.04 μm ($t=0$ h; prior to exposure to gastric media) to 5.46 ± 0.05 μm after 1 h incubation in gastric media, down to 3.57 ± 0.03 μm after 4 h in simulated intestinal medium. In contrast, bilosome vesicles significantly increase in diameter ($p<0.05$) from 6.19 ± 0.04 μm to 9.13 ± 0.31 μm when incubated in gastric medium for 1 h, yet they returned to their original vesicle size when placed back in simulated intestinal fluid SIF (to 6.11 ± 1.27 μm; $t=4$ h). Directly adding the vesicles into simulated intestinal fluid made no significant difference to the size of the niosomes or the bilosomes, demonstrating that the vesicle size was only influenced by the acidic gastric conditions (Wilkhu et al. 2013c).

When comparing the antigen retention of niosomes and bilosomes in GI tract conditions, there was no notable difference in antigen incorporation between niosomes and bilosomes (39 ± 3 vs. 33 ± 3 % for niosomes and bilosomes, respectively) on initial formulation. However after 15 min in gastric medium (pH 1.2) the niosome antigen retention decreased by approximately 10 % and then remained around 30 % level for up to 60 min in gastric media. In contrast, bilosomes were able to retain their antigen payload in the gastric media (Wilkhu et al. 2013a), suggesting the bile salt is able to enhance vesicle stability. However, in the intestinal media both the niosomes and bilosomes showed reduced antigen retention with only ~10–15 % of the antigen being retained suggesting the vesicles were less stable in SIF conditions compared to SGF conditions (Wilkhu et al. 2013a). These studies suggested that the incorporation of bile salts was advantageous in terms of offering protection through the gastric media.

6.9.4 *Biodistribution of Vesicle Systems After Oral Administration*

To investigate the biodistribution of our vesicular systems after immunisation we developed a dual-labelling system (Henriksen-Lacey et al. 2010) which allows us to follow the fate of both antigen and vesicle carriers after oral immunisation. Briefly, vesicles are prepared containing antigen radio-labelled with ^{125}I and vesicles are labelled with ^{3}H. To investigate the biodistribution after oral immunisation, mice were dosed with 200 μL of vesicle formulations, and at various time points, mice were euthanised and organs collected and analysed for both ^{125}I (to measure antigen) and ^{3}H (to measure vesicles). Through these investigations we demonstrated that formulating the antigen within bilosome vesicles increased antigen transit through the GI tract and antigen delivery to the target site by facilitating uptake via cells within the Peyer's patches (Wilkhu et al. 2013c) and that biodistribution was influenced by formulation attributes (Fig. 6.6). However when antigen is administered without a carrier system, only low levels of antigen are measured across the GI tract, with less than 10 % being detected in the small intestine after 1 h, suggesting antigen is either degraded and/or cleared quickly (Wilkhu et al. 2013c). In terms of antigen targeting to the site of action, antigen recovery at the Peyer's patches and mesenteric lymph tissue was significantly higher ($p<0.05$) when delivered using the bilosome carrier compared to antigen without a carrier system (Wilkhu et al. 2013c).

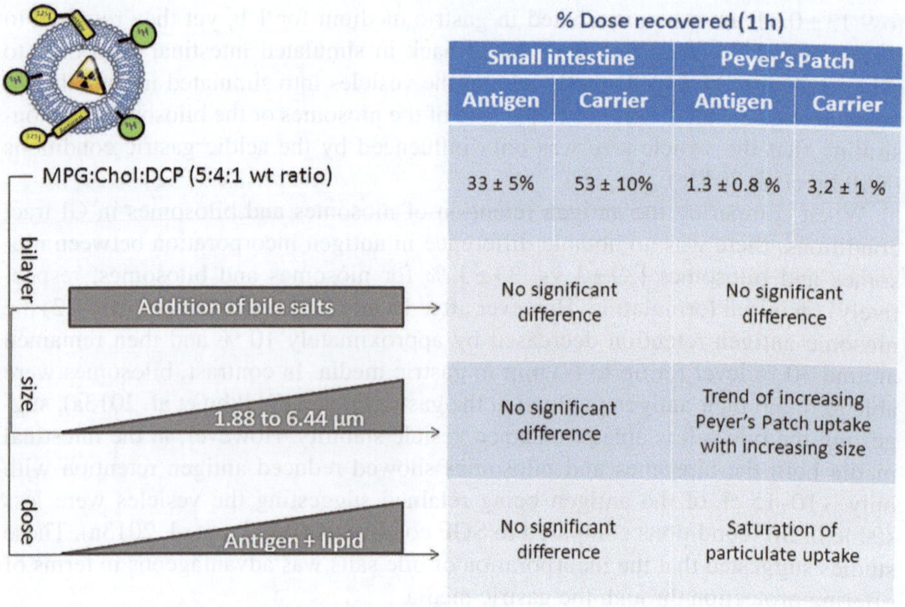

Fig. 6.6 The impact of vesicle vaccine design and dose on dose recovered in the small intestine and Peyer's Patch 1 h after administration

6.9.5 *Comparing the Biodistribution of Niosomes and Bilosomes*

When considering the impact of incorporating bile salts into the bilayer vesicles on their biodistribution after oral immunisation, our results (Wilkhu et al. 2013a) show that around 30 % of the initial dose was detected within the small intestine after 30 min with no significant difference between the two formulations. By 4 h, the levels in the small intestine had decreased with higher levels being detected in the cecum and colon, as would be expected for the formulations as they transit through the GI tract. However, again there was no significant difference in the profiles of niosomes and bilosomes ((Wilkhu et al. 2013a). Similarly, there was no significant difference in uptake of antigen and vesicle within the Peyer's patch or mesenteric lymph tissue between the niosomes and bilosomes at the time points measured. Thus whilst in vitro studies suggest the addition of bile salts improved vesicle stability, this did not translate to differences to in vivo transit.

6.9.5.1 The Impact of Vesicle Size on Biodistribution

As already noted, particle size is a critical factor in determining the fate of orally delivered particulates, both in terms of their uptake and the type of response the systems potentiate. Whilst three possible routes for GI uptake of small particles have been previously considered (intracellular uptake by enterocytes, intercellular/paracellular transfer and uptake via the M cells of the Peyer's patch), it is now mainly agreed that particulate uptake in mammals is chiefly via the M cells of Peyer's patches (Lavelle et al. 1995). Studies by Eldridge et al. (1990) on vesicle size uptake in the Peyer's patches show that uptake is dependent on both the vesicle size and hydrophobicity (Eldridge et al. 1990).

From our studies, using MPG, cholesterol and DCP at a molar ratio of 5:4:1, respectively, with 100 mM bile salts produced vesicles of the desired size and surface charge for oral immunisation without the need for further processing. However to consider the impact of size, we prepared bilosomes at 1.88±0.4 μm (via size reduction through probe sonication) and compared their biodistribution in mice after oral administration with larger (6.44±0.5 μm) vesicles (Wilkhu et al. 2013c). The antigen and vesicle recovery data suggests that after 30 min the majority of both formulations were present within the small intestine of mice. Considering the target site uptake of antigen and bilosomes into the Peyer's patch and mesenteric lymph tissue, significantly higher ($p<0.01$) levels of the larger vesicles were found in the Peyer's patches compared to the smaller vesicle size formulation (Wilkhu et al. 2013c). However, this did not translate into significantly increased levels of bilosomes or antigen in the MLN.

These findings were comparable to studies by Ebel (1990); in their studies the larger (9 μm) polystyrene latex beads were retained within the Peyer's patches with no presence in the MLN, whereas the smaller 2 μm particles were more notable in the MLN (Ebel 1990). Similarly, Eldridge et al (1990) used PLA microparticles to demonstrate that Peyer's patch uptake was restricted to particles less than 10 μm and microspheres less than 5 μm were transported through the efferent lymphatics within macrophages. These studies also suggested that that this pattern of absorption and redistribution may determine the type of immune response elicited by the vaccines, with microspheres below 5 μm inducing a predominantly circulating antibody response, whilst those above this size would stimulate a mucosal (IgA) immune response (Eldridge et al. 1990).

6.9.5.2 The Impact of Lipid Dose on Antigen Delivery

To consider if the uptake of bilosomes after oral administration was dose dependent, we investigated the lipid and antigen uptake based on four increasing doses: (1) antigen 180 μg/mL, lipid 27 mg/mL, (2) antigen 90 μg/mL, lipid 13.5 mg/mL, (3) antigen 45 μg/mL, lipid 6.75 mg/mL and (4) antigen 22.5 μg/mL, lipid 3.375 mg/ mL.

Uptake of vesicles and antigen at the Peyer's patches and mesenteric lymph tissue was considered (Wilkhu et al. 2013a). Our studies show that there was no significant difference in the percentage of antigen or vesicle recovery between different dose concentrations in the organs collected after a 30 min period and total percentage recovery of antigen was comparable (40–70 %) between all doses administered based on the initial dose. This suggesting that the clearance rate and the gastric emptying time were not dose dependent over the range tested (Wilkhu et al. 2013a). However, when considering concentrations of lipid and antigen uptake rather than percentage dose, there was a general trend of increased concentrations of antigen and carrier within both the Peyer's patches and the mesentery lymph tissue as the dose increases up to 90 μg/mL antigen, 13.5 mg/mL lipid. This suggests that increasing the dose of a vaccine can improve delivery to the Peyer's patches and mesentery lymphatics. However, a saturation point may be reached. This would suggest that uptake at the target site is dose-limited and therefore the lipid-antigen dose ratio may be a key factor in vaccine efficacy (Wilkhu et al. 2013a).

6.9.6 Options for Preparation of Bilosomes

Depending on the method of preparation, NISVs consist of one or more bilayer membranes (lamellae) from small unilamellar vesicles (SUV) with a diameter of about 75 to 100 nm, to large unilamellar vesicles (LUV) which can be several microns in size, and MLV which vary in size from anywhere above 100 nm to several microns. There are a wide range of methods that can be used to prepare niosomes, and nearly all of which are adaptations from the methods used to prepare liposomes (Chap. 5). The method of choice, combined with the surfactant types used, drug/antigen attributes and concentration will all contribute to the vesicle size, surface characteristics, loading efficacy, and release characteristics of entrapped drug or antigen. In addition to the method of preparation, the nature of the solute and hydration temperature also affect entrapment efficiency (Uchegbu and Florence 1995). However many of the methods used involve high process temperatures which can be detrimental to thermosensitive antigens or where the antigen is added to preformed vesicles resulting in increased surface adsorption of antigen which exposes antigen to the extensive milieu of the GI tract. Thus, we have recently developed a method of producing niosome vesicles which offers thermostability and allows the use of thermolabile antigens by lowering the process temperatures and the addition of the antigen stage in the production of the niosome vesicles (Wilkhu et al. 2013a). Briefly, appropriate amounts of MPG, Chol, and DCP are melted by heating at 120 °C for 10 min with occasional mixing. Using the molten lipids at 120 °C, an emulsion is created by the addition of the pre-incubated (30 °C) antigen buffered stock solution and homogenised. Once homogenisation has finished, the bilosome formulation is cooled to 30 °C in an incubator/shaker at 220 rpm By employing lower processing temperatures of 30 °C for the antigen it is protected from prolonged temperature stresses (Wilkhu et al. 2013a). To consider the location of the antigen and the ability of the vesicles to protect antigen from enzymatic degradation, both niosomes and bilosomes were subjected to protein (trypsin) digestion. Results showed that after incubation with trypsin, both formulations show low antigen loss (~5 %) suggesting that in both systems the antigen is predominately located within the vesicles (and hence protected from protease digestion).

Interestingly, by considering the melting points of the components individually, and combined at a 5:4:1 MPG:Chol:DCP weight ratio, we have shown that lower (90 °C) than previously reported (120–140 °C) temperatures could be adopted to produce molten surfactants for the production of niosomes (Wilkhu et al. 2014). This is advantageous for surfactant stability; whilst thermogravimetric studies showed that the individual surfactants (MPG, Chol and DCP) were stable to above 200 °C, the 5:4:1 MPG:Chol:DCP weight ratio mixture show ~2 % surfactant degradation at 140 °C, compared to 0.01 % was measured at 90 °C. Furthermore, vesicles formed at this lower temperature offered comparable characteristics to vesicles prepared using higher temperatures commonly reported in literature. In the formation of niosome vesicles, cholesterol played a key role. Langmuir monolayer studies demonstrated that intercalation of cholesterol in the monolayer did not occur in the MPG:Chol:DCP (5:4:1 weight ratio) mixture. However cholesterol may support bilayer assembly, with molecular simulation studies demonstrating that vesicles cannot be built without the addition of cholesterol, and higher concentrations of cholesterol (5:4:1 vs. 5:2:1, MPG:Chol:DCP) decreasing the time required for niosome assembly (Wilkhu et al. 2014).

6.9.7 *Scale Up and Suitability as a Vaccine Product*

For niosomes and bilosomes to be effective as a vaccine formulation suitable for application in public health strategies, formulations must be able to be prepared in a large scale and in a stable format. Using the above outlined method for the preparation of niosomes and bilosomes, scale up is straight forward; within the laboratory we have been able to quickly scale up production to 500 mL batches which offer reproducible vesicle characteristics (data not shown) showing this method is suitable for larger scale production. To enhance the shelf-life of these vesicles, we freeze-dried the vesicle suspensions. Lyophilisation was performed with the Virtis Advantage (Bio Pharma) freeze dryer. The samples to be freeze dried were stored at −70 °C and the freeze drying protocol was set as: primary drying to occur at −40 °C for 35 h, secondary drying at 20 °C for 10 h with a condenser temperature set at −75 °C. Initially, the lyophilisation procedure was based on a freeze drying protocol we developed for liposomes (Mohammed et al. 2006). As the preparation of vesicles for oral vaccines used higher lipid concentrations within the formulations compared to liposome vesicles the lyophilisation cycle was further optimised for drying times. This was carried out as presented in Fig. 6.7 where thermocouples were added to vials containing the vesicle suspension; the shelf temperature was kept constant at −40 °C and the change in temperature based on the thermocouples recorded the temperatures with and without a cryoprotectant. Figure 6.7 shows that there was no deviation in temperature change between 30 and 75 h of primary drying, implying that the temperature within the lipid cake is constant and that a primary drying cycle within this time frame could be effectively implemented. Figure 6.7 also demonstrates that the addition of a cryoprotectant (200 mM sucrose) allows primary drying to take place at a much higher temperature (−20 °C) and the vesicle attributes in terms of size and surface charge were retained throughout the process. Thereby we are able to produce an oral vaccine formulation using niosome-based vesicles which is effective in animal studies and that can be produced in a stable freeze-dried format.

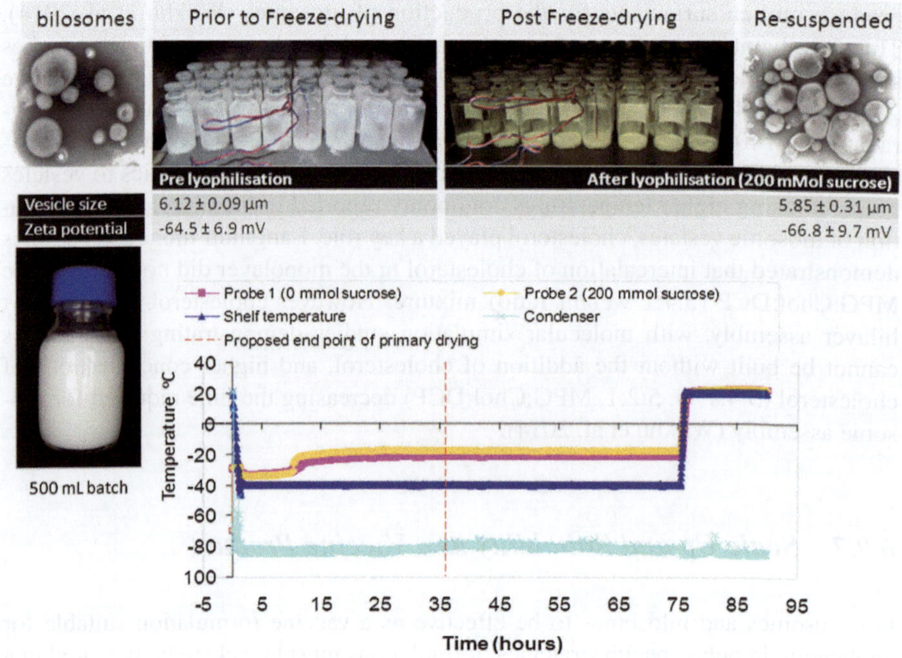

Fig. 6.7 Freeze-drying of vesicular vaccine suspensions with and without sucrose

6.10 Summary

Overall, there are a wide range of particulate systems that may be considered for the oral delivery of vaccines, all with associated advantages and disadvantages. However all must offer safe transit through the GI tract and promote appropriate uptake and stimulation of the required immune cells. Our work has focused on the development of niosome and bilosome systems and we have shown that by appropriate surfactant selection choice and optimisation of the physicochemical parameters of these vesicles we are able to produce vesicles entrapping influenza antigen that can promote protective immune responses in animal studies and these vesicles can be prepared using an easy-to-scale-up process in a stable freeze-dried format.

References

Allison AG, Gregoriadis G (1974) Liposomes as immunological adjuvants. Nature 252(5480):252

Almeida J, Edwards DC, Brand C, Heath T (1975) Formation of virosomes from influenza subunits and liposomes. Lancet 306:899–901

Aramaki Y, Tomizawa H, Hara T, Yachi K, Kikuchi H, Tsuchiya S (1993) Stability of liposomes in vitro and their uptake by rat Peyer's patches following oral administration. Pharm Res 10:1228–1231

Audran R, Peter K, Dannull J, Men Y, Scandella E, Groettrup M, Gander B, Corradin G (2003) Encapsulation of peptides in biodegradable microspheres prolongs their MHC class-I presentation by dendritic cells and macrophages in vitro. Vaccine 21:1250–1255

Azmin MN, Florence AT, Handjani-Vila RM, Stuart JF, Vanlerberghe G, Whittaker JS (1985) The effect of non-ionic surfactant vesicle (niosome) entrapment on the absorption and distribution of methotrexate in mice. J Pharm Pharmacol 37:237–242

Baillie AJ, Florence AT, Hume LR, Muirhead GT, Rogerson A (1985) The preparation and properties of niosomes–non-ionic surfactant vesicles. J Pharm Pharmacol 37:863–868

Baldwin SL, Bertholet S, Reese VA, Ching LK, Reed SG, Coler RN (2012) The importance of adjuvant formulation in the development of a tuberculosis vaccine. J Immunol 188: 2189–2197

Bangham AD, Horne RW (1964) Negative staining of phospholipids and their structural modification by surface-active agents as observed in the electron microscope. J Mol Biol 8:660–668

Baudner BC, O'Hagan DT (2010) Bioadhesive delivery systems for mucosal vaccine delivery. J Drug Target 18:752–770

Baxter D (2007) Active and passive immunity, vaccine types, excipients and licensing. Occup Med 57:552–556

Bennett E, Mullen AB, Ferro VA (2009) Translational modifications to improve vaccine efficacy in an oral influenza vaccine. Methods 49:322–327

Bilsborough J, Viney JL (2004) In vivo enhancement of dendritic cell function. Ann N Y Acad Sci 1029:83–87

Breau LM, Mcgrath PJ, Craig KD, Santor D, Cassidy KL, Reid GJ (2001) Facial expression of children receiving immunizations: a principal components analysis of the child facial coding system. Clin J Pain 17:178–186

Brewer JM, Alexander J (1994) Studies on the adjuvant activity of non-ionic surfactant vesicles: adjuvant-driven Igg2a production independent of MHC control. Vaccine 12(7):613–619

Burgdorf S, Kautz A, Bohnert V, Knolle PA, Kurts C (2007) Distinct pathways of antigen uptake and intracellular routing in Cd4 and Cd8 t cell activation. Science 316:612–616

Carafa, M., E. Santucci, et al. (1998). "Preparation and properties of new unilamellar non-ionic/ionic surfactant vesicles." International Journal of Pharmaceutics 160(1): 51–59

Chadwick S, Kriegel C, Amiji M (2010) Nanotechnology solutions for mucosal immunization. Adv Drug Deliv Rev 62:394–407

Chambers MA, Wright DC, Brisker J, Williams A, Hatch G, Gavier-Widen D, Hall G, Marsh PD, Glyn Hewinson R (2004) A single dose of killed mycobacterium bovis Bcg in a novel class of adjuvant (novasome) protects guinea pigs from lethal tuberculosis. Vaccine 22:1063–1071

Chen H, Torchilin V, Langer R (1996) Polymerized liposomes as potential oral vaccine carriers: stability and bioavailability. J Control Release 42:263–272

Chen Y, Lu Y, Chen J, Lai J, Sun J, Hu F, Wu W (2009) Enhanced bioavailability of the poorly water-soluble drug fenofibrate by using liposomes containing a bile salt. Int J Pharm 376: 153–160

Clark MA, Jepson MA, Hirst BH (2001) Exploiting M cells for drug and vaccine delivery. Adv Drug Deliv Rev 50:81–106

Conacher M, Alexander J, Brewer JM (2001) Oral immunisation with peptide and protein antigens by formulation in lipid vesicles incorporating bile salts (bilosomes). Vaccine 19:2965–2974

Conway A, Madrigal-Estebas L, Mcclean S, Brayden D, Mills K (2001) Protection against Bordetella pertussis infection following parenteral or oral immunization with antigens entrapped in biodegradable particles: effect of formulation and route of immunization on induction of Th1 and Th2 cells. Vaccine 19:1940–1950

Coombes JL, Powrie F (2008) Dendritic cells in intestinal immune regulation. Nat Rev Immunol 8:435–446

De Las Heras AI, Rodríguez Saint-Jean S, Pérez-Prieto SI (2010) Immunogenic and protective effects of an oral DNA vaccine against infectious pancreatic necrosis virus in fish. Fish Shellfish Immunol 28:562–570

Des Rieux A, Fievez V, Garinot M, Schneider YJ, Preat V (2006) Nanoparticles as potential oral delivery systems of proteins and vaccines: a mechanistic approach. J Control Release 116:1–27

Di Fabio JL, De Quadros C (2001) Considerations for combination vaccine development and use in the developing world. Clin Infect Dis 33:S340–S345

Ebel JP (1990) A method for quantifying particle absorption from the small intestine of the mouse. Pharm Res 7:848–851

Eldridge, J. H., J. A. Meulbroek, et al. (1989). "Vaccine-containing biodegradable microspheres specifically enter the gut-associated lymphoid tissue following oral administration and induce a disseminated mucosal immune response." Adv Exp Med Biol 251: 191–202

Eldridge JH, Hammond CJ, Meulbroek JA, Staas JK, Gilley RM, Tice TR (1990) Controlled vaccine release in the gut-associated lymphoid tissues. I. Orally administered biodegradable microspheres target the Peyer's patches. J Control Release 11:205–214

Fattal, E., S. Pecquet, et al. (2002). "Biodegradable microparticles for the mucosal delivery of antibacterial and dietary antigens." Int J Pharm 242(1-2): 15–24

Felgner PL, Gadek TR, Holm M, Roman R, Chan HW, Wenz M, Northrop JP, Ringold GM, Danielsen M (1987) Lipofection: a highly efficient, lipid-mediated DNA-transfection procedure. Proc Natl Acad Sci USA 84:7413–7417

Fine PE, Carneiro IA (1999) Transmissibility and persistence of oral polio vaccine viruses: implications for the global poliomyelitis eradication initiative. Am J Epidemiol 150:1001–1021

Florence AT (2005) Nanoparticle uptake by the oral route: fulfilling its potential? Drug Discov Today Technol 2:75–81

Freitas C, Müller RH (1998) Effect of light and temperature on zeta potential and physical stability in solid lipid nanoparticle (Sln™) dispersions. Int J Pharm 168:221–229

Gaucher G, Satturwar P, Jones MC, Furtos A, Leroux JC (2010) Polymeric micelles for oral drug delivery. Eur J Pharm Biopharm 76:147–158

Gelperina S, Kisich K, Iseman MD, Heifets L (2005) The potential advantages of nanoparticle drug delivery systems in chemotherapy of tuberculosis. Am J Respir Crit Care Med 172:1487–1490

González-Rodríguez M, Rabasco A (2011) Charged liposomes as carriers to enhance the permeation through the skin. Expert Opin Drug Deliv 8:857–871

Graham JE, Borda-Rodriguez A, Huzair F, Zinck E (2012) Capacity for a global vaccine safety system: the perspective of national regulatory authorities. Vaccine 30:4953–4959

Gregoriadis G, Ryman BE (1971) Liposomes as carriers of enzymes: a new approach to the treatment of storage diseases. Biochem J 124(5):58

Gregoriadis G, Mccormack B, Obrenovic M, Saffie R, Zadi B, Perrie Y (1999) Vaccine entrapment in liposomes. Methods 19:156–162

Handjani-Vila RM, Ribier A, Rondot B, Vanlerberghie G (1979) Dispersions of lamellar phases of non-ionic lipids in cosmetic products. Int J Cosmet Sci 1:303–314

Hanes J, Chiba M, Langer R (1995) Polymer microspheres for vaccine delivery. Pharm Biotechnol 6:389–412

Heinsbroek E, Ruitenberg EJ (2010) The global introduction of inactivated polio vaccine can circumvent the oral polio vaccine paradox. Vaccine 28:3778–3783

Henriksen-Lacey M, Bramwell V, Perrie Y* (2010) Radiolabelling of antigen and liposomes for vaccine biodistribution studies. Pharmaceutics 2:91–104

Holmgren J, Czerkinsky C (1992) Cholera as a model for research on mucosal immunity and development of oral vaccines. Curr Opin Immunol 4:387–391

Holmgren J, Czerkinsky C (2005) Mucosal immunity and vaccines. Nat Med 11:S45–S53

Holmgren J, Harandi AM, Czerkinsky C (2003) Mucosal adjuvants and anti-infection and anti-immunopathology vaccines based on cholera toxin, cholera toxin b subunit and CpG DNA. Expert Rev Vaccines 2:205–217

Hu S, Niu M, Hu F, Lu Y, Qi J, Yin Z, Wu W (2013) Integrity and stability of oral liposomes containing bile salts studied in simulated and ex vivo gastrointestinal media. Int J Pharm 441(1–2):693–700

Hull HF, Minor PD (2005) When can we stop using oral poliovirus vaccine? J Infect Dis 192:2033–2035

Hussain N, Jaitley V, Florence AT (2001) Recent advances in the understanding of uptake of microparticulates across the gastrointestinal lymphatics. Adv Drug Deliv Rev 50:107–142

Igartua M, Hernandez R, Esquisabel A, Gascon A, Calvo M, Pedraz J (1998) Enhanced immune response after subcutaneous and oral immunization with biodegradable PLGA microspheres. J Control Release 56:63–73

Israelachvili JN, Mitchell DJ (1975) A model for the packing of lipids in bilayer membranes. Biochim Biophys Acta 389(1):13–19

Israelachvili JN, Mitchell DJ, And Ninham BW (1977) Theory of self-assembly of lipid bilayers and vesicles. Biochim Biophys Acta 470(2):185–201

Iwasaki A, Kelsall BL (1999) Freshly isolated Peyer's patch, but not spleen, dendritic cells produce interleukin 10 and induce the differentiation of T helper type 2 cells. J Exp Med 190:229–239

Izadpanah A, Dwinell MB, Eckmann L, Varki NM, Kagnoff MF (2001) Regulated Mip-3alpha/Ccl20 production by human intestinal epithelium: mechanism for modulating mucosal immunity. Am J Physiol Gastrointest Liver Physiol 280:G710–G719

Jacob SS, Cherian S, Sumithra TG, Raina OK, Sankar M (2013) Edible vaccines against veterinary parasitic diseases—current status and future prospects. Vaccine 31:1879–1885

Kagnoff MF, Eckmann L (1997) Epithelial cells as sensors for microbial infection. J Clin Investig 100:6–10

Kane A, Lloyd J, Zaffran M, Simonsen L, Kane M (1999) Transmission of hepatitis B, hepatitis C and human immunodeficiency viruses through unsafe injections in the developing world: model-based regional estimates. Bull World Health Organ 77:801–807

Kiyono H, Fukuyama S (2004) Nalt- versus Peyer's-patch-mediated mucosal immunity. Nat Rev Immunol 4:699–710

Kraehenbuhl JP, Neutra MR (2000) Epithelial M cells: differentiation and function. Annu Rev Cell Dev Biol 16:301–332

Kumar VV (1991) Complementary molecular shapes and additivity of the packing parameter of lipids. Proc Natl Acad Sci USA 88(2):444–448

Langer R, Cleland JL, Hanes J (1997) New advances in microsphere-based single-dose vaccines. Adv Drug Deliv Rev 28:97–119

Langridge WH (2000) Edible vaccines. Sci Am 283:66–71

Lasic DD, Martin FJ, Gabizon A, Huang SK, Papahadjopoulos D (1991) Sterically stabilized liposomes: a hypothesis on the molecular origin of the extended circulation times. Biochim Biophys Acta 1070:187–192

Lavelle EC, Sharif S, Thomas NW, Holland J, Davis SS (1995) The importance of gastrointestinal uptake of particles in the design of oral delivery systems. Adv Drug Deliv Rev 18:5–22

Levine MM (2000) Immunization against bacterial diseases of the intestine. J Pediatr Gastroenterol Nutr 31:336–355

Lycke N (2004) From toxin to adjuvant: the rational design of a vaccine adjuvant vector, CTA1-DD/ISCOM. Cell Microbiol 6:23–32

Mahato RI, Narang AS, Thoma L, Miller DD (2003) Emerging trends in oral delivery of peptide and protein drugs. Crit Rev Ther Drug Carrier Syst 20:153–214

Mann JFS, Ferro VA, Mullen AB, Tetley L, Mullen M, Carter KC, Alexander J, Stimson WH (2004) Optimisation of a lipid based oral delivery system containing A/Panama influenza haemagglutinin. Vaccine 22:2425–2429

Mann JFS, Scales HE, Shakir E, Alexander J, Carter KC, Mullen AB, Ferro VA (2006) Oral delivery of tetanus toxoid using vesicles containing bile salts (bilosomes) induces significant systemic and mucosal immunity. Methods 38:90–95

Manosroi A, Wongtrakul P, Manosroi J, Sakai H, Sugawara F, Yuasa M, Abe M (2003) Characterization of vesicles prepared with various non-ionic surfactants mixed with cholesterol. Colloids Surf B Biointerfaces 30:129–138

Mbow ML, De Gregorio E, Valiante NM, Rappuoli R (2010) New adjuvants for human vaccines. Curr Opin Immunol 22:411–416

Minor, P. (2009). "Vaccine-derived poliovirus (VDPV): Impact on poliomyelitis eradication." Vaccine 27(20): 2649–2652
Mitragotri S (2005) Immunization without needles. Nat Rev Immunol 5:905–916
Mohammed AR, Bramwell VW, Coombes AGA, And Perrie Y (2006) Lyophilisation and sterilisation of liposomal vaccines to produce stable and sterile products. Methods 40(1):30–38
Mowat MA (2003) Anatomical basis of tolerance and immunity to intestinal antigens. Nature 3:331–341
Neutra MR, Kozlowski PA (2006) Mucosal vaccines: the promise and the challenge. Nat Rev Immunol 6:148–158
Ning M, Guo Y, Pan H, Yu H, And Gu Z (2005) Niosomes with sorbitan monoester as a carrier for vaginal delivery of insulin: studies in rats. Drug Deliv 12(6):399–407
Ning J-F, Zhu W, Xu J-P, Zheng C-Y, Meng X-L (2009) Oral delivery of DNA vaccine encoding VP28 against white spot syndrome virus in crayfish by attenuated Salmonella typhimurium. Vaccine 27:1127–1135
Niu M, Lu Y, Hovgaard L, Guan P, Tan Y, Lian R, Qi J, Wu W (2012) Hypoglycemic activity and oral bioavailability of insulin-loaded liposomes containing bile salts in rats: the effect of cholate type, particle size and administered dose. Eur J Pharm Biopharm 81:265–272
Norris DA, Puri N, Sinko PJ (1998) The effect of physical barriers and properties on the oral absorption of particulates. Adv Drug Deliv Rev 34:135–154
O'Hagan D, Rahman D, Mcgee J, Jeffery H, Davies M, Williams P, Davis S, Challacombe S (1991) Biodegradable microparticles as controlled release antigen delivery systems. Immunology 73:239–242
Ogra PL (2010) Ageing and its possible impact on mucosal immune responses. Ageing Res Rev 9:101–106
O'Hagan DT, Illum L (1990) Absorption of peptides and proteins from the respiratory tract and the potential for development of locally administered vaccine. Crit Rev Ther Drug Carrier Syst 7:35–97
Okada JI, Cohen S, Langer R (1995) In vitro evaluation of polymerized liposomes as an oral drug delivery system. Pharm Res 12:576–582
Perrie Y, Obrenovic M, Mccarthy D, Gregoriadis G (2002) Liposome (lipodine)-mediated DNA vaccination by the oral route. J Liposome Res 12:185–197
Pizza M, Giuliani MM, Fontana MR, Monaci E, Douce G, Dougan G, Mills KH, Rappuoli R, Del Giudice G (2001) Mucosal vaccines: non toxic derivatives of LT and CT as mucosal adjuvants. Vaccine 19:2534–2541
Plant A, Williams NA (2004) Modulation of the immune response by the cholera-like enterotoxins. Curr Top Med Chem 4:509–519
Rajesh Kumar S, Ishaq Ahmed VP, Parameswaran V, Sudhakaran R, Sarath Babu V, Sahul Hameed AS (2008) Potential use of chitosan nanoparticles for oral delivery of DNA vaccine in Asian sea bass (Lates calcarifer) to protect from Vibrio (Listonella) anguillarum. Fish Shellfish Immunol 25:47–56
Rappuoli R, Pizza M, Douce G, Dougan G (1999) Structure and mucosal adjuvanticity of cholera and Escherichia coli heat-labile enterotoxins. Immunol Today 20:493–500
Russell-Jones GJ (2000) Oral vaccine delivery. J Control Release 65:49–54
Santiago (1995) Vaccine design: the subunit and adjuvant approach. In: Powell MF, Newman MJ (eds). Plenum Press, London
Sarti F, Perera G, Hintzen F, Krotti K, Jarageorgiou V, Kammona O, Kiparissides C, Berkop-Schnurch A (2011) In vivo evidence of oral vaccination with PLGA nanoparticles containing the immunostimulant monophosphoryl lipid A. Biomaterials 32(16):4052–4057
Schijns VEJC (2000) Immunological concepts of vaccine adjuvant activity: commentary. Curr Opin Immunol 12:456–463
Schubert R, Jaroni H, Schoelmerich J, Schmidt KH (1983) Studies on the mechanism of bile salt-induced liposomal membrane damage. Digestion 28:181–190

Shakweh M, Besnard M, Nicolas VR, Fattal E (2005) Poly(lactide-co-glycolide) particles of different physicochemical properties and their uptake by Peyer's patches in mice. Eur J Pharm Biopharm 61:1–13

Shukla A, Khatri K, Gupta PN, Goyal AK, Mehta A, Vyas SP (2008) Oral immunization against hepatitis B using bile salt stabilized vesicles (bilosomes). J Pharm Pharm Sci 11:59–66

Srinivas S, Kumar YA, Hemanth A, And Anitha M (2010) Preparation and evaluation of niosomes containing aceclofenac. Dig J Nanomater Biostruct 5(1):249–254

Stanford M, Whittall T, Bergmeier LA, Lindblad M, Lundin S, Shinnick T, Mizushima Y, Holmgren J, Lehner T (2004) Oral tolerization with peptide 336-351 linked to cholera toxin B subunit in preventing relapses of uveitis in Behcet's disease. Clin Exp Immunol 137:201–208

Storni T, Kundig T, Senti G, Johansen P (2005) Immunity in response to particulate antigen-delivery systems. Adv Drug Deliv Rev 57:333–355

Tabata Y, Ikada Y (1988) Effect of the size and surface charge of polymer microspheres on their phagocytosis by macrophage. Biomaterials 9:356–362

Tamura, S. I. and T. Kurata (2000). "A proposal for safety standards for human use of cholera toxin (or Escherichia coli heat-labile enterotoxin) derivatives as an adjuvant of nasal inactivated influenza vaccine." Jpn J Infect Dis 53(3): 98–106

Tomizawa H, Aramaki Y, Fujii Y, Kara T, Suzuki N, Yachi K, Kikuchi H, Tsuchiya S (1993) Uptake of phosphatidylserine liposomes by rat Peyer's patches following intraluminal administration. Pharm Res 10:549–552

Tritto E, Mosca F, De Gregorio E (2009) Mechanism of action of licensed vaccine adjuvants. Vaccine 27:3331–3334

Troy SB, Ferreyra Reyes L, Huang C, Mahmud N, Lee YJ, Canizales Quintero S, Flaster H, Baez Saldana R, Garcia-Garcia L, Maldonado Y (2011) Use of a novel real-time PCR assay to detect oral polio vaccine (OPV) shedding and reversion in stool and sewage samples after a mexican national immunization day. J Clin Microbiol

Uchegbu IF, Vyas SP (1998) Non-ionic surfactant based vesicles (niosomes) in drug delivery. Int J Pharm 172(1–2):33–70

Uchegbu IF, Florence AT (1995) Non-ionic surfactant vesicles (niosomes): physical and pharmaceutical chemistry. Adv Colloid Interface Sci 58:1–55

Uchegbu IF, Double JA, Turton JA, Florence AT (1995) Distribution, metabolism and tumoricidal activity of doxorubicin administered in sorbitan monostearate (Span 60) niosomes in the mouse. Pharm Res 12(7):1019–1024

Wang D, Xu J, Feng Y, Liu Y, Mchenga SSS, Shan F, Sasaki J-I, Lu C (2010) Liposomal oral DNA vaccine (mycobacterium DNA) elicits immune response. Vaccine 28:3134–3142

Webster DE, Gahan ME, Strugnell RA, Wesselingh SL (2003) Advances in oral vaccine delivery options: what is on the horizon? Am J Drug Deliv 1:227–240

Wilkhu J, Mcneil SE, Kirby DJ, Perrie Y (2011) Formulation design considerations for oral vaccines. Ther Deliv 2:1141–1164

Wilkhu J, Mcneil S, Anderson D, Perrie Y (2013a) Consideration of the efficacy of non-ionic vesicles in the targeted delivery of oral vaccines. Drug Deliv Transl Res 4:233–245

Wilkhu J, Vangala A, Mohammed A, Perrie Y (2013b) Designing nonionic surfactant vesicles for the delivery of antigens for systemic and alternative delivery routes. In: Flower DR, Perrie Y (eds) Immunomic discovery of adjuvants and candidate subunit vaccines. Springer, New York

Wilkhu JS, Mcneil SE, Anderson DE, Perrie Y (2013c) Characterization and optimization of bilosomes for oral vaccine delivery. J Drug Target 21:291–299

Wilkhu JS, Ouyang D, Kirchmeier MJ, Anderson DE, Perrie Y (2014) Investigating the role of cholesterol in the formation of non-ionic surfactant based bilayer vesicles: thermal analysis and molecular dynamics. Int J Pharm 461(1–2):331–341

Yoshida H, Lehr CM, Kok W, Junginger HE, Verhoef JC, Bouwstra JA (1992) Niosomes for oral delivery of peptide drugs. J Control Release 21:145–153

Chapter 7
Cubosomes: Structure, Preparation and Use as an Antigen Delivery System

Shakila B. Rizwan and Ben J. Boyd

7.1 Introduction

Certain amphiphilic lipids and surfactants are able to self-assemble into highly ordered structures in an aqueous environment, with long-range order in one, two or three dimensions (3D) and short-range disorder at atomic distances (Quantan et al. 2004; Yano et al. 2005). Consequently, these structures have properties intermediate between those of solid crystals and isotropic liquids and are referred to as lyotropic liquid crystals or mesophases.

A unique feature of lyotropic systems is that in the absence of any physical or chemical changes, they remain thermodynamically stable in excess solvent. In contrast, other self-assembled lipid systems such as micelles dissociate into monomers upon dilution. This particular feature has led to significant research efforts to utilize these mesophases for various applications and some examples include (1) a platform for crystallization of membrane proteins (Cherezov et al. 2006), (2) delivery of food actives (Amar-Yuli et al. 2009) and (3) delivery of drugs (Rizwan et al. 2010).

This chapter is in three parts; firstly an overview of the self-assembling properties of lipids is provided with a view to introduce the reader to polymorphism in lipid–water systems. We then discuss the use of lyotropic mesophases for drug delivery, with a focus on the bicontinuous cubic phase (V_2). The remainder of the chapter will then focus on cubosomes, dispersions of the V_2 phase and their potential application in the field of vaccine delivery.

S.B. Rizwan (✉)
New Zealand's National School of Pharmacy, University of Otago, Dunedin, New Zealand
e-mail: shakila.rizwan@otago.ac.nz

B.J. Boyd
Faculty of Pharmacy and Pharmaceutical Sciences, Monash University, Melbourne, VIC, Australia
e-mail: ben.boyd@monash.edu

C. Foged et al. (eds.), *Subunit Vaccine Delivery*, Advances in Delivery Science and Technology, DOI 10.1007/978-1-4939-1417-3_7

7.2 Polymorphism in Lyotropic Liquid Crystal Systems

The spectrum of lyotropic self-assemblies that can be formed on exposure of lipids to aqueous environments range from simple micelles arranged in a cubic or hexagonal packing to flat lipid bilayer structures such as the lamellar phase (Lα), and to more complex non-lamellar structures composed of a lipid bilayer that have greater intrinsic curvature, such as the inverted hexagonal (H) and cubic (V) phase (Fig. 7.1). It is important to note that not all self-assemblies may be observed in any given lipid–water system. Lipid self-assembly is critical for several biological processes (Luzzati 1997; Lindblom and Rilfors 1989). The lamellar phase, in combination with membrane proteins and fatty acids provides the basic building block for all biological membranes. Non-lamellar structures have also been shown to play crucial roles in cellular processes such as cell stress, starvation and fusion (Colotto and Epand 1997; Almsherqi et al. 2006; Deng et al. 2002).

The rich array of polymorphism is primarily governed by the concentration of the solvent, geometric properties of the amphiphile and temperature. Exposure to a polar solvent causes the amphiphiles to position themselves in such a way to minimize the free energy of the system, where the polar solvent penetrates between the amphiphilic molecules exposing the hydrophilic parts to the aqueous environment and causes the hydrophobic parts to be sheltered from the solvent. This phenomenon is commonly referred to as the hydrophobic effect (Pratt 1985; Kaasgaard and Drummond 2006).

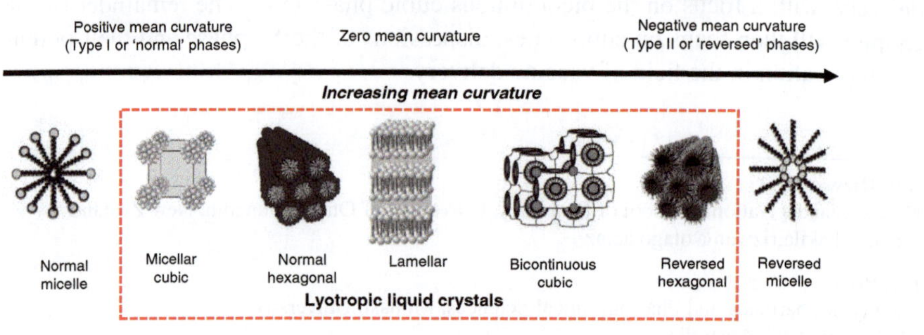

Fig. 7.1 Schematic representation of common liquid crystalline phases formed by self-assembly of amphiphiles in water. Figure modified after (Garti et al. 2012; Shearman et al. 2006)

In addition to the hydrophobic effect as a driving force for self-assembly, a number of packing constraints also play a critical role in the resulting structures and are determined by the geometric properties of the amphiphilic molecule, specifically packing and curvature. The critical packing parameter (*P*) described by Israelachvili (Israelachvili et al. 1976), is given by the equation:

$$P = \frac{v}{al} \tag{7.1}$$

where v and l denote the volume and the length of the hydrophobic chain and a is the optimal surface area of the polar head group. This equation provides a useful measure of aggregation morphology. The packing parameter is useful for predicting which phases may be preferentially formed for a given lipid. It connects the molecular shape and properties of a given lipid to its favoured curvature at the lipid–water interface, and therefore the topology and shape of the resulting aggregate as shown in Fig. 7.1 (Shearman et al. 2006). Curvature towards the chain region is conventionally designated as the normal (oil-in-water) or Type 1 phase where $P<1$, whilst curvature towards the water region is denoted as the reverse (water in oil) or Type 2 phase where $P>1$. Planar or lamellar phase structures are evident when $P=1$.

7.3 Bicontinuous Cubic Phase

The most common non-lamellar mesophase with a curved interface that is observed in lipid–solvent systems is the inverted or reversed hexagonal phase (H_2) (Seddon and Templer 1993). It consists of densely packed, water-filled cylindrical micelles arranged in a continuous matrix of fluid hydrocarbon chains (Fig. 7.1). In contrast, the inverted bicontinuous cubic phase (V_2) consists of a single continuous lipid bilayer, separating two non-intersecting water channels. The topology of the V_2 phase is categorized into the gyroid, diamond and primitive structures based on concepts of differential geometry and minimal surfaces and are associated with the space groups Ia3d, Pn3m and Im3m, respectively (Hyde 2001) (Fig. 7.2).

Various classes of synthetic lipids have been investigated, however, only a handful are capable of forming stable inverted mesophases (Fig. 7.3) (Boyd et al. 2006; Fong et al. 2007; Hato and Minamikawa 1996). A large amount of research in the literature is centred on unsaturated mono- and diglycerides, particularly monoolein and mixtures of monoolein with other lipids or its structural derivatives (Clogston and Caffrey 2005; Lara et al. 2005; Chang and Bodmeier 1997). These lipids are commonly used as emulsifying agents and food additives (Amar-Yuli et al. 2009; Ganem-Quintanar et al. 2000). However, the propensity, for example of monoolein and oleic acid to undergo esterase catalyzed hydrolysis can limit their applications in vivo. There is significant research effort to increase the repertoire of lipids that are able to form stable non-lamellar mesophases in water as alternatives to fatty-acid-based materials such as monoolein. One such alternative is phytantriol, an

Fig. 7.2 Schematic representations of the inverse bicontinuous cubic phases: Ia3d (gyroid), Pn3m (diamond) and Im3m (primitive) phase. Individual lipids are shown as ball-stick figures, whilst the regions filled with green and red colour represent water. Figure modified from (Caffrey 2000)

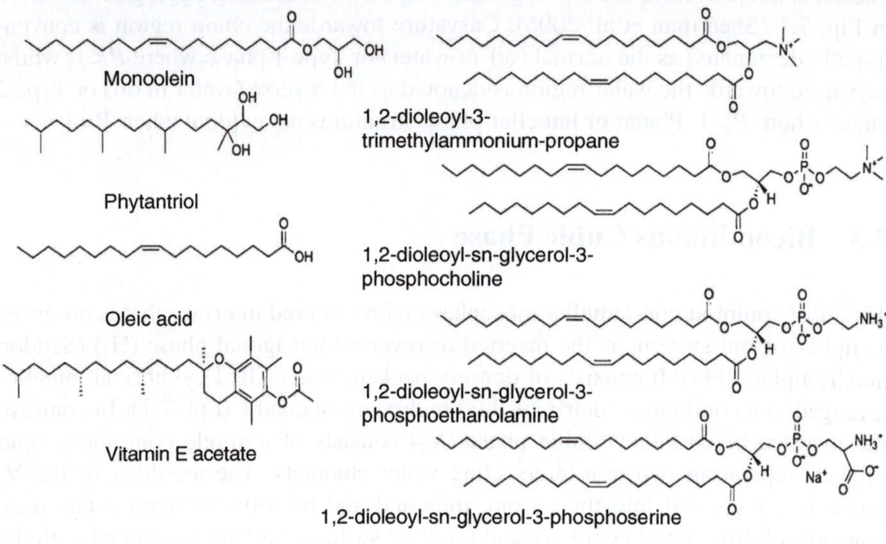

Fig. 7.3 Chemical structures of lipids known to form lyotropic mesophases. Figure modified after (Mulet et al. 2013)

additive used in cosmetics for improving moisture retention (Ribier and Biatry 1998; Barauskas and Landh 2003; Rizwan et al. 2009). Both phytantriol and glyceryl monooleate (GMO) differ structurally (Fig. 7.3), however, interestingly they both display similar phase behaviour, forming V_2 phases in excess water at low temperatures and transforming to the H_2 at higher temperatures (Barauskas et al. 2005a; Dong et al. 2006).

The nanostructure and thermodynamic stability of lyotropic mesophases has stimulated intense research, particularly towards exploring the potential of these lipid systems for various pharmaceutical applications. Macroscopically, the inverted mesophases are extremely viscous, almost solid-like, materials with a large specific

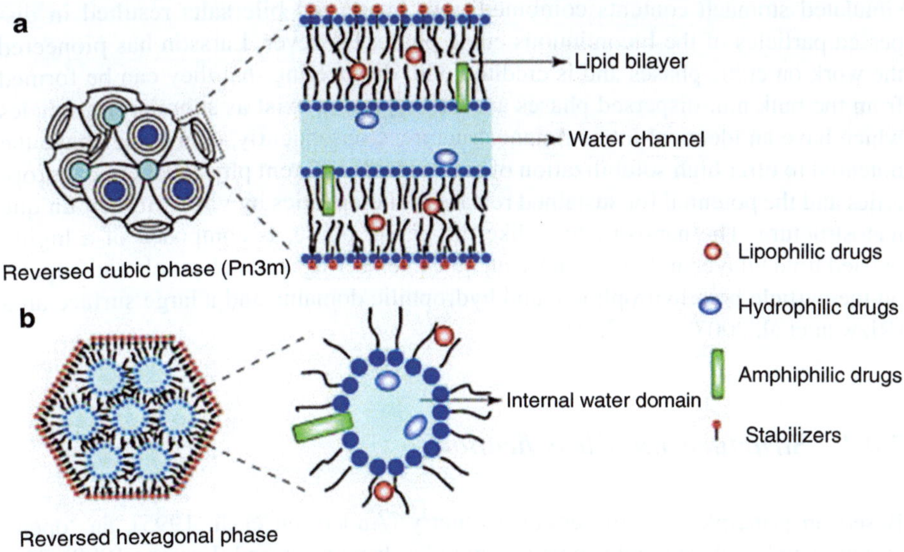

Fig. 7.4 Structures of (**a**) reversed bicontinuous cubic and (**b**) hexagonal phase showing the possible locations of drugs. Figure reproduced from (Guo et al. 2010)

surface area that can solubilize materials of varying polarity and provide a slow release matrix for drugs (Fig. 7.4) (Rizwan et al. 2009; Boyd et al. 2000; Clogston and Caffrey 2005). Retarded release of entrapped bioactives from the V_2 phase is a consequence of diffusion through the tortuous matrix.

In addition to the properties of the bioactive itself, factors believed to be regulating the release kinetics are the pore size and tortuosity of the water channels and the stiffness and high viscosity of the mesophase. However, the viscous nature of non-lamellar phases limit their pharmaceutical applications. To overcome this, the thermodynamic stability of these mesophases in excess water is exploited. Non-lamellar liquid crystals in equilibrium with excess water can be dispersed into submicron-sized particles which retain the unique microstructure of their respective 'parent' phase (Fig. 7.2) and are sometimes collectively referred to as non-lamellar liquid crystalline nanoparticles (LCPs). Dispersion of the H_2 and V_2 phases are known as hexosomes and cubosomes, respectively, analogous to liposomes, dispersions of the Lα phase.

7.4 Cubosomes

The term 'cubosome' was first mentioned in a review on cubic lipid/water phases by Kåre Larsson in the 1980s (Larsson 1983). Previous studies dating back prior to Larsson's description of cubosomes do exist. Patton and Carey (Patton and Carey 1979) in 1979 described their observation in studies involving fat digestion, where simulated stomach contents combined with lipase and bile salts resulted in dispersed particles of the bicontinuous cubic phase. However, Larsson has pioneered the work on cubic phases and is credited with discovering that they can be formed from the bulk non-dispersed phases and on dispersion exist as submicron particles which have an identical internal nanostructure. Consequently, cubosomes have the potential to offer high solubilization of actives with different physicochemical properties and the potential for sustained release of therapeutics by virtue of their unique nanostructure. The nanostructure, like the parent phase, is composed of a highly twisted lipid bilayer and two congruent, non-intersecting water channels and imparts on the particle both hydrophobic and hydrophilic domains and a large surface area (Rizwan et al. 2007) (Fig. 7.1).

7.4.1 Structure and Classification

Based on principles of differential geometry (Andersson et al. 1995), an 'open' structure and a 'closed' cubosome structure has been proposed (Larsson 1999). The two aqueous channels are in contact with the external environment in the open structure, whilst the closed cubosome has one water channel open towards the external environment with the other compartment closed in relation to the outside. The closed cubosome was proposed as the more stable structure of the two forms; however, recent studies strongly support the open model (Rizwan et al. 2007; Tilley et al. 2013). Like the bulk parent cubic phase, cubosomes are also classified into gyroid (Ia3d), primitive (Im3m) or diamond (Pn3m).

7.4.2 Components of Cubosomes

Various lipids have been used to prepare cubosomes and other LCPs [for an extensive summary the reader is referred to a review by Yaghmur (Yaghmur and Glatter 2008)]. Though the bulk mesophases are thermodynamically stable, when dispersed, the particles are not kinetically stable from a colloidal stability perspective, and tend to aggregate due to the exposure of the hydrophobic domains to the external aqueous environment (Dan and Poo 2004). In order to circumvent aggregation, stabilizing agents are required (Almgren 2003; Larsson 1999; Dan and Poo 2004). The main role of the stabilizing agent is to provide a steric or electrostatic barrier to prevent close particle contact (Nakano et al. 2001). The choice of a stabilizing or dispersing agent is crucial and should participate in the lipid–water assembly without disrupting the cubic liquid crystallinity of the structure (Dan and Poo 2004). Larsson's early cubosomes studies used bile salts to stabilize cubic phase dispersions (Larsson 1989).

Pluronics, in particular F 127, are the most commonly used stabilizing agents and even considered to be the 'gold standard.' Pluronics are a group of water soluble, self-assembling triblock co-polymers composed of polyethylene oxide (PEO) and polypropylene oxide (PPO) in a PEO–PPO–PEO configuration, where the PPO portion gives the polymer hydrophobic characteristics and the PEO portion is responsible for the hydrophilic character. The stabilizing effects of F 127 is thought to be brought about by the adsorption or incorporation of the hydrophobic PPO block onto the surface of the particles, whilst the hydrophilic PEO portion extends out into the aqueous environment to provide steric shielding (Dan and Poo 2004). The structure type of the dispersions can be influenced by the proportion of F 127 used in relation to the particular liquid crystal forming lipid (Almgren et al. 1996; Gustafsson et al. 1997; Larsson 2000). In GMO formulations, at relatively low F 127 concentrations (3 %), the D-type cubic phase is dominant, with minimal presence of the P-type cubic phase. Increases in the polymer concentration lead to an increase in the portion of the P-type cubic phase (Nakano et al. 2001; Almgren et al. 1996; Gustafsson et al. 1997; Larsson 2000). This has been attributed to the preferential location of F 127 on the surface of the particles at low concentrations, so very little polymer is thought to be involved with the internal cubic structure. However, when the F 127 is in excess, the surface is saturated and the polymer can incorporate within the lipid bilayer. As a consequence a transition from the Im3m structure to the Pn3m cubic structure is observed in monoolein–water system. In contrast, the nanostructure of phytantriol-based cubosomes appears to be unaffected by high concentrations of F127 (Dong et al. 2006).

Recently, Chong et al. investigated the ability of a wide range of non-ionic molecules to stabilize dispersions of phytantriol and GMO and compared them to F 127 (Chong et al. 2011). Interestingly, they showed that the poly(ethyleneoxide) stearate class of stabilizers, in particular Myrj 59® (100 poly(ethyleneoxide) units), was more effective in stabilizing phytantriol cubosomes when compared to the gold-standard F 127. The reason for the improved stability remains unclear.

7.4.3 Production of Cubosomes

Current research efforts, in addition to increasing the repository of available liquid crystal forming lipids, are focused on formulation of cubosomes. Although great progress has been made, there is still no consensus on an optimal method of cubosome production. Stability, biocompatibility and optimal drug release still remain unresolved. Generally, cubosomes reported in the literature have been produced by one of the following methods:

Application of high energy methods such as ultrasonication, microfluidization and homogenization of the viscous cubic phase in the presence of excess water (Larsson 1989; Gabizon et al. 2004; Almgren et al. 1996; Gustafsson et al. 1997).

Dry lipid films of lipid/stabilizer are produced, rehydrated and subsequently dispersed in excess water and fragmented into cubosomes using mechanical mixing methods such as homogenization (Nakano et al. 2002).

Dry powder precursors are prepared and form cubosomes upon hydration with a solvent (Spicer et al. 2002).

Microfluidization of a lipid/stabilizer mix followed by heat treatment at elevated temperature and subsequent cooling resulting in liquid crystalline dispersions with a narrow size distribution (Barauskas et al. 2005a; Wörle et al. 2006a).

Mixtures of liquid crystal forming lipid in ethanol or other organic solvents (liquid precursors) are dispersed in excess water (or the solvent of choice) resulting in the spontaneous formation of cubosomes (Rizwan et al. 2011; Spicer et al. 2001; Zheng et al. 2003; Chung et al. 2002).

7.4.4 Characterization of Non-lamellar Liquid Crystalline Dispersions

The physicochemical properties of LCPs have been characterized using various techniques. Amar-Yuli et al. (2009) recently grouped these techniques into two categories, *direct techniques* and *indirect techniques*. Direct techniques include small angle X-ray and neutron scattering and optical and electron microscopy. Indirect techniques include spectroscopy, including nuclear magnetic resonance, dynamic light scattering and rheology and provide supplementary information. Selected techniques will now be discussed in more detail.

7.4.4.1 Electron Microscopy

Cryogenic transmission electron microscopy (cryo-TEM) allows direct visualization of samples in the hydrated state through vitrification in a thin film suspended between polymer coated grids (Almgren et al. 1996). This is preferred over conventional (negative staining) TEM where samples are dried on carbon grids prior to viewing under the microscope, due to problems associated with dehydration. Cryo-TEM provides direct visualization and verification of lattice symmetry and is a powerful complementary technique to scattering data. The combination of cryo-TEM and scattering is considered the gold standard for characterizing the structure type of non-lamellar liquid crystalline dispersions. Cubosomes are generally recognized as cubic faceted particles as shown in Fig. 7.5.

The advantage of cryogenic field emission scanning electron microscopy (cryo-FESEM) as a complementary microscopy technique for investigating the nanostructure of non-lamellar mesophases has been recently reported (Boyd et al. 2007; Rizwan et al. 2007). Cryo-FESEM allows dispersions to be viewed in a frozen, close-to-natural state. Results summarized in Fig. 7.6 support the descriptions of the nanostructure of LCPs, particularly cubosomes based on differential geometry, where a single continuous lipid bilayer is contorted such that it divides space into two congruent and non-intersecting water channels. The nanostructure of the dispersions was also comparable to the microstructure of the non-dispersed phases determined by cryo-FESEM (Rizwan et al. 2007).

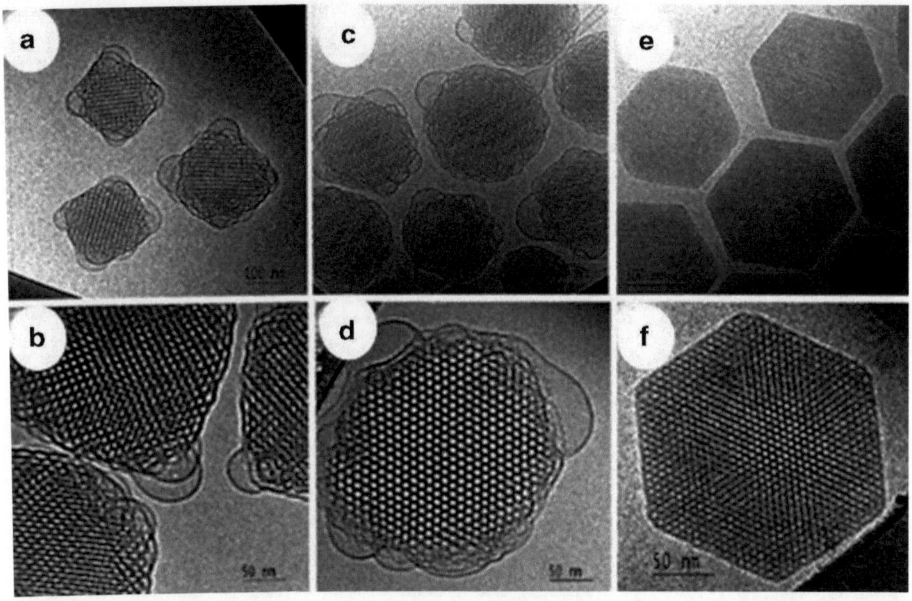

Fig. 7.5 Cryo-TEM images of non-lamellar nanoparticles with a reversed bicontinuous (**a–d**) and reversed hexagonal (**e, f**) lyotropic liquid crystal phases. Figure adapted from (Barauskas et al. 2005b)

One of the limitations of using a scanning microscopy technique to probe the structure of submicron particles when compared to microscopy in the transmission mode is the lower resolution [in the range of 50–100 nm (Pawley 1997)] and the other is the potential formation of ice crystals during sample transfer. Ice crystals are often large and can cover areas of interest. In addition to large ice crystals, frozen condensed water droplets can also make data interpretation difficult and misleading. Plunge freezing in liquid propane has been shown to reduce ice crystal size to below the resolution of the microscope. Additionally, samples are generally sublimed for a few minutes to remove unwanted surface ice prior to coating (Krauel et al. 2007).

7.4.4.2 X-ray Scattering

Scattering techniques are crucial in assigning unambiguously the structure of the mesophase of interest. Three different types of radiation are typically used in scattering studies; light, X-ray and neutron. Discussions will be restricted to X-ray scattering as this is the most frequently used approach for characterising liquid crystalline systems.

Typically small angle X-ray scattering (SAXS) experiments are used to measure the intensity of scattered X-rays at small angles to probe structure at the mesoscale,

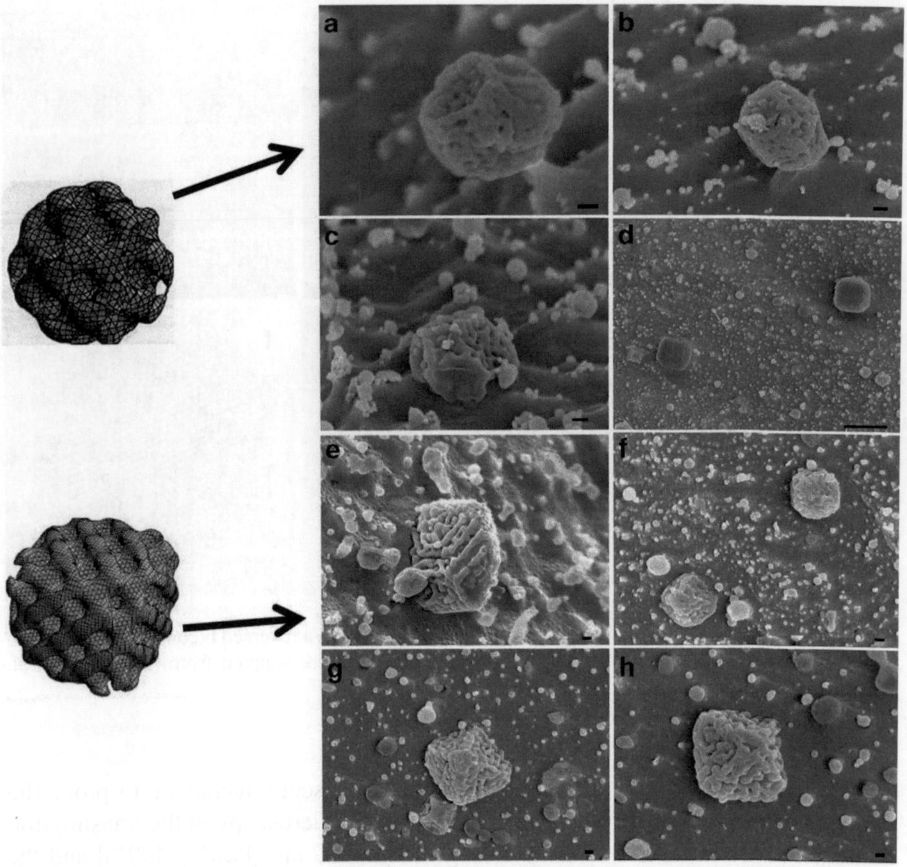

Fig. 7.6 Representative cryo-FESEM micrographs of cubosomes dispersions with spherical (**a**–**c**) or cubic morphology (**d**–**h**) compared with the proposed mathematical models based on differential geometry by Andersson et al. The *bar* represents 100 nm

which can be expressed by the length of the scattering vector q (Bergman et al. 2000):

$$q = \left(\frac{4\pi}{\lambda}\right)\sin\frac{\theta}{2} \tag{7.2}$$

where λ is the wavelength and θ the scattering angle. Typical values for the spacing of X-ray reflections characteristic of selected liquid crystalline mesophases are summarized in Table 7.1. Characteristic diffraction patterns are generated from an ordered microstructure due to specific repeat distances of the associated interlayer spacings, d. Furthermore, the mean lattice parameter, a, can be calculated from d.

Table 7.1 Selected lyotropic mesophases and their corresponding peak ratios observed using SAXS

Mesophase	Descriptor	Peak spacing ratios in q space
Lamellar	Lα	1:2:3:4:5:6:7...
Inverse hexagonal	H_{II}	$1:\sqrt{3}:\sqrt{4}:\sqrt{7}:\sqrt{9}:\sqrt{12}:\sqrt{13}...$
Bicontinuous cubic	P (Im3m)	$1:\sqrt{2}:\sqrt{4}:\sqrt{6}:\sqrt{8}:\sqrt{10}:\sqrt{12}...$
	D (Pn3m)	$1:\sqrt{2}:\sqrt{3}:\sqrt{4}:\sqrt{6}:\sqrt{8}:\sqrt{9}...$
	G (Ia3d)	$1:\sqrt{6}:\sqrt{8}:\sqrt{14}:\sqrt{16}:\sqrt{20}:\sqrt{22}...$

Table compiled from Hyde (2001) and Wörle et al. (2006b)

The value for the interlayer spacings d is calculated using (7.3) (Dong et al. 2006) and is related to (7.2) by:

$$d = \frac{2\pi}{q} \tag{7.3}$$

SAXS is an indispensible tool for identifying the various mesophases; however, it is not without some drawbacks. Problems associated with SAXS include weak reflections when acquired using a lab source, especially in dispersed liquid crystalline systems due to their small size and potentially non-uniform crystallographic microstructure. Furthermore, some systems may exhibit two or more co-existing mesophases and assignment of peaks to specific space groups becomes difficult (Amar-Yuli et al. 2009).

7.4.5 *Cubosomes for Vaccine Delivery*

Lipid-based particulate carriers have long been known to have the potential to increase the amount of antigen reaching antigen-presenting cells (APCs) and to stimulate a stronger immune response compared to antigen alone (Gregoriadis 1990). Furthermore, co-delivery of antigen and adjuvant(s) such as pattern-recognition receptors (PRRs) is also possible within a particulate system. This facilitates concurrent antigen processing and presentation as well as signalling via the relevant PRR pathway, and has been demonstrated as an effective strategy for increasing antigen delivery to APCs and expansion of effector T-cells (Kaisho and Akira 2002; Schlosser et al. 2008).

Liposomes, dispersions of lamellar liquid crystals, have been extensively studied as drug and vaccine carriers (Myschik et al. 2009). In contrast, research in the area of non-lamellar dispersions for vaccine delivery is limited. Given their unique nanostructure, cubosomes are flexible in the types of antigens and adjuvants that can be incorporated. Furthermore, encapsulation within the complex cubosome structure may also offer protection of the active against rapid degradation (Barauskas et al. 2005a). Cubosomes are also stable over time and upon dilution. These attributes make cubosomes an attractive candidate for delivery of subunit vaccines.

Cubosomes have been shown to successfully encapsulate a high percentage (as compared to liposomes) and show sustained release of the model antigen ovalbumin (OVA) in vitro (Rizwan et al. 2011). Whilst in vitro studies provide useful predictions of release kinetics, the correlation with antigen release in vivo is multifactorial and uncertain. To confirm this, in vivo release kinetics of OVA from various formulations were investigated. This was achieved by determining proliferation of carboxyfluorescein diacetate succinimidyl ester (CFSE)-labelled $CD8^{+}$ and $CD4^{+}$ T-cells at various time points following immunization. The continued expansion of CFSE-labelled T-cells from vaccinated animals indicated that OVA in cubosomes was delivered in a sustained manner and therefore was available for processing and presentation for at least 14 days (Rizwan et al. 2013), providing a clear in vitro and in vivo correlation.

In order to generate antigen-specific responses, APCs such as dendritic cells (DCs) need to recognize and take up particles containing the antigen of interest, process it and present it in the context of major histocompatibility complex (MHC) molecules for recognition by naïve T-cells, leading to subsequent T-cell proliferation (Cools et al. 2007). In addition, the APCs must become activated, up-regulating the expression of co-stimulatory molecules and cytokines (Lee and Iwasaki 2007; Cools et al. 2007). Cubosomes were subsequently modified to include the toll-like receptor (TLR) agonists, monophosphoryl lipid A (MPL) and imiquimod. All cubosome formulations were investigated in vitro in bone marrow-derived dendritic cells (BMDCs). The formulations were taken up by BMDCs in a lipid concentration-dependent manner. Co-stimulatory molecules, CD86 and MHC class-II, crucial for expansion of T-cells (Cools et al. 2007) were also expressed on the surface of activated BMDCs in a concentration-dependent manner. The greatest up-regulation of co-stimulatory molecules was observed when DCs were incubated with imiquimod + MPL cubosomes. Subsequently, cellular and humoral responses in mice were investigated after vaccination with cubosomes and comparable liposomal formulations. It was found that cubosomes containing the adjuvants imiquimod and MPL were more efficient at inducing antigen-specific immune responses than liposomes. Furthermore, cubosomes were more efficient at generating antigen-specific cellular responses and were equally as effective in generating humoral responses when compared to alum, the most widely used vaccine adjuvant, and to liposomes (Rizwan et al. 2013).

Needle-free immunization using cubosomes has also been investigated (Rattanapak et al. 2012). Transcutaneous immunization (TCI) is a promising vaccination strategy for obvious economic and social reasons. However, from a scientific view point, this approach is particularly attractive due to the abundance of APCs residing in the skin. The main obstacle for TCI, however, is the delivery of the vaccine through the stratum corneum to the APCs that reside in the deeper skin layers. Rattanapak et al. used a novel approach, where cubosomes and microneedles (MCNs) were utilized as a synergistic approach for vaccine delivery through the skin (Rattanapak et al. 2013). Initially, permeation of various lipid-based particulate formulations (with or without MCN pretreatment) through stillborn piglet skin was investigated in vitro under occlusive conditions using Franz diffusion cells

(Rattanapak et al. 2012). Intriguingly, cubosomes showed superior skin retention compared with liposomes and transfersomes. Subsequently, the combined approach with cubosomes + MCN was applied. This approach leads to significantly improved vaccine uptake into skin as compared to cubosomes alone. Vaccine antigen appeared to be preferentially taken up by a subpopulation of skin APC known as dermal DCs. Interestingly delivery of vaccine using MCNs in conjunction with cubosomes appeared to stimulate a $CD8^{+}$ T-cell biased immune response, crucial for effective therapeutic vaccines (van Duikeren et al. 2012).

7.5 Conclusions

Cubosomes represent an intriguing new delivery system for various applications. In the context of vaccine delivery, although the research field is still at early stages, cubosomes have been already shown to have the potential to encapsulate a high antigen load and can be modified to include adjuvants such as PRRs. They are recognized by processional APCs, with the antigen being processed and presented via the elusive MHC class I pathway in vitro and in vivo. They are also well tolerated in vivo and lead to greater expansion of T-cells compared to liposomes. However, a lot remains to be understood about these novel lipid-based carrier systems before they can be used in a clinical setting.

References

Almgren M (2003) Alexander lecture 2003; cubosomes, vesicles, and perforated bilayers in aqueous systems of lipids, polymers, and surfactants. Aust J Chem 56:959–970

Almgren M, Edwards K, Gustafsson J (1996) Cryo-transmission electron microscopy of thin vitrified samples. Curr Opin Colloid Interface Sci 1(2):270–278

Almsherqi ZA, Kohlwein SD, Deng Y (2006) Cubic membranes: a legend beyond the Flatland of cell membrane organization. J Cell Biol 173(6):839–844. doi:10.1083/jcb.200603055

Amar-Yuli I, Libster D, Aserin A, Garti N (2009) Solubilization of food bioactives within lyotropic liquid crystalline mesophases. Curr Opin Colloid Interface Sci 14(1):21–32. doi:10.1016/j.cocis.2008.02.001

Andersson S, Jacob M, Lidin S, Larsson K (1995) Structure of the cubosome—a closed lipid bilayer aggregate. Z Kristallographie 210:315–318

Barauskas J, Landh T (2003) Phase behaviour of the phytantriol/water system. Langmuir 19(23):9562–9565

Barauskas J, Johnsson M, Joabsson F, Tiberg F (2005a) Cubic phase nanoparticles (cubosome): principles for controlling size, structure, and stability. Langmuir 21:2569–2577

Barauskas J, Johnsson M, Tiberg F (2005b) Self-assembled lipid superstructures: beyond vesicles and liposomes. Nano Lett 5:1615–1619

Bergman A, Fritz G, Glatter O (2000) Solving the generalized indirect Fourier transformation (GIFT) by Boltzmann simplex simulated annealing (BSSA). J Appl Crystallogr 33:1212–1216

Boyd BJ, Drummond CJ, Krodkiewska I, Grieser F (2000) How chain length, headgroup polymerization, and anomeric configuration govern the thermotropic and lyotropic liquid crystalline phase behavior and the air-water interfacial adsorption of glucose-based surfactants. Langmuir 16:7359–7367

Boyd B, Whittaker D, Khoo S, Davey G (2006) Lyotropic liquid crystalline phases formed from glycerate surfactants as sustained release drug delivery systems. Int J Pharm 309:216–226

Boyd BJ, Rizwan SB, Dong Y, Hook S, Rades T (2007) Self-assembled geometric liquid-crystalline nanoparticles imaged in three dimensions: hexosomes are not necessarily flat hexagonal prisms. Langmuir 23(25):12461–12464

Caffrey M (2000) A lipid's eye view of membrane protein crystallization in mesophases. Curr Opin Struct Biol 10(4):486–497. doi:10.1016/S0959-440X(00)00119-6

Chang C, Bodmeier R (1997) Swelling of and drug release from monoglyceride-based drug delivery systems. J Pharm Sci 86(6):747–752

Cherezov V, Clogston J, Papiz MZ, Caffrey M (2006) Room to move: crystallizing membrane proteins in swollen lipidic mesophases. J Mol Biol 357(5):1605–1618. doi:10.1016/j.jmb.2006.01.049

Chong JYT, Mulet X, Waddington LJ, Boyd BJ, Drummond CJ (2011) Steric stabilisation of self-assembled cubic lyotropic liquid crystalline nanoparticles: high throughput evaluation of triblock polyethylene oxide-polypropylene oxide-polyethylene oxide copolymers. Soft Matter 7(10):4768–4777. doi:10.1039/C1SM05181D

Chung H, Kim J, Kwon I, Jeong S (2002) Self-assembled "nanocubicle" as a carrier for peroral insulin delivery. Diabetologica 45:448–451

Clogston J, Caffrey M (2005) Controlling release from the lipidic cubic phase. Amino acids, peptides, proteins and nucleic acid. J Control Release 107:97–111

Colotto A, Epand RM (1997) Structural study of the relationship between the rate of membrane fusion and the ability of the fusion peptide of influenza virus to perturb bilayers. Biochemistry 36(25):7644–7651. doi:10.1021/bi970382u

Cools N, Ponsaerts P, Van Tendeloo VFI, Berneman ZN (2007) Balancing between immunity and tolerance: an interplay between dendritic cells, regulatory T cells, and effector T cells. J Leukoc Biol 82:1365–1374

Dan Y, Poo M-M (2004) Spike timing-dependent plasticity of neural circuits. Neuron 44(1):23–30

Deng Y, Kohlwein SD, Mannella CA (2002) Fasting induces cyanide-resistant respiration and oxidative stress in the amoeba Chaos carolinensis: implications for the cubic structural transition in mitochondrial membranes. Protoplasma 219(3–4):160–167. doi:10.1007/s007090200017

Dong Y, Larson I, Hanley T, Boyd BJ (2006) Bulk and dispersed aqueous phase behavior of phytantriol: effect of vitamin E acetate and F127 polymer on liquid crystal structure. Langmuir 22:9512–9518

Fong C, Wells D, Krodkiewska I, Booth J, Hartley PJ (2007) Synthesis and mesophases of glycerate surfactants. J Phys Chem B 111:1384–1392

Gabizon A, Shmeeda H, Horowitz AT, Zalipsky S (2004) Tumor cell targeting of liposome-entrapped drugs with phospholipid-anchored folic acid-PEG conjugates. Adv Drug Deliv Rev 56:1177–1192

Ganem-Quintanar A, Quintanar-Guerrero D, Buri P (2000) Monoolein: a review of the pharmaceutical applications. Drug Dev Ind Pharm 26(8):809–820

Garti N, Libster D, Aserin A (2012) Lipid polymorphism in lyotropic liquid crystals for triggered release of bioactives. Food Funct 3(7):700–713

Gregoriadis G (1990) Immunological adjuvants: a role for liposomes. Immunol Today 11:89–97. doi:10.1016/0167-5699(90)90034-7

Guo C, Wang J, Cao F, Lee RJ, Zhai G (2010) Lyotropic liquid crystal systems in drug delivery. Drug Discov Today 15(23–24):1032–1040. doi:10.1016/j.drudis.2010.09.006

Gustafsson J, Ljusberg-Wahren H, Almgren M, Larsson K (1997) Submicron particles of reversed lipid phases in water stabilized by a nonionic amphiphilic polymer. Langmuir 13:6964–6971

Hato M, Minamikawa H (1996) The effects of oligo saccharide stereochemistry on the physical properties of aqueous synthetic lipids. Langmuir 12:1658–1665

Hyde S (2001) Identification of lyotropic liquid crystal mesophases, Chap 16. In: Holmberg K (ed) Handbook of applied surface and colloid chemistry. Wiley, New York, pp 299–331

Israelachvili JN, Mitchell DJ, Ninham BW (1976) Theory of self-assembly of hydrocarbon amphiphiles into micelles and bilayers. J Chem Soc Faraday Trans 2 72:1525–1568

Kaasgaard T, Drummond CJ (2006) Ordered 2-D and 3-D nanostructured amphiphile self-assembly materials stable in excess solvent. Phys Chem Chem Phys 8:4957–4975

Kaisho T, Akira S (2002) Toll-like receptors as adjuvant receptors. Biochim Biophys Acta 1589:1–13

Krauel K, Girvan L, Hook S, Rades T (2007) Characterisation of colloidal drug delivery systems from the naked eye to Cryo-FESEM. Micron 38:796–803

Lara MG, Bentley MV, Collet JH (2005) In vitro drug release mechanism and drug loading studies of cubic phase gels. Int J Pharm 293:241–250

Larsson K (1983) Two cubic phases in monoolein-water system. Nature 304(5927):664

Larsson K (1989) Cubic lipid-water phases: structures and biomembrane aspects. J Phys Chem 93:7304–7314

Larsson K (1999) Colloidal dispersions of ordered lipid-water phases. J Dispers Sci Technol 20:27–34

Larsson K (2000) Aqueous dispersions of cubic lipid-water phases. Curr Opin Colloid Interface Sci 5:64–69

Lee HK, Iwasaki A (2007) Innate control of adaptive immunity: dendritic cells and beyond. Semin Immunol 19:48–55

Lindblom G, Rilfors L (1989) Cubic phases and isotropic structures formed by membrane lipids—possible biological relevance. Biochim Biophys Acta 988:221–256

Luzzati V (1997) Biological significance of lipid polymorphism: the cubic phases. Curr Opin Struct Biol 7:661–668

Mulet X, Boyd BJ, Drummond CJ (2013) Advances in drug delivery and medical imaging using colloidal lyotropic liquid crystalline dispersions. J Colloid Interface Sci 393:1–20. doi:10.1016/j.jcis.2012.10.014

Myschik J, Rades T, Hook S (2009) Advances in lipid-based subunit vaccine formulations. Curr Immunol Rev 5:42–48

Nakano M, Sugita A, Matsuoka H, Handa T (2001) Small-angle x-ray scattering and 13C NMR investigation on the internal structure of "cubosomes". Langmuir 17(13):3917–3922

Nakano M, Teshigawara T, Sugita A, Leesajakul W, Taniguchi A, Kamo T, Matsuoka H, Handa T (2002) Dispersions of liquid crystalline phases of the monoolein/oleic acid/pluronic F127 system. Langmuir 18:9283–9288

Patton JS, Carey MC (1979) Watching fat digestion. Science 204(4389):145

Pawley J (1997) The development of field-emission scanning electron microscopy for imaging biological surfaces. Scanning 19:324–336

Pratt L (1985) Theory of hydrophobic effects. Annu Rev Phys Chem 36:433–449

Quantan N, Spicer J, Plunkett T, Pandha H (2004) Cellular immunotherapy for cancer: current concepts and clinical perspectives scientific basis and approaches for therapeutic cancer vaccines: Part 1. Clin Oncol 16:356–365

Rattanapak T, Young K, Rades T, Hook S (2012) Comparative study of liposomes, transfersomes, ethosomes and cubosomes for transcutaneous immunisation: characterisation and in vitro skin penetration. J Pharm Pharmacol 64(11):1560–1569. doi:10.1111/j.2042-7158.2012.01535.x

Rattanapak T, Birchall J, Young K, Ishii M, Meglinski I, Rades T, Hook S (2013) Transcutaneous immunization using microneedles and cubosomes: mechanistic investigations using optical coherence tomography and two-photon microscopy. J Control Release 172(3):894–903. doi:10.1016/j.jconrel.2013.08.018

Ribier A, Biatry B (1998) Cosmetic or dermatological composition in the form of an aqueous and stable dispersion of cubic gel particles based on phytanetriol and containing a surface-active agent which has a fatty chain, as dispersing and stabilizing agent. US Patent

Rizwan SB, Dong Y, Boyd BJ, Rades T, Hook S (2007) Characterisation of bicontinuous cubic liquid crystalline systems of phytantriol and water using cryo field emission scanning electron microscopy (cryo FESEM). Micron 38(5):478–485

Rizwan SB, Hanley T, Boyd BJ, Hook S (2009) Liquid crystalline systems of phytantriol and glyceryl monooleate containing a hydrophilic protein: characterisation, swelling and release kinetics. J Pharm Sci 98(11):4191–4204

Rizwan SB, Boyd BJ, Rades T, Hook S (2010) Bicontinuous cubic liquid crystals as sustained delivery systems for peptides and proteins. Expert Opin Drug Deliv 7(10):1133–1144

Rizwan SB, Assmus D, Boehnke A, Hanley T, Boyd BJ, Rades T, Hook S (2011) Preparation of phytantriol cubosomes by solvent precursor dilution for the delivery of protein vaccines. Eur J Pharm Biopharm 79(1):15–22. doi:10.1016/j.ejpb.2010.12.034

Rizwan SB, McBurney WT, Young K, Hanley T, Boyd BJ, Rades T, Hook S (2013) Cubosomes containing the adjuvants imiquimod and monophosphoryl lipid A stimulate robust cellular and humoral immune responses. J Control Release 165(1):16–21. doi:10.1016/j.jconrel.2012.10.020

Schlosser E, Mueller M, Fischer S, Basta S, Busch DH, Gander B, Groettrup M (2008) TLR ligands and antigen need to be coencapsulated into the same biodegradable microsphere for the generation of potent cytotoxic T lymphocyte responses. Vaccine 26:1626–1637

Seddon JM, Templer RH (1993) Cubic phases of self-assembled amphiphilic aggregates. Philos Trans R Soc Lond A 344(1672):377–401

Shearman GC, Templer RH, Seddon JM (2006) Inverse lyotropic phases of lipids and membrane curvature. J Phys Condens Matter 18:S1105–S1124

Spicer P, Hayden K, Lynch ML, Ofori-Boateng A, Burns JL (2001) Novel process for producing cubic liquid crystalline nanoparticles (cubosomes). Langmuir 17(19):5748–5756

Spicer P, Small WB II, Lynch ML, Burns JL (2002) Dry powder precursors of cubic liquid crystalline nanoparticles (cubosome). J Nanopart Res 4:297–311

Tilley AJ, Drummond CJ, Boyd BJ (2013) Disposition and association of the steric stabilizer Pluronic® F127 in lyotropic liquid crystalline nanostructured particle dispersions. J Colloid Interface Sci 392:288–296. doi:10.1016/j.jcis.2012.09.051

van Duikeren S, Fransen MF, Redeker A, Wieles B, Platenburg G, Krebber W-J, Ossendorp F, Melief CJM, Arens R (2012) Vaccine-induced effector-memory CD8+ T cell responses predict therapeutic efficacy against tumors. J Immunol 189(7):3397–3403. doi:10.4049/jimmunol.1201540

Wörle G, Siekmann B, Bunjes H (2006a) Effect of drug loading on the transformation of vesicular into cubic nanoparticles during heat treatment of aqueous monoolein/poloxamer dispersions. Eur J Pharm Biopharm 63:128–133

Wörle G, Siekmann B, Koch M, Bunjes H (2006b) Transformation of vesicular into cubic nanoparticles by autoclaving of aqueous monoolein/poloxamer dispersions. Eur J Pharm Sci 27:44–53

Yaghmur A, Glatter O (2008) Characterization and potential applications of nanostructured aqueous dispersions. Adv Colloid Interface Sci 147–148:333–342

Yano A, Onouka A, Asahi-Ozaki Y, Imai S, Hanada N, Miwa Y, Nisizawa T (2005) An ingenious design for peptide vaccines. Vaccine 23:2322–2326

Zheng L, Um J, Chung H, Kwon I, Li G, Jeong S (2003) Microstructure of dispersed colloidal particles of a bilayer cubic phase. J Dispers Sci Technol 24(1):123–128

Chapter 8
ISCOMs as a Vaccine Delivery System

Hanne M. Nielsen, Henriette B. Hübschmann, and Thomas Rades

8.1 Introduction

Vaccination utilizing adjuvants based on particles ranging from relatively small particles such as virus-like (VLP) particles or larger particles as liposomes and emulsion droplets is a feasible way to stimulate specific immune responses. Immune stimulating complexes (ISCOMs) can be categorized as small 40–60 nm lipid-based particles that have shown potential as adjuvants and carriers for antigens aiming at prophylactic or therapeutic vaccination. Both cellular and humoral immune responses have been reported after vaccination with antigens and ISCOM adjuvants (Sun et al. 2009; Morelli et al. 2012); some of which are in clinical trials (Hook and Rades 2013). Immune stimulation has been observed after administration by injection, via administration to mucosal sites and after cutaneous application (Morein et al. 1984; Sjölander et al. 1998; Pearse and Drane 2005; Sun et al. 2009; Alving et al. 2012; Morelli et al. 2012). The adjuvant particles are formed in solution by self-assembly at well-defined ratios of phospholipid, saponin, and cholesterol. In aqueous dispersion, they appear as cage-like structures with a hollow centre. The state-of-the-art with regards to formulation design, characterization, and assessment of the mechanisms of action for ISCOMs are summarized and discussed along with addressing the different routes of administration and the future perspectives of using ISCOMs as vaccine adjuvants.

H.M. Nielsen (✉) • H.B. Hübschmann • T. Rades
Department of Pharmacy, Faculty of Health and Medical Sciences, University of Copenhagen, Universitetsparken 2, Copenhagen Ø 2100, Denmark
e-mail: hanne.morck@sund.ku.dk; thomas.rades@sund.ku.dk; henriettehenriettehenriette@gmail.com

C. Foged et al. (eds.), *Subunit Vaccine Delivery*, Advances in Delivery Science and Technology, DOI 10.1007/978-1-4939-1417-3_8

8.2 Characteristics

ISCOMs, similar to VLPs (Chap. 9), are self-assembling particles of a size comparable to most typical viruses; and are a type of vaccine adjuvant and delivery system that is attracting continuous attention for development of vaccines (Scheerlinck and Greenwood 2008). The resemblance to viruses in terms of geometry has been proposed to be a beneficial feature, as ISCOMs like VLPs may easily be recognized and taken up by antigen-presenting cells (APCs) due to their composition, size, and surface structure (Scheerlinck and Greenwood 2008), followed by processing and presentation of the antigen incorporated into the vaccine leading to induction of an immune response (Morein et al. 1984). ISCOMs may be comparable to other phospholipid-based vesicle adjuvants, as they have a hollow (albeit open) core, such as liposomes, niosomes, flexosomes, vesosomes, exosomes, and ethosomes although these are usually larger in size being in the size range of 100 nm and larger. Likewise, emulsions and other lipid-based nanoparticles with potential as future adjuvants (like solid lipid nanoparticles and cubosomes) usually appear larger (Nordly et al. 2009).

Microscopic characterization of ISCOMs dispersed in aqueous medium by cryotransmission electron microscopy (TEM) displays spherical hollow particles in the size range of 40–60 nm (Fig. 8.1a) with the presumed 3D hollow and open structure depicted in Fig. 8.1b. Recently, small angle X-ray scattering (SAXS) was employed to describe ISCOMs in suspension confirming the organization of the constituents in the expected structure (Fig. 8.1c) (Pedersen et al. 2012). Still, only qualified theories exist as to how the single constituents self-assemble and are organized to form spherical hollow particles. One hypothesis is that the constituents are placed in stacks with the hydrophobic parts of the molecules facing the interior of the particle bilayers and the more hydrophilic parts oriented towards the aqueous dispersion medium inside and outside the particle structure (Kersten et al. 1991; Kersten and Crommelin 1995). This hypothesis has been used to explain the organization of different types of ISCOM particles, irrespective of whether their net surface charge is negative or positive, as illustrated in Fig. 8.1d (Lendemans et al. 2005). The particles are thus organized into hollow structures with both locally charged areas and lipophilic bilayers, with which the antigens of choice may interact, and are shaped and stabilized by hydrophobic interactions, electrostatic repulsion, steric factors, and possibly hydrogen bonds (Kersten et al. 1991; Lendemans et al. 2005).

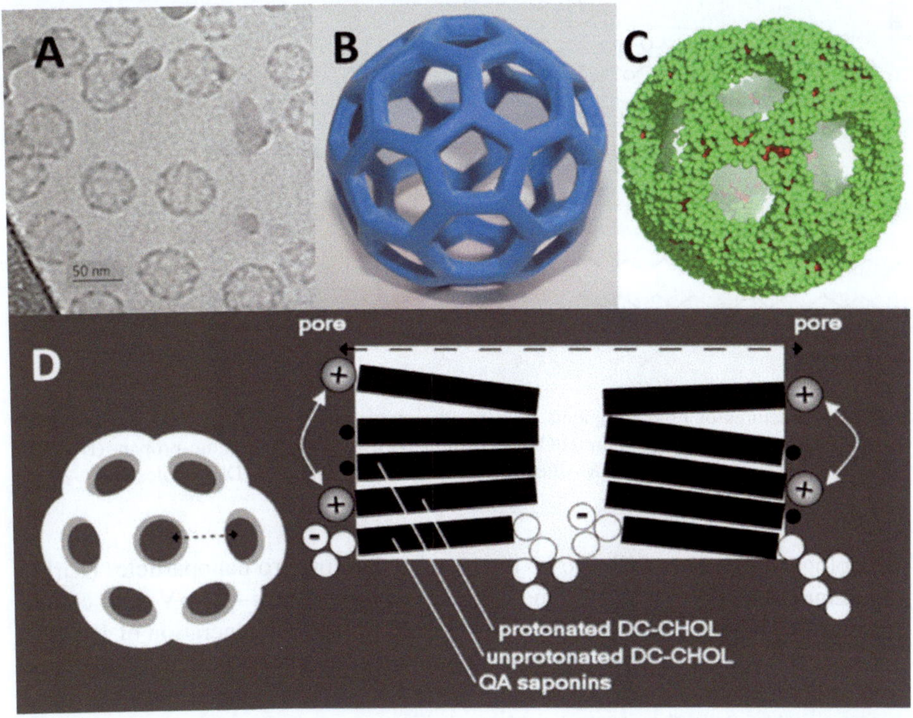

Fig. 8.1 (**a**) Cryo-TEM image of ISCOMs in suspension (Pedersen et al. 2012), (**b**) schematic 3D-model of the ISCOM cage-like structure, (**c**) structure of ISCOMs as derived from SAXS analysis (Pedersen et al. 2012), (**d**) proposed molecular alignment of components in lipid bilayers (Lendemans et al. 2005). *Reprinted with permissions from Elsevier* (*Figure 8.1A and 8.1C*) *and from Wiley* (*Figure 8.1D*)

8.3 Lipid Components

By formulation design and exchange of the traditionally used excipients, new generations of ISCOMs have appeared with slightly improved safety as well as immunostimulatory profiles. Such optimizations of the ISCOM properties have been done by partly or fully exchanging the neutral cholesterol component with the positively charged 3β-[*N*-(*N'*,*N'*-dimethylaminoethane)-carbamoyl](DC)-cholesterol (Kirkby and Samuelsen 2006; Lendemans et al. 2005) or by exchanging the zwitterionic phospholipids with the cationic dioleoyl-trimethyl-ammonium-propane (DOTAP) (Lendemans et al. 2007) (Fig. 8.2), to decrease the anionic surface charge (Posintros) or to provide them with a cationic surface charge (PLUSCOMs). The net cationic surface charge of PLUSCOMs (approximately +25 mV) (McBurney et al. 2008) will increase the likeliness of (electrostatic) interaction with antigens usually possessing a pI value below 7.4, and thus carrying an overall negative charge at

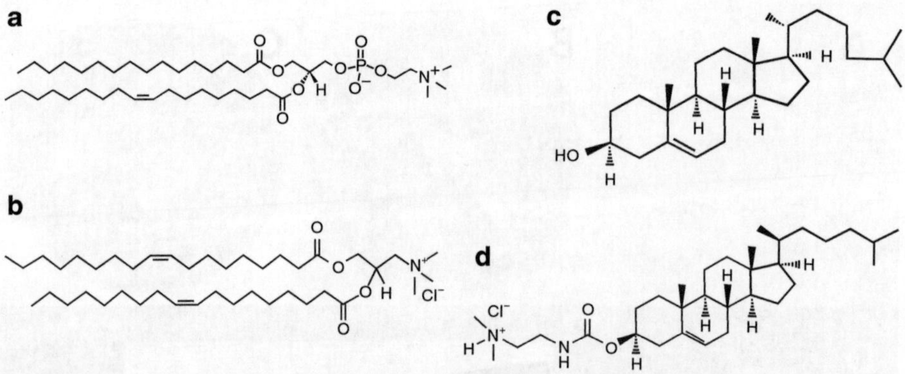

Fig. 8.2 Structures of lipid excipients used for the preparation of ISCOMs. (**a**) 1-palmitoyl-2-oleoyl-*sn*-glycero-3-phosphocholine (POPC), (**b**) dioleoyl-trimethyl-ammonium-propane (DOTAP), (**c**) cholesterol, and (**d**) 3β-[*N*-(*N'*,*N'*-dimethylaminoethane)-carbamoyl]((DC)-cholesterol)

physiological pH. This will also be the case for the Posintro nanoparticles (Kirkby and Samuelsen 2006), which are net anionic (approximately −30 mV), but contain a higher degree of cationic charges as compared to the first generation of ISCOMs with a net charge of −40 mV. Incorporation of other components with lipidic backbone structures into the ISCOM particle could also prove beneficial for the physical stability and/or the effect of the vaccine.

The presence of phospholipids in the ISCOMs has been reported to be important for the formation of the cage-like geometry (Myschik et al. 2006), yet also argued not to be crucial for particle formation, although important for the incorporation or association of antigens into the structure (Lövgren and Morein 1988). These discrepancies reflect the delicate balance between the molar ratios of the constituents that must be finely tuned when preparing ISCOMs by one of the various preparation methods, as will be described below. Also, the presence of cholesterol seems to be crucial in order to assemble the ISCOM structure together with the saponin adjuvant (Lövgren and Morein 1988).

8.4 Immunostimulating Component

In order to improve the efficacy of the formulation, also the immunogenic component may be changed. Saponins are natural products, which are surface active, negatively charged, possess strong adjuvant properties (Dalsgard 1974; Alving et al. 2012) and are used as the main adjuvant in the preparation of ISCOMs. The crude saponin mixture used for first generation ISCOMs is obtained from extracts from the *Quillaja saponaria* tree and is now partly purified to give the currently most often used, although still complex mixture, Quil A, or completely purified to obtain one of the main and most safe components, QS-21 from the extracts (Fig. 8.3)

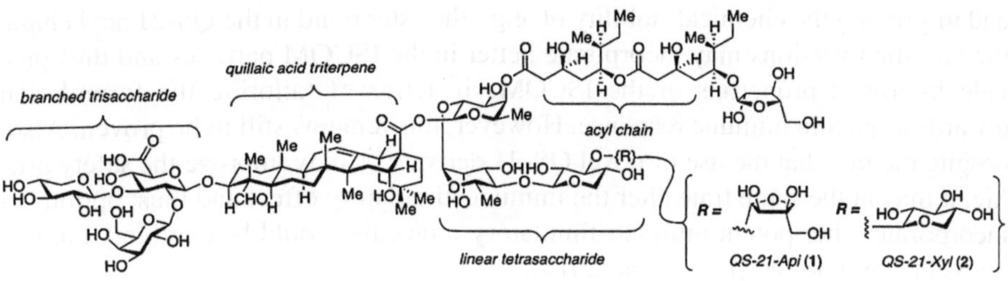

Fig. 8.3 Molecular structure of QS-21. *Reprinted with permission from Chea et al.* 2012. *Copyright (2012) American Chemical Society*

(Hook and Rades 2013). The QS-21 fraction has also been used as an adjuvant together with other adjuvants leading to non-ISCOM type structures. Examples of these are the adjuvants AS01, AS02, and AS15 from GSK, all containing QS21 and monophosphoryl lipid A (MPL) and one also CpG (AS15), in liposomal (AS01, AS15) or oil-in-water emulsion formulations (AS02). Structurally, these saponins consist of a rigid lipophilic backbone and two large polar head groups and belong to the group of bola surfactants. They are characterized as triterpene glycosides, and the presence of the glucoronic acid present in the sugar units of the molecule is the main contributor to the negative charge. Their surfactant properties seem to be important for their role in the structure formation of the ISCOMs. Saponins have the ability to bind cholesterol, which is the reason for their known effect on cell plasma membranes mediating lysis at high concentrations due to cholesterol depletion, but may also contribute to the stabilization of the ISCOM structure, which at the same time reduces the side effects of the saponin upon injection (Pham et al. 2006). Although incorporation of the saponins into the ISCOMs has been shown to reduce the cytotoxic effect of the compound (Cox et al. 1998; Kamstrup et al. 2000), the high content of the immune stimulating and negatively charged saponin is responsible for the often overall negative charge of the resulting ISCOM nanoparticles. The unique capacity to stimulate both the production of T-lymphocytes as well as to stimulate a Th1-based immune response makes saponins ideal adjuvants in therapeutic as well as prophylactic subunit vaccines (Sun et al 2009).

The structure–activity relationship of the saponins in terms of adjuvanticity is influenced by the hydrophilic sugar side chains and the hydrophobic aglycone backbone, but it is also thought to be related to the aldehyde groups present in the lipophilic backbone of the molecules or to the acyl residue bearing the aglycone (Sun et al. 2009; Soltysik et al. 1995). Yet, the overall adjuvant mechanism is not completely understood. Modifications in the acyl backbone (Wang et al. 2013) have, however, been shown to induce specific alterations in the antibody and cytotoxic T-cell responses as well as in the hemolytic activity of QS-21 variants (Chea et al. 2012). Also, synthetic versions of QS-21 including carbohydrate modifications of the apiose and xylose moieties along with acyl chain modifications have been shown to have an impact on the immunological response and on how well the QS-21 derivative is tolerated (Chea et al. 2012). Besides altering the properties described above and improving the chemical stability of, e.g., the ester bond in the QS-21 acyl chain, the modified versions may incorporate better in the ISCOM particles and thus provide improved properties of the ISCOMs in terms of tailoring the formulation towards a specific immune response. However, this remains still to be proven. Also, despite the fact that the use of novel QS-21 derivatives may improve the safety profile, it may at the same time alter the immunostimulatory effect and thus, options to incorporate other potent immunostimulatory molecules could be explored, but has not yet been reported (Brito et al. 2013).

8.5 Antigen Component

Due to the hollow geometry, the lipid bilayers and the presence of anionic (and sometimes cationic) patches in the ISCOM, the antigen may be enclosed in the pores, in the interior of the particle or closely associated to the surface of the particle, but cannot as such be encapsulated in the ISCOM structure like it is the case for, e.g., liposomes. The complexation with the ISCOM may thus be mediated by both electrostatic as well as hydrophobic interactions depending on the properties of the antigen. As mentioned above, a way to enhance the interaction between the adjuvant and the usually negatively charged antigen may be to modify the ISCOM to carry more positive charges, and it was indeed demonstrated that when using (partially) positively charged ISCOMs (e.g., PLUSCOMs), a high association of (negatively charged) antigen was achieved, likely due to electrostatic interactions between the ISCOM particles and the antigen (McBurney et al. 2008). Further, modification of the ISCOM by incorporating molecules that may bind directly with a given antigen is a strategy to improve the loading of hydrophilic peptides or proteins in ISCOM-adjuvanted vaccines (Andersson et al. 2000; Cruz-Bustos et al. 2012). Some antigens derived from membrane spanning proteins may by nature contain hydrophobic domains that are likely to interact with the hydrophobic parts of the ISCOM, and some antigens, such as tetanus toxoid, can be partly unfolded to expose hydrophobic patches (Morein et al. 1990) that can interact with the ISCOM lipid bilayers. Another viable approach is to conjugate lipophilic moieties to antigens to promote hydrophobic interactions with the ISCOM bilayers. An example of the latter approach includes conjugation of palmitic acid to ovalbumin (Könnings et al. 2002). However, care should be taken that this does not compromise the antigenicity of the molecule.

In efficacy studies, the antigen is usually co-administered with the adjuvant and detailed systematic studies regarding the localization of the antigen on or in the ISCOM structure prior to and after administration remains still to be reported. Electron microscopy does not provide sufficient resolution unless a thick antigen corona covers the ISCOMs, thus possible quantitative adsorption experiments along with biophysical analysis by, e.g., isothermal calorimetry may be used to describe the interaction between ISCOMs and antigens. Also, as recently demonstrated, modelling of SAXS data might provide valuable information about the localization and the amount of antigen interacting with the ISCOM particles (Pedersen et al. 2012).

8.6 ISCOM Terminology

The terminology used in literature for describing the various ISCOM nanoparticles is depending on the specific composition, and to some extent reflects their properties. In Table 8.1, the most often used terminologies are described. The differences mainly lie in whether a crude saponin mixture or the different fractions of the Quil A components are used resulting in different preparations that may be recommended for different species depending on the species sensitivity to the saponin component (Fossum et al. 2014).

Table 8.1 Overview of the composition and properties of different ISCOM-based adjuvants and vaccines

Terminology	Composition	Properties	References
ISCOM	Used to describe the original technology for the adjuvant consisting of Quil A, cholesterol, and phospholipids; which has subsequently been optimized. Currently used as a general term describing the self-assembled structures (Fig. 8.1) used for vaccination purposes, i.e., with or without content of antigen	Hollow structure, size: 40–50 nm, zeta potential: −10 to −40 mV (depending on preparation procedure)	Morein et al. (1984); Brito et al. (2013); Hook and Rades (2013)
ISCOM-Matrix	Often used to describe the adjuvant nanoparticle without antigen content	As ISCOM	
ISCOMATRIX	Preformed adjuvant with different ratio of a more purified Quil A as opposed to the original ISCOM technology. Without content of antigen	As ISCOM	Vujanic et al. (2010); Brito et al. (2013); DiStefano et al. (2013)
MATRIX-M	Two different fractions of Quil A as immunopotentiators	As ISCOM	Fossum et al. (2014)
MATRIX-Q	Quil A incorporated as immunopotentiator	As ISCOM	Fossum et al. (2014)
Posintro	Positively charged DC-cholesterol partly exchanged with cholesterol. Less negative than ISCOM prepared under same conditions (−40 mV)	Size: 40–50 nm Zeta potential: −30 mV	Madsen et al. (2009); Madsen et al. (2010)
PLUSCOM	Cholesterol fully exchanged with DC-cholesterol	Size: 40–50 nm Zeta potential: +25 mV	Lendemans et al. (2005); McBurney et al. (2008)
ISCOPREP saponin	Specific type of saponin for use in the preparation of ISCOMs or ISCOMATRIX	n.a.	DiStefano et al. (2013)

DC-cholesterol = 3β-[*N*-(*N*′,*N*′-dimethylaminoethane)-carbamoyl]-cholesterol
Most names are registered trademarks and some commercially available

8.7 Preparation and Characterization of ISCOM-Based Vaccines

The properties of ISCOM adjuvants or ISCOM-based particles depend on a variety of factors; some of which are described below.

8.7.1 Preparation Methods

The methods used for preparation of ISCOMs are similar to the methods used for preparation of liposomes, and include dialysis, ultracentrifugation, lipid film hydration, freeze drying, and ethanol or ether injection. The methodologies and their pros and cons are thoroughly described elsewhere (Hook and Rades 2013; Sun et al. 2009) and will not be further addressed in the present chapter. However, briefly it should be mentioned that although each of the mentioned methods results in colloidal dispersions, the main differences lie in (a) the time needed for reaching equilibrium mainly due to the need for removing the lipid solubilizing surfactant, (b) the sample yield, and (c) the homogeneity of the resulting dispersion. Overall, despite the relatively long dialysis time needed to remove the solubilizing surfactant, the dialysis method is often preferred due to the high homogeneity of the resulting ISCOMs.

8.7.2 Component Ratio

The need for including both phospholipid and cholesterol for the formation of ISCOMs is well recognized (Kersten et al. 1991; Myschik et al. 2006) although it has been extensively debated. The type of the three components, the exact ratio as well as the preparation method and conditions used determine the properties of the resulting self-assembled dispersed structures (Myschik et al. 2006; Hook and Rades 2013). Moreover, the total concentration of lipids is important for the outcome of the preparation. As mentioned, the molecular ratio of the different lipids and the saponin are important for the preparation of the ISCOMs. Thus, the narrow window of molar ratios that lead to ISCOM formation under the specified conditions must be identified, optimally by constructing ternary phase diagrams (Hook and Rades 2013). Briefly, as described in literature, in the absence of the saponin, liposomes are formed, whereas even relatively low amounts of saponin induce the formation of ISCOM structures. If the concentration of the saponin is too low, the dispersion will appear as a mixture of liposomes with ISCOM and/or ring-like micelles. Worm-like micelles and helical structures are typically formed if only cholesterol and saponin are present. Also, as the critical micellar concentration (CMC) of saponin is low compared to the large concentrations required for ISCOM preparation (e.g., the CMC for Quil A was determined to 0.03 % (Özel et al. 1989)), the appearance of saponin micelles may occur in the sample (Madsen et al. 2009). Further, lamellar structures are dominant if no or only a very little fraction of cholesterol is present.

The optimal window for ISCOM preparation may thus more realistically be described as the ratios where mainly ISCOM structures are present after the complete preparation process rather than the more unlikely case of obtaining a sample with only ISCOM structures present. It should be emphasized that the theoretical ratios used may not represent the final ratios in the ISCOMs. As an example, the preparation of ISCOMs and Posintro by the dialysis method (Höglund et al. 1989) was done with an initial weight ratio of 5:1:1 for Quil A, POPC, and cholesterol and when exchanging some of the cholesterol with DC-cholesterol, quantitative lipid analysis showed that the relative amount of DC-cholesterol compared to the other lipids in the resulting particles was much higher than theoretically expected, due to loss during the dialysis process (Madsen, et al. 2010). In a study by Behboudi et al. (1995), it was also shown that for five different types of ISCOMs, the measured amount of lipids, especially the phospholipid, was much lower than the theoretical value, whereas the amount of saponin in most cases was close to the expected value.

Although the colloidal stability of some ISCOM dispersions has been reported to be longer than a year, the colloidal stability is critical to consider when handling ISCOM-based vaccines. Especially as it is a self-assembled particle in an equilibrium state, the colloidal stability of the dispersion will depend on the surrounding conditions, e.g., the storage medium and temperature, and changes to this as for example mixing and diluting with an antigen solution prior to use.

8.7.3 *Structural Characterization*

As the sample stability is dependent on, e.g., the ISCOM concentration and the ionic strength of the dispersion medium, care must be taken when preparing samples for analysis and when interpreting results obtained on for example diluted samples. Dilution of samples may be needed for proper determination of the size by using dynamic laser scattering (DLS) or of the zeta potential by using laser Doppler electrophoresis, which is often measured on diluted samples in low-ionic strength buffer. However, the size and shape characteristics may change upon storage or dilution, as results of *in vitro* studies have demonstrated that with higher dilution, the structure changes from ISCOM structures to more liposomal-like structures indicating a diffusion of the saponin out of the ISCOM (Lendemans et al. 2006). As an expression of negative surface charge density, the mobility of ISCOM particles has been monitored by titration with a cationic polymer, which clearly showed the change in surface charge density when incorporating the positively charged DC-cholesterol (Madsen et al. 2010).

8.7.4 *Size and Structure*

For confirmation of size and structural properties, transmission electron microscopy (TEM) or cryo-TEM has traditionally been applied elucidating ISCOMs as spherical particles in the size range of 40–60 nm and composed of ring-like subunits and a hollow centre (Özel et al. 1989; Kersten et al. 1991). The pores of around 7–10 nm in the structure (Kersten et al. 1988; Özel et al. 1989) has also been indicated by freeze fracture electron microscopy.

Recently, SAXS was applied on an undiluted ISCOM sample prepared by dialysis with a resulting size of 43.9 ± 0.2 nm and a polydispersity index of 0.14, measured by DLS, indicating a rather narrow and homogeneous size distribution of the ISCOM particles. Based on Monte Carlo simulation integrations, a novel modelling method was developed and implemented in order to describe the obtained SAXS data. The sample clearly showed a more polydisperse distribution with three types of perforated bilayer vesicles; namely icosahedral (29 nm), football (49 nm), and tennis ball (38 nm) structures. The predominant species that was named the tennis ball structure, accounted for 76–79 % of the ISCOMs in the dispersion by number and mass fractions, respectively. Modeling of these ISCOMs showed 20 pores per tennis ball of a diameter of 5–6 nm and a lipid bilayer membrane thickness of 4.6 nm (Pedersen et al. 2012), corresponding to the general perception of the ISCOM pore size. The structures determined by SAXS were very similar to structures observed in cryo-TEM images on the same batch (Fig. 8.4) strongly indicating that SAXS may be used to model ISCOM structures.

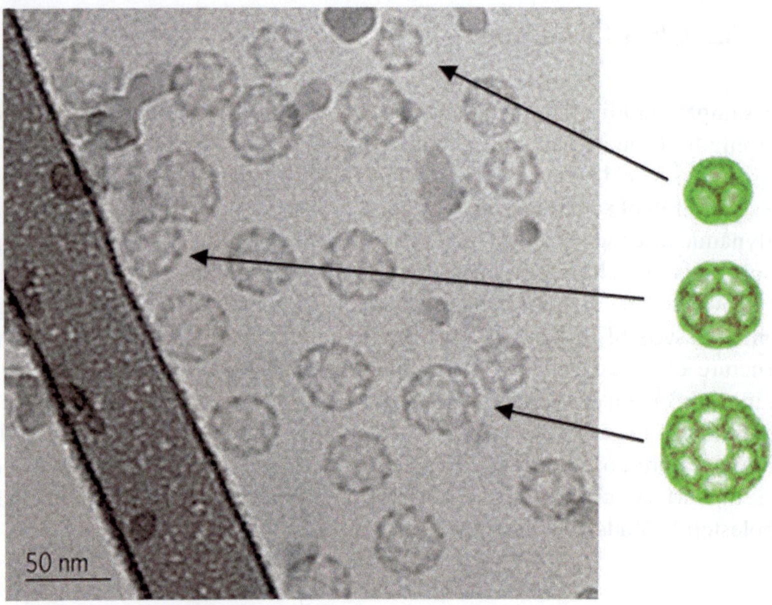

Fig. 8.4 Cryo-TEM image of dispersed ISCOMs (scale bar 50 nm) and SAXS-derived suggested structures indicating three different populations of ISCOMs in the sample (not to scale). *Reprinted from Pedersen* et al. (2012) *with permission from Elsevier*

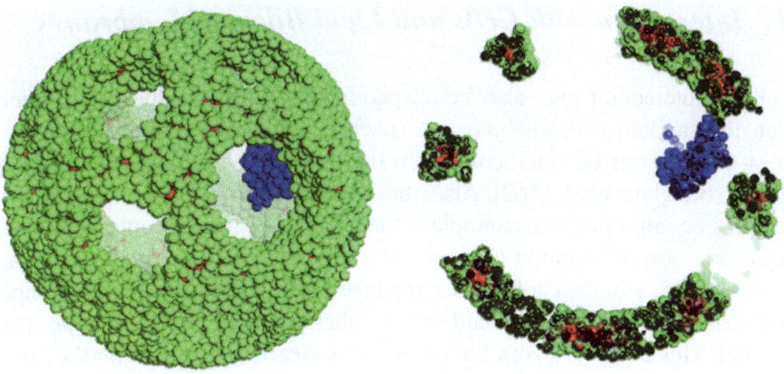

Fig. 8.5 Localization of tetanus toxoid on the predominant ISCOM species (38 nm, pore size 5–6 nm) derived from SAXS analysis. Surface representation (left) and cross-section (right) *Reprinted from Pedersen* et al. (2012) *with permission from Elsevier*

8.7.5 *Association of Antigen*

Co-administration of antigens to preformed ISCOM-based adjuvants likely broadens the use of the adjuvants as compared to ISCOM-based vaccines with antigens incorporated or chemically bound to the surface. In both cases, however, the stability and the localization of the antigen are considered key for the efficacy of the vaccine (Brito et al. 2013). How and to which extent the antigens may incorporate into the ISCOMs is speculated to depend strongly on the properties of both components. Only rarely, the ISCOMs with associated antigens are distinguishable from the ISCOMs without the antigen present, as visualized by electron microscopy (Barr and Mitchell 1996), which indicates that only a low number of antigen molecules may (partly) be incorporated in the adjuvant system (Hook and Rades 2013). This hypothesis was confirmed based on modeling of data obtained by using SAXS, as it was evident that only one molecule on average of the tetanus toxoid antigen monomer associated to one ISCOM structure with a size of 38 nm, which did not lead to a detectable change in the size as measured by using DLS. Further, and surprisingly, it was indicated that the tetanus toxoid was located just below the membrane inside the particles (Fig. 8.5). Thus, scattering may provide a useful tool to predict the further information on the interaction of specific antigens with ISCOMs.

8.8 Administration and Mechanisms of Action

The exact mode of action for ISCOMs to induce an immune response is not completely understood, yet at the cellular level, the endocytotic uptake in APCs is stimulated by the particulate nature of the adjuvant/drug delivery system (Kersten and Crommelin 2003), which is likely to be dependent on the interaction with the plasma membrane of the cell.

8.8.1 Interaction with Cells and Lipid Bilayer Membranes

The cellular interaction may also be unspecifically increased due to the interaction between the saponin carbohydrate and specific receptors on dendritic cells (DCs) (Jiang et al. 1995) or by interaction with the cholesterol component of the plasma membrane (Bangham et al. 1962). Also, the overall charge and thus composition of the particle may be important. An example is the clear increase in the interaction between net negatively charged stratum corneum-like liposomes and the Posintro, which has (theoretically) 25 % of the cholesterol exchanged with DC-cholesterol as compared to ISCOMs without DC-cholesterol and thus a higher degree of cationic charges (Madsen et al. 2010). This entropy-driven interaction was clearly dependent on the content of DC-cholesterol in the ISCOMs and resulted in interference with the lipid bilayer. In addition, the uptake specificity and kinetics may be modulated by formulation design to target, e.g., B-cells, by incorporating specific receptor ligands (Lycke 2004; Helgeby et al. 2006). The uptake and resulting specific cytokine responses may be dependent not only on the incorporated or co-administered antigen, but also on the adjuvant/carrier composition as well as the route and mode of administration.

8.8.2 Injection of ISCOM-Based Vaccines

The structure of the nanoparticles will inevitably be affected by the administration due to dilution or interaction with the surrounding biological matrix. Upon either subcutaneous (s.c.) or intramuscular (i.m.) injection, the ISCOMs may form a depot at the injection site and by this attract APCs although the results of some studies claim that the particles quickly disappear from the site of injection (Pearse and Drane 2005; Morein and Bengtsson 1998). Given the charged properties of ISCOMs, it is likely that some aggregation occurs as reported (Henriksen-Lacey et al. 2010) for the somewhat larger liposome adjuvant CAF01 currently in clinical trials, followed by disintegration and diffusion of individual components from the depot over time. The kinetics of the depot formation and disassembly will depend on the specific formulation and site of injection. However, it has been demonstrated that the size of nanoparticles in a range from 25 to 100 nm is a prerequisite for their ability to be transported via the lymphatic capillaries to the draining lymph nodes after injection, thus targeting lymph node-residing DCs (Reddy et al. 2007). This corresponds with reports that the DCs in the lymphoid organs and the spleen have been shown to be a target after s.c. or intraperitoneal injection of ISCOMs in mice (Sjölander et al. 1996, 1997).

Recent studies have shown promising results using ISCOMs as adjuvants administered by injection; one being a phase 1 clinical study in healthy adults demonstrating efficacy of influenza vaccination (Fries et al. 2013). Further, chickens were efficiently vaccinated by i.m. administration resulting in increased levels of antigen-specific intestinal IgA and CD4 and CD8 positive intestinal intraepithelial T-lymphocytes after a subsequent oral challenge with the antigen (Zhang et al. 2014).

8.8.3 *Non-injectable Administration of ISCOM-Based Vaccines*

Non-injectable administration of vaccines constitutes a more patient friendly, more convenient, and potentially also safer alternative to i.m. and s.c. vaccination strategies. At the same time it provides the possibility for induction of a local immune response at the site of dosing, e.g., to obtain a higher mucosal IgA response after dosing to mucosal sites. Mucosal administration of nanoparticle vaccines includes dosing primarily via the airway and oral routes, which are also the primary sites of infection.

Oral administration and single-dose vaccines have long been desired, yet major challenges remain to formulate a vaccine that is effectively delivered to the target, the gut-associated lymphoid tissue (GALT) within a time frame ensuring sufficient colloidal stability of the drug delivery system and also the appropriate chemical stability of the antigen or subunit antigen in the harsh environment of the gastrointestinal tract. Although the oral mucosa in general is considered to be relatively immune tolerant rather than mediating immune responses (Scheerlinck and Greenwood 2008) it is intriguing to aim for an oral vaccine with ISCOMs and indeed some are tested after oral administration (Gregory et al. 2013; Mowat et al. 1999).

Immunization via the airways may be achieved via the nose- or bronchial-associated lymphoid tissue (NALT and BALT, respectively) and ISCOMs are also a realistic option to be applied for this route of vaccination. Administration via the nose was recently demonstrated to be effective in boosting an existing immunity in draining nasal lymph nodes, whereas pulmonary administration induced strong immune responses in both the lung lavage as well as in the blood (Vujanic et al. 2012). Pulmonary administration of the ISCOMATRIX™ was also shown effective for influenza vaccination (Vujanic et al. 2010). Also, for vaccination against respiratory syncytial virus, the particle size of the nanoparticles was found to significantly influence the immune response (Mottram et al. 2007), which should attract attention for the development of future vaccines.

Transcutaneous immunization by cutaneous application of ISCOM-based vaccines has also been investigated (Combadiere and Mahe 2008). This is mediated by the fact that the strongly immune competent Langerhans cells (LC) are present in high numbers in the epidermis and thus covering a large area underneath the skin surface (Huang 2007). Upon stimulation, these LCs, and activated DCs residing in the dermis, migrate to the lymph nodes resulting in cellular immune responses and antibody production resulting in also mucosal immunity (Frech et al. 2008). Penetration of the adjuvant and an antigen through the outermost layer of the skin, the stratum corneum, constitutes a delivery challenge and considerable efforts are put into creation of novel devices and strategies for expanding the repertoire of skin-breaching modalities, such as the use of microneedles (Bal et al. 2010). Also, the development of novel adjuvants suitable for transcutaneous immunization is a focus area, and it was demonstrated that the application of Posintro particles to human skin *in vitro* significantly enhanced the penetration of an incorporated dye into stratum corneum, and that the application of ISCOMs using a hydrogel patch resulted in ultrastructural changes in the human stratum corneum (Madsen et al. 2009). Previously, indications that the hair follicles may be a route of entry for ISCOM-based vaccines were given when fluorophore-labeled Posintro particles were observed to localize in the hair follicles of mouse skin after cutaneous application *in vivo* (Madsen 2010). Especially, since the appearance of LC protrusions is pronounced close to the hair follicles, this finding is valuable.

8.9 Summary and Perspectives

Since the first description of the potential of ISCOMs as adjuvants by Morein et al. (1984), the technology has matured significantly, and several vaccines with this adjuvant have been tested in clinical trials (Hook and Rades 2013). Extensive modification of the basic technology was first made to mitigate the toxicity due to the presence of the saponin, which included use of purified fractions of the crude mixture of saponins, a different Quil A-to-lipid ratio (Brito et al. 2013) as well as synthetic, more specific immunopotentiators prepared from chemical modifications of the saponin skeleton. Recent mechanistic insight into the influence of the glycol moieties on the adjuvant is opening new perspectives for improved design of carbohydrate-based vaccines (Berti and Adamo 2013) and by use of the progressing biosynthetic technologies, novel glycoconjugates may be pursued for use in future ISCOM-based vaccines.

Research is ongoing with regards to optimizing the formulation design of the ISCOM-based adjuvants and vaccines, especially on the control of the type and specificity, as well as efficacy of the immune response generated, which seem to depend on both the properties of the nanoparticles, the administration route, the dose and dosing regimen as well as the antigen used. The stability and localization of the co-delivered antigens are key factors for the concept to be a success, and since the performance of ISCOM-based vaccines appears to be partially dependent on antigen association (Brito et al. 2013) various approaches to improve the association and binding of the antigens to the particles have been described. The successful use of for example Matrix-M as an adjuvant simply admixed with the antigen, poses the question as to what extent incorporation of the antigen into the ISCOM structure is really needed (Bengtsson et al. 2011). The concept of using preformed ISCOM adjuvants to which the antigen of choice is added prior to use seems therefore promising both from a manufacturing perspective as well as with regards to broader application ranges. However, also in this case it is of importance to investigate the antigen-to-particle association behavior and the importance of this to the efficacy of the vaccine. Many open or only partially answered questions still remain in order to fully understand and develop ISCOM adjuvants and vaccines further: The molecular level mechanism of ISCOM adjuvanticity is still not well understood, nor is the fate of ISCOMS after administration. Also the question of what importance the different colloidal structures, which are found when slightly modifying the component ratios in ISCOM formulations, are for adjuvanticity or immunogenicity needs to be further investigated. The question regarding in which cases co-localization of the antigen and the ISCOM is advantageous and in which cases co-administration is sufficient should be investigated further. Finally, alternative application routes need to be explored and exploited in more detail. These are only a few of the remaining challenges that inspire current application of ISCOM as adjuvants and require future research. Despite these open questions, as ISCOM-based vaccines are generally well tolerated and only inducing minor local side effects upon injection, they are likely to be a part of the future adjuvants and vaccines also for, e.g., cancer vaccines and for both human and veterinary use.

References

Alving CR, Peachman KK, Rao M, Reed SG (2012) Adjuvants for human vaccines. Curr Opin Immunol 24:310–315

Andersson C, Sandberg L, Wernérus H, Johansson M, Lövgren-Bengtsson K, Ståhl S (2000) Improved systems for hydrophobic tagging of recombinant immunogens for efficient iscom preparation. J Immunol Methods 238(1–2):181–193

Bal SM, Ding Z, van Riet E, Jiskoot W, Bouwstra JA (2010) Advances in transcutaneous vaccine delivery: do all ways lead to Rome? J Control Release 148:266–282

Bangham AD, Horne RW, Glauert AM, Dinglejt LJA (1962) Action of saponin on biological cell membranes. Nature 196:952–955

Barr IG, Mitchell GF (1996) ISCOMs (immunostimulating complexes): the first decade. Immunol Cell Biol 74:8–25

Behboudi S, Morein B, Rönnberg B (1995) Isolation and quantification of Quillaja saponaria Molina saponins and lipids in iscom-matrix and iscoms. Vaccine 13:1690–1696

Bengtsson KL, Morein B, Osterhaus ADME (2011) ISCOM technology-based Matrix M™ adjuvant: success in future vaccines relies on formulation. Expert Rev Vaccines 10:401–403

Berti F, Adamo R (2013) Recent mechanistic insights on glycoconjugate vaccines and future perspectives. ACS Chem Biol 8:1653–1663

Brito LA, Malyala P, O'Hagan DT (2013) Vaccine adjuvant formulations: a pharmaceutical perspective. Semin Immunol 25:130–145

Chea EK, Fernández-Tejada A, Damani P, Adams MM, Gardner JR, Livingston PO, Ragupathi G, Gin DY (2012) Synthesis and preclinical evaluation of QS-21 variants leading to simplified vaccine adjuvants and mechanistic probes. J Am Chem Soc 134:13448–13457

Combadiere B, Mahe B (2008) Particle-based vaccines for transcutaneous vaccination. Comp Immunol Microbiol Infect Dis 31:293–315

Cox JC, Sjölander A, Barr IG (1998) ISCOMs and other saponin based adjuvants. Adv Drug Deliv Rev 32(3):247–271

Cruz-Bustos T, González-González G, Morales-Sanfrutos J, Megía-Frenández A, Santoyo-González F, Osuna A (2012) Functionalization of immunostimulating complexes (ISCOMs) with lipid vinyl sulfones and their application in immunological techniques and therapy. Int J Nanomedicine 7:5941–5956

Dalsgard K (1974) Saponin adjuvants. 3. Isolation of a substance form Quillaja saponaria Molina with adjuvant activity in food-and-mouth disease vaccines. Arch Gesamte Virsuforschung 44:243–254

DiStefano D, Antonello JM, Bett AJ, Medi MB, Casimiro D, ter Meulen J (2013) Immunogenicity of a reduced-dose whole killed rabies vaccine is significantly enhanced by ISCOMATRIX™ adjuvant, Merck amorphous aluminium hydroxylphosphate sulfate (MAA) or a synthetic TLR9 agonist in rhesus macaques. Vaccine 31:4888–4893

Fossum C, Hjertner B, Ahlberg V, Charerntantanakul W, McIntosh K, Fuxler L, Balagunaseelan N, Wallgren P, Bengtsson KL (2014) Early inflammatory response to the saponin adjuvant Matrix-M in the pig. Vet Immunol Immunopathol 158:53–61

Frech SA, Dupont HL, Bourgeois AL, McKenzie R, Belkind-Gerson J, Figueroa JF, Okhuysen PC, Guerrero NH, Martinez-Sandoval FG, Meléndez-Romero JH, Jiang ZD, Halpern J, Torres OR, Hoffman AS, Villar CP, Kassem RN, Flyer DC, Andersen BH, Kazempour K, Breisch SA, Glenn GM (2008) Use of a patch containing heat-labile toxin from *Escherichia coli* against travellers' diarrhea: a phase II, randomized, double-blind, placebo-controlled field trial. Lancet 371:2019–2025

Fries LF, Smith GE, Glenn GM (2013) A recombinant viruslike particle influenza A (H7N9) vaccine. N Engl J Med 369:2564–2566

Gregory AE, Titball R, Williamson D (2013) Vaccine delivery using nanoparticles. Front Cell Infect Microbiol 3:1–13

Helgeby A, Robson NC, Donachie AM, Beacock-Sharp H, Lövgren K, Schön K, Mowat A, Lycke NY (2006) The combined CTA1-DD/ISCOM adjuvant vector promotes priming of mucosal and systemic immunity to incorporated antigens by specific targeting of B-cells. J Immunol 15:3697–3706

Henriksen-Lacey M, Bramwell VW, Agger EM, Andersen P, Perrie Y (2010) Liposomes based on dimethyldioctadecylammonium promote a depot effect and enhance immunogenicity of soluble antigen. J Control Release 142:180–186

Höglund S, Dalsgaard K, Lovgren K, Sundquist B, Osterhaus A, Morein B (1989) ISCOMs and immunostimulation with viral antigens. Subcell Biochem 15:39–68

Hook S, Rades T (2013) Immune stimulating complexes (ISCOMs) and Quil-A containing particulate formulations as vaccine delivery systems. In: Flower DR, Perrie Y (eds) Immunomic discovery of adjuvants and candidate subunit vaccines. Immunomics reviews, vol 5. Springer, New York, pp 233–261. doi:10.1007/978-1-4614-5070-2_12

Huang CM (2007) Topical vaccination: the skin as a unique portal to adaptive immune responses. Semin Immunopathol 29:71–80

Jiang W, Swiggard WJ, Heufler C, Peng M, Mirza A, Steinman RM, Nussenzweig MC (1995) The receptor DEC-205 expressed by dendritic cells and thymic epithelial cells is involved in antigen processing. Nature 375:151–155

Kamstrup S, San MR, Doberti A, Grande H, Dalsgaard K (2000) Preparation and characterisation of quillaja saponin with less heterogeneity than Quil-A. Vaccine 18:2244–2249

Kersten GF, Crommelin DJ (1995) Liposomes and ISCOMs as vaccine formulations. Biochim Biophys Acta 1241:117–138

Kersten GF, Crommelin DJ (2003) Liposomes and ISCOMs. Vaccine 21:915–920

Kersten GF, Teerlink T, Derks HJ, Verkleij AJ, van Wezel TL, Crommelin DJ, Beuvery EC (1988) Incorporation of the major outer membrane protein of Neisseria gonorrhoeae in saponin-lipid complexes (iscoms): chemical analysis, some structural features, and comparison of their immunogenicity with three other antigen delivery systems. Infect Immun 56:432–438

Kersten GF, Spiekstra A, Beuvery EC, Crommelin DJ (1991) On the structure of immune-stimulating saponin-lipid complexes (iscoms). Biochim Biophys Acta 1062:165–171

Kirkby N, Samuelsen P (2006) Composition for vaccination. US 2006/0147509 A1, 6 July 2006

Könnings C, Copland MJ, Davies NM, Rades T (2002) A method for the incorporation of ovalbumin into immune stimulating complexes prepared by the hydration method. Int J Pharm 241:385–389

Lendemans DG, Myschik J, Hook S, Rades T (2005) Cationic cage-like complexes formed by DC-cholesterol, Quil-A, phospholipid. J Pharm Sci 94:1794–1807

Lendemans DG, Egert AM, Myschik J, Hook S, Rades, T (2006) On the dilution behaviour of immuno-stimulating complexes (ISCOMs). Pharmazie 61:689–695

Lendemans DG, Egert AM, Hook S, Rades T (2007) Cage-like complexes formed by DOTAP, Quil-A and cholesterol. Int J Pharm 332:192–195

Lövgren K, Morein B (1988) The requirement of lipids for the formation of immunostimulating complexes (iscoms). Biotechnol Appl Biochem 10:161–172

Lycke N (2004) From toxin to adjuvant: the rational design of a vaccine adjuvant vector, CTA1-DD/ISCOM. Cell Microbiol 6:23–32

Madsen HB (2010) Penetration and interaction of ISCOM nanoparticles with the skin. PhD thesis, University of Copenhagen. ISBN 978-87-92199-87-4

Madsen HB, Ifversen P, Madsen F, Brodin B, Hausser I, Nielsen HM (2009) In vitro cutaneous application of ISCOMs on human skin enhances delivery of hydrophobic model compounds through the stratum corneum. AAPS J 4:728–739

Madsen HB, Arboe-Andersen HM, Rozlosnik N, Madsen F, Ifversen P, Kasimova MR, Nielsen HM (2010) Investigation of the interaction between modified ISCOMs and stratum corneum lipid model systems. Biochim Biophys Acta 1798:1779–1789

McBurney WT, Lendemans DG, Myschik J, Hennessy T, Rades T, Hook S (2008) *In vivo* activity of cationic immune stimulating complexes (PLUSCOMs). Vaccine 26:4549–4556

Morein B, Bengtsson KL (1998) Functional aspects of iscoms. Immunol Cell Biol 76:295–299

Morein B, Sundquist B, Höglund S, Dalsgaard K, Osterhaus A (1984) ISCOM, a novel structure for antigenic presentation of membrane proteins from enveloped viruses. Nature 308:457–460

Morein B, Ekström J, Lövgren K (1990) Increased immunogenicity of a non-amphipathic protein (BSA) after infusion into iscoms. J Immunol Methods 128:177–181

Morelli AB, Becher D, Koernig S, Silva A, Drane D, Maraskovsky E (2012) ISCOMATRIX: a novel adjuvant for use in prophylactic and therapeutic vaccines against infectious diseases. J Med Microbiol 61:935–943

Mottram PL, Leong D, Crimeen-Irwin B, Gloster S, Xiang SD, Meanger J, Ghildyal R, Vardaxis N, Plebanski M (2007) Type 1 and 2 immunity following vaccination is influenced by nanoparticle size: formulation of a model vaccine for respiratory syncytial virus. Mol Pharm 4:73–84

Mowat AM, Smith RE, Donachie AM, Furrie E, Grdic D, Lycke N (1999) Oral vaccination with immune stimulating complexes. Immunol Lett 65:133–140

Myschik J, Lendemans DG, McBurney WT, Demana PH, Hook S, Rades T (2006) On the preparation, microscopic investigation and application of ISCOMs. Micron 37:724–734

Nordly P, Madsen HB, Nielsen HM, Foged C (2009) Status and future prospects of lipid-based particulate delivery systems as vaccine adjuvants and their combination with immunostimulators. Exp Opin Drug Deliv 6:1–16

Özel M, Hoglund S, Gelderblom HR, Morein B (1989) Quaternary structure of the immunostimulating complex (iscom). J Ultrastruct Mol Struct Res 102:240–248

Pearse MJ, Drane D (2005) ISCOMATRIX adjuvant for antigen delivery. Adv Drug Deliv Rev 57:465–474

Pedersen JS, Oliveira CLP, Hübschmann HB, Arleth L, Manniche S, Kirkby N, Nielsen HM (2012) Structure of ISCOM matrices and ISCOMs in suspension determined by small-angle X-ray scattering. Biophys J 102:2372–2380

Pham HL, Ross BP, McGeary RP, Shaw PN, Hewavitharana AK, Davies NM (2006) Saponins from Qillaja saponina Molina: isolation, characterization and ability to form immune stimulatory complexes (ISCOMs). Curr Drug Deliv 3:389–397

Reddy ST, van der Vlies AJ, Simeoni E, Angeli V, Randolph GJ, O'Neil CP, Lee LK, Swartz MA, Hubbell JA (2007) Exploiting lymphatic transport and complement activation in nanoparticle vaccines. Nat Biotechnol 25:1159–1164

Scheerlinck J-PY, Greenwood DLV (2008) Virus-sized vaccine delivery systems. Drug Discov Today 13:882–887

Sjölander A, Bengtsson KL, Johansson M, Morein B (1996) Kinetics, localization and isotype profiling of antibody responses to immune stimulating complexes (iscoms) containing human influenza virus envelope glycoproteins. Scand J Immunol 43:164–172

Sjölander A, Bengtsson KL, Morein B (1997) Kinetics, localization and cytokine profile of T-cell responses to immune stimulating complexes (iscoms) containing human influenza virus envelope glycoproteins. Vaccine 15:1030–1038

Sjölander A, Cox JC, Barr IG (1998) ISCOMs: an adjuvant with multiple functions. J Leukoc Biol 64:713–723

Soltysik S, Wu J-Y, Recchia J, Wheeler DA, Newman MJ, Coughlin RT, Kensil CR (1995) Structure/function studies of QS-21 adjuvant: assessment of triterpene aldehyde and glucuronic acid roles in adjuvant function. Vaccine 13:1403–1410

Sun H-X, Xie Y, Ye Y-P (2009) ISCOMs and ISCOMATRIX™. Vaccine 27:4388–4401

Vujanic A, Wee JLK, Snibson KJ, Edwards S, Pearse M, Quinn C, Moloney M, Taylor S, Scheerlinck J-PY, Sutton P (2010) Combined mucosal and systemic immunity following pulmonary delivery of ISCOMATRIX™ adjuvanted recombinant antigens. Vaccine 28: 2593–2597

Vujanic A, Sutton P, Snibson KL, Yen H-H, Scheerlinck J-PY (2012) Mucosal vaccination: lung versus nose. Vet Immunol Immunopathol 148:172–177

Wang P, Dai Q, Thogaripally P, Zhang P, Michalek SM (2013) Synthesis of QS-21 based immunoadjuvants. J Org Chem 78:11525–11534

Zhang Y, Wang L, Ruan W, Zhang J, Yao P, Zhou S, An J (2014) Immunization with recombinant 3-1E protein in AbISCO®-300 adjuvant induced protective immunity against Eimeria acervulina infection in chickens. Exp Parasitol 141:75–81

Chapter 9
Virus-Like Particles, a Versatile Subunit Vaccine Platform

Braeden Donaldson, Farah Al-Barwani, Vivienne Young, Sarah Scullion, Vernon Ward, and Sarah Young

9.1 Introduction

Novel vaccine development requires a balance between eliciting a potent immune response while limiting the unintentional induction of hypersensitivity and off-target effects. Virus-like particles (VLPs) are a form of subunit vaccine consisting of self-assembling shells derived from virus capsid proteins. Due to the absence of viral genomic material, VLPs are rendered non-replicative and non-infectious, enhancing their safety profile. In comparison to other subunit vaccines, the resemblance of VLPs to their corresponding native virus provides enhanced immunogenicity and specificity. VLP capsid proteins retain their natural structural conformation, harbouring undamaged antigenic motifs in a more immunologically relevant state than an inactivated virus vaccine. VLPs can also resemble a live attenuated virus without replicative or infectious capacity due to structural similarity and utilisation of similar processing pathways. In general, VLPs are considered significantly safer than many other virally derived vaccines by avoiding potential hazards such as attenuated virus reversion or incomplete inactivation.

Author contributed equally with all other contributors

B. Donaldson • F. Al-Barwani
Department of Pathology, School of Medicine, University of Otago, Dunedin, New Zealand

Department of Microbiology and Immunology, School of Medical Sciences, University of Otago, Dunedin, New Zealand

V. Young • S. Scullion • V. Ward
Department of Microbiology and Immunology, School of Medical Sciences, University of Otago, Dunedin, New Zealand

S. Young (✉)
Department of Pathology, Dunedin School of Medicine, University of Otago, PO Box 56, Dunedin, New Zealand
e-mail: sarah.young@otago.ac.nz

C. Foged et al. (eds.), *Subunit Vaccine Delivery*, Advances in Delivery Science and Technology, DOI 10.1007/978-1-4939-1417-3_9

The first VLP identified and studied was isolated from patients infected with Hepatitis B virus (HBV) in 1968 (Bayer et al. 1968). Due to the ability of some viral capsid proteins to spontaneously form stable particles, many viruses produce VLPs as a natural by-product of their infection cycle. Engerix (GlaxoSmithKline) was the first human VLP-based vaccine, licensed in 1989 for vaccination against HBV. Since then a number of VLP vaccines have been approved for clinical use, including Epaxal (Crucell) for Hepatitis A virus (HAV), Recombivax (Merck), Hepavax (Crucell), and many others for HBV, Gardasil (Merck), and Cervarix (GlaxoSmithKline) for human papillomavirus (HPV), and Inflexal V (Crucell) for Influenza. Bolstered by the success of these vaccines many new VLPs are being developed, with a selection of examples summarised in Table 9.1. In addition, VLPs have also been produced from various non-human mammalian viruses, primarily for vaccination of livestock. Examples include porcine circovirus (Kim et al. 2002), bovine rotavirus (Rodriguez-Limas et al. 2011), chicken anaemia virus (Noteborn et al. 1998; Koch et al. 1995), SARS coronavirus (Liu et al. 2011), Nipah virus (Walpita et al. 2011), and swine vesicular stomatitis virus (Ko et al. 2005).

9.2 VLP Structural Conformation

Spontaneous polymerisation of a range of viral capsid proteins can yield VLPs with authentic geometric symmetry, usually icosahedral, spherical or rod-like in shape, depending on the source virus. VLPs can be generally categorised into groups based on their structural complexity, including single-protein non-enveloped (e.g. VLPs derived from caliciviruses (Jiang et al. 1992), papillomaviruses (Kirnbauer et al. 1992), and parvoviruses (Lopez de Turiso et al. 1992)), multi-protein non-enveloped (e.g. VLPs derived from infectious bursal disease virus (Kibenge et al. 1999), poliovirus (Brautigam et al. 1993), and reoviruses (French et al. 1990; French and Roy 1990)) and enveloped VLPs (e.g. VLPs derived from Hantaan virus (Betenbaugh et al. 1995), hepatitis C virus (Baumert et al. 1998), influenza A (Latham and Galarza 2001), and retroviruses (Yamshchikov et al. 1995)) as illustrated in Fig. 9.1. While single-protein VLPs have a relatively simple structure, multi-protein VLPs can contain unique structural features such as several distinct capsid layers. For example, expression of various combinations of the VP2, VP4, VP6, and VP7 capsid proteins of rotavirus can produce stable VLPs with double or even triple capsid layers (Crawford et al. 1994; Sabara et al. 1991).

Multi-protein VLPs can also be produced from variant copies of the same protein derived from different viral strains. These mosaic VLPs efficiently confer protection against several strains of the same virus (Buonamassa et al. 2002). An alternative means of increasing VLP versatility is through the incorporation of antigens from heterologous sources. Chimeric VLPs contain antigenic material from a target source supported by a stable VLP framework. These antigens can be inserted as peptides into the VLP capsid protein or substructural secondary VLP proteins, or covalently coupled to the surface of VLP. Chimeric VLPs have an extensive range of potential applications, and will be discussed later in this chapter. Enveloped

Table 9.1 Examples of VLPs licensed or in clinical trials

Vaccine target	Company/Institution	Administration (adjuvant)	Expression system	VLP platform	Antigen	References
Respiratory syncytial virus	Novavax	IM (aluminum phosphate)	Insect (Sf-9 cells)	RSV	RSV-F protein	Glenn et al. (2013)
Norwalk virus	LigoCyte Pharmaceuticals	IN (±MPL, chitosan, mannitol, sucrose)	Insect (Sf-9 cells)	NV	NV capsid protein	El-Kamary et al. (2010)
Hepatitus B (Engerix-B)	GSK	IM (aluminum hydroxide)	Yeast (*S. cerevisiae*)	HBV	HBV surface antigen	Keating and Noble (2003)
Human papilloma virus (Gardasil)	Merck	IM (aluminum hydroxyphosphate sulphate)	Yeast (*S. cerevisiae*)	HPV	HPV6/11/16/18L1	Romanowski (2011)
Human papilloma virus (Cervarix)	GSK	IM (aluminum hydroxide & MPL)	Insect (Hi Five cells)	HPV	HPV16/18 LI	Romanowski (2011)
Alzheimer's disease (CAD106)	Cytos Biotechnology/ Novartis	SC, IM (unspecified)	Bacteria (*E. coli*)	Qβ (carrier)	Ab 1-6	Lemere and Masliah (2010)
Human parvovirus B19 (VAI-VP705)	NHI/Meridian Life Science	IM (MF59)	Insect (Sf-9 cells)	B19	B19 VP1, VP2	Bernstein et al. (2011)
Human immunodeficiency virus	British Biotech Pharmaceuticals/ NAID	SC/IM, IM (aluminum hydroxide) Oral or rectal (none)	Yeast (*S. cerevisiae*)	Ty p1 (carrier)	HIV-1 Gag p17/p24	Peters et al. (1997)
Rabies	Thomas Jefferson University	Oral (none)	Plant (Tg spinach)	Alfalfa mosaic virus (carrier)	Rabies GP/NP	Yusibov et al. (2002)
Influenza	Novavax	IM (none)	Insect (Sf-9 cells)	Influenza virus	A/California/04/09 (H1N1) HA, NA	Lopez-Macias et al. (2011)
Influenza	Medicago	IM (none)	Plant (transient *N. benthamiana*)	Influenza virus	A/California/04/09 (H1N1) HA, NA	Medicago
Breast Cancer	Pevion Biotech	IM (none)	Cell-free	Influenza virosome (carrier)	Her2/neu	Wiedermann et al. (2010)
Hepatitus A (Eqaxal)	Crucell	IM (none)	Cell-free	Influenza virosome (carrier)	Inactivated Hepatitus A vaccine	Bovier (2008)

Adapted from Kushnir et al. (2012). *Abbreviations*: *GP* glycoprotein, *HA* haemagglutinin, *HBV*, hepatitis B virus, *HPV* human papillomavirus, *IM* intramuscular, *IN* intra-nasal, *NA* neuraminidase, *NP* nucleocapsid protein, *NV* Norwalk virus, *RSV* respiratory syncytial virus, *SC* sub-cutaneous

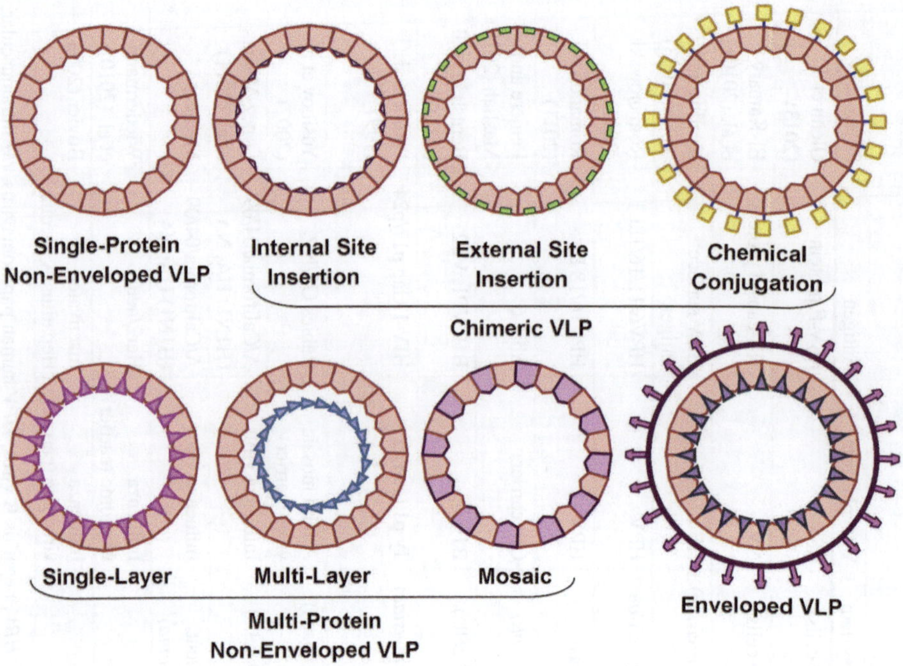

Fig. 9.1 VLP Structure. VLPs can be categorised based on characteristic structural features such as capsid protein composition, encapsulation inside a lipid bilayer envelope, and incorporation of antigens by recombinant insertion or chemical conjugation. Additional combinations other than those illustrated also exist, such as multi-protein chimeric VLPs and enveloped mosaic or chimeric VLPs

VLPs consist of either a single-protein or multi-protein VLPs encapsulated in a lipid bilayer captured from the cell membrane. Co-expression of haemagglutinin (HA), neuraminidase (NA), matrix protein M1, and ion channel protein M2 from influenza virus produces enveloped VLPs with the same size and morphology as native influenza virions, including the characteristic surface spikes HA and NA (Latham and Galarza 2001). The lipid bilayer of enveloped VLPs can also support the incorporation of transmembrane anchored proteins from multiple viral strains (enveloped mosaic VLPs) or even heterologous pathogens (enveloped chimeric VLPs) (Buonaguro et al. 2001; Halsey et al. 2008; Visciano et al. 2011). VLP structural complexity appears to have few limitations, with intriguing novel constructs still frequently theorised and investigated.

9.3 Production of VLPs

VLPs are a natural by-product produced during the infection cycle of certain viruses (Bayer et al. 1968). The same characteristics that benefit efficient virus reproduction, such as spontaneously polymerising capsid proteins, also promote the formation of VLPs; however, the isolation of VLPs produced from virally infected cells is not an efficient means of purification. An expansive range of protein expression systems have been developed for a variety of applications, and can be effectively commandeered for the production and purification of high quality VLPs. Recombinant expression of viral capsid proteins through tailored expression systems can also enable the production of VLPs from viruses not routinely cultured in laboratories. Common VLP expression systems include bacteria, yeast, insect cell lines, mammalian cell lines, plants, and cell-free cultures. Each expression system has its benefits and pitfalls as outlined in Table 9.2 (Rebeaud and Bachmann 2012). While most VLPs can be produced in multiple expression systems, the quaternary structural conformation of the capsid proteins produced can vary due to differences in post-translational modifications such as phosphorylation and glycosylation. This can have significant effects on the immunogenicity of VLPs, as these modifications are often essential for eliciting the desired immune response.

Escherichia coli (*E. coli*) has long been a primary laboratory workhorse bacterium, facilitating the expression and purification of recombinant proteins through plasmid transformation or bacteriophage vector delivery. Expression in *E. coli* is often preferred when producing small proteins with limited post-translational modifications; however, larger proteins with post-translational modifications require a more complex expression system (e.g. Chinese hamster ovary (CHO) mammalian cell line). The presence of endotoxins during downstream purification also presents a significant challenge for vaccine development from a bacterial expression system. Each VLP is unique, with optimal expression identified through trial and error by comparing the translated products of multiple expression systems. For example, Rabbit Haemorrhagic Disease Virus (RHDV) VLPs can be optimally produced by expressing the RHDV VP60 capsid protein in *Spodoptera frugiperda* (SF) cells using a recombinant baculovirus vector (Young et al. 2004, 2006; Peacey et al. 2007). Icosahedral T=3 VLPs with a diameter of around 40 nm spontaneously form when VP60 is expressed in SF21 cells. Each VLP contains 180 copies of the VP60 capsid protein, representing a relatively simple single-protein non-enveloped VLP. These VLPs structurally resemble native RHDV virions, as illustrated in Fig. 9.2 (Katpally et al. 2010; Wang et al. 2013).

Production of VLPs in transgenic plants (e.g. tobacco, potato, tomato) is a relatively new concept with interesting applications. Expression of recombinant proteins in plants is achieved through transgene insertion into the nuclear or plastid genome, or using plant viral vectors. While plant cells do not have a mammalian-like post-translational modification system, plant-specific glycosylation can have an immunostimulatory effect. Some examples of VLPs produced in a transgenic plant system include Norwalk virus (Tacket 2007; Tacket et al. 2000), HIV-1 (Scotti et al. 2009), and influenza virus VLPs (Medicago). Another recently developed expression platform is the cell-free system. This usually consists of extracts from *E. coli* or yeast cells, and was developed primarily to enable the production of viral capsid proteins which have toxic intermediate protein forms. Development of VLPs containing unnatural amino acids (UAAs) has also been achieved using the cell-free system. The non-replenishing nature of a cell-free system renders this method highly demanding with some scalability limitations. The influenza vaccine

Table 9.2 VLP expression systems

	Advantages	Limitations
Bacteria (e.g. *Escherichia coli*) 	• Rapid cell growth	• No post-translational modification
	• Highest yield	• Limited applications for mammalian VLPs
	• Low production cost	• May form inclusion bodies
	• Scalable	• Requires removal of endotoxins
Yeast (e.g. *Saccharomyces cerevisiae*) 	• Rapid cell growth	• Limited post-translational modification
	• High yield	• May form inclusion bodies
	• Low production cost	
	• Scalable	
	• Already has some regulatory approval	
Insect cells/Baculovirus (e.g. *Spodoptera frugiperda*) 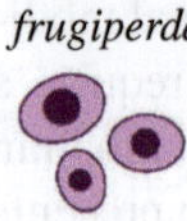	• Average cell growth	• Requires removal of baculovirus proteins
	• High yield	• May form inclusion bodies
	• Scalable	
	• Complex post-translational modification	
	• Formation of multi-protein VLP	
Plant cells (e.g. *Nicotiana* sp.) 	• Rapid production	• Limited post-translational modification
	• Low production cost	• Relatively new system
	• Scalable	
Mammalian cells (e.g. Chinese hamster ovary cells) 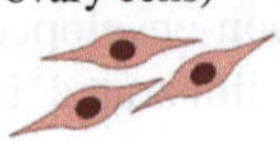	• Scalable	• Slow growth
	• Complex post-translational modification	• Low yield
	• Formation of multi-protein	• Demanding culture conditions
	• VLP	• High production cost
		• Potential infectious contamination
Cell free 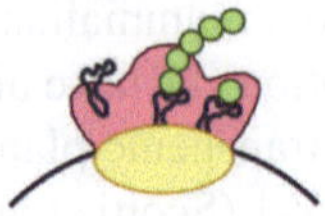	• Almost exclusive production of target protein	• Very high production cost
	• Limited cellular contaminants	• Limited scalability
	• Enables production of VLPs containing non-natural amino acids or toxic protein intermediates	• Relatively new system, not well characterised

Adapted from Rebeaud and Bachmann (2012)

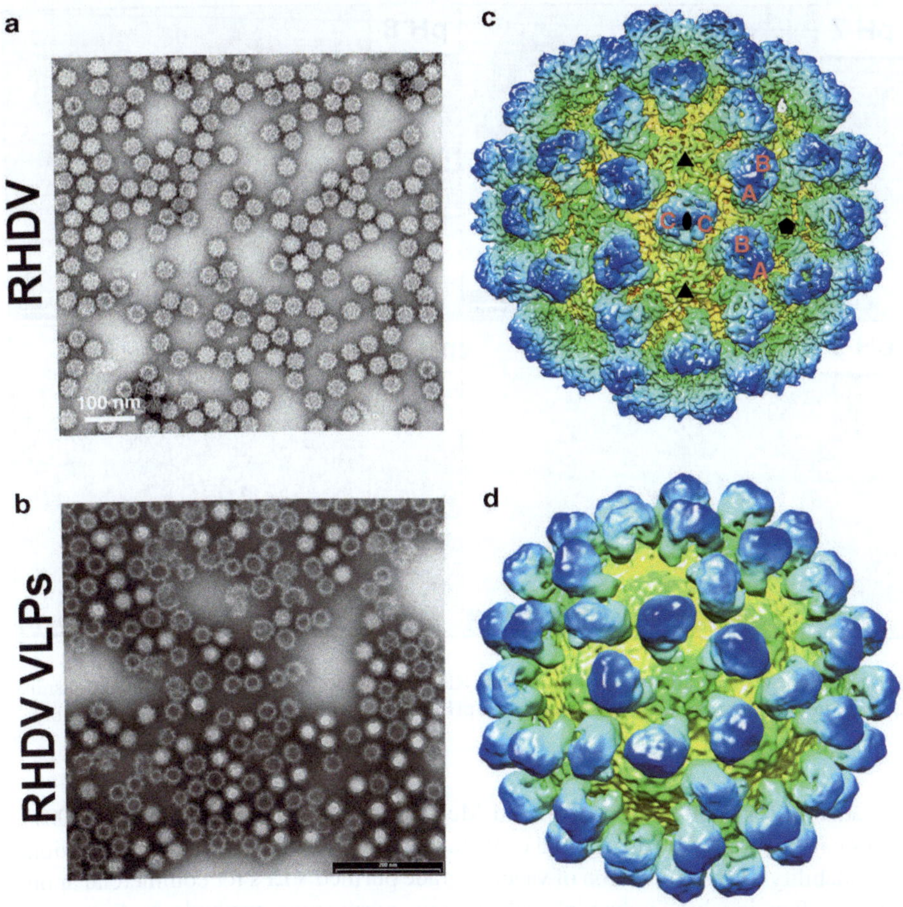

Fig. 9.2 Comparison of RHDV and RHDV VLP structure. RHDV VLPs expressed in insect cells visibly share structural characteristics with the native virus as viewed by transmission electron microscopy (**a**, **b**) and 3D modelling from cryo-electron microscopy and crystallography (**c**, **d**). (**a**, **c**) Adapted from Wang et al. (2013). (**d**) Generously supplied by Thomas J. Smith (Katpally et al. 2010)

Inflexal V (Herzog et al. 2009), and the hepatitis A vaccine Epaxal (Bovier 2008) (Crucell) are two commercialised VLP vaccines that consist of virosomes produced in a cell-free expression system.

Following production, VLPs must be isolated from the expression system and purified to sufficient quality for downstream applications. VLPs are usually isolated through a combination of cell lysis, removal of cellular debris, VLP concentration, and selective purification. Some mammalian and insect cell lines secrete VLPs into the supernatant, negating the necessity for cell lysis (Vicente et al. 2011). Resilient cells (e.g. bacteria, plant cells) may require more robust mechanical manipulation such as ultrasonication, compression, abrasion, repeated freeze/thawing, or enzymatic

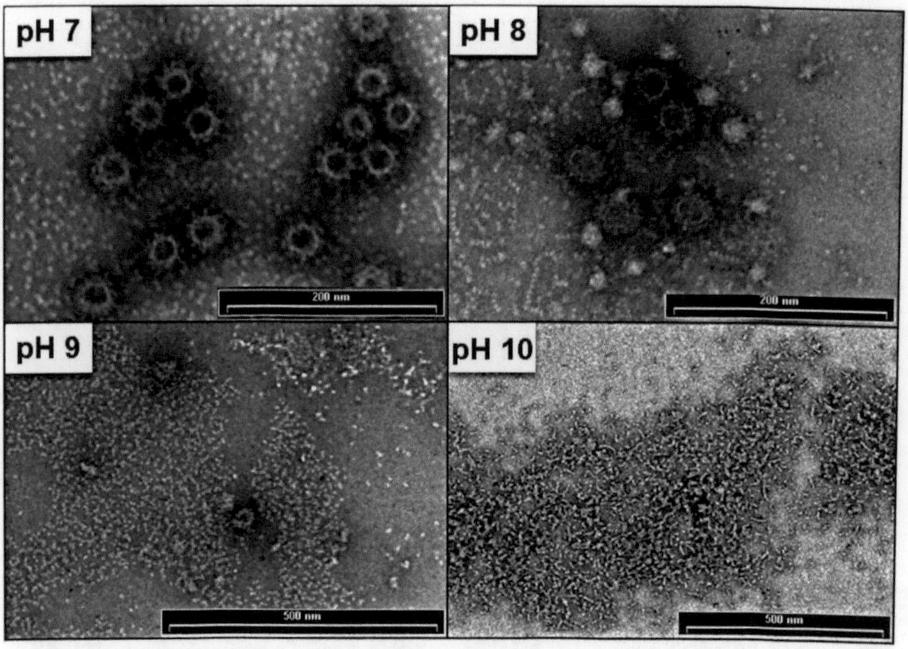

Fig. 9.3 RHDV VLP stability at an alkaline pH. RHDV VLPs are visibly perturbed under transmission electron microscopy in an elevated solvent pH, resulting in irreversible particle disassembly

degradation for VLP release (Cull and McHenry 1990; Salazar and Asenjo 2007). VLPs can be purified by differential centrifugation; however, this can pose limitations on scalability. GMP production of vaccine-grade purified VLPs for commercial applications often involves industrial-scale protein purification methods such as size-exclusion, ion exchange, or affinity chromatography columns (Vicente et al. 2011). Optimal solvent conditions must also be identified to maintain VLP stability. Solution pH is critical, as some VLPs irreversibly denature and disassemble beyond a specific pH range. For example, RHDV VLPs deteriorate in an alkaline environment, with complete VLP disruption above pH 9 (Fig. 9.3). VLP solubility is another important consideration as some VLPs may aggregate at higher concentrations, forming an insoluble precipitate.

9.4 VLPs as an Antigen Scaffold

VLPs are an incredibly versatile vaccination tool. In addition to harbouring multiple copies of immunologically recognisable antigens from their source virus, VLPs can also be used as a nanoparticulate delivery vector for heterogenous antigenic molecules. Each VLP can be considered a polymerised protein subunit vaccine, supporting modifications such as recombinant peptide insertion and chemical conjugation of peptides, proteins, lysate, carbohydrates, and lipoproteins to form chimeric VLPs. Some VLPs can support insertion of short peptide sequences at specific sites in their structural capsid proteins without impairing VLP formation. For example, RHDV VP60 is known to retain its ability to spontaneously form VLPs despite recombinant insertion of peptides at the N-terminus, C-terminus, or at amino acid residue 306 (Crisci et al. 2009); however, these sites have restrictions on inserted peptide length and residue sequence. The N-terminus of RHDV VP60 can support an insertion of <33 amino acids (Peacey et al. 2007), while other viral proteins such as polyomavirus VP1 can support insertions ranging from 9 to 120 amino acids (Eriksson et al. 2011; Lasickiene et al. 2012; Mazeike et al. 2012; Middelberg et al. 2011). Inclusion of secondary capsid proteins, such as polyomavirus VP2, with a limited contribution to VLP stability, can support large insertions including truncated proteins (Tegerstedt et al. 2005). HBV core antigen (HBcAg) has been one of the most popular VLP forming viral capsid proteins for recombinant insertion of immunogenic epitopes. Notable recombinant HBc vaccines include Malariavax, which targets *Plasmodium falciparum* (*P. falciparum*); the protozoan responsible for malaria (Gregson et al. 2008), and a pan-influenza A vaccine against the influenza matrix protein 2 (M2e) ectodomain universally conserved between strains (Fiers et al. 2009). The phase I clinical trial of the recombinant influenza A M2e vaccine ACAM-FLU-A reported a 90 % seroconversion rate amongst participants after two vaccinations, with no severe adverse side-effects observed (Fiers et al. 2009).

Recombinant non-mammalian VLPs can also be used to deliver immunogenic peptides derived from mammalian pathogens. Various plant viruses have demonstrated promise as recombinant antigen scaffolds, including alfalfa mosaic virus (AlMV) VLP containing epitopes from HIV-1 or rabies virus (Yusibov et al. 2002), cowpea mosaic virus (CPMV) VLP containing epitopes from *Bacillus anthracis* (Phelps et al. 2007) or HPV16 (Matic et al. 2011) and tobacco mosaic virus (TMV) VLP containing epitopes for *P. falciparum* (Turpen et al. 1995). The potential applications of chimeric VLPs extend far beyond the incorporation of epitopes from mammalian pathogens. VLPs containing auto-antigens naturally present in humans have been harnessed for a variety of novel roles, such as targeting angiotensin II to combat hypertension (Ambuhl et al. 2007; Maurer and Bachmann 2010), targeting Aβ protein to treat Alzheimer's disease (Zamora et al. 2006; Wiessner et al. 2011; Chackerian et al. 2006), or interrupting cytokine signalling pathways (Spohn et al. 2008; Link and Bachmann 2010). VLPs that stimulate an immune response against nicotine have even been investigated for their potential to break smoking addiction (Maurer et al. 2005; Cornuz et al. 2008). The therapeutic application of VLPs can also be used to develop novel vaccines for cancer. For example, recombinant insertion of a truncated form of Her2 protein into polyomavirus VP1/VP2 VLP confers protection against Her2 positive mammary carcinoma (Tegerstedt et al. 2005).

The selected site of recombinant insertion in chimeric VLP provides an inherent limitation on vaccine applications. Recognition of native unprocessed antigen is an essential component of B cell activation in the humoral immune system leading to antibody production. Internal site insertion prevents such interactions, and restricts these sites to insertion of antigenic epitopes that utilise intracellular processing pathways. External site insertion is required for effective B cell activation and depending on the selected VLP this method of chimeric VLP design can have significantly increased complexity. While internal site insertions are typically cloned onto an internal terminus of the VLP capsid protein, external site insertion requires in-depth understanding of the capsid protein quaternary structure and the interactions between each protein subunit to identify the optimal site for insertion. For example, the RHDV VP60 capsid protein is arranged with its N-terminus facing internally and C-terminus facing externally; however, the quaternary structural conformation has the C-terminus folded backwards into the central protein bulk. Amino acid 306 of RHDV VP60 capsid protein was instead identified as a potential external insertion site for exposure of immunogenic peptides tailored for B cell recognition (Crisci et al. 2009).

An alternative means of avoiding the limitations imposed by recombinant peptide insertion is to utilise VLPs as a particulate scaffold for chemical conjugation of antigenic molecules such as peptides, proteins, lysate, carbohydrates, and lipoproteins. Advancement in coupling chemistry continues to expand the list of viable conjugation candidates, primarily limited by their effects on the resulting particles size, solubility, and processing. Conjugation uses a chemical linker as a bridge, commonly conjugating proteins by acylation of amino groups, alkylation of sulfhydryl groups, or the activation of carboxylic acid residues. Some of the most commonly used protein coupling chemistries are illustrated in Fig. 9.4 (Smith et al. 2013). Azide-alkyne click chemistry (Patel and Swartz 2011) and biotin–streptavidin complexes (Chackerian et al. 2008) are also used in VLP conjugation. Heterobifunctional crosslinkers have two reactive groups enabling efficient protein crosslinking; however, this has the unfortunate side-reaction of non-specific conjugation, which can result in protein aggregation in a complex solution such as tumour lysate. Bioorthogonal crosslinkers are a specific alternative that enables selective conjugation, limiting side-reactions by targeting chemical motifs absent in biological systems (e.g. phosphines).

Chemical conjugation of immunogenic peptides is a relatively simple alternative to recombinant insertion, using commercial protein crosslinkers such as Sulfo-SMCC (Thermo Fisher Scientific Inc., Rockford, IL, USA). Sulfo-SMCC is a heterobifunctional crosslinker which permits a stepwise conjugation process targeting the amino groups on the VLP surface (NHS-ester) and the sulfhydryl group (maleimide group) from a cysteine residue introduced at the N-terminus of a target epitope (Peacey et al. 2008). Direct comparison between recombinant insertion and chemical conjugation of an $H2k^b$-restricted peptide from lymphocytic choreomeningitis virus (LCMV) gp33 in chimeric RHDV VLP indicated that the recombinant form induced a superior in vivo cytotoxic response (Li et al. 2013). Remarkably both the coupled and recombinant forms induced 20 % and 50 % tumour-free survival, respectively, following a single vaccination without adjuvant in mice grafted with Lewis' lung carcinoma tumours expressing gp33. The addition of a 4 week vaccination boost with VLP and adjuvant increased tumour-free survival to 70 % in mice vaccinated with the recombinant gp33 RHDV VLP (Li et al. 2013).

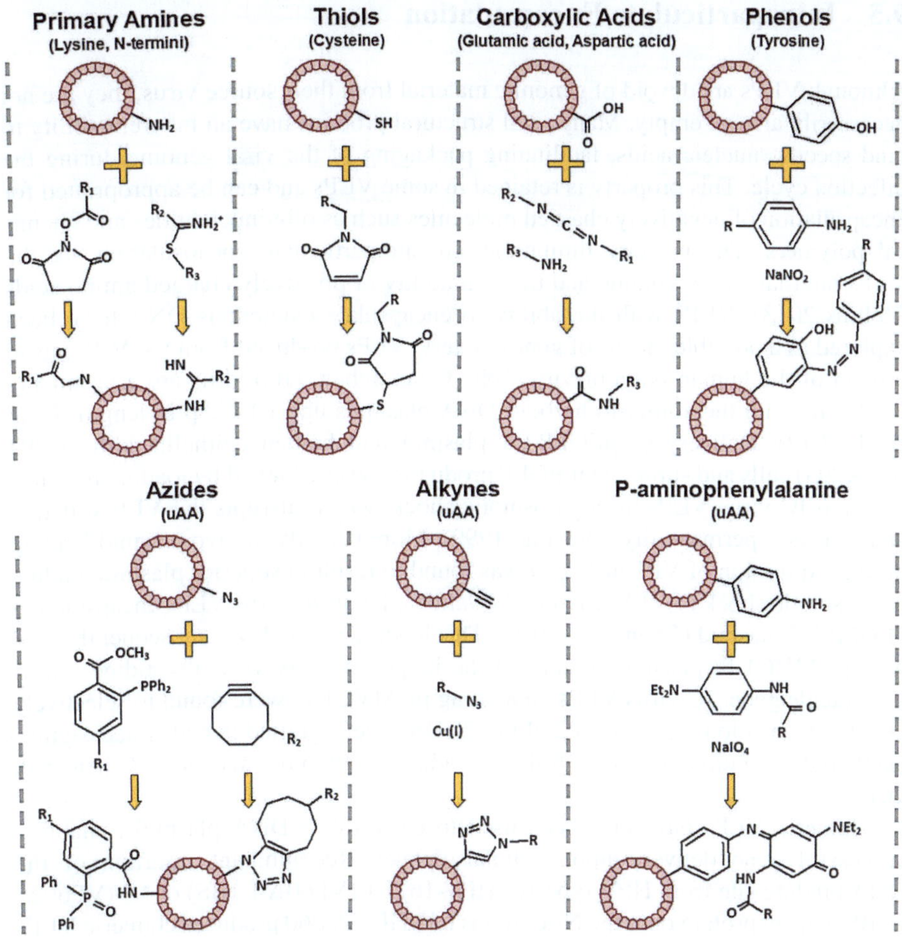

Fig. 9.4 Chemical conjugation onto the surface of VLPs. Various molecules can be coupled onto the surface of VLPs through chemical conjugation, targeting primary amines (e.g. lysine, N-termini), thiols (e.g. cysteine), carboxylic acids (e.g. glutamic acid, aspartic acid), and phenols (e.g. tyrosine). Unnatural amino acids (UAAs) incorporated into the structure of VLPs can also be used as potential chemical conjugation targets by introducing reactive azide or alkyne groups. Adapted from Smith et al (2013)

Conjugation of large proteins or cell lysate onto the surface of VLPs enables the induction of an immune response without identification of specific immunogenic peptides. Tumour lysates derived from the MART-1 expressing melanoma cell line Mel888 conjugated onto RHDV VLP provides more efficient lysate delivery to dendritic cell antigen processing compartments and enhances MART-1-specific CD8+ T cell stimulation compared to lysate alone (Win et al. 2012). The use of tumour lysate in VLP vaccines has potential applications for personalised anti-cancer immunotherapy.

9.5 Intraparticulate Encapsulation

Although VLPs are devoid of genomic material from their source virus, they are not necessarily always empty. Many viral structural proteins have an inherent ability to bind specific nucleic acids, facilitating packaging of the viral genome during the infection cycle. This property is retained in some VLPs and can be appropriated for encapsulation of negatively charged molecules such as oligonucleotides and chemical polymers. The primary limitations on intraparticulate encapsulation are the VLPs internal cavity volume and the availability of positively charged amino acids (Zeltins 2013). VLPs with the ability to encapsulate exogenous DNA have been explored as a possible means of gene delivery. VLPs produced from the VP1 capsid protein of the human polyomavirus John Cunningham (JC) virus are a promising candidate, with the ability to harbour DNA plasmids up to 14 kbp in length (Fang et al. 2012). Delivery of pEGFP-C1 plasmid into human epithelial kidney 293 (HEK293) cells and subsequent EGFP production was achieved by loading the plasmid into JC virus VLPs through osmotic shock, which disrupts the VLP structure and increases permeability (Ou et al. 1999). More recently in vivo plasmid loading during expression of VP1 in *E. coli* was found to result in superior plasmid loading over osmotic shock or VLP reassembly, validated with JC virus VLPs encapsulating pEGFP-N3 plasmid (Chen et al. 2010). This loading method was subsequently used with pUMVC1-tk plasmid containing the herpes simplex virus thymidine kinase (tk) suicide gene. JC virus VLPs containing pUMVC1-tk were found to selectively target human colon carcinoma (COLO-320 HSR) cells grafted in nude mice, significantly reducing tumour volume following administration of ganciclovir (Chen et al. 2010).

Chimeric VLPs have also been used to encapsulate DNA plasmids, enabling successful gene delivery upon cellular uptake. Recombinant insertion of the DNA binding site from HPV type 16 (HPV-16) L1 (VP60Δ-L1BS) or L2 (VP60Δ-L2BS) capsid protein onto the N-terminus of RHDV VP60 produces chimeric VLPs with the capacity to encapsulate pCMV-β in vitro through VLP reassembly. β-Galactosidase expression was identified in a range of cell lines (Cos-7, R17, HuH-7, and $CaCo_2$) treated with VP60Δ-L1BS containing pCMV-β plasmid (El Mehdaoui et al. 2000). It is possible that any VLP that supports internal site insertion may be capable of DNA plasmid encapsulation and delivery by incorporation of an appropriate DNA binding site. VLPs have even been used to induce the expression of their own structural capsid protein, enhancing immunogenicity by mimicking viral replication in vivo (Pichlmair et al. 2010). A variety of molecules other than nucleic acids have also been successfully encapsulated into VLPs, including enzymes into Qβ VLPs (Fiedler et al. 2010), polymerase into rotavirus VLPs (Boudreaux et al. 2013), and fluorophores into cucumber mosaic virus VLPs (Lu et al. 2012). Recombinant insertion of a heterodimeric coiled-coil amino acid motif at the N-terminus of a capsid protein from cowpea chlorotic mottle virus (CCMV) produced CCMV VLPs capable of encapsulating up to 15 EGFP proteins (Minten et al. 2009). Encapsulating exogenous proteins inside VLPs is an interesting concept with multiple potential therapeutic applications.

9.6 VLP Vaccine Immunogenicity

The immune response to VLP vaccines is primarily determined by particle size, capsid structure, and innate immunity activation. Particles larger than 200 nm require transportation to the lymph nodes by peripheral antigen presenting cells (APCs) at the vaccination site (e.g. macrophages, dendritic cells) (Manolova et al. 2008). VLPs are usually below 200 nm, facilitating free circulation and drainage directly to lymph nodes in addition to peripheral APC transportation. Once in the lymph node, free VLPs are internalised by resident APCs. The repetitive structure of VLPs promotes efficient uptake by APCs through mechanisms such as phagocytosis and macropinocytosis (Win et al. 2011; Xiang et al. 2006; Scheerlinck and Greenwood 2008). Due to VLPs consisting of native viral capsid proteins, they often contain receptor binding motifs that also enable receptor-mediated endocytosis. An example of this is the binding of haemagglutinin in influenza VLPs to the sialylose receptor on the cell surface (Pan et al. 2010).

The processing of VLPs by APCs can lead to the stimulation of both the humoral and cell-mediated arms of the immune system. Historically vaccines have been designed to target the initiation of humoral immunity, generating vaccine-specific antibodies. While this type of immune response is essential for the clearance of extracellular pathogens, establishing protection against intracellular conditions such as viral infections and cancer often requires the generation of cell-mediated immunity. The ability of VLPs to stimulate both arms of the immune system is advantageous, as the combination of immune responses often work synergistically to ensure disease clearance and the initiation of effective immunological memory. The immune response to VLPs is summarised in Fig. 9.5.

Following internalisation VLPs are degraded and processed through the exogenous antigen processing pathway with immunogenic peptides loaded onto MHC class II (MHC-II) molecules for surface presentation. Processed antigens are recognised by $CD4^+$ T helper cells, resulting in activation and release of immunostimulatory cytokines, such as Interleukin (IL) 4, IL-5, which are essential for B cell activation or IL-2 and IFNγ, which are essential for induction of cell-mediated immunity. For successful activation of the humoral immune system, B cells must also interact with native antigen through B cell receptors (BCRs). The repetitive nature of VLPs promotes crosslinking of BCRs, enhancing activation and promoting memory-cell formation. This stimulation is notably superior to soluble peptides or non-repetitive proteins (Jegerlehner et al. 2007; Bachmann and Jennings 2010). Antibodies produced in response to VLPs can neutralise native virions to prevent active infection, and memory-cell formation promotes long-term immunity. Notable examples of licensed VLP vaccines that successfully induce high titres of long-lasting neutralising antibodies include the Hepatitis B vaccine Engerix (Keating and Noble 2003) and the HPV vaccines Cervarix and Gardasil (Romanowski 2011).

Some VLPs are also cross-presented by APCs, primarily $CD8^+$ dendritic cells, facilitating loading of exogenous antigen onto MHC class I (MHC-I). With additional stimulation from co-stimulatory molecules (e.g. CD40, CD80/86), MHC-I

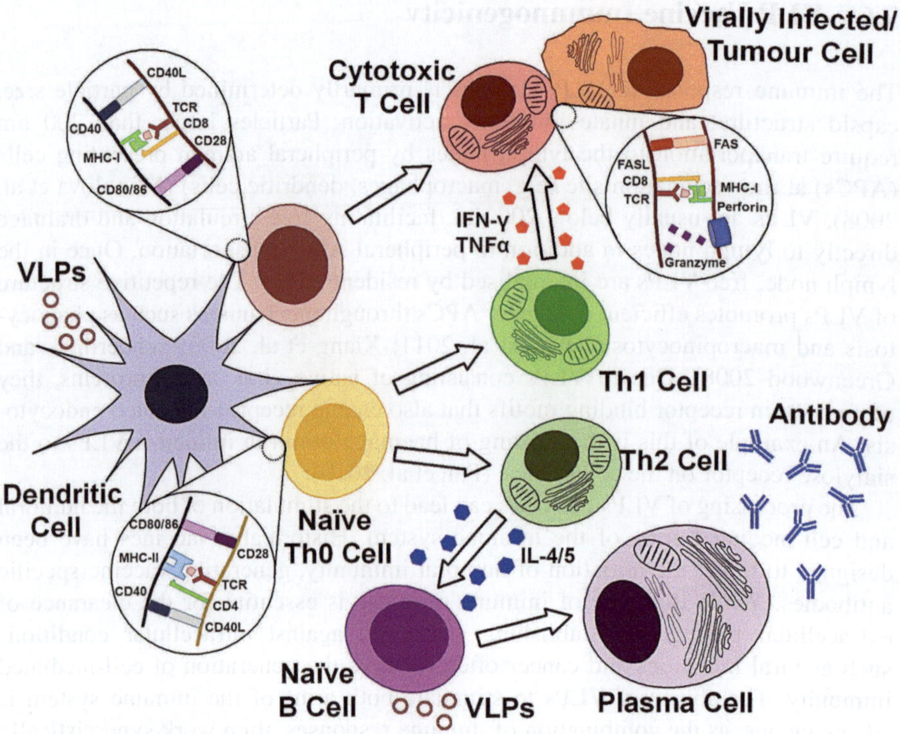

Fig. 9.5 Antigen presenting cell (APC) cross-presentation pathways. APCs can cross-present exogenous antigens onto MHC-I through pathways including: (**a**) Gap junctions; (**b**) Endosome-to-cytosol; (**c**) ER-Endosome fusion; (**d**) Receptor recycling; and (**e**) Exosomes. Peptides derived from RHDV VLPs are known to be crosspresented through the receptor recycling pathway (Win et al. 2011). Adapted from Groothuis and Neefjes (2005)

loaded with immunogenic peptides can activate $CD8^+$ T cells to initiate a cell-mediated immune response. Cytotoxic T lymphocyte (CTL) activation is advantageous for clearance of intracellular conditions such as viral infection or cancer. For example, RHDV VLPs are processed through the receptor-recycling pathway for cross-presentation. Immunogenic peptides are loaded onto MHC-I when phagolysosomes containing degraded VLPs fuse with endosomes containing MHC-I molecules recycling from the cell surface (Win et al. 2011). The endosomes return to the cell surface, displaying new MHC–peptide complexes on the cell membrane as illustrated in Fig. 9.6.

The site of vaccine administration can play an important role in VLP immunogenicity. As the mucosal membranes (e.g. respiratory tract, digestive tract, urogenital tract) are a common route of infection, generation of mucosal immunity can provide early protection against pathogenic intrusion. A number of preclinical VLP vaccine trials have demonstrated that VLPs have the ability to initiate potent mucosal immunity. For example, intranasal vaccination with cholera toxin B conjugated to simian immunodeficiency virus VLPs significantly increased the

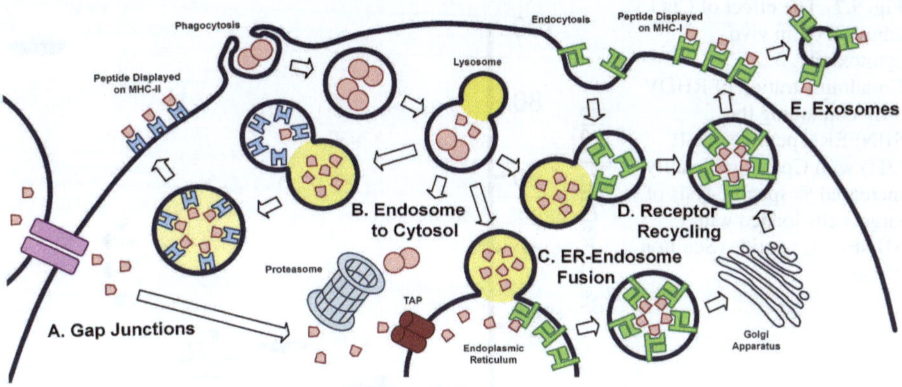

Fig. 9.6 The immune response to VLPs. VLPs have the capacity to stimulate both the cell-mediated and humoral immune system. The desired immune response can be promoted through adjuvant selection

levels of antibody detected in the mucosae compared to unconjugated cholera toxin B (Kang et al. 2003).

Depending on the protein expression system used, some VLPs can contain exogenous immunogenic molecules such as bacterial RNA which promote APC activation through stimulation of pattern recognition receptors such as toll-like receptors (TLRs) (Rebeaud and Bachmann 2012). This can have an adjuvanting effect due to activation of the innate immune system; however, we have found that successful vaccination can be achieved in the absence of these adjuvanting molecules (Li et al. 2013). RHDV VLPs produced in insect cell cultures do not stimulate innate cells cultured in vitro, but effectively activate the cell-mediated and humoral immune systems in vivo. Despite the success of VLP vaccination alone, the inclusion of an adjuvant can still be beneficial in driving specific immune pathway activation and leading to further enhancement of immune stimulation.

A number of adjuvants have been used in humans; these include mineral salts (e.g. aluminum hydroxide, aluminum phosphate, and aluminum hydroxy phosphates), emulsions (e.g. MF59, AS01, and AS02), and microbial derivatives (monophosphoryl lipid A and CpG) (Rappuoli et al. 2011). Traditionally adjuvants were used to enhance vaccine immunogenicity by stimulating the humoral immune system to induce antibody production and isotype-switching. This was thought to be facilitated by formation of a slow-release antigen depot. Modern adjuvants are designed to directly target immune activation and to tailor vaccination to drive a specific immune response, enabling selective stimulation of the cell-mediated immune system. TLR agonists such as unmethylated CpG DNA can induce a potent cell-mediated response by stimulating upregulated expression of co-stimulatory molecules in innate immune cells. We have found that the addition of CpG to recombinant RHDV VLP containing the SIINFEKL peptide derived from the model antigen OVA (VLP.OTI) led to an enhancement in the generation of SIINFEKL-specific cytotoxic T cells (Fig. 9.7) (Scullion 2012). This association was also confirmed using the Lewis' lung carcinoma model with recombinant gp33 RHDV VLP (Li et al. 2013).

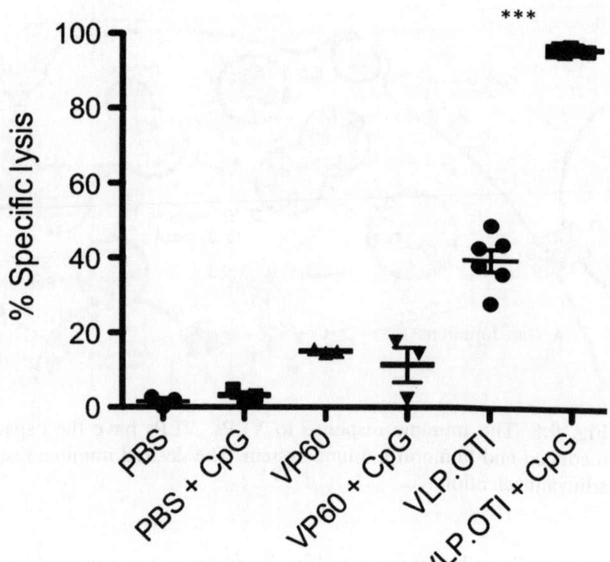

Fig. 9.7 The effect of CpG adjuvant on in vivo cytotoxicity. Co-administration of RHDV VLP containing the SIINFEKL peptide (VLP.OTI) with CpG significantly increased % specific lysis of target cells loaded with SIINFEKL peptide (Scullion 2012)

Co-delivery of vaccine adjuvants and antigens to cells through physical association or linkage leads to enhanced immunogenicity and a reduction in adjuvant-related side effects. Viruses often bind molecules that facilitate entry into host cells. RHDV VLPs retain the ability of its native virus to bind carbohydrate moieties, providing a useful mechanism for modification and vaccine enhancement (Ruvoen-Clouet et al. 2000; McKee et al. 2012). α-Galactosylceramide is a glycolipid adjuvant which was found to directly associate with RHDV VLP, forming a composite particle. α-Galactosylceramide activates natural killer-like T cells, leading to the enhancement of both the innate and acquired immune responses (Bendelac et al. 2007). Prophylactic vaccination with recombinant gp33. RHDV VLP and α-galactosylceramide led to the generation of gp33-specific T cells and enhanced protection against subcutaneous tumour challenge. The adjuvanting effect of α-galactosylceramide was increased >10-fold when delivered as a composite particle compared to the co-delivery of unassociated α-galactosylceramide and recombinant gp33.RHDV VLP (McKee et al. 2012).

9.7 Conclusion

The primary requirements of a successful vaccine include safety, reliability, and efficacy. Along with other subunit vaccines VLPs have an impeccable safety record, with some vaccines already approved for routine use. The absence of the viral genome renders VLPs incapable of infection or replication, preventing inadvertent reversion as has occurred with attenuated viruses. Extensive purification of VLPs produced using protein expression systems helps to limit contamination with substances which might induce allergies or undesirable side effects. Each new VLP is extensively characterised during development, containing immunogenic epitopes in a more natural state than inactivated virus vaccines. The particulate nature of VLPs also provides some protection from degradation. VLPs are uniform in structure, providing consistency between preparations and a reliable vaccine. VLPs are readily processed by the immune system and can stimulate cytotoxic and humoral immune responses. VLPs can provide protection against their source virus, or they can harbour exogenous antigens. This unprecedented versatility in vaccine design enables the utilisation of VLP vaccines in a wide range of applications including prophylactic viral vaccines, therapeutic cancer vaccines, and even gene delivery. VLP vaccines have only begun to unveil their true potential as a versatile subunit vaccine platform.

References

Ambuhl PM, Tissot AC, Fulurija A, Maurer P, Nussberger J, Sabat R, Nief V, Schellekens C, Sladko K, Roubicek K, Pfister T, Rettenbacher M, Volk HD, Wagner F, Muller P, Jennings GT, Bachmann MF (2007) A vaccine for hypertension based on virus-like particles: preclinical efficacy and phase I safety and immunogenicity. J Hypertens 25(1):63–72. doi:10.1097/HJH.0b013e32800ff5d6

Bachmann MF, Jennings GT (2010) Vaccine delivery: a matter of size, geometry, kinetics and molecular patterns. Nat Rev Immunol 10(11):787–796. doi:10.1038/nri2868

Baumert TF, Ito S, Wong DT, Liang TJ (1998) Hepatitis C virus structural proteins assemble into viruslike particles in insect cells. J Virol 72(5):3827–3836

Bayer ME, Blumberg BS, Werner B (1968) Particles associated with Australia antigen in the sera of patients with leukaemia, Down's syndrome and hepatitis. Nature 218(5146):1057–1059

Bendelac A, Savage PB, Teyton L (2007) The biology of NKT cells. Annu Rev Immunol 25:297–336. doi:10.1146/annurev.immunol.25.022106.141711

Bernstein DI, El Sahly HM, Keitel WA, Wolff M, Simone G, Segawa C, Wong S, Shelly D, Young NS, Dempsey W (2011) Safety and immunogenicity of a candidate parvovirus B19 vaccine. Vaccine 29(43):7357–7363. doi:10.1016/j.vaccine.2011.07.080

Betenbaugh M, Yu M, Kuehl K, White J, Pennock D, Spik K, Schmaljohn C (1995) Nucleocapsid- and virus-like particles assemble in cells infected with recombinant baculoviruses or vaccinia viruses expressing the M and the S segments of Hantaan virus. Virus Res 38(2–3):111–124

Boudreaux C, Vile D, Gilmore B, Tanner J, Kelly D, McDonald S (2013) Rotavirus core shell subdomains involved in polymerase encapsidation into virus-like particles. J Gen Virol 94(Pt 8):1818–1826. doi:10.1099/vir.0.052951-0

Bovier PA (2008) Recent advances with a virosomal hepatitis A vaccine. Expert Opin Biol Ther 8(8):1177–1185. doi:10.1517/14712598.8.8.1177

Brautigam S, Snezhkov E, Bishop DH (1993) Formation of poliovirus-like particles by recombinant baculoviruses expressing the individual VP0, VP3, and VP1 proteins by comparison to particles derived from the expressed poliovirus polyprotein. Virology 192(2):512–524. doi:10.1006/viro.1993.1067

Buonaguro L, Buonaguro FM, Tornesello ML, Mantas D, Beth-Giraldo E, Wagner R, Michelson S, Prevost MC, Wolf H, Giraldo G (2001) High efficient production of Pr55(gag) virus-like particles expressing multiple HIV-1 epitopes, including a gp120 protein derived from an Ugandan HIV-1 isolate of subtype A. Antiviral Res 49(1):35–47

Buonamassa DT, Greer CE, Capo S, Yen TS, Galeotti CL, Bensi G (2002) Yeast coexpression of human papillomavirus types 6 and 16 capsid proteins. Virology 293(2):335–344. doi:10.1006/viro.2001.1289

Chackerian B, Rangel M, Hunter Z, Peabody DS (2006) Virus and virus-like particle-based immunogens for Alzheimer's disease induce antibody responses against amyloid-beta without concomitant T cell responses. Vaccine 24(37–39):6321–6331. doi:10.1016/j.vaccine.2006.05.059

Chackerian B, Durfee MR, Schiller JT (2008) Virus-like display of a neo-self antigen reverses B cell anergy in a B cell receptor transgenic mouse model. J Immunol 180(9):5816–5825

Chen LS, Wang M, Ou WC, Fung CY, Chen PL, Chang CF, Huang WS, Wang JY, Lin PY, Chang D (2010) Efficient gene transfer using the human JC virus-like particle that inhibits human colon adenocarcinoma growth in a nude mouse model. Gene Ther 17(8):1033–1041. doi:10.1038/gt.2010.50

Cornuz J, Zwahlen S, Jungi WF, Osterwalder J, Klingler K, van Melle G, Bangala Y, Guessous I, Muller P, Willers J, Maurer P, Bachmann MF, Cerny T (2008) A vaccine against nicotine for smoking cessation: a randomized controlled trial. PLoS One 3(6):e2547. doi:10.1371/journal.pone.0002547

Crawford SE, Labbe M, Cohen J, Burroughs MH, Zhou YJ, Estes MK (1994) Characterization of virus-like particles produced by the expression of rotavirus capsid proteins in insect cells. J Virol 68(9):5945–5952

Crisci E, Almanza H, Mena I, Cordoba L, Gomez-Casado E, Caston JR, Fraile L, Barcena J, Montoya M (2009) Chimeric calicivirus-like particles elicit protective anti-viral cytotoxic responses without adjuvant. Virology 387(2):303–312. doi:10.1016/j.virol.2009.02.045

Cull M, McHenry CS (1990) Preparation of extracts from prokaryotes. Methods Enzymol 182:147–153

El Mehdaoui S, Touze A, Laurent S, Sizaret PY, Rasschaert D, Coursaget P (2000) Gene transfer using recombinant rabbit hemorrhagic disease virus capsids with genetically modified DNA encapsidation capacity by addition of packaging sequences from the L1 or L2 protein of human papillomavirus type 16. J Virol 74(22):10332–10340

El-Kamary SS, Pasetti MF, Mendelman PM, Frey SE, Bernstein DI, Treanor JJ, Ferreira J, Chen WH, Sublett R, Richardson C, Bargatze RF, Sztein MB, Tacket CO (2010) Adjuvanted intranasal Norwalk virus-like particle vaccine elicits antibodies and antibody-secreting cells that express homing receptors for mucosal and peripheral lymphoid tissues. J Infect Dis 202(11):1649–1658. doi:10.1086/657087

Eriksson M, Andreasson K, Weidmann J, Lundberg K, Tegerstedt K, Dalianis T, Ramqvist T (2011) Murine polyomavirus virus-like particles carrying full-length human PSA protect BALB/c mice from outgrowth of a PSA expressing tumor. PLoS One 6(8):e23828. doi:10.1371/journal.pone.0023828

Fang CY, Lin PY, Ou WC, Chen PL, Shen CH, Chang D, Wang M (2012) Analysis of the size of DNA packaged by the human JC virus-like particle. J Virol Methods 182(1–2):87–92. doi:10.1016/j.jviromet.2012.03.017

Fiedler JD, Brown SD, Lau JL, Finn MG (2010) RNA-directed packaging of enzymes within virus-like particles. Angew Chem Int Ed Engl 49(50):9648–9651. doi:10.1002/anie.201005243

Fiers W, De Filette M, El Bakkouri K, Schepens B, Roose K, Schotsaert M, Birkett A, Saelens X (2009) M2e-based universal influenza A vaccine. Vaccine 27(45):6280–6283. doi:10.1016/j.vaccine.2009.07.007

French TJ, Roy P (1990) Synthesis of bluetongue virus (BTV) corelike particles by a recombinant baculovirus expressing the two major structural core proteins of BTV. J Virol 64(4):1530–1536

French TJ, Marshall JJ, Roy P (1990) Assembly of double-shelled, viruslike particles of bluetongue virus by the simultaneous expression of four structural proteins. J Virol 64(12):5695–5700

Glenn GM, Smith G, Fries L, Raghunandan R, Lu H, Zhou B, Thomas DN, Hickman SP, Kpamegan E, Boddapati S, Piedra PA (2013) Safety and immunogenicity of a Sf9 insect cell-derived respiratory syncytial virus fusion protein nanoparticle vaccine. Vaccine 31(3):524–532. doi:10.1016/j.vaccine.2012.11.009

Gregson AL, Oliveira G, Othoro C, Calvo-Calle JM, Thorton GB, Nardin E, Edelman R (2008) Phase I trial of an alhydrogel adjuvanted hepatitis B core virus-like particle containing epitopes of Plasmodium falciparum circumsporozoite protein. PLoS One 3(2):e1556. doi:10.1371/journal.pone.0001556

Groothuis TA, Neefjes J (2005) The many roads to cross-presentation. J Exp Med 202(10):1313–1318. doi:10.1084/jem.20051379

Halsey RJ, Tanzer FL, Meyers A, Pillay S, Lynch A, Shephard E, Williamson AL, Rybicki EP (2008) Chimaeric HIV-1 subtype C Gag molecules with large in-frame C-terminal polypeptide fusions form virus-like particles. Virus Res 133(2):259–268. doi:10.1016/j.virusres.2008.01.012

Herzog C, Hartmann K, Kunzi V, Kursteiner O, Mischler R, Lazar H, Gluck R (2009) Eleven years of Inflexal V-a virosomal adjuvanted influenza vaccine. Vaccine 27(33):4381–4387. doi:10.1016/j.vaccine.2009.05.029

Jegerlehner A, Maurer P, Bessa J, Hinton HJ, Kopf M, Bachmann MF (2007) TLR9 signaling in B cells determines class switch recombination to IgG2a. J Immunol 178(4):2415–2420

Jiang X, Wang M, Graham DY, Estes MK (1992) Expression, self-assembly, and antigenicity of the Norwalk virus capsid protein. J Virol 66(11):6527–6532

Kang SM, Yao Q, Guo L, Compans RW (2003) Mucosal immunization with virus-like particles of simian immunodeficiency virus conjugated with cholera toxin subunit B. J Virol 77(18):9823–9830

Katpally U, Voss NR, Cavazza T, Taube S, Rubin JR, Young VL, Stuckey J, Ward VK, Virgin HW 4th, Wobus CE, Smith TJ (2010) High-resolution cryo-electron microscopy structures of murine norovirus 1 and rabbit hemorrhagic disease virus reveal marked flexibility in the receptor binding domains. J Virol 84(11):5836–5841. doi:10.1128/JVI.00314-10

Keating GM, Noble S (2003) Recombinant hepatitis B vaccine (Engerix-B): a review of its immunogenicity and protective efficacy against hepatitis B. Drugs 63(10):1021–1051

Kibenge FS, Qian B, Nagy E, Cleghorn JR, Wadowska D (1999) Formation of virus-like particles when the polyprotein gene (segment A) of infectious bursal disease virus is expressed in insect cells. Can J Vet Res 63(1):49–55

Kim Y, Kim J, Kang K, Lyoo YS (2002) Characterization of the recombinant proteins of porcine circovirus type2 field isolate expressed in the baculovirus system. J Vet Sci 3(1):19–23

Kirnbauer R, Booy F, Cheng N, Lowy DR, Schiller JT (1992) Papillomavirus L1 major capsid protein self-assembles into virus-like particles that are highly immunogenic. Proc Natl Acad Sci U S A 89(24):12180–12184

Ko YJ, Choi KS, Nah JJ, Paton DJ, Oem JK, Wilsden G, Kang SY, Jo NI, Lee JH, Kim JH, Lee HW, Park JM (2005) Noninfectious virus-like particle antigen for detection of swine vesicular disease virus antibodies in pigs by enzyme-linked immunosorbent assay. Clin Diagn Lab Immunol 12(8):922–929. doi:10.1128/CDLI.12.8.922-929.2005

Koch G, van Roozelaar DJ, Verschueren CA, van der Eb AJ, Noteborn MH (1995) Immunogenic and protective properties of chicken anaemia virus proteins expressed by baculovirus. Vaccine 13(8):763–770

Kushnir N, Streatfield SJ, Yusibov V (2012) Virus-like particles as a highly efficient vaccine platform: diversity of targets and production systems and advances in clinical development. Vaccine 31(1):58–83. doi:10.1016/j.vaccine.2012.10.083

Lasickiene R, Gedvilaite A, Norkiene M, Simanaviciene V, Sezaite I, Dekaminaviciute D, Shikova E, Zvirbliene A (2012) The use of recombinant pseudotype virus-like particles harbouring inserted target antigen to generate antibodies against cellular marker p16INK4A. ScientificWorldJournal 2012:263737. doi:10.1100/2012/263737

Latham T, Galarza JM (2001) Formation of wild-type and chimeric influenza virus-like particles following simultaneous expression of only four structural proteins. J Virol 75(13):6154–6165. doi:10.1128/JVI.75.13.6154-6165.2001

Lemere CA, Masliah E (2010) Can Alzheimer disease be prevented by amyloid-beta immunotherapy? Nat Rev Neurol 6(2):108–119. doi:10.1038/nrneurol.2009.219

Li K, Peers-Adams A, Win SJ, Scullion S, Wilson M, Young VL, Jennings P, Ward VK, Baird MA, Young SL (2013) Antigen incorporated in virus-like particles is delivered to specific dendritic cell subsets that induce an effective antitumor immune response in vivo. J Immunother 36(1):11–19. doi:10.1097/CJI.0b013e3182787f5e

Link A, Bachmann MF (2010) Immunodrugs: breaking B- but not T-cell tolerance with therapeutic anticytokine vaccines. Immunotherapy 2(4):561–574. doi:10.2217/imt.10.30

Liu YV, Massare MJ, Barnard DL, Kort T, Nathan M, Wang L, Smith G (2011) Chimeric severe acute respiratory syndrome coronavirus (SARS-CoV) S glycoprotein and influenza matrix 1 efficiently form virus-like particles (VLPs) that protect mice against challenge with SARS-CoV. Vaccine 29(38):6606–6613. doi:10.1016/j.vaccine.2011.06.111

Lopez de Turiso JA, Cortes E, Martinez C, Ruiz de Ybanez R, Simarro I, Vela C, Casal I (1992) Recombinant vaccine for canine parvovirus in dogs. J Virol 66(5):2748–2753

Lopez-Macias C, Ferat-Osorio E, Tenorio-Calvo A, Isibasi A, Talavera J, Arteaga-Ruiz O, Arriaga-Pizano L, Hickman SP, Allende M, Lenhard K, Pincus S, Connolly K, Raghunandan R, Smith G, Glenn G (2011) Safety and immunogenicity of a virus-like particle pandemic influenza A (H1N1) 2009 vaccine in a blinded, randomized, placebo-controlled trial of adults in Mexico. Vaccine 29(44):7826–7834. doi:10.1016/j.vaccine.2011.07.099

Lu X, Thompson JR, Perry KL (2012) Encapsidation of DNA, a protein and a fluorophore into virus-like particles by the capsid protein of cucumber mosaic virus. J Gen Virol 93(Pt 5):1120–1126. doi:10.1099/vir.0.040170-0

Manolova V, Flace A, Bauer M, Schwarz K, Saudan P, Bachmann MF (2008) Nanoparticles target distinct dendritic cell populations according to their size. Eur J Immunol 38(5):1404–1413. doi:10.1002/eji.200737984

Matic S, Rinaldi R, Masenga V, Noris E (2011) Efficient production of chimeric human papillomavirus 16L1 protein bearing the M2e influenza epitope in Nicotiana benthamiana plants. BMC Biotechnol 11:106. doi:10.1186/1472-6750-11-106

Maurer P, Bachmann MF (2010) Immunization against angiotensins for the treatment of hypertension. Clin Immunol 134(1):89–95. doi:10.1016/j.clim.2009.06.003

Maurer P, Jennings GT, Willers J, Rohner F, Lindman Y, Roubicek K, Renner WA, Muller P, Bachmann MF (2005) A therapeutic vaccine for nicotine dependence: preclinical efficacy, and phase I safety and immunogenicity. Eur J Immunol 35(7):2031–2040. doi:10.1002/eji.200526285

Mazeike E, Gedvilaite A, Blohm U (2012) Induction of insert-specific immune response in mice by hamster polyomavirus VP1 derived virus-like particles carrying LCMV GP33 CTL epitope. Virus Res 163(1):2–10. doi:10.1016/j.virusres.2011.08.003

McKee SJ, Young VL, Clow F, Hayman CM, Baird MA, Hermans IF, Young SL, Ward VK (2012) Virus-like particles and alpha-galactosylceramide form a self-adjuvanting composite particle that elicits anti-tumor responses. J Control Release 159(3):338–345. doi:10.1016/j.jconrel.2012.02.015

Medicago I News Release 17 June, 2013. Medicago receives authorization from health Canada and commences phase ii clinical trial for an H5N1 vaccine

Middelberg AP, Rivera-Hernandez T, Wibowo N, Lua LH, Fan Y, Magor G, Chang C, Chuan YP, Good MF, Batzloff MR (2011) A microbial platform for rapid and low-cost virus-like particle and capsomere vaccines. Vaccine 29(41):7154–7162. doi:10.1016/j.vaccine.2011.05.075

Minten IJ, Hendriks LJ, Nolte RJ, Cornelissen JJ (2009) Controlled encapsulation of multiple proteins in virus capsids. J Am Chem Soc 131(49):17771–17773. doi:10.1021/ja907843s

Noteborn MH, Verschueren CA, Koch G, Van der Eb AJ (1998) Simultaneous expression of recombinant baculovirus-encoded chicken anaemia virus (CAV) proteins VP1 and VP2 is required for formation of the CAV-specific neutralizing epitope. J Gen Virol 79(Pt 12):3073–3077

Ou WC, Wang M, Fung CY, Tsai RT, Chao PC, Hseu TH, Chang D (1999) The major capsid protein, VP1, of human JC virus expressed in Escherichia coli is able to self-assemble into a capsid-like particle and deliver exogenous DNA into human kidney cells. J Gen Virol 80(Pt 1):39–46

Pan YS, Wei HJ, Chang CC, Lin CH, Wei TS, Wu SC, Chang DK (2010) Construction and characterization of insect cell-derived influenza VLP: cell binding, fusion, and EGFP incorporation. J Biomed Biotechnol 2010:506363. doi:10.1155/2010/506363

Patel KG, Swartz JR (2011) Surface functionalization of virus-like particles by direct conjugation using azide-alkyne click chemistry. Bioconjug Chem 22(3):376–387. doi:10.1021/bc100367u

Peacey M, Wilson S, Baird MA, Ward VK (2007) Versatile RHDV virus-like particles: incorporation of antigens by genetic modification and chemical conjugation. Biotechnol Bioeng 98(5):968–977. doi:10.1002/bit.21518

Peacey M, Wilson S, Perret R, Ronchese F, Ward VK, Young V, Young SL, Baird MA (2008) Virus-like particles from rabbit hemorrhagic disease virus can induce an anti-tumor response. Vaccine 26(42):5334–5337. doi:10.1016/j.vaccine.2008.07.074

Peters BS, Cheingsong-Popov R, Callow D, Foxall R, Patou G, Hodgkin K, Weber JN (1997) A pilot phase II study of the safety and immunogenicity of HIV p17/p24:VLP (p24-VLP) in asymptomatic HIV seropositive subjects. J Infect 35(3):231–235

Phelps JP, Dang N, Rasochova L (2007) Inactivation and purification of cowpea mosaic virus-like particles displaying peptide antigens from Bacillus anthracis. J Virol Methods 141(2):146–153. doi:10.1016/j.jviromet.2006.12.008

Pichlmair A, Habjan M, Unger H, Weber F (2010) Virus-like particles expressing the nucleocapsid gene as an efficient vaccine against Rift Valley fever virus. Vector Borne Zoonotic Dis 10(7):701–703. doi:10.1089/vbz.2009.0248

Rappuoli R, Mandl CW, Black S, De Gregorio E (2011) Vaccines for the twenty-first century society. Nat Rev Immunol 11(12):865–872. doi:10.1038/nri3085

Rebeaud F, Bachmann M (2012) Virus-like particles as efficient delivery platform to induce a potent immune response. In: Baschieri S (ed) Innovation in vaccinology: from design, through to delivery and testing. Springer Science + Buisness Media, Dordrecht, pp 87–122

Rodriguez-Limas WA, Tyo KE, Nielsen J, Ramirez OT, Palomares LA (2011) Molecular and process design for rotavirus-like particle production in Saccharomyces cerevisiae. Microb Cell Fact 10:33. doi:10.1186/1475-2859-10-33

Romanowski B (2011) Long term protection against cervical infection with the human papillomavirus: review of currently available vaccines. Hum Vaccin 7(2):161–169

Ruvoen-Clouet N, Ganiere JP, Andre-Fontaine G, Blanchard D, Le Pendu J (2000) Binding of rabbit hemorrhagic disease virus to antigens of the ABH histo-blood group family. J Virol 74(24):11950–11954

Sabara M, Parker M, Aha P, Cosco C, Gibbons E, Parsons S, Babiuk LA (1991) Assembly of double-shelled rotaviruslike particles by simultaneous expression of recombinant VP6 and VP7 proteins. J Virol 65(12):6994–6997

Salazar O, Asenjo JA (2007) Enzymatic lysis of microbial cells. Biotechnol Lett 29(7):985–994. doi:10.1007/s10529-007-9345-2

Scheerlinck JP, Greenwood DL (2008) Virus-sized vaccine delivery systems. Drug Discov Today 13(19–20):882–887. doi:10.1016/j.drudis.2008.06.016

Scotti N, Alagna F, Ferraiolo E, Formisano G, Sannino L, Buonaguro L, De Stradis A, Vitale A, Monti L, Grillo S, Buonaguro FM, Cardi T (2009) High-level expression of the HIV-1 Pr55gag polyprotein in transgenic tobacco chloroplasts. Planta 229(5):1109–1122. doi:10.1007/s00425-009-0898-2

Scullion S (2012) Investigating cytotoxic T cell responses to RHDV virus-like particles: a thesis submitted for the degree of Master of Science at the University of Otago, Dunedin, New Zealand. MSc, University of Otago

Smith MT, Hawes AK, Bundy BC (2013) Reengineering viruses and virus-like particles through chemical functionalization strategies. Curr Opin Biotechnol 24(4):620–626. doi:10.1016/j.copbio.2013.01.011

Spohn G, Keller I, Beck M, Grest P, Jennings GT, Bachmann MF (2008) Active immunization with IL-1 displayed on virus-like particles protects from autoimmune arthritis. Eur J Immunol 38(3):877–887. doi:10.1002/eji.200737989

Tacket CO (2007) Plant-based vaccines against diarrheal diseases. Trans Am Clin Climatol Assoc 118:79–87

Tacket CO, Mason HS, Losonsky G, Estes MK, Levine MM, Arntzen CJ (2000) Human immune responses to a novel norwalk virus vaccine delivered in transgenic potatoes. J Infect Dis 182(1):302–305. doi:10.1086/315653

Tegerstedt K, Lindencrona JA, Curcio C, Andreasson K, Tullus C, Forni G, Dalianis T, Kiessling R, Ramqvist T (2005) A single vaccination with polyomavirus VP1/VP2Her2 virus-like particles prevents outgrowth of HER-2/neu-expressing tumors. Cancer Res 65(13):5953–5957. doi:10.1158/0008-5472.CAN-05-0335

Turpen TH, Reinl SJ, Charoenvit Y, Hoffman SL, Fallarme V, Grill LK (1995) Malarial epitopes expressed on the surface of recombinant tobacco mosaic virus. Biotechnology 13(1):53–57

Vicente T, Roldao A, Peixoto C, Carrondo MJ, Alves PM (2011) Large-scale production and purification of VLP-based vaccines. J Invertebr Pathol 107(Suppl):S42–S48. doi:10.1016/j.jip.2011.05.004

Visciano ML, Diomede L, Tagliamonte M, Tornesello ML, Asti V, Bomsel M, Buonaguro FM, Lopalco L, Buonaguro L (2011) Generation of HIV-1 virus-like particles expressing different HIV-1 glycoproteins. Vaccine 29(31):4903–4912. doi:10.1016/j.vaccine.2011.05.005

Walpita P, Barr J, Sherman M, Basler CF, Wang L (2011) Vaccine potential of Nipah virus-like particles. PLoS One 6(4):e18437. doi:10.1371/journal.pone.0018437

Wang X, Xu F, Liu J, Gao B, Liu Y, Zhai Y, Ma J, Zhang K, Baker TS, Schulten K, Zheng D, Pang H, Sun F (2013) Atomic model of rabbit hemorrhagic disease virus by cryo-electron microscopy and crystallography. PLoS Pathog 9(1):e1003132. doi:10.1371/journal.ppat.1003132

Wiedermann U, Wiltschke C, Jasinska J, Kundi M, Zurbriggen R, Garner-Spitzer E, Bartsch R, Steger G, Pehamberger H, Scheiner O, Zielinski CC (2010) A virosomal formulated Her-2/neu multi-peptide vaccine induces Her-2/neu-specific immune responses in patients with metastatic breast cancer: a phase I study. Breast Cancer Res Treat 119(3):673–683

Wiessner C, Wiederhold KH, Tissot AC, Frey P, Danner S, Jacobson LH, Jennings GT, Luond R, Ortmann R, Reichwald J, Zurini M, Mir A, Bachmann MF, Staufenbiel M (2011) The second-generation active Abeta immunotherapy CAD106 reduces amyloid accumulation in APP transgenic mice while minimizing potential side effects. J Neurosci 31(25):9323–9331. doi:10.1523/JNEUROSCI.0293-11.2011

Win SJ, Ward VK, Dunbar PR, Young SL, Baird MA (2011) Cross-presentation of epitopes on virus-like particles via the MHC I receptor recycling pathway. Immunol Cell Biol 89(6):681–688. doi:10.1038/icb.2010.161

Win SJ, McMillan DG, Errington-Mais F, Ward VK, Young SL, Baird MA, Melcher AA (2012) Enhancing the immunogenicity of tumour lysate-loaded dendritic cell vaccines by conjugation to virus-like particles. Br J Cancer 106(1):92–98. doi:10.1038/bjc.2011.538

Xiang SD, Scholzen A, Minigo G, David C, Apostolopoulos V, Mottram PL, Plebanski M (2006) Pathogen recognition and development of particulate vaccines: does size matter? Methods 40(1):1–9. doi:10.1016/j.ymeth.2006.05.016

Yamshchikov GV, Ritter GD, Vey M, Compans RW (1995) Assembly of SIV virus-like particles containing envelope proteins using a baculovirus expression system. Virology 214(1):50–58. doi:10.1006/viro.1995.9955

Young SL, Buchan G, Slobbe LJ, Williman J, Wilson M, Ward V (2004) Novel vaccine formulations to enhance Type 1 immune responses. Immunology 1:555–563

Young SL, Wilson M, Wilson S, Beagley KW, Ward V, Baird MA (2006) Transcutaneous vaccination with virus-like particles. Vaccine 24(26):5406–5412. doi:10.1016/j.vaccine.2006.03.052

Yusibov V, Hooper DC, Spitsin SV, Fleysh N, Kean RB, Mikheeva T, Deka D, Karasev A, Cox S, Randall J, Koprowski H (2002) Expression in plants and immunogenicity of plant virus-based experimental rabies vaccine. Vaccine 20(25–26):3155–3164

Zamora E, Handisurya A, Shafti-Keramat S, Borchelt D, Rudow G, Conant K, Cox C, Troncoso JC, Kirnbauer R (2006) Papillomavirus-like particles are an effective platform for amyloid-beta immunization in rabbits and transgenic mice. J Immunol 177(4):2662–2670

Zeltins A (2013) Construction and characterization of virus-like particles: a review. Mol Biotechnol 53(1):92–107. doi:10.1007/s12033-012-9598-4

Chapter 10
Polymeric Particulates for Subunit Vaccine Delivery

Thomas Schuster, Martin Nussbaumer, Patric Baumann, Nico Bruns, Wolfgang Meier, and Anja Car

10.1 Introduction to Polymeric Particulates

Nanoparticles are defined as particles with a diameter in the sub-micrometer range that exhibit properties not found in the same bulk material. This classical definition is not very specific but is compatible across different areas of science.

The basic concept of using nanoparticles for drug delivery appeared in the late 1950s (Holdermann and Greiling 1954), followed by other articles, which addressed infections such as tetanus, diphtheria, and other diseases that require multiple doses of vaccines (Birrenbach 1973). The vaccines described in the articles were based on micellar formulations (Kreuter 2007). Since then and due to progress in pharmaceutical technology and immunology, various types of particulates have been developed as delivery platforms (Rice-Ficht et al. 2010; Moon et al. 2012; Naahidi et al. 2013; Leleux and Roy 2013). The diversity of structures ranges from micro- and nanoparticles, nanogels, micro- and nanocapsules to dendrimers, vesicles, and micelles, as illustrated in Fig. 10.1.

Author contributed equally with all other contributors

T. Schuster • M. Nussbaumer • P. Baumann • W. Meier • A. Car (✉)
Department of Chemistry, University of Basel,
Klingelbergstrasse 80, 4056 Basel, Switzerland
e-mail: thomas.schuster@yahoo.fr; martin.nussbaumer@unibas.ch;
patric.baumann@unibas.ch; wofgang.meier@unibas.ch; anja.car@unibas.ch

N. Bruns
Department of Chemistry, University of Basel,
Klingelbergstrasse 80, 4056 Basel, Switzerland

Adolphe Merkle Institute, University of Fribourg,
Rte de l'ancienne Papeterie, CP 209, 1723 Marly 1, Switzerland
e-mail: nico.bruns@unifr.ch

C. Foged et al. (eds.), *Subunit Vaccine Delivery*, Advances in Delivery Science and Technology, DOI 10.1007/978-1-4939-1417-3_10

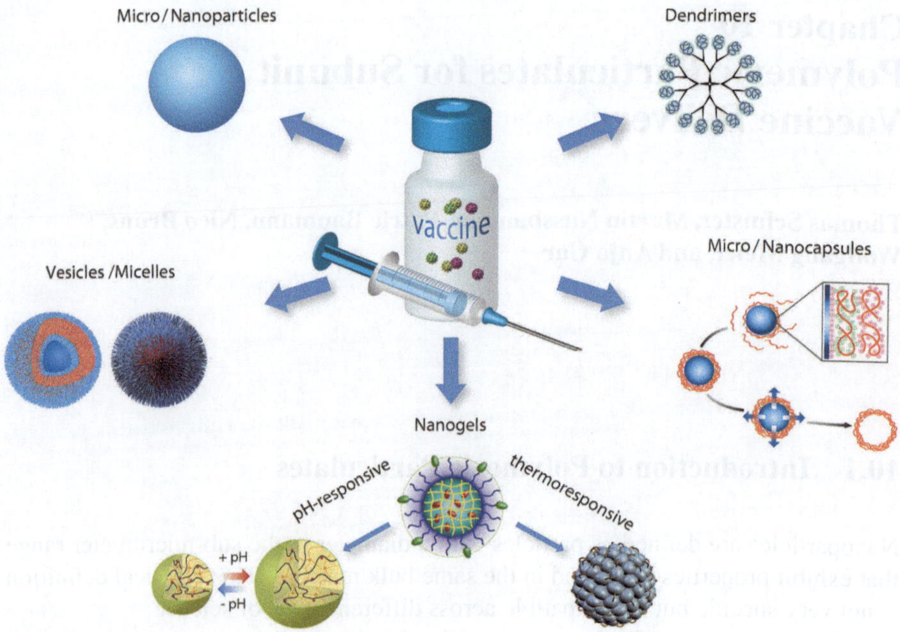

Fig. 10.1 Overview of currently available delivery systems based on polymeric particulates. Adapted from (Chacko et al. 2012; Wu and Zhou 2010; Sadler and Tam 2002; De Geest et al. 2007) with permission

Within the past few years, the number of articles published on subunit vaccines (antigen, protein, peptides, deoxyribonucleic acid (DNA), etc.) has increased significantly, and so have reviews dealing with specific delivery platforms. The first review published 20 years ago by Couvreur et al. focused on the progress and development of nanoparticulate systems for delivery of peptides and proteins (Couvreur and Puisieux 1993). Furthermore, in 2006 Pinto Reis et al. summarized the advances on biomedical applications and the status of peptide delivery systems (Pinto Reis et al. 2006). More recently, an excellent review compiling advances on vaccine delivery system using different polymeric nanocarriers was published (Correia-Pinto et al. 2013). Although 247 nanomedicine products (as for January 2013) are presently approved or in one of the three clinical phases, it should be clearly noted that none of these products are polymeric particulates applied in vaccine delivery. Rather, these systems are used as standard drug delivery platforms (Etheridge et al. 2013). Vaccine delivery using nanoparticles is still in its infancy and requires overcoming some challenges (e.g., prevention of toxic side effects by high-efficiency of encapsulation of the bioactive material, eases of administration, and introduction of an antigen-specific immune response) before proceeding from fundamental research to clinical applications (De Temmerman et al. 2011; Kamaly et al. 2012). For example, small molecules such as peptides acting as antigens are not recognized by the immune system, i.e., they are not immunostimulating, and therefore they rely solely on a vaccine delivery platform (Janeway et al. 2004). Thus, a suitable delivery platform has to combine high loading efficiency of antigens/adjuvants with a controlled release mechanism and a versatile chemistry of targeting ligands on the surface (Li and Mooney 2013).

Various delivery platforms based on polymeric particulates and their current status as vaccine delivery systems is discussed in this chapter.

10.2 Micro- and Nanoparticles

Currently, the most commonly used polymers in controlled drug release are poly(lactic-co-glycolic acid) (PLGA), poly(lactic acid) (PLA), poly(glutamic acid) (PGA), polycaprolactone, *N*-(2-hydroxypropyl)-methacylate copolymers (HPMA) and polyamido acids. PLGA, PLA, and PGA are biocompatible, biodegradable, and approved for biomedical application by the U.S. Food and Drug Administration (FDA) (Panyam and Labhasetwar 2003; Uto et al. 2013). PLGA particles, used in subunit vaccines delivery, can be prepared by various methods including spray drying, phase separation, solvent evaporation, and solvent extraction (Byrd et al. 2005). The first step of preparation includes the formation of a primary water-in-oil emulsion from an aqueous antigen solution and a solubilized organic polymer. The second step, which differs for each of the methods listed above, yields dispersed oil-containing microparticles. Several PLGA-based particles have been specifically designed to obtain an initial burst release and to guarantee that the encapsulation process provides inert and insulated environmental conditions for the active compound (Sanchez et al. 1996; Sah et al. 1995; Youan et al. 2001). This design is based on the combination of two microparticle formulations with carefully selected copolymer composition and preparation parameters, and enables a single dose vaccination, which mimics the effect of a booster dose (Sanchez et al. 1996; O'Hagan et al. 1998; Crotts and Park 1998). In this context, polyesters are often used as they undergo hydrolysis after parenteral administration, and the biocompatible breakdown products (e.g., lactic and glycolic acid) are easily cleared from the body (Andriano et al. 1999). In addition, early release of the immunogenic cargo before internalization in antigen-presenting cells (APC) is reduced by a slow degradation rate (Hamdy et al. 2011). Moreover, it has been shown that micro- and nanoparticles exhibit advantageous properties such as passive dendritic cell (DC) targeting, high storage stability, ease of scale-up, controlled release, and delivery of antigens, which are afterwards presented by both major histocompatibility complex (MHC) class I and II pathways (Men et al. 1999; Shen et al. 2006). All of these properties make them an attractive delivery system for tumor vaccines (Mueller et al. 2012). PLGA particulates have also been investigated for use in vaccinations against, e.g., tetanus, malaria, human immunodeficiency virus (HIV) and hepatitis B (Saroja et al. 2011). Several examples of vaccine formulations have been selected from a large number of publications in order to demonstrate the progress in this area.

In the 1990s, a single dose administration of tetanus toxoid-loaded PLGA microparticles was investigated by Men and coworkers (Men et al. 1995), who demonstrated that this delivery system elicits similar or superior T-cell and antibody response compared to alum formulations. The same group later investigated a PLGA particulate delivery system loaded with malaria-specific cytotoxic T-lymphocyte CTL peptides, which was capable of inducing antigen-specific CTL responses in vivo (Men et al. 1997). They showed that injection of mice with a short CTL epitope—microencapsulated or adsorbed on empty PLGA—enhanced specific CTL responses comparable to those obtained with incomplete Freund's adjuvant. Similar studies were performed by Igartua et al. where PLGA microspheres were loaded with SPf66 (synthetic malarial antigen) (Igartua et al. 2008). The aim was to evaluate the effect of γ-irradiation on the biopharmaceutical properties of particles. It was demonstrated that subcutaneous administration of irradiated and nonirradiated microspheres into mice induced a similar immune response (IgG, IgG1, IgG2a levels), to the one observed when SPf66 was emulsified with Freund's complete adjuvant. Preclinical studies by Hamdy et al. showed that PLGA particles were efficiently taken up in vitro by murine DCs (Elamanchili et al. 2007). The same study also demonstrated that the Toll-like receptor ligand (TLR) 4 was significantly more effective at inducing maturation of DCs in vitro when provided in PLGA particles rather than in the soluble form (Hamdy et al. 2008).

Experiments carried out by Uchida et al. using ovalbumin (OVA) entrapped in PLGA particles (Uchida et al. 1996) showed that only one subcutaneous administration was necessary to generate a maximal OVA-specific antibody response. Furthermore, PLGA-OVA encapsulated particles were more effective at inducing responses than complete Freund's adjuvant combined with soluble OVA. Similar results were obtained by a different group using bovine serum albumin (BSA) with the immunogen entrapped in PLGA microparticles (Igartua et al. 1997, 1998). Interestingly, it was reported that PLGA microparticles decorated with anti-CD40 antibody and human IgG immobilized on the surface induced enhanced maturation and activation of DCs (Kempf et al. 2003).

Another important aspect to be considered is the stability of the encapsulated cargo. It has been reported that some protein antigens aggregate or degrade upon entrapment or release from the matrix. These problems have been partially solved by optimizing manufacturing parameters, through the addition of stabilizers and by avoiding exposure of proteins to organic solvents (Crotts and Park 1998; Zhu et al. 2000; Tamber et al. 2005).

Although much progress has been made and encouraging results have been reported, none of the vaccine delivery systems investigated have yet been approved for clinical use. This is most probably due to the fact that the co-delivery of antigens and adjuvants is very limited. Furthermore, the slow release mechanism of the antigen from the particles is possibly not favorable to induce a strong immune response. In addition to the aforementioned reasons, technological and economic issues also play an important role (Johansen et al. 2007). However, in contrast to vaccine delivery, various drug delivery systems based on PLGA, PLA, and PGA particulates have been developed, clinically approved, and commercialized. They are applied, for instance, in cancer treatment, acromegaly, periodontal disease, and schizophrenia (Danhier et al. 2012; Choi et al. 2012).

In addition to PLGA, other materials such as chitosan (CS) and its derivatives have been applied in immunotherapy. In general, CS has been in use for decades in biomedical applications due to its non-toxicity, biodegradability, and abundance in nature (the polymer is derived from chitin, a component of exoskeletons of crustaceans and insects) (Pangburn et al. 1984). Alonso and coworkers were among the first to study different CS systems for protein delivery, with focus on nasal and ocular administration (Calvo et al. 1997; Fernandez-Urrusuno et al. 1999; De Campos et al. 2001; Vila et al. 2004; Angelatos et al. 2006). More recently, this group demonstrated the possibility of intranasal delivery of hepatitis B antigens using different CS-based nanocarriers (Prego et al. 2010). In parallel, the group of Kawahima, Pan, and Leong demonstrated the potential of CS particulates for oral and intestinal delivery. Moreover, nanoparticles synthesized by salt-induced complex coacervation of cDNA and polycations such as gelatin and chitosan were evaluated as gene delivery vehicles for a variety of cell lines (Leong et al. 1998; Kawashima et al. 2000; Pan et al. 2002).

Another interesting platform, based on acetalated dextran (Ac-DEX), was described by the group of Fréchet for small interfering RNA (siRNA) delivery (Cui et al. 2012). Such an acid-sensitive, biocompatible, and biodegradable microparticulate delivery system demonstrated efficient gene knockdown in HeLa-luc cells while exhibiting minimal toxicity (Cohen et al. 2011). Since controlled, nonviral, intracellular delivery of genetic material for vaccines remains a major challenge, the authors studied the effect of co-encapsulating DNA and antigens in the Ac-DEX delivery system in vivo. A cytotoxicity assay proved that co-encapsulation of a model antigen protein (OVA) and an immunostimulatory agent (CpG DNA) in particles led to superior CTL activity when compared to particles co-administrated with adjuvant in the soluble form (Beaudette et al. 2009). Another study on protective immunity utilizing an Ac-DEX delivery system have demonstrated that, depending on the degree and type of acetal modification, antigen presentation pathways (MHC class I and MHC class II) can also be tuned (Broaders et al. 2009). Furthermore, blends of Ac-DEX and poly(β-amino ester) particles, prepared by the double emulsion technique with controlled degradation kinetics and surface functionality, were successfully utilized for gene delivery (Cohen et al. 2010). In other studies, Ac-DEX particles were modified with mannose in order to enhance internalization and activation of APCs (Cui et al. 2011).

10.3 Hydrogel Nanoparticles (Nanogels)

Hydrogel nanoparticles or nanogels are cross-linked polymeric particles composed of amphiphilic or polyionic polymers (Oh et al. 2008). The materials used are generally biocompatible and possess moderate mechanical properties (Raemdonck et al. 2009). Their tunable size, the large surface area available for bioconjugation, their injectability and the possibility of achieving prolonged release make them interesting delivery platforms (Carvalho et al. 2011; Wu et al. 2012). Nanogels made from natural polymers (e.g., alginate) and synthetic hydrogel nanoparticles based on poly(vinyl alcohol) (PVA), poly(ethylene oxide) (PEO), poly(ethyleneimine) (PEI), poly(vinyl pyrrolidone) (PVP), and poly(*N*-isopropyl acrylamide) (PNIPAm) have been widely investigated in drug delivery applications (Vinogradov et al. 2002). The interior of the swollen networks provides a space for the incorporation of bioactive molecules (small molecule drugs, proteins, DNA, etc.), which can be released from the polymeric matrix under specific conditions (Kabanov and Vinogradov 2009). Regardless of the type of polymer, the release of the active compound from hydrogel nanoparticles follows a complex mechanism and depends on three main factors: (1) drug diffusion, (2) hydrogel matrix swelling (degree of cross-linking), and (3) the chemical affinity between drug and matrix (Dorwal 2012). The first hydrogel nanoparticles used for vaccination purposes were prepared by Kreuter et al., who synthesized poly(methyl methacrylate) (PMMA) particles by polymerizing methyl methacrylate monomers in the presence of influenza virus or by addition of the virus to preprepared PMMA (Kreuter and Speiser 1976). The results showed that the simultaneously polymerized system of methylacrylate monomers and viruses was considerably more effective than the simple addition of viruses to pre-polymerized polymer. Antibody responses were measured in mice and guinea pigs, the morphology of particles was investigated using electron microscopy and the immunological reactivity of the particles was studied using hemagglutination and antibody-binding assays. At the same time, the Sjoholm group explored polyacrylamide particles cross-linked with *N*, *N*′-methylene-bis-acrylamide and incorporated human serum albumin (HSA) (Ekman et al. 1976). Recently, Fréchet and coworkers cross-linked polyacrylamide hydrogel particles with acid-sensitive moieties and co-administrated them with immunostimulatory DNA. They paid particular attention to the impact of particle size on the activation of T-cells (Cohen et al. 2009). In contrast to previous reports, the authors suggested that there was no significant difference in the magnitude of T-cell activation between particles of micro- and nanometer size carrying protein antigens.

In addition to research on polyacrylamide hydrogel particles, the group of Sjoholm extensively studied polyacryldextran and polyacryl starch during the 1980s in order to improve the biodegradability of these nanoparticles. Enzyme kinetics, release profiles, surface localization, and heat stability were investigated. Additionally, the degradation of microparticles in serum and in the target organelles, the lysosomes, was tested in vitro (Edman et al. 1980; Artursson et al. 1984). This was followed by the encapsulation of HSA and mouse serum albumin (MSA) antigens, and the investigation of their ability to stimulate an immune response in mice. In the case of MSA entrapped in particles, no detectable response was obtained after administration, while encapsulated HSA induced a dose-dependent immune response. However, injection of free HSA or in combination with empty microparticles did not cause any response (Artursson et al. 1985).

A decade later, in the 1990s, the group of Akiyoshi described different polysaccharides nanogels, which were further developed for the delivery of proteins such as insulin, interleukin 12, and HER2 (Akiyoshi et al. 1991; Akiyoshi et al. 1993; Hirakura et al. 2010). This work triggered research interest towards polysaccharide-based nanogels and their chemical and surface variations for protein and DNA delivery. Polysaccharide-based vaccine delivery systems obtained from CS and dextran sulfate (DEXS) loaded with capsid protein of HIV-1 were studied, and a specific immune response was observed in mice with high production of antibodies (Weber et al. 2010). Other groups focused on targeting particular DC receptors using dextran-based nanogels, which were surface decorated with alginate or mannosylated alginate carrying model antigen OVA. It was demonstrated that the delivery of OVA using nanogels with mannose surface decoration was superior to free OVA for the induction of interferon-γ production by T-lymphocytes (Thomann-Harwood et al. 2013).

In the same context, alginate and its derivatives have been thoroughly studied. This polymer is polyionic and can form intermolecular electrostatic bonds with multivalent cations. Hence, polyvalent cations (e.g., Ca^{2+}, Zn^{2+}) induce cross-linking and lead to the formation of particles. The incorporation of model proteins, such as BSA, human hemoglobin, and *Helicobacter pylori* antigen, into these matrices has been investigated (Leonard et al. 2004; George and Abraham 2006), and the results suggested that alginate particulate delivery platforms have potential as vaccine nanocarriers.

Responsive systems that release active compounds upon environmental changes, such as pH-responsive nanogels for endosomal release and intracellular delivery, are of particular interest. The ability of pH-sensitive nanogels to escape from endosomes, kill the cells, and migrate to adjacent cells (in a virus-like way) is very promising for vaccine delivery (Lee et al. 2008). A detailed discussion of pH-responsive systems can be found in a recent review (Ferreira et al. 2013). Thermo-responsive nanogels are mostly based on PNIPAm and are most frequently used for conventional drug delivery. A general overview about past, present, and future prospects regarding nanogels can be found elsewhere (Amidi et al. 2010; Garcia-Fuentes and Alonso 2012).

10.4 Micro- and Nanocapsules

Micro- and nanocapsules are hollow or filled spherically shaped structures that can be used to encapsulate pharmaceuticals (Nill 2005). Polypeptide assemblies such as bio-nanocapsules (BNCs) from the hepatitis B virus surface antigen (HBsAg) (Yamada et al. 2001; Jung et al. 2011) or poly-γ-glutamic acid capsules also fall under this definition (Rhie et al. 2003; Schneerson et al. 2003; Wang et al. 2004; Chabot et al. 2004; Joyce et al. 2006). Although these natural structures are fascinating, their further description is beyond the focus of this chapter.

The invention of the layer-by-layer (LbL) microencapsulation technique in the late 1990s by Decher inspired scientists to apply this technique to vaccine delivery (Decher 1997). Iterative and alternating adsorption of oppositely charged polyelectrolytes, i.e., polycations and polyanions, was used to build-up multilayers on colloidal substrates, thereby forming polyelectrolyte microcapsules (Donath et al. 1998;

Sukhorukov et al. 1998; Caruso et al. 1998) as carrier systems for vaccines (De Geest et al. 2009; De Temmerman et al. 2011;). A variety of polyanion/polycation pairs have been used, poly(styrene sulfonate)/ poly(allylamine hydrochloride) (PSS/PAH) (Donath et al. 1998; Sukhorukov et al. 1998; Caruso et al. 1998), dextran sulfate/poly-L-arginine (DEXS/pARG) (De Geest et al. 2006), DNA /poly-L-lysine or poly-L-glutamic acid (DNA/PLL or PGA), (Johnston et al. 2005; Johnston and Caruso 2007) and poly(methacrylic acid)/poly(vinylpyrrolidone) (PMA/PVPON) to name a few (De Geest et al. 2009; Chong et al. 2009). These examples include non-degradable (e.g., PSS/PAH) and biodegradable (e.g., DEXS/pARG) polyelectrolytes, which can be selected according to specific requirements.

The choice of polymer defines the surface chemistry, which strongly influences the interaction with the targeted APC (De Rose et al. 2008). As template, i.e., 3D structure for polyelectrolyte adsorption, several materials have been used such as dextran-hydroxyethyl methacrylate microgels copolymerized with dimethylaminoethyl methacrylate (DEX-HEMA-DMAEMA) (De Geest et al. 2005), SiO_2 (Chong et al. 2009), and $CaCO_3$ particles (De Koker et al. 2007; De Temmerman et al. 2011). The latter can be used to incorporate the antigen within the particles, which are dissolved upon addition of EDTA after LbL deposition, and thus combine minimal stress for the antigens while obtaining high loading efficiencies (De Temmerman et al. 2011). Capsules do not have to be loaded with biological entities before the polyelectrolyte multilayer is established; however, "post-loading" the capsules after removal of the scarifying template may affect their integrity (De Geest et al. 2006).

Different release mechanisms have been reported based on these materials. Polyelectrolyte capsules with HEMA microgel cores release their payload through biodegradation of the gel, which also causes rupture of the multilayer shell (De Geest et al. 2005; De Geest et al. 2009). Biodegradable DEXS/pARG capsules are efficiently taken up by DCs through the macropinocytotic route and release their payload upon intracellular degradation (De Koker et al. 2007; De Koker et al. 2009a). This system was modified with immunopotentiators to further activate DCs. Thus, a pulmonary adaptive immune response was stimulated, which was characterized by induction of a strong Th17-polarized response (De Koker et al. 2009b; De Temmerman et al. 2011; De Temmerman et al. 2012). PMA_{SH}/PVPON polyelectrolyte capsules, which are controllable in their size and peptide vaccine loading, were designed to disintegrate and release the peptide vaccine by rupture of disulfide linkages upon exposure to reductive conditions. It was shown that they were internalized into APCs, released, and presented their payload within an MHC class I context to elicit an immune response (Chong et al. 2009; De Rose et al. 2008). Furthermore, whole encapsulated OVA was internalized by mouse APCs, which led to the presentation of OVA epitopes and subsequent activation of OVA-specific $CD4^+$ and $CD8^+$ T-cells in vitro, and greater proliferation in vivo (Sexton et al. 2009).

A relatively novel approach is DNA vaccination, where a plasmid encoding a protein from a pathogen is introduced into human cells. Once the plasmid is inside the cells, the pathogen's protein is produced, recognized as foreign and displayed on the host cells surface, which activates the immune system. Drawbacks of this approach are the poor targeting of APCs and low immunogenicity. Microcapsules have been tested using this approach (Jewell and Lynn 2008), where efficient delivery and processing of DNA, i.e., production of sufficient amount of encoded antigen, is of key importance (Saurer et al. 2009). In these systems, DNA is used as natural polyanion in the LbL technique, which leads to reduced enzymatic degradation of the polynucleotides (Briones et al. 2001). Erodible microcapsules with well-controlled DNA loading, accomplished through the LbL technique, were useful in transporting the payload into a mouse macrophage cell line, and led to a gradual release of plasmid DNA under physiological conditions (Saurer et al. 2009).

10.5 Dendrimers

The word "dendrimer" arises from the Greek words *dendron*, "tree" and *meros*, "part." The core in the center of the structure has at least one branching point and three anchoring groups. Thus, in second generations, at least three monomers can be incorporated. With every additional generation, the molecular mass and the number of surface groups increase by a factor of two. The stepwise synthesis allows highly symmetric structures with a well-defined molecular weight to be produced. The branches of the polymer arrange in a way that the dendrimer possesses a globular structure. In contrast to linear or randomly branched polymers, they are spatially structured. Dendrimeric polymers were first reported by Vögtle et al. (Buhleier et al. 1978). In the past years, the synthesis of dendrimers evolved to allow the production of larger molecules (Denkewalter et al. 1981; Tomalia et al. 1985; Hawker and Frechet 1990).

A high degree of branching leads to high density of functionalities on the surface, and therefore to a multivalent structure. With higher dendrimer generation numbers, the dendrimer properties become dominated by the end groups and its size, while the core is shielded (Heegaard et al. 2010). Additionally, a high degree of functionalities at the end groups allows various modifications, which makes dendrimers attractive as vaccine delivery systems.

Traditionally, antigens are attached to small proteins in order to increase their immunogenicity (Van Regenmortel et al. 1988). This effect can also be achieved by attaching antigens to dendrimers, with the advantage of well-defined molecular structures provided by the dendrimer. Such systems with a high number of T- and B-cell epitopes also enhance the immunostimulatory effect through multiple interactions with cells of the immune system (de Oliveira et al. 1994; Wang et al. 1995; Cavenaugh et al. 2003; Sadler and Tam 2002; Iglesias et al. 2005; Fujita and Taguchi 2011; Tarradas et al. 2012).

In 1988, Tam and coworkers developed the so-called multiple antigenic peptide (MAP) system, the first immunostimulatory dendrimer-like system (Posnett et al. 1988; Tam 1988). The dendritic part of the MAP was built from lysine. MAP systems are asymmetrical and possess a wedge-like shape, because the α- and ε-amines of the core-lysine act as an anchor point for the next generation, and the carboxylic group of the core-lysine is modified with other amino acids or lipids. Frequently, the second- or third-generation lysine dendrimers are used, presenting either four or eight anchor points for small molecular weight antigens, e.g., short peptides (Heegaard et al. 2010). The simple design and ease of synthesis has contributed to the success of this system (Crespo et al. 2005), and MAP dendrimers are promising candidates for vaccine delivery against AIDS, (Defoort et al. 1992) foot and mouth disease (de Oliveira et al. 2002), influenza (Zhao et al. 2010), swine fever (Tarradas et al. 2012), or malaria (Nardin et al. 2001; Kublin et al. 2002). However, when some of these vaccines were tested in Phase I clinical trial, it was reported that simple linear systems worked as well (Calvo-Calle et al. 2006). While some groups reported that adjuvants are needed to make MAP systems more efficient and that best results were obtained with saponin adjuvants (Moreno et al. 1999), others reported that MAPs exhibit self-adjuvanticity (Tam 1996; Olive et al. 2003;).

Compared to the lysine-based MAP systems, immunostimulatory systems based on other monomers are less common. Dendrimers used as vaccines often present specific carbohydrates on their surface. Carbohydrates found at the outer cell membrane are important in a number of biological processes, such as intercellular interactions and cell recognition, which is of particular interest for vaccines. Anomalous patterns of carbohydrates are present, for example, in cancer cells (Shiao and Roy 2012; Baek et al. 2001; Roy and Baek 2003). Carbohydrates not only represent interesting antigens for cancer vaccines but also for vaccines against viruses, which are often heavily glycosylated (Niederhafner et al. 2008; Šebestík et al. 2012). Dendrimers decorated with carbohydrates can exhibit adjuvant properties (Sorensen et al. 2011), e.g., potentiation of immunogenicity can be achieved by conjugation of lectin-binding carbohydrates to antigenic-peptide modified poly(amidoamine) (PAMAM) dendrimers. This potentiation can be partially ascribed to increased binding and recognition of antigens by DCs (Sheng et al. 2008). Another example of self-adjuvanting systems are phosphoric acid-capped dendrimers, which target and activate monocytes (Poupot et al. 2006). Dendrimer nanoparticles can also be formulated by conjugation of a B-cell epitope to a hydrophobic dendrimer. The resulting amphiphiles can form immunogenic nanoparticles, which have an epitope outer layer and a polymer core (Skwarczynski et al. 2010).

Furthermore, dendrimers have been utilized to deliver DNA vaccines. Daftarian et al. conjugated MHC class II-targeting peptides onto a generation 5 (G5)-PAMAM), and reported a high transfection efficiency, specific targeting to the APCs and increased immunogenicity (Daftarian et al. 2011).

10.6 Polymersomes and Micelles

Block copolymers with amphiphilic properties can be synthesized using monomers with different polarities (i.e., hydrophobic/hydrophilic). The groups of Discher and Eisenberg discovered that amphiphilic block copolymers self-assemble in aqueous solution into vesicular structures (polymersomes) (Discher et al. 1999; Discher and Eisenberg 2002). Polymersomes have an aqueous cavity surrounded by a hydrophobic membrane. Drugs or antigens can be encapsulated in the hydrophilic core or hydrophobic interior of the polymersome membrane without destroying their functionality (Christian et al. 2009; Grzelakowski et al. 2009). In contrast to vesicles, micelles are smaller and have a hydrophobic core surrounded by a hydrophilic corona. According to vaccine delivery platform demands, where hydrophilic (antigen) and hydrophobic (adjuvants) cargos are required to be encapsulated simultaneously, the vesicular morphology is beneficial.

Polymersomes are tunable in size, and can be transported via the blood stream or into cells via endocytosis. An additional benefit of polymersomes is that different ligands can be attached onto the surface to yield targeted delivery systems (Egli et al. 2011a, b). Triggerable polymersome systems can also be created for controlled release (Johansen et al. 2007; Cabane et al. 2011; Du et al. 2012). Such systems were developed to transport siRNA into cells; the release was controlled through the incorporation of pH-triggerable linkers, which were cleaved inside the cells due to the acidic environment in lysosomes and caused the polymersomes to burst (Christian et al. 2009). The use of cationic diblock copolymers allows electrostatic interactions with DNA and RNA molecules, and therefore a higher loading efficiency compared to passive encapsulation (Korobko et al. 2005; Korobko et al. 2006). In contrast to liposomes, polymersomes offer a stable and chemically more tunable structure to protect the antigen or DNA/RNA against degradation processes (Hao et al. 2006). In the past few years, there have been countless reports on polymersome and micellar formulations for drug delivery and medical application (Brinkhuis et al. 2011; Renggli et al. 2011). Therefore, the examples discussed herein are restricted to those with a direct application to vaccination.

The group of Kataoka was one of the first to use polyionic complexed micelles for vaccine delivery (Harada and Kataoka 1995). They used poly(ethylene glycol)-poly(L-lysine) block copolymers (PEG-PLL) to form micelles, and exploited the reversible electrostatic interaction between DNA and polymer for self-assembly (Katayose and Kataoka 1997). The DNA was trapped within the core of the micelle surrounded by PEG, thereby protecting the DNA from chemical or enzymatic denaturation. Using a similar polymersome, Cheng et al. showed that a number of different proteins can be loaded into vesicles and released (Cheng et al. 2011). This provided a universally applicable platform for vaccine delivery, which does not need to be adapted to particular antigens.

Christian et al. designed a vaccine system by conjugating a TAT peptide to the exterior of the polymersome (Christian et al. 2007). Characterization showed that such surface modification increased the uptake to around 70,000 vesicles per cell. Loading of such targeted polymersomes with antigens or siRNA would thus be suitable for vaccine delivery. Due to the biocompatibility and biodegradability of most of the polymers used for polymersome formation, the resulting vaccine delivery systems would hold potential for clinical trials. Liposome-based antigen delivery is reportedly more immunogenic than the antigen alone (Childers et al. 2000), as liposomes mimic the uptake of viruses and bacteria better than the antigen itself (Brewer et al. 2004); hence, polymersomes may exhibit similar effects (Christian et al. 2007). One possible strategy to control the release, while benefiting from the higher stability of the polymersomes, would be to insert the antigen in the membrane of the vesicles in a way that could be presented to the immune system. The first studies made in this direction are the integration of membrane proteins into polymeric vesicles, and their functionalization (Stoenescu et al. 2004; Egli et al. 2011a, b). However and to the best of our knowledge, no such experiments have been carried out for vaccine delivery.

10.7 Conclusions

Scientists are trying to successfully increase efficacy of vaccines against severe diseases like AIDS, malaria, Leishmaniasis, or cancer. In general, the objectives of vaccine development are to make them safer, more effective, easy to administrate, and cheaper. Towards obtaining these goals, new drugs and delivery platforms are explored. Although some polymeric particulate platforms have already been approved and are commercialized as drug delivery applications, vaccine delivery is currently in its infancy. Each of the presented particulate systems show advantages and drawbacks, but fail to fully comply with general requirements like biocompatibility, non-toxicity, high encapsulation efficiency, versatile surface chemistry, and smart behavior. It will take a few more years of research to obtain synthetic-based carriers that completely fulfill these requirements.

Acknowledgments This work was supported by the Swiss National Science Foundation (SNSF), NCCR Nanosciences, Marie Curie Actions-Intra European fellowship (IEF) (p.n. 301398) (A. C.), the Holcim Stiftung (N. B.), and the Gerbert Rüf Stiftung (GRS-048/11) (T. S.). We also thank Gesine Gunkel and Elisa Nogueira for editorial help and Ruth Pfalzberger for her help with the preparation of the graphic.

References

Akiyoshi K, Yamaguchi S, Sunamoto J (1991) Self-aggregates of hydrophobic polysaccharide derivatives. Chem Lett 7:1263–1266. doi:10.1246/cl.1991.1263

Akiyoshi K, Deguchi S, Moriguchi N, Yamaguchi S, Sunamoto J (1993) Self-aggregates of hydrophobized polysaccharides in water. Formation and characteristics of nanoparticles. Macromolecules 26(12):3062–3068. doi:10.1021/ma00064a011

Amidi M, Mastrobattista E, Jiskoot W, Hennink WE (2010) Chitosan-based delivery systems for protein therapeutics and antigens. Adv Drug Deliv Rev 62(1):59–82. doi:10.1016/j.addr.2009.11.009

Andriano KP, Tabata Y, Ikada Y, Heller J (1999) In vitro and in vivo comparison of bulk and surface hydrolysis in absorbable polymer scaffolds for tissue engineering. J Biomed Mater Res. 48(5):602–612. doi: 10.1002/(sici)1097-4636(1999)48:5<602::aid-jbm3>3.0.co;2-6

Angelatos AS, Katagiri K, Caruso F (2006) Bioinspired colloidal systems via layer-by-layer assembly. Soft Matter 2(1):18–23. doi:10.1039/b511930h

Artursson P, Edman P, Laakso T, Sjoeholm I (1984) Characterization of polyacryl starch microparticles as carriers for proteins and drugs. J Pharm Sci 73(11):1507–1513. doi:10.1002/jps.2600731103

Artursson P, Edman P, Sjoeholm I (1985) Biodegradable microspheres. II: immune response to a heterologous and an autologous protein entrapped in polyacryl starch microparticles. J Pharmacol Exp Ther 234(1):255–260

Baek M-G, Rittenhouse-Olson K, Roy R (2001) Synthesis and antibody binding properties of glycodendrimers bearing the tumor related T-antigen. Chem Commun 3:257–258

Beaudette TT, Bachelder EM, Cohen JA, Obermeyer AC, Broaders KE, Frechet JM, Kang ES, Mende I, Tseng WW, Davidson MG, Engleman EG (2009) In vivo studies on the effect of co-encapsulation of CpG DNA and antigen in acid-degradable microparticle vaccines. Mol Pharm 6(4):1160–1169. doi:10.1021/mp900038e

Birrenbach G (1973) Über Mizellpolymerisate, mögliche Einschlussverbindungen (Nanokapseln) und deren Eigenung als Adjuvantien. ETH, Zürich

Brewer JM, Pollock KGJ, Tetley L, Russell DG (2004) Vesicle size influences the trafficking, processing, and presentation of antigens in lipid vesicles. J Immunol 173(10):6143–6150

Brinkhuis RP, Rutjes FPJT, van Hest JCM (2011) Polymeric vesicles in biomedical applications. Polym Chem 2(7):1449–1462

Briones M, Singh M, Ugozzoli M, Kazzaz J, Klakamp S, Ott G, O'Hagan D (2001) The preparation, characterization, and evaluation of cationic microparticles for DNA vaccine delivery. Pharm Res 18(5):709–712

Broaders KE, Cohen JA, Beaudette TT, Bachelder EM, Frechet JM (2009) Acetalated dextran is a chemically and biologically tunable material for particulate immunotherapy. Proc Natl Acad Sci U S A 106(14):5497–5502. doi:10.1073/pnas.0901592106

Buhleier E, Wehner W, Vogtle F (1978) Cascade-chain-like and nonskid-chain-like syntheses of molecular cavity topologies. Synthesis-Stuttgart 2:155–158

Byrd W, de Lorimier A, Zheng Z-R, Cassels FJ (2005) Microencapsulated subunit vaccine approach to enterotoxigenic Escherichia coli and other mucosal pathogens. Adv Drug Deliv Rev 57(9):1362–1380. doi:10.1016/j.addr.2005.01.014

Cabane E, Malinova V, Menon S, Palivan CG, Meier W (2011) Photoresponsive polymersomes as smart, triggerable nanocarriers. Soft Matter 7(19):9167–9176

Calvo P, Remunan-Lopez C, Vila-Jato JL, Alonso MJ (1997) Chitosan and chitosan/ethylene oxide-propylene oxide block copolymer nanoparticles as novel carriers for proteins and vaccines. Pharm Res 14(10):1431–1436

Calvo-Calle JM, Oliveira GA, Watta CO, Soverow J, Parra-Lopez C, Nardin EH (2006) A linear peptide containing minimal T- and B-cell epitopes of plasmodium falciparum circumsporozoite protein elicits protection against transgenic sporozoite challenge. Infect Immun 74(12): 6929–6939. doi:10.1128/iai.01151-06

Caruso F, Caruso RA, Moehwald H (1998) Nanoengineering of inorganic and hybrid hollow spheres by colloidal templating. Science 282(5391):1111–1114. doi:10.1126/science.282.5391.1111

Carvalho V, Castanheira P, Madureira P, Ferreira SA, Costa C, Teixeira JP, Faro C, Vilanova M, Gama M (2011) Self-assembled dextrin nanogel as protein carrier: controlled release and biological activity of IL-10. Biotechnol Bioeng 108(8):1977–1986. doi:10.1002/bit.23125

Cavenaugh J, Wang H-K, Hansen C, Smith R, Herron J (2003) How well can a T-cell epitope replace its parent carrier protein? A dose-response study. Pharm Res 20(4):591–596. doi:10.1023/a:1023242631413

Chabot DJ, Scorpio A, Tobery SA, Little SF, Norris SL, Friedlander AM (2004) Anthrax capsule vaccine protects against experimental infection. Vaccine 23(1):43–47. doi:10.1016/j.vaccine.2004.05.029

Chacko RT, Ventura J, Zhuang J, Thayumanavan S (2012) Polymer nanogels: a versatile nanoscopic drug delivery platform. Adv Drug Deliv Rev 64(9):836–851. doi:10.1016/j.addr.2012.02.002

Cheng R, Meng F, Ma S, Xu H, Liu H, Jing X, Zhong Z (2011) Reduction and temperature dual-responsive crosslinked polymersomes for targeted intracellular protein delivery. J Mater Chem 21(47):19013–19020

Childers NK, Miller KL, Tong G, Llarena JC, Greenway T, Ulrich JT, Michalek SM (2000) Adjuvant activity of monophosphoryl lipid A for nasal and oral immunization with soluble or liposome-associated antigen. Infect Immun 68(10):5509–5516. doi:10.1128/iai.68.10.5509-5516.2000

Choi J-S, Seo K, Yoo J-W (2012) Recent advances in PLGA particulate systems for drug delivery. J Pharmaceut Invest 42(3):155–163. doi:10.1007/s40005-012-0024-5

Chong S-F, Sexton A, De Rose R, Kent SJ, Zelikin AN, Caruso F (2009) A paradigm for peptide vaccine delivery using viral epitopes encapsulated in degradable polymer hydrogel capsules. Biomaterials 30(28):5178–5186. doi:10.1016/j.biomaterials.2009.05.078

Christian NA, Milone MC, Ranka SS, Li G, Frail PR, Davis KP, Bates FS, Therien MJ, Ghoroghchian PP, June CH, Hammer DA (2007) Tat-functionalized near-infrared emissive polymersomes for dendritic cell labeling. Bioconjug Chem 18(1):31–40

Christian DA, Cai S, Bowen DM, Kim Y, Pajerowski JD, Discher DE (2009) Polymersome carriers: from self-assembly to siRNA and protein therapeutics. Eur J Pharm Biopharm 71(3): 463–474

Cohen JA, Beaudette TT, Tseng WW, Bachelder EM, Mende I, Engleman EG, Frechet JM (2009) T-cell activation by antigen-loaded pH-sensitive hydrogel particles in vivo: the effect of particle size. Bioconjug Chem 20(1):111–119. doi:10.1021/bc800338n

Cohen JA, Beaudette TT, Cohen JL, Broaders KE, Bachelder EM, Frechet JM (2010) Acetal-modified dextran microparticles with controlled degradation kinetics and surface functionality for gene delivery in phagocytic and non-phagocytic cells. Adv Mater 22(32):3593–3597. doi:10.1002/adma.201000307

Cohen JL, Schubert S, Wich PR, Cui L, Cohen JA, Mynar JL, Frechet JM (2011) Acid-degradable cationic dextran particles for the delivery of siRNA therapeutics. Bioconjug Chem 22(6):1056–1065. doi:10.1021/bc100542r

Correia-Pinto JF, Csaba N, Alonso MJ (2013) Vaccine delivery carriers: insights and future perspectives. Int J Pharm 440(1):27–38. doi:10.1016/j.ijpharm.2012.04.047

Couvreur P, Puisieux F (1993) Nano- and microparticles for the delivery of polypeptides and proteins. Adv Drug Deliv Rev 10(2–3):141–162. doi:10.1016/0169-409x(93)90046-7

Crespo L, Sanclimens G, Pons M, Giralt E, Royo M, Albericio F (2005) Peptide and amide bond-containing dendrimers. Chem Rev 105(5):1663–1681. doi:10.1021/cr030449l

Crotts G, Park TG (1998) Protein delivery from poly(lactic-co-glycolic acid) biodegradable microspheres: release kinetics and stability issues. J Microencapsul 15(6):699–713. doi:10.3109/02652049809008253

Cui L, Cohen JA, Broaders KE, Beaudette TT, Frechet JM (2011) Mannosylated dextran nanoparticles: a pH-sensitive system engineered for immunomodulation through mannose targeting. Bioconjug Chem 22(5):949–957. doi:10.1021/bc100596w

Cui L, Cohen JL, Chu CK, Wich PR, Kierstead PH, Frechet JM (2012) Conjugation chemistry through acetals toward a dextran-based delivery system for controlled release of siRNA. J Am Chem Soc 134(38):15840–15848. doi:10.1021/ja305552u

Daftarian P, Kaifer AE, Li W, Blomberg BB, Frasca D, Roth F, Chowdhury R, Berg EA, Fishman JB, Al Sayegh HA, Blackwelder P, Inverardi L, Perez VL, Lemmon V, Serafini P (2011) Peptide-conjugated PAMAM dendrimer as a universal DNA vaccine platform to target antigen-presenting cells. Cancer Res 71(24):7452–7462. doi:10.1158/0008-5472.can-11-1766

Danhier F, Ansorena E, Silva JM, Coco R, Le Breton A, Preat V (2012) PLGA-based nanoparticles: an overview of biomedical applications. J Control Release 161(2):505–522. doi:10.1016/j.jconrel.2012.01.043

De Campos AM, Sanchez A, Alonso MJ (2001) Chitosan nanoparticles: a new vehicle for the improvement of the delivery of drugs to the ocular surface. Application to cyclosporin A. Int J Pharm 224(1–2):159–168. doi:10.1016/s0378-5173(01)00760-8

De Geest BG, Dejugnat C, Sukhorukov GB, Braeckmans K, De Smedt SC, Demeester J (2005) Self-rupturing microcapsules. Adv Mater 17:2357–2361

De Geest BG, Vandenbroucke RE, Guenther AM, Sukhorukov GB, Hennink WE, Sanders NN, Demeester J, De Smedt SC (2006) Intracellularly degradable polyelectrolyte microcapsules. Adv Mater 18(8):1005–1009. doi:10.1002/adma.200502128

De Geest BG, Sanders NN, Sukhorukov GB, Demeester J, De Smedt SC (2007) Release mechanisms for polyelectrolyte capsules. Chem Soc Rev 36(4):636–649. doi:10.1039/b600460c

De Geest BG, De Koker S, Demeester J, De Smedt SC, Hennink WE (2009) Pulsed in vitro release and in vivo behavior of exploding microcapsules. J Control Release 135(3):268–273. doi:10.1016/j.jconrel.2009.01.017

De Koker S, De Geest BG, Cuvelier C, Ferdinande L, Deckers W, Hennink WE, De Smedt S, Mertens N (2007) In vivo cellular uptake, degradation, and biocompatibility of polyelectrolyte microcapsules. Adv Funct Mater 17(18):3754–3763. doi:10.1002/adfm.200700416

De Koker S, De Geest BG, Singh SK, De Rycke R, Naessens T, Van Kooyk Y, Demeester J, De Smedt SC, Grooten J (2009a) Polyelectrolyte microcapsules as antigen delivery vehicles to dendritic cells: uptake, processing, and cross-presentation of encapsulated antigens. Angew Chem Int Ed 48(45):8485–8489. doi:10.1002/anie.200903769

De Koker S, Naessens T, De Geest BG, Bogaert P, Demeester J, De Smedt S, Grooten J (2009b) Biodegradable polyelectrolyte microcapsules: antigen delivery tools with Th17 skewing activity after pulmonary delivery. J Immunol 184(1):203–211. doi:10.4049/jimmunol.0803591

de Oliveira GA, Clavijo P, Nussenzweig RS, Nardin EH (1994) Immunogenicity of an alum-adsorbed synthetic multiple-antigen peptide based on B- and T-cell epitopes of the Plasmodium falciparum CS protein: possible vaccine application. Vaccine 12(11):1012–1017. doi:http://dx.doi.org/10.1016/0264-410X(94)90337-9

de Oliveira E, Villén J, Giralt E, Andreu D (2002) Synthetic approaches to multivalent lipopeptide dendrimers containing cyclic disulfide epitopes of foot-and-mouth disease virus. Bioconjug Chem 14(1):144–152. doi:10.1021/bc025577f

De Rose R, Zelikin AN, Johnston APR, Sexton A, Chong S-F, Cortez C, Mulholland W, Caruso F, Kent SJ (2008) Binding, internalization, and antigen presentation of vaccine-loaded nanoengineered capsules in blood. Adv Mater 20(24):4698–4703. doi:10.1002/adma.200801826

De Temmerman ML, Rejman J, Demeester J, Irvine DJ, Gander B, De Smedt SC (2011) Particulate vaccines: on the quest for optimal delivery and immune response. Drug Discov Today 16(13–14):569–582. doi:10.1016/j.drudis.2011.04.006

De Temmerman M-L, Rejman J, Lucas B, Vandenbroucke RE, Libert C, Demeester J, De Smedt SC (2012) Modulation of dendritic cells by lipid grafted polyelectrolyte microcapsules. Adv Funct Mater 22(20):4236–4243. doi:10.1002/adfm.201103151

Decher G (1997) Fuzzy nanoassemblies: toward layered polymeric multicomposites. Science 277(5330):1232–1237. doi:10.1126/science.277.5330.1232

Defoort JP, Nardelli B, Huang W, Ho DD, Tam JP (1992) Macromolecular assemblage in the design of a synthetic AIDS vaccine. Proc Natl Acad Sci U S A 89(9):3879–3883. doi:10.1073/pnas.89.9.3879

Denkewalter RG, Kolc J, Lukasavage WJ (1981) Macromolecular highly branched homogeneous compound based on lysine units. Application: US Patent 1979-27622 4289872

Discher DE, Eisenberg A (2002) Polymer vesicles. Science 297(5583):967–973

Discher BM, Won YY, Ege DS, Lee JCM, Bates FS, Discher DE, Hammer DA (1999) Polymersomes: tough vesicles made from diblock copolymers. Science 284(5417):1143–1146

Donath E, Sukhorukov GB, Caruso F, Davis SA, Mohwald H (1998) Novel hollow polymer shells by colloid-templated assembly of polyelectrolytes. Angew Chem Int Ed 37(16):2202–2205

Dorwal D (2012) Nanogels as novel and versatile pharmaceuticals. Int J Pharm Pharm Sci 4(3):67–74

Du Y, Chen W, Zheng M, Meng F, Zhong Z (2012) pH-sensitive degradable chimaeric polymersomes for the intracellular release of doxorubicin hydrochloride. Biomaterials 33(29): 7291–7299

Edman P, Ekman B, Sjoholm I (1980) Immobilization of proteins in microspheres of biodegradable polyacryldextran. J Pharm Sci 69(7):838–842

Egli S, Nussbaumer MG, Balasubramanian V, Chami M, Bruns N, Palivan C, Meier W (2011a) Biocompatible functionalization of polymersome surfaces: a new approach to surface immobilization and cell targeting using polymersomes. J Am Chem Soc 133(12):4476–4483

Egli S, Schlaad H, Bruns N, Meier W (2011b) Functionalization of block copolymer vesicle surfaces. Polymers 3(4):252–280. doi:10.3390/polym3010252
Ekman B, Lofter C, Sjoholm I (1976) Incorporation of macromolecules in microparticles: preparation and characteristics. Biochemistry 15(23):5115–5120. doi:10.1021/bi00668a026
Elamanchili P, Lutsiak CME, Hamdy S, Diwan M, Samuel J (2007) "Pathogen-mimicking" nanoparticles for vaccine delivery to dendritic cells. J Immunother 30(4):378–395. doi:10.1097/CJI.1090b1013e31802cf31803e31803
Etheridge ML, Campbell SA, Erdman AG, Haynes CL, Wolf SM, McCullough J (2013) The big picture on nanomedicine: the state of investigational and approved nanomedicine products. Nanomedicine 9(1):1–14. doi:10.1016/j.nano.2012.05.013
Fernandez-Urrusuno R, Calvo P, Remunan-Lopez C, Vila-Jato JL, Alonso MJ (1999) Enhancement of nasal absorption of insulin using chitosan nanoparticles. Pharm Res 16(10):1576–1581
Ferreira SA, Gama FM, Vilanova M (2013) Polymeric nanogels as vaccine delivery systems. Nanomedicine 9(2):159–173. doi:10.1016/j.nano.2012.06.001
Fujita Y, Taguchi H (2011) Current status of multiple antigen-presenting peptide vaccine systems: application of organic and inorganic nanoparticles. Chem Cent J 5(1):1–8. doi:10.1186/1752-153x-5-48
Garcia-Fuentes M, Alonso MJ (2012) Chitosan-based drug nanocarriers: where do we stand? J Control Release 161(2):496–504. doi:10.1016/j.jconrel.2012.03.017
George M, Abraham TE (2006) Polyionic hydrocolloids for the intestinal delivery of protein drugs: alginate and chitosan—a review. J Control Release 114(1):1–14. doi:10.1016/j.jconrel.2006.04.017
Grzelakowski M, Onaca O, Rigler P, Kumar M, Meier W (2009) Immobilized protein-polymer nanoreactors. Small 5(22):2545–2548
Hamdy S, Molavi O, Ma Z, Haddadi A, Alshamsan A, Gobti Z, Elhasi S, Samuel J, Lavasanifar A (2008) Co-delivery of cancer-associated antigen and Toll-like receptor 4 ligand in PLGA nanoparticles induces potent CD8+ T cell-mediated anti-tumor immunity. Vaccine 26(39):5046–5057. doi:10.1016/j.vaccine.2008.07.035
Hamdy S, Haddadi A, Hung RW, Lavasanifar A (2011) Targeting dendritic cells with nano-particulate PLGA cancer vaccine formulations. Adv Drug Deliv Rev 63(10–11):943–955. doi:10.1016/j.addr.2011.05.021
Hao J, Kwissa M, Pulendran B, Murthy N (2006) Peptide crosslinked micelles: a new strategy for the design and synthesis of peptide vaccines. Int J Nanomedicine 1(1):97–103
Harada A, Kataoka K (1995) Formation of polyion complex micelles in an aqueous milieu from a pair of oppositely-charged block-copolymers with poly(ethylene glycol) segments. Macromolecules 28(15):5294–5299
Hawker CJ, Frechet JMJ (1990) Preparation of polymers with controlled molecular architecture. A new convergent approach to dendritic macromolecules. J Am Chem Soc 112(21):7638–7647. doi:10.1021/ja00177a027
Heegaard PMH, Boas U, Sorensen NS (2010) Dendrimers for vaccine and immunostimulatory uses. A review. Bioconjug Chem 21(3):405–418. doi:10.1021/bc900290d
Hirakura T, Yasugi K, Nemoto T, Sato M, Shimoboji T, Aso Y, Morimoto N, Akiyoshi K (2010) Hybrid hyaluronan hydrogel encapsulating nanogel as a protein nanocarrier: new system for sustained delivery of protein with a chaperone-like function. J Control Release 142(3):483–489. doi:10.1016/j.jconrel.2009.11.023
Holdermann K, Greiling W (1954) Im Banne der Chemie-Carl Bosch, Leben und Werk, 2nd edn. Econ-Verlag, Düsseldorf
Igartua M, Hernandez RM, Esquisabel A, Gascon AR, Calvo MB, Pedraz JL (1997) Influence of formulation variables on the in-vitro release of albumin from biodegradable microparticulate systems. J Microencapsul 14(3):349–356. doi:10.3109/02652049709051138
Igartua M, Hernandez RM, Esquisabel A, Gascon AR, Calvo MB, Pedraz JL (1998) Stability of BSA encapsulated into PLGA microspheres using PAGE and capillary electrophoresis. Int J Pharm 169(1):45–54. doi:10.1016/s0378-5173(98)00101-x

Igartua M, Hernandez RM, Rosas JE, Patarroyo ME, Pedraz JL (2008) Gamma-irradiation effects on biopharmaceutical properties of PLGA microspheres loaded with SPf66 synthetic vaccine. Eur J Pharm Biopharm 69(2):519–526. doi:10.1016/j.ejpb.2007.12.014

Iglesias E, Aguilar JC, Cruz LJ, Reyes O (2005) Broader cross-reactivity after conjugation of V3 based multiple antigen peptides to HBsAg. Mol Immunol 42(1):99–104. doi: http://dx.doi.org/10.1016/j.molimm.2004.06.038

Janeway CA, Travers P, Walport M, Shlomchik M (2004) Immunobiology: the immune system in health and disease, 6th edn. Churchill Livingstone, Edinburgh

Jewell CM, Lynn DM (2008) Multilayered polyelectrolyte assemblies as platforms for the delivery of DNA and other nucleic acid-based therapeutics. Adv Drug Deliv Rev 60(9):979–999. doi:10.1016/j.addr.2008.02.010

Johansen P, Martinez Gomez JM, Gander B (2007) Development of synthetic biodegradable microparticulate vaccines: a roller coaster story. Expert Rev Vaccines 6(4):471–474. doi:10.1586/14760584.6.4.471

Johnston AP, Caruso F (2007) Exploiting the directionality of DNA: controlled shrinkage of engineered oligonucleotide capsules. Angew Chem Int Ed Engl 46(15):2677–2680. doi:10.1002/anie.200605135

Johnston APR, Read ES, Caruso F (2005) DNA multilayer films on planar and colloidal supports: sequential assembly of like-charged polyelectrolytes. Nano Lett 5(5):953–956. doi:10.1021/nl050608b

Joyce J, Cook J, Chabot D, Hepler R, Shoop W, Xu Q, Stambaugh T, Aste-Amezaga M, Wang S, Indrawati L, Bruner M, Friedlander A, Keller P, Caulfield M (2006) Immunogenicity and protective efficacy of Bacillus anthracis Poly-γ-D-glutamic acid capsule covalently coupled to a protein carrier using a novel triazine-based conjugation strategy. J Biol Chem 281(8):4831–4843. doi:10.1074/jbc.M509432200

Jung J, Iijima M, Yoshimoto N, Sasaki M, Niimi T, Tatematsu K, Jeong S-Y, Choi EK, Tanizawa K, S-i K (2011) Efficient and rapid purification of drug- and gene-carrying bio-nanocapsules, hepatitis B virus surface antigen L particles, from Saccharomyces cerevisiae. Protein Expr Purif 78(2):149–155. doi:10.1016/j.pep.2011.04.008

Kabanov AV, Vinogradov SV (2009) Nanogels as pharmaceutical carriers: finite networks of infinite capabilities. Angew Chem Int Ed Engl 48(30):5418–5429. doi:10.1002/anie.200900441

Kamaly N, Xiao Z, Valencia PM, Radovic-Moreno AF, Farokhzad OC (2012) Targeted polymeric therapeutic nanoparticles: design, development and clinical translation. Chem Soc Rev 41(7):2971–3010. doi:10.1039/c2cs15344k

Katayose S, Kataoka K (1997) Water-soluble polyion complex associates of DNA and poly(ethylene glycol)-poly(L-lysine) block copolymer. Bioconjug Chem 8(5):702–707

Kawashima Y, Yamamoto H, Takeuchi H, Kuno Y (2000) Mucoadhesive DL-lactide/glycolide copolymer nanospheres coated with chitosan to improve oral delivery of elcatonin. Pharm Dev Technol 5(1):77–85. doi:10.1081/pdt-100100522

Kempf M, Mandal B, Jilek S, Thiele L, Voros J, Textor M, Merkle HP, Walter E (2003) Improved stimulation of human dendritic cells by receptor engagement with surface-modified microparticles. J Drug Target 11(1):11–18. doi:10.1080/1061186031000072978

Korobko AV, Jesse W, van der Maarel JRC (2005) Encapsulation of DNA by cationic diblock copolymer vesicles. Langmuir 21(1):34–42

Korobko AV, Backendorf C, van der Maarel JRC (2006) Plasmid DNA encapsulation within cationic diblock copolymer vesicles for gene delivery. J Phys Chem B 110(30):14550–14556

Kreuter J (2007) Nanoparticles—a historical perspective. Int J Pharm 331(1):1–10. doi:10.1016/j.ijpharm.2006.10.021

Kreuter J, Speiser PP (1976) In vitro studies of poly(methyl methacrylate) adjuvants. J Pharm Sci 65(11):1624–1627. doi:10.1002/jps.2600651115

Kublin JG, Lowitt MH, Hamilton RG, Oliveira GA, Nardin EH, Nussenzweig RS, Schmeckpeper BJ, Diggs CL, Bodison SA, Edelman R (2002) Delayed-type hypersensitivity in volunteers immunized with a synthetic multi-antigen peptide vaccine (PfCS-MAP1NYU) against Plasmodium falciparum sporozoites. Vaccine 20(13–14):1853–1861. doi:http://dx.doi.org/10.1016/S0264-410X(02)00037-3

Lee ES, Kim D, Youn YS, Oh KT, Bae YH (2008) A virus-mimetic nanogel vehicle. Angew Chem Int Ed Engl 47(13):2418–2421. doi:10.1002/anie.200704121

Leleux J, Roy K (2013) Micro and nanoparticle-based delivery systems for vaccine immunotherapy: an immunological and materials perspective. Adv Healthc Mater 2(1):72–94. doi:10.1002/adhm.201200268

Leonard M, De Boisseson MR, Hubert P, Dalencon F, Dellacherie E (2004) Hydrophobically modified alginate hydrogels as protein carriers with specific controlled release properties. J Control Release 98(3):395–405. doi:10.1016/j.jconrel.2004.05.009

Leong KW, Mao HQ, Truong-Le VL, Roy K, Walsh SM, August JT (1998) DNA-polycation nanospheres as non-viral gene delivery vehicles. J Control Release 53(1–3):183–193. doi:10.1016/s0168-3659(97)00252-6

Li WA, Mooney DJ (2013) Materials based tumor immunotherapy vaccines. Curr Opin Immunol 25(2):238–45. doi:10.1016/j.coi.2012.12.008

Men Y, Thomasin C, Merkle HP, Gander B, Corradin G (1995) A single administration of tetanus toxoid in biodegradable microspheres elicits T cell and antibody responses similar or superior to those obtained with aluminum hydroxide. Vaccine 13(7):683–689

Men Y, Tamber H, Audran R, Gander B, Corradin G (1997) Induction of a cytotoxic T lymphocyte response by immunization with a malaria specific CTL peptide entrapped in biodegradable polymer microspheres. Vaccine 15(12/13):1405–1412. doi:10.1016/s0264-410x(97)00047-9

Men Y, Audran R, Thomasin C, Eberl G, Demotz S, Merkle HP, Gander B, Corradin G (1999) MHC class I- and class II-restricted processing and presentation of microencapsulated antigens. Vaccine 17(9–10):1047–1056. doi:10.1016/s0264-410x(98)00321-1

Moon JJ, Huang B, Irvine DJ (2012) Engineering nano- and microparticles to tune immunity. Adv Mater 24(28):3724–3746. doi:10.1002/adma.201200446

Moreno CA, Rodriguez R, Oliveira GA, Ferreira V, Nussenzweig RS, Moya Castro ZR, Calvo-Calle JM, Nardin E (1999) Preclinical evaluation of a synthetic Plasmodium falciparum MAP malaria vaccine in Aotus monkeys and mice. Vaccine 18(1–2):89–99. doi: http://dx.doi.org/10.1016/S0264-410X(99)00184-X

Mueller M, Reichardt W, Koerner J, Groettrup M (2012) Coencapsulation of tumor lysate and CpG-ODN in PLGA-microspheres enables successful immunotherapy of prostate carcinoma in TRAMP mice. J Control Release 162(1):159–166. doi:10.1016/j.jconrel.2012.06.015

Naahidi S, Jafari M, Edalat F, Raymond K, Khademhosseini A, Chen P (2013) Biocompatibility of engineered nanoparticles for drug delivery. J Control Release 166(2):182–194. doi:10.1016/j.jconrel.2012.12.013

Nardin EH, Calvo-Calle JM, Oliveira GA, Nussenzweig RS, Schneider M, Tiercy J-M, Loutan L, Hochstrasser D, Rose K (2001) A totally synthetic polyoxime malaria vaccine containing plasmodium falciparum B cell and universal T cell epitopes elicits immune responses in volunteers of diverse HLA types. J Immunol 166(1):481–489

Niederhafner P, Reiniš M, Šebestík J, Ježek J (2008) Glycopeptide dendrimers, part III—a review: use of glycopeptide dendrimers in immunotherapy and diagnosis of cancer and viral diseases. J Pept Sci 14(5):556–587. doi:10.1002/psc.1011

Nill K (2005) Glossary of biotechnology and nanobiotechnology terms, 4th ed by Taylor & Francis Group, LLC

Oh JK, Drumright R, Siegwart DJ, Matyjaszewski K (2008) The development of microgels/nanogels for drug delivery applications. Prog Polym Sci 33(4):448–477. doi:10.1016/j.progpolymsci.2008.01.002

O'Hagan D, Singh M, Gupta RK (1998) Poly(lactide-co-glycolide) microparticles for the development of single-dose controlled-release vaccines. Adv Drug Deliv Rev 32(3):225–246. doi:10.1016/s0169-409x(98)00012-x

Olive C, Batzloff M, Horváth A, Clair T, Yarwood P, Toth I, Good MF (2003) Potential of lipid core peptide technology as a novel self-adjuvanting vaccine delivery system for multiple different synthetic peptide immunogens. Infect Immun 71(5):2373–2383. doi:10.1128/iai.71.5.2373-2383.2003

Pan Y, Li Y-J, Zhao H-Y, Zheng J-M, Xu H, Wei G, Hao J-S, Cui F-D (2002) Bioadhesive polysaccharide in protein delivery system: chitosan nanoparticles improve the intestinal absorption of insulin in vivo. Int J Pharm 249(1–2):139–147. doi:10.1016/s0378-5173(02)00486-6

Pangburn SH, Trescony PV, Heller J (1984) Partially deacetylated chitin: its use in self-regulated drug delivery systems. Chitin, Chitosan, Relat Enzymes, [Proc Jt US-Jpn Semin Adv Chitin, Chitosan, Relat Enzymes] 3–19

Panyam J, Labhasetwar V (2003) Biodegradable nanoparticles for drug and gene delivery to cells and tissue. Adv Drug Deliv Rev 55(3):329–347. doi:10.1016/s0169-409x(02)00228-4

Pinto Reis C, Neufeld RJ, Ribeiro AJ, Veiga F (2006) Nanoencapsulation II. Biomedical applications and current status of peptide and protein nanoparticulate delivery systems. Nanomedicine 2(2):53–65. doi:10.1016/j.nano.2006.04.009

Posnett DN, McGrath H, Tam JP (1988) A novel method for producing anti-peptide antibodies. Production of site-specific antibodies to the T cell antigen receptor beta-chain. J Biol Chem 263(4):1719–1725

Poupot M, Griffe L, Marchand P, Maraval A, Rolland O, Martinet L, L'Faqihi-Olive F-E, Turrin C-O, Caminade A-M, Fournié J-J, Majoral J-P, Poupot R (2006) Design of phosphorylated dendritic architectures to promote human monocyte activation. FASEB J 20(13):2339–2351. doi:10.1096/fj.06-5742com

Prego C, Paolicelli P, Diaz B, Vicente S, Sanchez A, Gonzalez-Fernandez A, Alonso MJ (2010) Chitosan-based nanoparticles for improving immunization against hepatitis B infection. Vaccine 28(14):2607–2614. doi:10.1016/j.vaccine.2010.01.011

Raemdonck K, Demeester J, De Smedt S (2009) Advanced nanogel engineering for drug delivery. Soft Matter 5(4):707. doi:10.1039/b811923f

Renggli K, Baumann P, Langowska K, Onaca O, Bruns N, Meier W (2011) Selective and responsive nanoreactors. Adv Funct Mater 21(7):1241–1259

Rhie G-E, Roehrl MH, Mourez M, Collier RJ, Mekalanos JJ, Wang JY (2003) A dually active anthrax vaccine that confers protection against both bacilli and toxins. Proc Natl Acad Sci U S A 100(19):10925–10930. doi:10.1073/pnas.1834478100

Rice-Ficht AC, Arenas-Gamboa AM, Kahl-McDonagh MM, Ficht TA (2010) Polymeric particles in vaccine delivery. Curr Opin Microbiol 13(1):106–112. doi:10.1016/j.mib.2009.12.001

Roy R, Baek M-G (2003) Multivalent breast cancer T-antigen markers scaffolded onto PAMAM dendrimers. In: Yuan CL, Reiko TL (eds) Methods in enzymology, vol 362. Academic Press, New York, pp 240–249. doi:http://dx.doi.org/10.1016/S0076-6879(03)01017-6

Sadler K, Tam JP (2002) Peptide dendrimers: applications and synthesis. Rev Mol Biotechnol 90(3–4):195–229. doi:http://dx.doi.org/10.1016/S1389-0352(01)00061-7

Sah H, Toddywala R, Chien YW (1995) Continuous release of proteins from biodegradable microcapsules and in vivo evaluation of their potential as a vaccine adjuvant. J Control Release 35(2–3):137–144. doi:10.1016/0168-3659(95)00029-8

Sanchez A, Gupta RK, Alonso MJ, Siber GR, Langer R (1996) Pulsed controlled-release system for potential use in vaccine delivery. J Pharm Sci 85(6):547–552. doi:10.1021/js960069y

Saroja C, Lakshmi P, Bhaskaran S (2011) Recent trends in vaccine delivery systems: a review. Int J Pharm Investig 1(2):64–74. doi:10.4103/2230-973X.82384

Saurer EM, Jewell CM, Kuchenreuther JM, Lynn DM (2009) Assembly of erodible, DNA-containing thin films on the surfaces of polymer microparticles: toward a layer-by-layer approach to the delivery of DNA to antigen-presenting cells. Acta Biomater 5(3):913–924. doi:10.1016/j.actbio.2008.08.022

Schneerson R, Kubler-Kielb J, Liu T-Y, Dai Z-D, Leppla SH, Yergey A, Backlund P, Shiloach J, Majadly F, Robbins JB (2003) Poly(γ-D-glutamic acid) protein conjugates induce IgG antibodies in mice to the capsule of Bacillus anthracis: a potential addition to the anthrax vaccine. Proc Natl Acad Sci U S A 100(15):8945–8950. doi:10.1073/pnas.1633512100

Šebestík J, Reiniš M, Ježek J (2012) Vaccines and immunomodulation. In: Biomedical Applications of peptide-, glyco- and glycopeptide dendrimers, and analogous dendrimeric structures. Springer, Vienna, pp 199–207. doi:10.1007/978-3-7091-1206-9_22

Sexton A, Whitney PG, Chong S-F, Zelikin AN, Johnston APR, De Rose R, Brooks AG, Caruso F, Kent SJ (2009) A protective vaccine delivery system for in vivo t cell stimulation using nanoengineered polymer hydrogel capsules. ACS Nano 3(11):3391–3400. doi:10.1021/nn900715g

Shen H, Ackerman AL, Cody V, Giodini A, Hinson ER, Cresswell P, Edelson RL, Saltzman WM, Hanlon DJ (2006) Enhanced and prolonged cross-presentation following endosomal escape of exogenous antigens encapsulated in biodegradable nanoparticles. Immunology 117(1):78–88. doi:10.1111/j.1365-2567.2005.02268.x

Sheng K-C, Kalkanidis M, Pouniotis DS, Esparon S, Tang CK, Apostolopoulos V, Pietersz GA (2008) Delivery of antigen using a novel mannosylated dendrimer potentiates immunogenicity in vitro and in vivo. Eur J Immunol 38(2):424–436. doi:10.1002/eji.200737578

Shiao TC, Roy R (2012) Glycodendrimers as functional antigens and antitumor vaccines. New J Chem 36(2):324–339

Skwarczynski M, Zaman M, Urbani CN, Lin IC, Jia Z, Batzloff MR, Good MF, Monteiro MJ, Toth I (2010) Polyacrylate dendrimer nanoparticles: a self-adjuvanting vaccine delivery system. Angew Chem Int Ed 49(33):5742–5745. doi:10.1002/anie.201002221

Sorensen NS, Boas U, Heegaard PMH (2011) Enhancement of muramyldipeptide (MDP) immunostimulatory activity by controlled multimerization on dendrimers. Macromol Biosci 11(11):1484–1490. doi:10.1002/mabi.201100105

Stoenescu R, Graff A, Meier W (2004) Asymmetric ABC-triblock copolymer membranes induce a directed insertion of membrane proteins. Macromol Biosci 4:930–935. doi:10.1002/mabi.200400065

Sukhorukov GB, Donath E, Davis S, Lichtenfeld H, Caruso F, Popov VI, Mohwald H (1998) Stepwise polyelectrolyte assembly on particle surfaces: a novel approach to colloid design. Polym Adv Technol 9(10–11):759–767. doi:10.1002/(sici)1099-1581(1998100)9:10/11<759::aid-pat846>3.0.co;2-q

Tam JP (1988) Synthetic peptide vaccine design: synthesis and properties of a high-density multiple antigenic peptide system. Proc Natl Acad Sci U S A 85(15):5409–5413

Tam JP (1996) Recent advances in multiple antigen peptides. J Immunol Methods 196(1):17–32. doi:http://dx.doi.org/10.1016/0022-1759(96)00066-X

Tamber H, Johansen P, Merkle HP, Gander B (2005) Formulation aspects of biodegradable polymeric microspheres for antigen delivery. Adv Drug Deliv Rev 57(3):357–376. doi:10.1016/j.addr.2004.09.002

Tarradas J, Monsó M, Fraile L, de la Torre BG, Muñoz M, Rosell R, Riquelme C, Pérez LJ, Nofrarías M, Domingo M, Sobrino F, Andreu D, Ganges L (2012) A T-cell epitope on NS3 non-structural protein enhances the B and T cell responses elicited by dendrimeric constructions against CSFV in domestic pigs. Vet Immunol Immunopathol 150(1–2):36–46. doi:http://dx.doi.org/10.1016/j.vetimm.2012.08.006

Thomann-Harwood LJ, Kaeuper P, Rossi N, Milona P, Herrmann B, McCullough KC (2013) Nanogel vaccines targeting dendritic cells: contributions of the surface decoration and vaccine cargo on cell targeting and activation. J Control Release 166(2):95–105. doi:10.1016/j.jconrel.2012.11.015

Tomalia DA, Baker H, Dewald J, Hall M, Kallos G, Martin S, Roeck J, Ryder J, Smith P (1985) A new class of polymers: starburst-dendritic macromolecules. Polym J (Tokyo) 17(1):117–132. doi:10.1295/polymj.17.117

Uchida T, Yagi A, Oda Y, Goto S (1996) Microencapsulation of ovalbumin in poly(lactide-co-glycolide) by an oil-in-oil (o/o) solvent evaporation method. J Microencapsul 13(5):509–518. doi:10.3109/02652049609026036

Uto T, Toyama M, Nishi Y, Akagi T, Shima F, Akashi M, Baba M (2013) Uptake of biodegradable poly(γ-glutamic acid) nanoparticles and antigen presentation by dendritic cells in vivo. Res Immunol 3:1–9. doi:10.1016/j.rinim.2012.11.002

Van Regenmortel MHV, Briand JP, Muller S, Plaue S (1988) Laboratory techniques in biochemistry and molecular biology, vol. 19: synthetic polypeptides as antigens, Elsevier Science Publishers B. V. (Biomedical Devision)

Vila A, Sánchez A, Janes K, Behrens I, Kissel T, Jato JLV, Alonso MJ (2004) Low molecular weight chitosan nanoparticles as new carriers for nasal vaccine delivery in mice. Eur J Pharm Biopharm 57(1):123–131. doi:10.1016/j.ejpb.2003.09.006

Vinogradov SV, Bronich TK, Kabanov AV (2002) Nanosized cationic hydrogels for drug delivery: preparation, properties and interactions with cells. Adv Drug Deliv Rev 54(1):135–147. doi:10.1016/s0169-409x(01)00245-9

Wang R, Charoenvit Y, Corradin G, Porrozzi R, Hunter RL, Glenn G, Alving CR, Church P, Hoffman SL (1995) Induction of protective polyclonal antibodies by immunization with a Plasmodium yoelii circumsporozoite protein multiple antigen peptide vaccine. J Immunol 154(6):2784–2793

Wang TT, Fellows PF, Leighton TJ, Lucas AH (2004) Induction of opsonic antibodies to the γ-D-glutamic acid capsule of Bacillus anthracis by immunization with a synthetic peptide-carrier protein conjugate. FEMS Immunol Med Microbiol 40(3):231–237. doi:10.1016/s0928-8244(03)00366-3

Weber C, Drogoz A, David L, Domard A, Charles MH, Verrier B, Delair T (2010) Polysaccharide-based vaccine delivery systems: macromolecular assembly, interactions with antigen presenting cells, and in vivo immunomonitoring. J Biomed Mater Res A 93(4):1322–1334. doi:10.1002/jbm.a.32605

Wu W, Zhou S (2010) Hybrid micro-/nanogels for optical sensing and intracellular imaging. Nano Rev 1. doi:10.3402/nano.v1i0.5730

Wu Z, Zhang X, Guo H, Li C, Yu D (2012) An injectable and glucose-sensitive nanogel for controlled insulin release. J Mater Chem 22(42):22788. doi:10.1039/c2jm34082h

Yamada T, Iwabuki H, Kanno T, Tanaka H, Kawai T, Fukuda H, Kondo A, Seno M, Tanizawa K, Kuroda S (2001) Physicochemical and immunological characterization of hepatitis B virus envelope particles exclusively consisting of the entire L (pre-S1 + pre-S2 + S) protein. Vaccine 19(23–24):3154–3163. doi:10.1016/s0264-410x(01)00017-2

Youan BBC, Jackson TL, Dickens L, Hernandez C, Owusu-Ababio G (2001) Protein release profiles and morphology of biodegradable microcapsules containing an oily core. J Control Release 76(3):313–326. doi:10.1016/s0168-3659(01)00445-x

Zhao G, Lin Y, Du L, Guan J, Sun S, Sui H, Kou Z, Chan C, Guo Y, Jiang S, Zheng B-J, Zhou Y (2010) An M2e-based multiple antigenic peptide vaccine protects mice from lethal challenge with divergent H5N1 influenza viruses. Virol J 7(1):9

Zhu G, Mallery SR, Schwendeman SP (2000) Stabilization of proteins encapsulated in injectable poly(lactide-co-glycolide). Nat Biotechnol 18(1)

Chapter 11
Gels as Vaccine Delivery Systems

Sarah Gordon

11.1 Background

The majority of the gel systems presented in this chapter, and indeed the majority of gel systems utilised for the purposes of vaccine delivery, can be further defined as hydrogels. While numerous definitions for both gels and hydrogels can be found, in basic terms gels may be considered as semi-solid systems of two or more components in which a small amount of solid is dispersed in a comparatively larger quantity of liquid (Almdal et al. 1993; Gupta et al. 2002). Hydrogels may therefore be defined in simple terms as gel systems in which the liquid component is water. More specifically, hydrogels consist of networks of natural or synthetic hydrophilic polymers that are capable of taking up large amounts of water and swelling, while still maintaining their distinct three-dimensional structure (Gupta et al. 2002; Kashyap et al. 2005; Sood et al. 2013). Their high molecular permeability, low interfacial tension, and similar mechanical properties to physiological soft tissue (Gupta et al. 2002) make them attractive for drug and vaccine delivery alike, while the low polymer content of hydrogels (typically of the order of 1–20 %) generally means that the administration of such systems is unlikely to result in concentration-induced irritancy (Gupta et al. 2002; Ishihara et al. 2006).

In addition to these properties, the ability of hydrogels to facilitate sustained release of actives makes them systems of considerable interest for both drug and vaccine delivery. With respect to the latter, subunit antigens are well known to have a good safety profile but a poor inherent immunogenicity, which often necessitates the administration of multiple vaccine doses in order to induce an appropriate and effective immune response (Ellis 2001). The employment of immunopotentiating adjuvants and incorporation of antigens into particulate systems are discussed in

S. Gordon (✉)
Helmholtz Institute for Pharmaceutical Research Saarland (HIPS), Saarland University, Campus, Geb. A4.1, Saarbrücken 66123, Germany
e-mail: Sarah.Gordon@helmholtz-hzi.de

C. Foged et al. (eds.), *Subunit Vaccine Delivery*, Advances in Delivery Science and Technology, DOI 10.1007/978-1-4939-1417-3_11

detail in other chapters of this book as potential strategies to combat the poor immunogenicity of subunit vaccines; an additional strategy that has demonstrated considerable promise is the incorporation of antigen into a sustained release delivery system. The probability of inducing an effective immune response to a specific antigen appears to increase if release of antigen occurs in a sustained manner, due to the fact that antigen is present and available for interaction with immune cells for an extended period of time (Lofthouse 2002). By increasing the likelihood of a more effective immune response being induced, delivery systems that facilitate the sustained release of antigen may enable a reduction in the need for multiple immunisations, or potentially abolishment of this requirement altogether. Such an occurrence could in turn facilitate a decrease in expenses associated with immunisation programmes, and a corresponding increase in patient compliance with resulting simpler and cheaper immunisation courses (Zhao and Leong 1996).

11.2 Classification and Manufacture of Gel Systems for Vaccine Delivery

Not only may gels be defined in a variety of ways, but they may also be classified according to a number of different criteria. In this chapter, the mechanism of gelation will be used to broadly group gel systems employed for the purposes of vaccine delivery into three major categories: namely, gelation by precipitation, as a result of crosslinking, or due to solidification (Agarwal and Rupenthal 2013).

11.2.1 Gelation by Precipitation

11.2.1.1 Gelation by Precipitation: Thermosensitive, In Situ-Forming Gels

Thermosensitive, in situ-forming gels constitute a subclass of precipitating gel systems with particularly promising applications in the field of vaccine delivery. As their name suggests, such systems are polymeric solutions that transform into gel depots (gels) upon in vivo administration. While the exact mechanism driving this transformation may vary from polymer to polymer (and as such, will be discussed in more detail in the context of specific thermosensitive gel systems), gel formation generally occurs as a result of decreased polymer solubility in response to an increase in temperature from ambient to physiological levels (Ruel-Gariépy and Leroux 2004). Temperature change is considered to be a particularly advantageous stimulus for induction of gelation, as there is no reliance on the use of potentially harsh reagents such as organic solvents or copolymerisation agents (Jeong et al. 2002). The mild nature of thermosensitive gel formation is therefore associated with little risk of damage to the gel-incorporated active—a particularly important consideration in the context of vaccine delivery, where the active is often a protein or peptide. The fact that thermosensitive, in situ-forming gels are liquids at room temperature is also a considerable advantage, as this allows for incorporation of desired actives into thermosensitive solutions by a simple mixing process. Furthermore, there is no need for surgical implantation of such systems, as thermosensitive solutions can be easily injected; and, if the gel-forming polymer is biodegradable, subsequent surgical removal of the system is also unnecessary (Jeong et al. 2002).

Thermosensitive, in situ-forming gels can be produced from a wide range of polymers, which in turn are often further functionalised or copolymerised in order to achieve control over specific system properties such as precise gelation temperature and required polymer concentration. The following will focus on description of vaccine delivery systems formed from chitosan and poloxamer, two polymers which are often employed for the preparation of thermosensitive, in situ-forming gels.

Chitosan Gels

Chitosan is a polymer of *N*-acetylglucosamine and glucosamine which is derived from the partial deacetylation of chitin, a naturally occurring component of crustacean exoskeletons (Chenite et al. 2000). The term "chitosan" can be used to describe any member of this large group of chitin derivatives, having various molecular weights and degrees of deacetylation (Fig. 11.1) (Chenite et al. 2001). Chitosans collectively are ideal candidates for in vivo applications, being both biocompatible and biodegradable in nature and exhibiting low systemic and local toxicity (Agarwal and Rupenthal 2013; Qin et al. 2006). As is indicated in Fig. 11.1, amine groups present in the chitosan structure may be protonated under acidic conditions, rendering the polymer soluble. However, at neutral or alkaline pH levels, where a deprotonation of

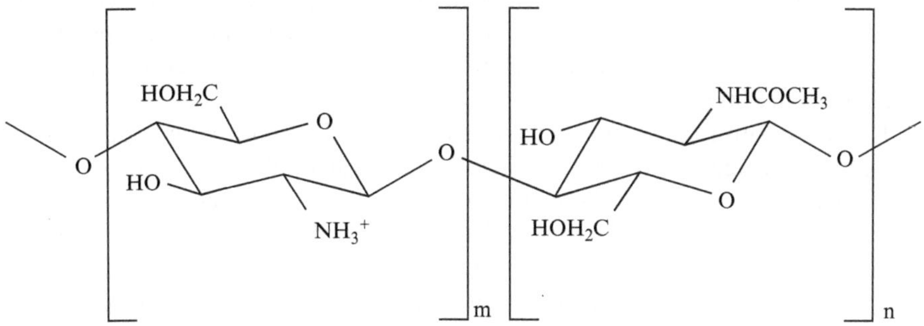

Fig. 11.1 Chemical structure of chitosan. Glucosamine and *N*-acetylglucosamine monomers are shown. In the case of the parent polymer chitin, $m \ll n$; the term "chitosan" is generally conferred on polymers with a degree of deacetylation >50 % (Cho et al. 2005), with degree of deacetylation given by $(m/(m+n))100$). Figure reproduced from Berger et al. (2005) with kind permission (Elsevier)

such groups occurs, a hydrated gel-like precipitate is formed (Ganguly and Dash 2004). Chitosan solutions therefore typically exhibit pH-dependent gel formation; however, if polyol salts such as glycerol phosphate are used as pH-altering agents, gel formation does not occur purely as a result of an increase in pH. Use of these agents instead leads to the conversion of purely pH-dependent gel-forming chitosan solutions into temperature-controlled, pH-dependent gel-forming solutions (Chenite et al. 2000; Mao et al. 2006). The mechanism by which addition of glycerol phosphate and other polyol salts to chitosan solutions facilitates temperature-controlled, pH-dependent gel formation is proposed to be the result of a number of processes. A decrease in electrostatic repulsion and increase in hydrogen bonding of chitosan chains due to neutralisation of chitosan charge, together with the occurrence of electrostatic interactions between ammonium groups of chitosan and phosphate groups of glycerol phosphate are considered to be responsible for the pH-dependent nature of chitosan gel formation; an increase in hydrophobic interactions between chitosan chains as a consequence of the structuring effect of the salt glycerol moiety on water is thought to be responsible for the temperature-controlled gelling property of such systems (Chenite et al. 2000; Cho et al. 2005).

In addition to having the general advantages of thermosensitive gel systems, such as existence as a solution at ambient temperature allowing for ease of incorporation of actives, chitosan gels have the additional benefit of a residual positive charge at physiological pH. This facilitates the ready entrapment of anionic moieties (including many antigens and adjuvants), and also allows for adhesive interactions with negatively charged sites on cell surfaces and body tissues. Chitosan has also been shown to attract and activate inflammatory cells such as neutrophils and macrophages upon in vivo administration (Peluso et al. 1994; Porporatto et al. 2005; VandeVord et al. 2002), and to exhibit an inherent immunogenicity (Bueter et al. 2011; Gordon et al. 2008; Li et al. 2008; Porporatto et al. 2003). Such properties may prove highly advantageous for a vaccine delivery system; however, the induction of severe inflammatory reactions due to chitosan administration has been demonstrated in some cases (Tomihata and Ikada 1997). Such adverse reactions may potentially be avoided through the use of highly deacetylated chitosans, which have been shown to form gels with a slower rate of degradation and a correspondingly lower level of reactogenicity than chitosans with a low degree of deacetylation (Chenite et al. 2000).

Thermosensitive chitosan gels have found particular application to date for the delivery of vaccines via the nasal route (Çokçalışkan et al. 2013; Günbeyaz et al. 2010; Wu et al. 2012b, c), where antigen residence time is often a limiting factor for vaccine efficacy. Wu et al. successfully formulated a thermosensitive gel composed of a quaternary ammonium derivative of chitosan (*N*-[(2-hydroxy-3-trimethylammonium)propyl]chitosan chloride) together with α,β-glycerophosphate, which proved to be a free-flowing solution at room temperature (allowing for incorporation of antigens) and gelled rapidly upon intranasal administration to mice (Wu et al. 2012b, c). Such a system facilitated an increase in antigen residence time as well as an enhancement of paracellular antigen transport, consistent with the well-known ability of chitosan to act as a permeation enhancer via alteration of tight junctions (Amidi et al. 2006; Prego et al. 2005). Potentiation of both mucosal and systemic immune responses was observed as a result of incorporation of an H5N1 antigen into the developed thermosensitive chitosan gel, together with promotion of a $CD8^+$ T-cell memory response in the nasal-associated lymphoid tissue (Wu et al. 2012b). Similarly, delivery of an envelope glycoprotein antigen derived from adenovirus-based Zaire Ebola virus within such a gel has been shown to result in the induction of both local respiratory and systemic immune responses (Wu et al. 2012c). Potentiation of humoral immunity together with an enhanced, Th1-biased cellular immune response relative to antigen in the absence of a gel component was induced; an acceptable biocompatibility of the gel formulation was also demonstrated, even following repeated administrations.

In addition to their use for intranasal vaccine delivery, thermosensitive chitosan hydrogels have also shown promise as injectable delivery systems. For example, a thermosensitive chitosan gel produced from chitosan and glycerophosphate and loaded with the model protein antigen ovalbumin has shown the ability to facilitate a sustained release of antigen in vitro, and to induce both the proliferation of $CD4^+$ T-cells and production of antigen-specific IgG as a result of subcutaneous administration to mice (Gordon et al. 2008). Similarly, gel delivery systems prepared from purified poly-*N*-acetylglucosamine have been investigated as injectable delivery systems for peptide- as well as DNA-based vaccines. These systems have shown the ability to induce greater antigen-specific T-cell responses than those produced by administration of antigen in the absence of gel, and to confer protection in tumour challenge studies (Cole et al. 1997; Maitre et al. 1999; Nguyen et al. 2001; Salem et al. 2010).

Poloxamer Gels

Poloxamers, often referred to by their common commercial name, Pluronics®, are triblock copolymers of poly(ethylene oxide)-*b*-poly(propylene oxide)-*b*-*poly*(ethylene oxide). Such copolymers have a non-ionic, amphiphilic surfactant character, and spontaneously form micelles in aqueous solutions due to the relatively hydrophilic character of poly(ethylene oxide) and comparatively more hydrophobic nature of poly(propylene oxide) (Klouda and Mikos 2008). Gelation of poloxamer solutions is facilitated by an increase in temperature, and is thought to occur as a result of association of poloxamer surfactant micelles. As the temperature of poloxamer solutions is increased and the critical gel-forming temperature (discussed further later) is approached, a dehydration of the more hydrophobic poly(propylene oxide) residues occurs, leading to an increasingly tight packing of poloxamer micelles (Cabana et al. 1997). The progressive condensation of micellar structures is accompanied by an increase in entanglement of poloxamer polymer chains, which ultimately leads to the induction of gel formation (Fig. 11.2) (Brown et al. 1991; Cabana et al. 1997; Hvidt et al. 1994). While the micellar structure of poloxamer gels confers a number of advantages on these systems, including the ability to encapsulate both hydrophilic and hydrophobic actives, it is also

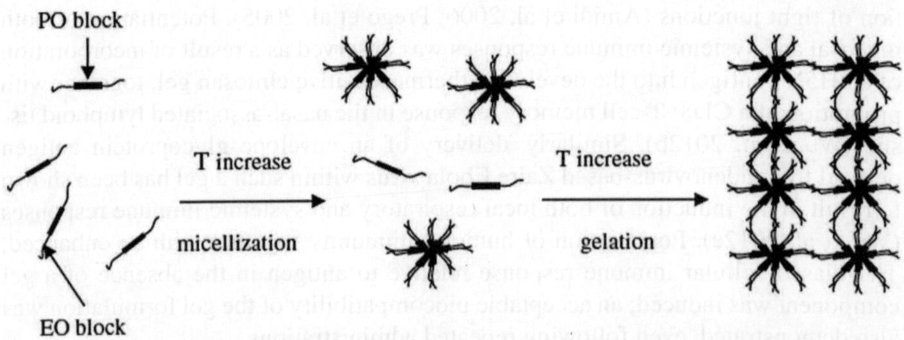

Fig. 11.2 Schematic representation of micelle and gel formation of poloxamer 407 in water. 'PO' and 'EO' denote poly(propylene oxide) and poly(ethylene oxide) blocks respectively, while 'T' signifies temperature. Figure reproduced from Dumortier et al. (2006) with kind permission (Springer Science+Business Media)

responsible for the generally weak mechanical strength of poloxamer gels and their typically short residence time (Chitkara et al. 2006; He et al. 2008). This is due to the fact that a rapid drop in the effective concentration of poloxamer occurs on exposure of poloxamer gels to excess (biological) fluid, causing micelles within the gel structure to fall apart (Chung et al. 2008). The effect of poloxamer dilution may be somewhat countered by the use of higher polymer concentrations, however given that a polymer concentration of at least 15 % w/w is generally needed to facilitate thermosensitive gel formation, further increases may lead to toxic effects (Agarwal and Rupenthal 2013). Poloxamers are additionally non-biodegradable, which may prove a hindrance to their use in vivo (Chitkara et al. 2006).

Poloxamer-based thermosensitive gels have long been investigated for their ability to act as gene and drug delivery systems, and in more recent years have begun to find application in the field of tissue engineering (Chung et al. 2008; Kabanov et al. 2002). Poloxamer-based thermosensitive gels commonly consisting of poloxamer P407 (commercially known as Pluronic® F127) have also been investigated for use as vaccine delivery systems, often in combination with additional polymer components in an effort to increase the resulting gel strength and stability (Kojarunchitt and Hook 2012). Such a system, consisting of P407 and used in combination with a reverse poloxamer (Pluronic® 25R4) in an attempt to enhance gel stability, was investigated by Kojarunchitt et al. (2011). An increased in vitro stability of the produced thermosensitive gel was achieved as a result of the combination of P407 with Pluronic® 25R4. Thermosensitive hydrogels consisting of P407 in combination with chitosan have also been investigated for their ability to act as effective vaccine delivery systems (Coeshott et al. 2004; Westerink et al. 2001). In one such case where delivery to mucosal surfaces was intended, chitosan was chiefly employed for its ability to act as a penetration enhancer rather than being utilised in an attempt to improve gel stability. In this instance, a combination of P407/chitosan gel incorporating tetanus toxoid as an antigen was able to enhance both local and systemic humoral immune responses relative to a formulation consisting of antigen in buffer alone, when administered intranasally to mice (Westerink et al. 2001).

The same combination P407/chitosan gel formulation has also been employed for the purposes of systemic vaccine delivery (Coeshott et al. 2004). In this case, tetanus toxoid (or alternatively, diphtheria toxoid or anthrax recombinant protective antigen) was added to P407 solutions in combination with either chitosan or immunopotentiating CpG motifs and administered subcutaneously to mice, upon which gel formation occurred. Administration of gel-based formulations resulted in long-lived and enhanced IgG antibody production relative to the administration of antigens and immunopotentiators in the absence of a gel component. A single dose of gel-based formulations also proved to be more immunogenic than several administrations of tetanus toxoid adsorbed to aluminium phosphate; tetanus toxoid delivered by P407 gels in combination with CpG was also seen to result in the initiation of superior immune responses in comparison to a formulation of tetanus toxoid together with CpG in incomplete Freund's adjuvant. Lethal challenge experiments further demonstrated that P407 gels incorporating tetanus toxoid and either chitosan or CpG were capable of conferring protective immunity.

Also worthy of mention is the existence of studies in which poloxamers are formulated in combination with the mucoadhesive polymers polycarbophil or poly(ethylene oxide) in order to produce thermosensitive, in situ-gelling systems for the delivery of plasmid DNA (Han et al. 2006; Oh et al. 2003; Park et al. 2002). When administered intranasally to mice, such systems have been shown to demonstrate an improvement in retention and a corresponding increase in the absorption of plasmid DNA (Park et al. 2002). Similar systems incorporating hepatitis B surface antigen have also been used successfully for the purposes of intravaginal vaccination (Oh et al. 2003). Administration of these in situ-gelling, mucoadhesive systems resulted in the production of IgA antibody in both vaginal secretions and saliva, and also induced systemic immunity as measured by production of serum IgG antibody.

PECE and PCEC Gels

The copolymers poly(ethylene glycol)-poly(ε-caprolactone)-poly(ethylene glycol) (PECE) (Gong et al. 2009) and poly(ε-caprolactone)-poly(ethylene glycol)-poly(ε-caprolactone) (PCEC) (Wu et al. 2012a) have also been utilised as the basis for thermosensitive, in situ gel-forming systems for vaccine delivery. Incorporation of basic fibroblast growth factor (bFGF) into a solution of PECE has been shown to be possible at ambient temperature, allowing for subcutaneous administration via simple injection. Equilibration of the injected solution to physiological temperature resulted in formation of an antigen gel depot. Robust and long-lived humoral immune responses were observed following such administration, with detectable levels of serum IgG persisting in immunised mice for as long as 14 weeks following administration of thermosensitive gel systems (Gong et al. 2009). Formulation of PCEC thermosensitive gels has been proposed to offer additional advantages over those prepared using PECE, including an improved biocompatibility and prolonged gel residence time (Wu et al. 2012a). Thermosensitive gels composed of PCEC have also been formulated incorporating bFGF, as well as mannan, in order to target the lectin receptor present on the surface of dendritic cells (DCs). Subcutaneous administration of such a PCEC-based gel also demonstrated the ability to induce appreciable humoral immune responses, resulting in production of serum IgG antibody in excess of that produced in response to vaccination with control formulations (Wu et al. 2012a).

11.2.1.2 Gelation by Precipitation: pH-Sensitive Gels

As alluded to above in the case of chitosan, gelation of polymer solutions may also occur in response to a change in pH. Polymeric macromolecules containing ionisable groups, also known as polyelectrolytes, behave in a similar manner to weak acids and bases, in that they exhibit different degrees of ionisation in solution dependent on solution pH. Such a characteristic proves very useful with respect to gel formation. Solutions of polymers which exhibit a decrease in solubility or a change in polymer chain arrangement with a change in pH can be used to form gel-based delivery systems; this behaviour may be tailored to occur upon a change to physiological pH, enabling such polymers to be used as a basis for in situ-forming gel delivery systems.

Carbopol (Poly-Acrylic Acid) Gels

Carbopols are a group of hydrophilic polyanionic polymers (or carbomers) consisting of cross-linked acrylic acid residues. Carbopol has been utilised widely as a basis for controlled release and bioadhesive delivery systems, and is often employed in veterinary vaccines in the form of suspensions (Dey et al. 2012; Krashias et al. 2010). Although a number of mechanisms are thought to contribute to carbopol gelation, this may be considered to occur chiefly in response to an increase in pH (Kumar and Himmelstein 1995). Dispersion of carbopol in water results in the formation of an acidic solution of partially coiled polymer chains; neutralisation of this solution by addition of a base confers a negative charge on carbopol polymer chains, which in turn leads to electrostatic repulsion, complete chain uncoiling, and the formation of a rigid gel structure (Felt et al. 2002).

Carbopol gels appear to have found particular application as vaccine delivery systems for intravaginal administration, possibly due to the already approved employment of carbopol in numerous products for vaginal use. Cranage et al. formulated a pre-neutralised carbopol gel containing an HIV-1 envelope glycoprotein antigen, and investigated the immune response resulting from topical administration of such a formulation in rabbits (Cranage et al. 2009). Administration of the carbopol gel-based formulation was well tolerated, and was considered to form a depot of antigen at the mucosal surface. One-off administration of such a system was seen to stimulate the production of both local (mucosal) and systemic (seral) IgG antibody, with further potentiation of antibody titres occurring with successive administrations of gel-incorporated antigen. Topical administration of a similar carbopol gel formulation in a more advanced animal model, namely macaques, was shown to be capable of B-cell priming, leading to antibody production after a subsequent intramuscular boost; in a similar vein, humoral immune responses were noted upon intramuscular priming followed by an intravaginally administered boost with gel-based antigen (Cranage et al. 2011). Carbopol gels incorporating HIV-1 envelope glycoprotein-based antigen have further been employed as vaccine delivery systems for subcutaneous administration in mice. Following delivery via such a route, humoral and cell-mediated immune responses comparable to those induced by antigen administered in complete Freund's adjuvant were noted (Krashias et al. 2010). Moreover, incorporation of purified influenza haemagglutinin into carbopol gels followed by subcutaneous administration has been shown to result in the production of high antigen-specific antibody titres and protection in lethal challenge experiments, demonstrating the versatility of carbopol gel delivery systems in facilitating the induction of effective immune responses to various antigens.

11.2.2 Gelation by Crosslinking

Crosslinking of polymer chains in order to form gels may occur via a number of different mechanisms, and instances of the use of gel-based vaccine delivery systems formed as a result of ion-mediated crosslinking or photopolymerisation may be found in the literature. Examples of systems relying on each of these gel-forming mechanisms are therefore mentioned below.

11.2.2.1 Alginate Gels

Alginates are naturally occurring biopolymers generally sourced from brown algae, or kelp. They are linear, non-branched polysaccharides, incorporating 1,4′-linked β-D-mannuronic acid and α-L-guluronic acid residues in varying amounts and sequences to form an overall block-like structure (Gombotz and Wee 1998). Alginate gel formation occurs as a result of ionic crosslinking, initiated by divalent cations such as Ca^{2+}. Incubation of alginate with Ca^{2+} in an aqueous solution leads to the formation of polymer chain-connecting ionic bridges and a stacking of guluronic acid residues within alginate polymer chains. This results in the formation of the so-called egg-box structure, which is characteristic of alginate gel networks (Agarwal and Rupenthal 2013; Gombotz and Wee 1998). While not widely employed in the field of vaccination, alginate gels have shown a promising ability to act as in situ-gelling vaccine delivery systems. Hori et al. demonstrated that an antigen-loaded, in situ-gelling alginate formulation could be prepared by mixing antigen-loaded DCs and Ca^{2+}-incorporating alginate microspheres with a bulk alginate solution (Hori et al. 2008). Mixing of formulation components followed directly by subcutaneous administration to mice resulted in gel depot formation, due to the liberation of Ca^{2+} from formulation microspheres and subsequent diffusion into the bulk alginate solution. Alginate gel depots formed in this manner were observed to act as sites of attraction for host DCs and T-cells; transport of formulation-incorporated DCs to host lymphatic tissues was also noted.

11.2.2.2 Modified Poly(Ethylene Glycol) Gels

Similarly to the case of alginate mentioned above, polymers capable of forming cross-linked gels in the presence of a photoinitiator have not been commonly employed as the basis for vaccine delivery systems to date. An example can be found in the area of veterinary vaccination however, in which photopolymerised poly(ethylene glycol)-cross-linked gels were investigated as systems for remote ballistic delivery not of a subunit vaccine antigen, but of a live, poorly bioactive Brucella vaccine (Christie et al. 2006). For this purpose, poly(ethylene glycol) modification was first carried out in order to incorporate degradable glycolide or lactide residues, which were functionalised with photopolymerisable methacrylate groups. The resulting polymer was then filled into commercially available "Biobullets", together with a photoinitiator (Irgacure®184) and polystyrene microspheres as a model for live Brucella bacteria. Crosslinking of modified poly(ethylene glycol) chains was then initiated by exposure of delivery systems to UV light, resulting in the formation of a gel. A sustained release of model polystyrene particles was observed from "Biobullet"-encased gels in vivo, demonstrating the ability of such a system to act as an antigen depot.

11.2.3 Gelation by Solidification

When employed as vehicles for vaccine delivery, solidifying gel systems have typically been formed from hot melts that solidify upon a reduction in temperature to physiological levels. Examples of carrageenan gels and surfactant-based organogels formed as a result of solidification and utilised in the field of vaccine delivery are detailed below.

11.2.3.1 Carrageenan Gels

Carrageenans are a group of hydrophilic, sulphated polysaccharides which are commonly employed in food products for their thickening, stabilising, and gelling properties. Carrageenans in general are soluble at high temperatures in excess of their melting temperature, and are capable of forming gel networks upon cooling (Tecante and del Carmen Núñez Santiago 2012). Carrageenan gel-based systems have been employed as vehicles for an oral coccidiosis vaccine for poultry (Danforth et al. 1997; Dasgupta and Lee 2000). Vaccine preparation involved dissolving a carrageenan gum at high temperatures, followed by cooling in order to allow for antigen incorporation. Gel formation was then facilitated by further cooling of the system. Such a vaccine was shown to be capable of inducing protective immune responses in vaccinated animals.

11.2.3.2 Water-in-Sorbitan-Monostearate Organogels

As detailed above in this chapter, a gel may be classified as a hydrogel when the liquid component of the system is water. The liquid component of a gel may also consist of an organic solvent however, in which case the gel may be designated as an organogel (Murdan 2005). While their applications have been far less commonly explored to date, instances of the employment of organogels as vaccine delivery systems may be found in literature. Murdan et al. demonstrated the production of such an organogel consisting of the non-ionic surfactant and principal gelation-inducing agent (or organogelator) sorbitan monostearate, together with polysorbate 20, and isopropyl myristate as the organic solvent (Murdan 2005). The gel system was produced by preparing a water-in-oil emulsion containing the above components at elevated temperatures, followed by cooling to room temperature in order to facilitate setting of the emulsion to an organogel. Gel formation in this instance was proposed to be due to a reduction in the oil solubility of sorbitan monostearate as a result of the decrease in temperature, leading to the formation and subsequent entanglement of surfactant aggregates (within which the aqueous phase is situated) and the ultimate formation of a gel network. Loading of such a system can be achieved by directly dissolving/dispersing the active in the solution prior to cooling in the case of hydrophobic entities, or dissolution of the active in an aqueous solvent followed by addition to the solution in the case of a hydrophilic entity. With respect to the latter case, incorporation of the model antigen bovine serum albumin into the aqueous phase of such an organogel has been successfully demonstrated. Intramuscular administration of the pre-formed gel to mice was shown to result in depot formation, and to facilitate a sustained release of the incorporated model antigen (Murdan et al. 1999a).

11.3 Characterisation of Gel Systems

While quantification of the immunogenicity of gel-based vaccine delivery systems is of clear and primary importance (as seen above in the context of various preclinical studies), additional properties of such systems may also be investigated as part of the characterisation process. Parameters of interest for further characterisation may vary depending on the polymer constituent of individual systems and the mechanism by which gelation occurs, however characterisation of certain gel properties may be considered to be of general relevance. Such properties include gel morphology, which may be visualised using electron microscopy techniques such as cryo-field emission scanning electron microscopy (cryo-FESEM, Fig. 11.3) or freeze-fracture transmission electron microscopy (freeze fracture-TEM). The degree of swelling of gel systems is also a parameter of general interest. Swelling behaviour may be characterised by determining changes in gel mass or volume over time, or by quantification of gel water content (Gupta et al. 2002). Determination of the degradation behaviour of gel systems may also be of value, together with assessment of the

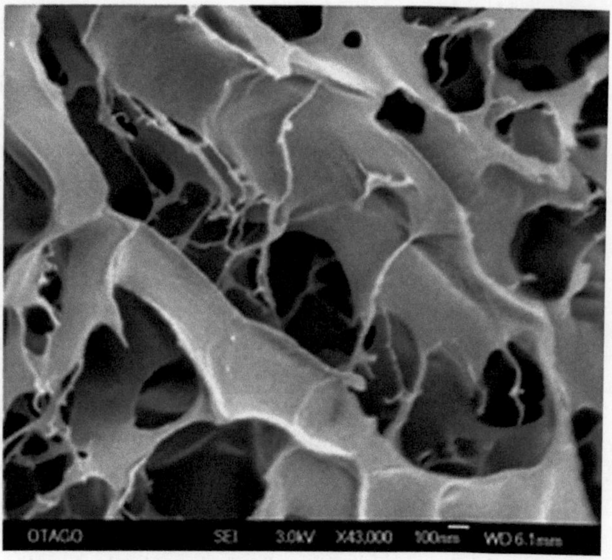

Fig. 11.3 Cryo-FESEM image illustrating the morphology of a chitosan gel containing the model antigen ovalbumin following immersion in phosphate-buffered saline at 37 °C. Figure reproduced from Gordon et al. (2008) with kind permission (John Wiley and Sons)

biocompatibility of the parent gel as well as any degradation products. Determination of the release kinetics of any gel-entrapped active is also of great importance; this may be carried out in either in vitro or in vivo environments (Gordon et al. 2008, 2010). Characterisation of the rheological behaviour of both solution and gel forms is also of great importance for the majority of gel-based vaccine delivery systems. Such information may for example be used to determine the pH at which the transition from solution to gel occurs in the case of pH-dependent gelling systems, or the temperature at which gelation occurs in the case of thermosensitive gelling systems. A case example of the rheological characterisation of a thermosensitive gel system is given by Kojarunchitt et al. (2011), where oscillatory shear measurements of the storage (elastic response) and loss (viscous response) moduli of a poloxamer P407 gel-based vaccine delivery system were conducted over a range of oscillatory frequencies and temperatures. Such measurements allowed for quantification of the complex viscosity of the gel system as a function of temperature, and subsequent determination of the solution-to-gel transition temperature. Further kinetic measurements of storage and loss moduli at the determined temperature of gelation were then carried out in order to give a more precise indication of the time required for gelation of the system to occur (Kojarunchitt et al. 2011; Winter and Chambon 1986).

11.4 Combination Systems: Gel-Incorporated Particulate Systems for Vaccine Delivery

The use of more complex gel-based vaccine delivery systems is also noted in the literature. Specifically, gel systems which incorporate antigen (and in some cases also adjuvant) associated with a particulate carrier are being increasingly utilised, in order to combine the advantages associated with both gel-based and particulate delivery. Careful fine-tuning of such systems is however extremely important in order to realise the full potential of such a combination strategy, and to ultimately demonstrate an immunological advantage. The stability of the particulate system within the gel matrix must be ensured; particle-related characteristics such as size, antigen loading, and gel incorporation concentration must also be optimised.

Thermosensitive, in situ-forming chitosan gels containing antigen and adjuvant in a number of different particulate carriers have been formulated and tested for their ability to stimulate an immune response. In particular, gel systems consisting of cationic liposomes and cubosomes encapsulating the model antigen ovalbumin together with the adjuvant Quil A have been prepared (Gordon et al. 2012). The incorporation of a particulate component into chitosan gel systems was noted to impact on the kinetics of antigen release from gels, as well as on the immune response resulting from subcutaneous administration of gel-based systems to mice (Gordon et al. 2012). This is consistent with previous observations related to changes in the characteristics of chitosan gel delivery systems upon addition of lipid-based particulate systems, such as an increase in gel strength and a prolongation of release of incorporated active (Ruel-Gariépy et al. 2002). Chitosan gels containing ovalbumin incorporated into silica nanoparticles in both the presence and absence of the adjuvant Quil A have also been formulated, and have been shown to stimulate both cell-mediated and humoral immune responses in vivo (Gordon et al. 2010). Poloxamer-based thermosensitive, in situ-forming gels have also been utilised as the basis for combination vaccine delivery systems, and have been formulated to contain human papilloma virus (HPV) virus-like particles (VLPs) together with cholera toxin as an additional adjuvant (Park et al. 2003). Intravaginal administration of such systems to mice induced a greater production of vaginal and salivary IgA and serum IgG antibody than that induced by adjuvanted HPV VLPs delivered in an aqueous vehicle.

The previously mentioned copolymers PCEC and PECE have also been employed in order to form combination vaccine delivery systems, as has the discussed water-in-sorbitan-monostearate organogel system. In the case of PCEC and PECE, PCEC nanoparticles loaded with human bFGF were incorporated into a PECE-based thermosensitive gel. Entirely consistent with the general properties of gel delivery systems, a prolonged release of antigen was noted from such combination formulations in comparison to release from PCEC nanoparticles alone; moreover, subcutaneous administration of combination systems to mice resulted in the production of potent and long-lived humoral immune responses, as well as indications of the induction of protective immunity (Wu et al. 2011). A strong humoral immune response was also noted as a result of vaccination of mice with the aforementioned water-in-sorbitan-monostearate organogel, used in combination with haemagglutinin-loaded niosomes (non-ionic surfactant vesicles) (Murdan et al. 1999b). This modified formulation was composed of a vesicle-in-water-in-oil gel, in which the aqueous phase containing antigen-loaded niosomes was dispersed within surfactant structures in the oil component. Interestingly, the level of antibody production induced by an organogel formulation containing non-particulate haemagglutinin was not significantly different to that produced in response to vaccination with the combination system, and in fact a greater efficacy at lower antigen dose was noted as a result of vaccination with the organogel system in the absence of a particulate component.

A combination system consisting of the in situ-gelling and mucoadhesive polymer polygalacturonic acid incorporating Norwalk virus VLPs has also been investigated by Velasquez et al. (2011): such a system was initially prepared as a liquid, which was then spray dried in order to produce a dry powder formulation, termed GelVac™. GelVac™ formulations were shown to be capable of inducing local and systemic antibody responses equal to or greater than those resulting from the administration of VLPs together with an adjuvant in an aqueous vehicle.

11.5 Clinical Studies

As is clearly evident, all above discussions of the application of gels as vaccine delivery systems have been in the context of preclinical research. Indeed, there is a paucity of clinical studies involving gels for the delivery of vaccines, indicating that considerable development is still necessary in order for the preclinical promise of gels for vaccine delivery to be realised in the clinical setting. In one of few examples, Lewis et al. conducted a Phase I, double-blind randomised controlled trial investigating the efficacy of a carbopol gel system containing an HIV-1 envelope protein antigen following repeated intravaginal administration (Lewis et al. 2011). The gel-based vaccine system was demonstrated to be well tolerated, but was not observed to induce any appreciable immune responses. An intranasal H5N1 influenza vaccine based on the above mentioned GelVac™ system is also currently being investigated in the context of Phase I clinical testing (ClinicalTrials.gov Identifier NCT01258062, http://clinicaltrials.gov/ct2/show/NCT01258062?term=GelVac&rank=1).

11.6 Outlook

This chapter has attempted to give a comprehensive overview of the body of literature currently available concerning the application of gels as vaccine delivery systems. Overall, the information currently available indicates that gel-based vaccine delivery systems are associated with numerous advantages. Gels have the potential to facilitate a sustained release of antigen, a property which has positive flow-on effects with respect to requirements for boosting and patient compliance. In many cases, gels also demonstrate excellent biocompatibility and biodegradability characteristics, and may also exhibit inherent biological activity due to their constituent polymers. In situ gel-forming systems (which exist as solutions at the benchtop and form gels upon administration and adjustment to the physiological environment) are also associated with significant advantages, including the ready incorporation of antigens and adjuvants, and ease of administration. Furthermore, the concept of combination systems in which antigen is associated with a particulate carrier prior to incorporation into a gel matrix holds great promise in the area of vaccine delivery, with such systems having the potential to considerably enhance or direct the resulting immune response. These advantages are clearly demonstrated in a considerable number of preclinical studies employing gel-based vaccine delivery systems; the translation of this promise into the clinical setting will be a process to be monitored with considerable interest in the coming years.

References

Agarwal P, Rupenthal ID (2013) Injectable implants for the sustained release of protein and peptide drugs. Drug Discov Today 18(7–8):337–349

Almdal K, Dyre J, Hvidt S, Kramer O (1993) Towards a phenomenological definition of the term 'gel'. Polym Gels Networks 1(1):5–17

Amidi M, Romeijn SG, Borchard G, Junginger HE, Hennink WE, Jiskoot W (2006) Preparation and characterization of protein-loaded N-trimethyl chitosan nanoparticles as nasal delivery system. J Control Release 111(1–2):107–116

Berger J, Reist M, Chenite A, Felt-Baeyens O, Mayer JM, Gurny R (2005) Pseudo-thermosetting chitosan hydrogels for biomedical application. Int J Pharm 288(1):17–25

Brown W, Schillen K, Almgren M, Hvidt S, Bahadur P (1991) Micelle and gel formation in a poly(ethylene oxide)/poly(propylene oxide)/poly(ethylene oxide) triblock copolymer in water solution: dynamic and static light scattering and oscillatory shear measurements. J Phys Chem 95(4):1850–1858

Bueter CL, Lee CK, Rathinam VAK, Healy GJ, Taron CH, Specht CA, Levitz SM (2011) Chitosan but not chitin activates the inflammasome by a mechanism dependent upon phagocytosis. J Biol Chem 286(41):35447–35455

Cabana A, Aït-Kadi A, Juhász J (1997) Study of the gelation process of polyethylene oxide$_a$–polypropylene oxide$_b$–polyethylene oxide$_a$ copolymer (Poloxamer 407) aqueous solutions. J Colloid Interface Sci 190(2):307–312

Chenite A, Chaput C, Wang D, Combes C, Buschmann MD, Hoemann CD, Leroux JC, Atkinson BL, Binette F, Selmani A (2000) Novel injectable neutral solutions of chitosan form biodegradable gels in situ. Biomaterials 21(21):2155–2161

Chenite A, Buschmann M, Wang D, Chaput C, Kandani N (2001) Rheological characterisation of thermogelling chitosan/glycerol-phosphate solutions. Carbohydr Polym 46(1):39–47

Chitkara D, Shikanov A, Kumar N, Domb AJ (2006) Biodegradable injectable in situ depot-forming drug delivery systems. Macromol Biosci 6(12):977–990

Cho J, Heuzey M-C, Bégin A, Carreau PJ (2005) Physical gelation of chitosan in the presence of β-glycerophosphate: the effect of temperature. Biomacromolecules 6(6):3267–3275

Christie RJ, Findley DJ, Dunfee M, Hansen RD, Olsen SC, Grainger DW (2006) Photopolymerized hydrogel carriers for live vaccine ballistic delivery. Vaccine 24(9):1462–1469

Chung HJ, Lee Y, Park TG (2008) Thermo-sensitive and biodegradable hydrogels based on stereocomplexed Pluronic multi-block copolymers for controlled protein delivery. J Control Release 127(1):22–30

Coeshott CM, Smithson SL, Verderber E, Samaniego A, Blonder JM, Rosenthal GJ, Westerink MAJ (2004) Pluronic® F127-based systemic vaccine delivery systems. Vaccine 22(19):2396–2405

Çokçalışkan C, Özyörük F, Gürsoy RN, Alkan M, Günbeyaz M, Arca HÇ, Uzunlu E, Şenel S (2013) Chitosan-based systems for intranasal immunization against foot-and-mouth disease. Pharm Dev Technol. doi:10.3109/10837450.2013.763263:1-8

Cole DJ, Gattoni-Celli S, McClay EF, Metcalf JS, Brown JM, Nabavi N, Newton DA, Woolhiser CB, Wilson MC, Vournakis JN (1997) Characterization of a sustained-release delivery system for combined cytokine/peptide vaccination using a poly-N-acetyl glucosamine-based polymer matrix. Clin Cancer Res 3(6):867–873

Cranage MP, Fraser CA, Stevens Z, Huting J, Chang M, Jeffs SA, Seaman MS, Cope A, Cole T, Shattock RJ (2009) Repeated vaginal administration of trimeric HIV-1 clade C gp140 induces serum and mucosal antibody responses. Mucosal Immunol 3(1):57–68

Cranage MP, Fraser CA, Cope A, McKay PF, Seaman MS, Cole T, Mahmoud AN, Hall J, Giles E, Voss G, Page M, Almond N, Shattock RJ (2011) Antibody responses after intravaginal immunisation with trimeric HIV-1CN54 clade C gp140 in Carbopol gel are augmented by systemic priming or boosting with an adjuvanted formulation. Vaccine 29(7):1421–1430

Danforth HD, Lee EH, Martin A, Dekich M (1997) Evaluation of a gel-immunization technique used with two different Immucox vaccine formulations in battery and floor-pen trials with broiler chickens. Parasitol Res 83(5):445–451

Dasgupta T, Lee EH (2000) A gel delivery system for coccidiosis vaccine: uniformity of distribution of oocysts. Can Vet J 41(8):613–616

Dey AK, Burke B, Sun Y, Hartog K, Heeney JL, Montefiori D, Srivastava IK, Barnett SW (2012) Use of a polyanionic carbomer, Carbopol971P, in combination with MF59, improves antibody responses to HIV-1 envelope glycoprotein. Vaccine 30(17):2749–2759

Dumortier G, Grossiord JL, Agnely F, Chaumeil JC (2006) A review of poloxamer 407 pharmaceutical and pharmacological characteristics. Pharm Res 23(12):2709–2728

Ellis RW (2001) Technologies for the design, discovery, formulation and administration of vaccines. Vaccine 19(17–19):2681–2687

Felt O, Einmahl S, Gurny R, Furrer P, Baeyens V (2002) Polymeric systems for ophthalmic drug delivery. In: Dumitriu S (ed) Polymeric biomaterials, revised and expanded, 2nd edn. Marcel Dekker, New York

Ganguly S, Dash AK (2004) A novel in situ gel for sustained drug delivery and targeting. Int J Pharm 276(1–2):83–92

Gombotz WR, Wee SF (1998) Protein release from alginate matrices. Adv Drug Deliv Rev 31(3):267–285

Gong C-Y, Shi S, Peng X-Y, Kan B, Yang L, Huang M-J, Luo F, Zhao X, Wei Y-Q, Qian Z-Y (2009) Biodegradable thermosensitive injectable PEG-PCL-PEG hydrogel for bFGF antigen delivery to improve humoral immunity. Growth Factors 27(6):377–383

Gordon S, Saupe A, McBurney W, Rades T, Hook S (2008) Comparison of chitosan nanoparticles and chitosan hydrogels for vaccine delivery. J Pharm Pharmacol 60(12):1591–1600

Gordon S, Teichmann E, Young K, Finnie K, Rades T, Hook S (2010) In vitro and in vivo investigation of thermosensitive chitosan hydrogels containing silica nanoparticles for vaccine delivery. Eur J Pharm Sci 41(2):360–368

Gordon S, Young K, Wilson R, Rizwan S, Kemp R, Rades T, Hook S (2012) Chitosan hydrogels containing liposomes and cubosomes as particulate sustained release vaccine delivery systems. J Liposome Res 22(3):193–204

Günbeyaz M, Faraji A, Özkul A, Puralı N, Şenel S (2010) Chitosan based delivery systems for mucosal immunization against bovine herpesvirus 1 (BHV-1). Eur J Pharm Sci 41(3–4): 531–545

Gupta P, Vermani K, Garg S (2002) Hydrogels: from controlled release to pH-responsive drug delivery. Drug Discov Today 7(10):569–579

Han I-K, Kim YB, Kang H-S, Sul D, Jung W-W, Cho HJ, Oh Y-K (2006) Thermosensitive and mucoadhesive delivery systems of mucosal vaccines. Methods 38(2):106–111

He C, Kim SW, Lee DS (2008) In situ gelling stimuli-sensitive block copolymer hydrogels for drug delivery. J Control Release 127(3):189–207

Hori Y, Winans AM, Huang CC, Horrigan EM, Irvine DJ (2008) Injectable dendritic cell-carrying alginate gels for immunization and immunotherapy. Biomaterials 29(27):3671–3682

Hvidt S, Joergensen EB, Brown W, Schillen K (1994) Micellization and gelation of aqueous solutions of a triblock copolymer studied by rheological techniques and scanning calorimetry. J Phys Chem 98(47):12320–12328

Ishihara M, Obara K, Nakamura S, Fujita M, Masuoka K, Kanatani Y, Takase B, Hattori H, Morimoto Y, Ishihara M, Maehara T, Kikuchi M (2006) Chitosan hydrogel as a drug delivery carrier to control angiogenesis. J Artif Organs 9(1):8–16

Jeong B, Kim SW, Bae YH (2002) Thermosensitive sol–gel reversible hydrogels. Adv Drug Deliv Rev 54(1):37–51

Kabanov AV, Batrakova EV, Alakhov VY (2002) Pluronic® block copolymers as novel polymer therapeutics for drug and gene delivery. J Control Release 82(2–3):189–212

Kashyap N, Kumar N, Kumar MNVR (2005) Hydrogels for pharmaceutical and biomedical applications. Crit Rev Ther Drug Carrier Syst 22(2):107–150

Klouda L, Mikos AG (2008) Thermoresponsive hydrogels in biomedical applications. Eur J Pharm Biopharm 68(1):34–45

Kojarunchitt T, Hook S (2012) Thermoresponsive chitosan and poloxamer 407 hydrogels for drug and vaccine delivery. In: Câmara FV, Ferreira LJ (eds) Hydrogels: synthesis, characterization and applications. Biochemistry research trends. Nova Biomedical, Waltham

Kojarunchitt T, Hook S, Rizwan S, Rades T, Baldursdottir S (2011) Development and characterisation of modified poloxamer 407 thermoresponsive depot systems containing cubosomes. Int J Pharm 408(1–2):20–26

Krashias G, Simon A-K, Wegmann F, Kok W-L, Ho L-P, Stevens D, Skehel J, Heeney JL, Moghaddam AE, Sattentau QJ (2010) Potent adaptive immune responses induced against HIV-1 gp140 and influenza virus HA by a polyanionic carbomer. Vaccine 28(13):2482–2489

Kumar S, Himmelstein KJ (1995) Modification of in situ gelling behavior of carbopol solutions by hydroxypropyl methylcellulose. J Pharm Sci 84(3):344–348

Lewis DJ, Fraser CA, Mahmoud AN, Wiggins RC, Woodrow M, Cope A, Cai C, Giemza R, Jeffs SA, Manoussaka M, Cole T, Cranage MP, Shattock RJ, Lacey CJ (2011) Phase I randomised clinical trial of an HIV-1_{CN54}, Clade C, trimeric envelope vaccine candidate delivered vaginally. PLoS One 6(9):e25165

Li H, Willingham SB, Ting JP-Y, Re F (2008) Cutting edge: inflammasome activation by alum and alum's adjuvant effect are mediated by NLRP3. J Immunol 181(1):17–21

Lofthouse S (2002) Immunological aspects of controlled antigen delivery. Adv Drug Deliv Rev 54(6):863–870

Maitre N, Brown JM, Demcheva M, Kelley JR, Lockett MA, Vournakis J, Cole DJ (1999) Primary T-cell and activated macrophage response associated with tumor protection using peptide/poly-N-acetyl glucosamine vaccination. Clin Cancer Res 5(5):1173–1182

Mao J, Kondu S, Ji H-F, McShane MJ (2006) Study of the near-neutral pH-sensitivity of chitosan/gelatin hydrogels by turbidimetry and microcantilever deflection. Biotechnol Bioeng 95(3):333–341

Murdan S (2005) Organogels in drug delivery. Expert Opin Drug Deliv 2(3):489–505

Murdan S, Bvd B, Gregoriadis G, Florence AT (1999a) Water-in-sorbitan monostearate organogels (water-in-oil gels). J Pharm Sci 88(6):615–619

Murdan S, Gregoriadis G, Florence AT (1999b) Sorbitan monostearate/polysorbate 20 organogels containing niosomes: a delivery vehicle for antigens? Eur J Pharm Sci 8(3):177–185

Nguyen CL, Bui JT, Demcheva M, Vournakis JN, Cole DJ, Gillanders WE (2001) Sustained release of granulocyte-macrophage colony-stimulating factor from a modular peptide-based cancer vaccine alters vaccine microenvironment and enhances the antigen-specific T-cell response. J Immunother 24(5):420–429

Oh Y-K, Park J-S, Yoon H, Kim C-K (2003) Enhanced mucosal and systemic immune responses to a vaginal vaccine coadministered with RANTES-expressing plasmid DNA using in situ-gelling mucoadhesive delivery system. Vaccine 21(17–18):1980–1988

Park J-S, Oh Y-K, Yoon H, Kim JM, Kim C-K (2002) In situ gelling and mucoadhesive polymer vehicles for controlled intranasal delivery of plasmid DNA. J Biomed Mater Res 59(1):144–151

Park J-S, Oh Y-K, Kang M-J, Kim C-K (2003) Enhanced mucosal and systemic immune responses following intravaginal immunization with human papillomavirus 16L1 virus-like particle vaccine in thermosensitive mucoadhesive delivery systems. J Med Virol 70(4):633–641

Peluso G, Petillo O, Ranieri M, Santin M, Ambrosic L, Calabró D, Avallone B, Balsamo G (1994) Chitosan-mediated stimulation of macrophage function. Biomaterials 15(15):1215–1220

Porporatto C, Bianco ID, Riera CM, Correa SG (2003) Chitosan induces different l-arginine metabolic pathways in resting and inflammatory macrophages. Biochem Biophys Res Commun 304(2):266–272

Porporatto C, Bianco ID, Correa SG (2005) Local and systemic activity of the polysaccharide chitosan at lymphoid tissues after oral administration. J Leukoc Biol 78(1):62–69

Prego C, García M, Torres D, Alonso MJ (2005) Transmucosal macromolecular drug delivery. J Control Release 101(1–3):151–162

Qin C, Gao J, Wang L, Zeng L, Liu Y (2006) Safety evaluation of short-term exposure to chitooligomers from enzymic preparation. Food Chem Toxicol 44(6):855–861

Ruel-Gariépy E, Leroux J-C (2004) In situ-forming hydrogels—review of temperature-sensitive systems. Eur J Pharm Biopharm 58(2):409–426

Ruel-Gariépy E, Leclair G, Hildgen P, Gupta A, Leroux JC (2002) Thermosensitive chitosan-based hydrogel containing liposomes for the delivery of hydrophilic molecules. J Control Release 82(2–3):373–383

Salem ML, Demcheva M, Gillanders WE, Cole DJ, Vournakis JN (2010) Poly-N-acetyl glucosamine gel matrix as a non-viral delivery vector for DNA-based vaccination. Anticancer Res 30(10):3889–3894

Sood N, Nagpal S, Nanda S, Bhardwaj A, Mehta A (2013) An overview on stimuli responsive hydrogels as drug delivery system. J Control Release. doi:10.1016/j.jconrel.2013.02.023

Tecante A, del Carmen Núñez Santiago M (2012) Solution properties of κ-carrageenan and its interaction with other polysaccharides in aqueous media. In: De Vicente J (ed) Rheology. InTech, Croatias

Tomihata K, Ikada Y (1997) In vitro and in vivo degradation of films of chitin and its deacetylated derivatives. Biomaterials 18(7):567–575

VandeVord PJ, Matthew HWT, DeSilva SP, Mayton L, Wu B, Wooley PH (2002) Evaluation of the biocompatibility of a chitosan scaffold in mice. J Biomed Mater Res 59(3):585–590

Velasquez LS, Shira S, Berta AN, Kilbourne J, Medi BM, Tizard I, Ni Y, Arntzen CJ, Herbst-Kralovetz MM (2011) Intranasal delivery of Norwalk virus-like particles formulated in an in situ gelling, dry powder vaccine. Vaccine 29(32):5221–5231

Westerink MAJ, Smithson SL, Srivastava N, Blonder J, Coeshott C, Rosenthal GJ (2001) ProJuvant™ (Pluronic F127®/chitosan) enhances the immune response to intranasally administered tetanus toxoid. Vaccine 20(5–6):711–723

Winter HH, Chambon F (1986) Analysis of linear viscoelasticity of a crosslinking polymer at the gel point. J Rheol 30(2):367–382

Wu QJ, Zhu XC, Xiao X, Wang P, Dk X, Gong CY, Wang YS, Yang L, Wei YQ (2011) A novel vaccine delivery system: Biodegradable nanoparticles in thermosensitive hydrogel. Growth Factors 29(6):290–297

Wu Q, Gong C, Shi S, Wang Y, Huang M, Yang L, Zhao X, Wei Y, Qian Z (2012a) Mannan loaded biodegradable and injectable thermosensitive PCL-PEG-PCL hydrogel for vaccine delivery. Soft Mater 10(4):472–486

Wu Y, Wei W, Zhou M, Wang Y, Wu J, Ma G, Su Z (2012b) Thermal-sensitive hydrogel as adjuvant-free vaccine delivery system for H5N1 intranasal immunization. Biomaterials 33(7): 2351–2360

Wu Y, Wu S, Hou L, Wei W, Zhou M, Su Z, Wu J, Chen W, Ma G (2012c) Novel thermal-sensitive hydrogel enhances both humoral and cell-mediated immune responses by intranasal vaccine delivery. Eur J Pharm Biopharm 81(3):486–497

Zhao Z, Leong KW (1996) Controlled delivery of antigens and adjuvants in vaccine development. J Pharm Sci 85(12):1261–1270

Chapter 12
Implants as Sustained Release Delivery Devices for Vaccine Antigens

Julia Engert

12.1 Background and History

Implant systems have been investigated for the controlled and sustained release of drugs for several decades. In particular, depot systems for highly active drugs delivering a low dose per day have been investigated. Implants are small, mainly cylindrical rods in which the drug is either embedded in a polymer matrix, or the drug-loaded matrix is coated with a polymer, or a combination of both. In the 1980s, Norplant was the first implant containing the contraceptive drug levonorgestrel, which consisted of either silastic capsules or covered rods (Peralta et al. 1995). Implantable systems have also been investigated in research areas such as controlled and sustained release of antibiotics, ocular treatment, or chemotherapeutics.

The use of implantable systems for vaccine delivery has, however, not been investigated in such depths. Implants could be a valuable delivery system for vaccines as they may release the antigen (and adjuvant) over a prolonged period of time, thereby reducing the need for frequent administrations. In addition, patient compliance may increase based on this therapy regime and the in vivo stability of the antigens may be enhanced. First reports of the advantages for the employment of implants as vaccine delivery systems were published as early as 1976 by Langer and Folkman. The authors reported the successful preparation of a polymer pellet made of ethylene-vinyl-copolymer, which was prepared by a molding procedure, that released the incorporated drug over >100 days with zero-order kinetics (Langer and Folkman 1976). One major obstacle during the production process was the incorporation of peptides or proteins into the matrix (Langer and Folkman 1976). Proteins are prone to chemical or physical degradation during manufacture, storage, and release due to their sensitive three-dimensional structure. However, only a few

J. Engert (✉)
Department of Pharmacy, Pharmaceutical Technology and Biopharmaceutics,
Ludwig-Maximilians-University Munich, Butenandtstr. 5, 81377 Munich, Germany
e-mail: julia.engert@cup.uni-muenchen.de

C. Foged et al. (eds.), *Subunit Vaccine Delivery*, Advances in Delivery Science and Technology, DOI 10.1007/978-1-4939-1417-3_12

years later Preis and Langer published a study where they investigated the use of polymer pellets loaded with bovine serum albumin (BSA), γ-globulin, or ribonuclease. These implants were administered to C57Bl/6 mice, and antibody IgG responses to the antigen were compared to the antigen administered in complete Freund's adjuvant (Preis and Langer 1979). In response to the extended antigen delivery facilitated by implant incorporation, antibody production increased and no induction of tolerance was observed (Preis and Langer 1979). In 1987, Wise and coworkers concluded in a review on implantable polymeric systems that biodegradable implants releasing antigens could be attractive devices in vaccine delivery (Wise et al. 1988). They proposed that it was conceivable to deliver two distinct pulses of antigen to sufficiently stimulate immune responses. In addition, they highlighted the possibility to further incorporate an adjuvant into the matrix to deliver both antigen and adjuvant at the same time. In the early 1990s the search for controlled release vaccine delivery systems mainly focused on biocompatible or biodegradable microparticles for various delivery routes (Aguado and Lambert 1992; O'Hagan et al. 1991), but research on implant systems continued at the same time. One obstacle all researchers were facing was the fear of inducing tolerance by delivering the antigen over a prolonged time period. Dixon and Maurer had observed that specific immunogenic unresponsiveness can be induced in rabbits after long-term administration of excess antigen (Dixon and Maurer 1955). In the early 1960s Dresser had reported that the delivery of non-particulate soluble antigen (bovine γ-globulin) caused immunological tolerance in mice (Dresser 1962). Furthermore, Ramsdell and Fowlkes had pointed out in 1992 that the persistence of antigen can lead to in vivo tolerance (Ramsdell and Fowlkes 1992). However later, Walduck and Opdebeeck found in their study that antibody responses were not significantly affected by the dose administered at either one time or by continuous delivery of an antigen (Walduck et al. 1998). In contrast to the proposed required pulsed delivery of antigen by Wise et al. (1988), they found that a continuous slow release of antigen was effective in stimulating immune responses (Walduck et al. 1998). Similarly, Khan et al. evaluated the use of cholesterol-lecithin implants loaded with BSA in mice, and showed that BSA-specific antibodies were produced in response to the BSA released from the implant (Khan et al. 1991). Investigations on cholesterol-lecithin implants containing the adjuvant Quil-A were first reported by Demana et al. in 2005, showing that immunostimulating complexes (ISCOMs) and related colloidal structures were released from compressed pellets (Demana et al. 2005). This research was further extended to a preparation of implants for an in vivo study, in which the implant induced comparable immune responses to two immunizations given by injection (Myschik et al. 2008a, c, d). The co-delivery of an antigen plus adjuvant from an implant was further explored using the small molecular weight immunopotentiator imiquimod and an α-galactosylceramide analogue (Myschik et al. 2008b). In summary, implants appear to have potential as controlled or sustained release vaccine delivery systems. However, the transfer from model antigens to true vaccines has not been performed so far. It still needs to be investigated in the future if real vaccines can be incorporated into implant matrices and how this type of delivery affects immune responses in vivo.

12.2 Types of Implants

Different types of implants can be prepared, depending on the manufacturing process and the manufacturing equipment. An overview is given in Fig. 12.1.

12.2.1 *Rod-Shaped Implants*

Mainly rod-shaped strands are described in literature, as these implants can be produced with varying diameters. In addition, these types of implants can be easily implanted subcutaneously through a trochar or a needle, depending on their diameter.

12.2.2 *Tablet-Shaped Implants*

Implants may be produced by direct compression of lipids, e.g., compressed by a hydraulic press. The resulting implants can therefore display a form that is similar to that of a tablet, or implants having a cylindrical shape can also be prepared. The use of tableting technology may be beneficial as this is a commonly used technique for the preparation of solid dosage forms.

12.2.3 *Single-layer or Multilayer Coated Implants*

Prepared implants may be coated with one or more layers in order to increase the diffusion layer, resulting in a slowing of drug/antigen release (Walduck et al. 1998; Lofthouse et al. 2002) by increasing the thickness of the diffusional layer.

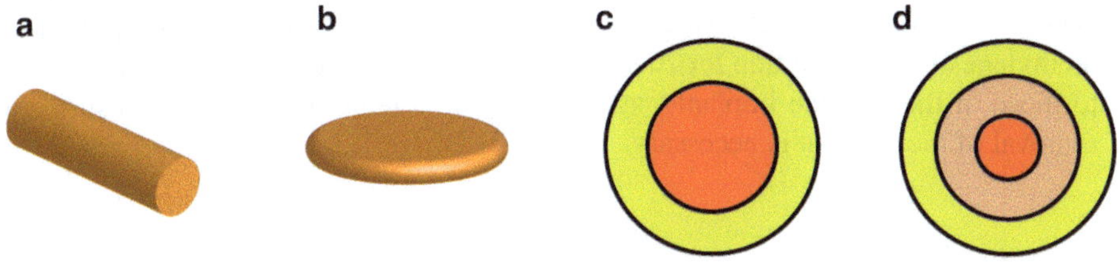

Fig. 12.1 Overview of different implant shapes. (**a**) Rod-shaped implant, (**b**) disc-shaped implant, (**c**) cross-section single-layer coated rod, (**d**) cross-section double-layer coated rod

12.3 Choice of Materials and Excipient(s)

For the preparation of implants, different techniques are available and their use is influenced by the type of material and/or excipients used for the implant matrix. Different materials have been utilized for the preparation of antigen-loaded implants. But all materials have certain advantages and disadvantages.

12.3.1 Silicone

Silicone can be used as a nonbiodegradable polymer for the preparation of silicone matrix implants or rod-shaped silicone implants. The material has been employed for a number of different model proteins and vaccines such as human serum albumin (HSA), interferon (INF), avidin, *Clostridium tetani*, or *Clostridium novyi* toxoids (Lofthouse et al. 2001, 2002; Kajihara et al. 2000, 2001; Kemp et al. 2002). The preparation usually requires the use of a catalyst and a cross-linking agent to ensure the formation of a stable silicone matrix. The mixture is pressed through an extruder equipped with a die, and the implants are then left to cure at room temperature for 3–4 days prior to use. The advantage of using silicone as the matrix former is the absence of pH or temperature effects during manufacture as well as the avoidance of organic solvents. The disadvantage of using silicone-based implants is their non-biodegradability, resulting in the need for surgical removal of the drug-depleted matrix.

12.3.2 Ethylene-Vinyl-Acetate Copolymer

Langer and Folkman (1976) developed an implant using ethylene-vinyl acetate copolymer as matrix and reported that this type of implant was inert, noninflammatory, and capable of releasing macromolecules for over 100 days in vitro and in vivo (Langer and Folkman 1976). The advantage of using this sort of polymer is its inertness as well as the manufacturing procedure. Here, dry protein powder was mixed with a small volume of methylene chloride containing the copolymer. After vacuum-drying the resulting pellets were cut into pieces and then dipped into a polymer-methylene chloride solution for 20 s. The disadvantages of these systems are similar to those of the silicone implants, as the material is nonbiodegradable, so surgical removal of the implant is necessary.

12.3.3 Collagen

Collagen minipellets have been employed by Lofthouse et al. to deliver avidin plus interleukin-1β or a clostridial vaccine to sheep and mice (Lofthouse et al. 2001). The authors were able to show that the cylindrical-shaped implants degraded within

35 days of implantation. Surprisingly, only a few studies can be found in literature using this type of polymer: the studies by Lofthouse and co-workers and a study by Higaki et al. using tetanus and diphtheria toxoids (Higaki et al. 2001). One more study by Ochiya et al. utilized a collagen minipellet for the controlled delivery of plasmid DNA to mice (Ochiya et al. 1999).

12.3.4 *Lipids*

Current research focuses on using lipids or lipid mixtures as matrix formers for the delivery of proteins, as it was shown that the use of triglycerides can overcome at least some of the problems that have been encountered when using polymers. Lipids are a promising alternative to polymer-based systems, as they can be relatively easily formed into microparticles or implants. It was furthermore shown that macromolecules such as INF-α can be incorporated into and released from implants almost exclusively in a monomeric form (Mohl and Winter 2004). Tripalmitin-based implants have also been developed and employed for the delivery of protein drugs such as insulin (Appel et al. 2006), interleukins (Koennings et al. 2006), lysozyme, and neurotropic factor (Koennings et al. 2007). Also, lipid implants prepared from cholesterol alone or mixtures of cholesterol and phospholipids, as well as implants made of triglycerides and blends of cholesterol and phospholipids (Guse et al. 2006) have been explored for the delivery of model antigens such as ovalbumin (OVA) and BSA (Khan et al. 1991, 1993; Myschik et al. 2008a, b, d), showing that the sustained release of antigen with or without the inclusion of an adjuvant is possible. The shift from polymeric matrix material to lipids can be explained by the fact that lipids are naturally occurring substances and therefore, a foreign body response is less likely to be raised by the immune system. However, even for triglyceride-cholesterol implants blended with phospholipids, an inflammatory reaction was reported when phospholipid content in the implants was increased to higher concentrations (Guse et al. 2006).

12.3.5 *Further Additives/Excipients*

In some cases, it is necessary to incorporate further additives and excipients into the implants to allow for ease of production and/or to modify the antigen release behavior.

12.3.5.1 Excipients

As proteins and peptides are molecules sensitive to changes in pH and susceptible to degradation upon heat, shear stress, or hydrolysis, the addition of further additives or excipients may be necessary to stabilize the protein or peptide. Cases in which this approach has been successfully employed have been reported in literature. Hydroxypropyl-β-cyclodextrin was used in the case of INF-α as a lyoprotectant during the freeze-drying step, when dry IFN-α powder was incorporated into a lipid implant (Schulze and Winter 2009). This excipient had been proven to stabilize the protein not only during the freeze-drying step, but also to increase storage stability of the protein in compressed lipid implants (Mohl and Winter 2004).

12.3.5.2 Pore-Forming Agents

Pore-forming agents may be added to the implant matrix in order to adjust the release profile of the protein. For instance, polyethylene glycol 6000 (PEG 6000) has been used as a pore former in lipid implants containing IFN-α (Herrmann et al. 2007b; Schulze and Winter 2009).

12.4 Manufacturing Processes

A number of different manufacturing processes can be used to prepare implants, including direct compression, molding/melting, casting, and extrusion. These processes are discussed further below.

12.4.1 Direct Compression

The easiest method to prepare an implant matrix is usually direct compression, meaning that a dry powder of the matrix plus any drug (which is also in a dry form) is compressed using a punch and die system in a hydraulic press by applying a defined mass (Myschik et al. 2008a). Alternatively, a single punch tableting machine can be employed (Khan et al. 1991; Cardamone et al. 1997). Direct compression has been utilized for the preparation of pure cholesterol implants (Opdebeeck and Tucker 1993), cholesterol/lecithin implants (Khan et al. 1991, 1993; Walduck et al. 1998), tristearin implants (Mohl and Winter 2004; Herrmann et al. 2007b), and glyceryl palmitostearate implants (Pongjanyakul et al. 2004), showing that this method can be employed for a variety of different excipients and excipient mixtures. The main disadvantage of direct compression is the limitation towards scale up of the production of implants. Most hydraulic presses are operated manually and therefore only production at a very small lab-scale can be achieved. In addition, flowability of the lipid powder may be poor, which results in unfavorable loading of the punch/die system, and necessitates the use of large amounts of lubricants in order to achieve good flowability of the powder. The addition of lubricants may in turn impact on the release kinetics of antigen from implants. Direct compression does however usually result in a continuous distribution of the drug in the matrix, as shown by Kreye et al. (2008).

12.4.2 Molding/Melting

Using a melt of the matrix former and employing subsequent controlled cooling has been less frequently reported in the literature. Pongajanyakul et al. utilized polyethylene tubes with an internal diameter of 2.5 mm filled with a suspension of lysozyme in molten glyceryl palmitostearate (Pongjanyakul et al. 2004). Similarly, Yamagata et al. heated a lipidic matrix former above its T_m to subsequently force this molten mixture through an 14G stainless steel needle (Yamagata et al. 2000). This molding technique was used for IFN-α as a model protein, but has so far not been reported for antigens. This method is limited to materials that do not show polymorphic transition and therefore different physicochemical characteristics. During this process, some lipids may undergo polymorphic changes that during storage result in transformation into a more stable crystalline form (Kreye et al. 2008). These changes may have a profound impact on the release characteristics of incorporated drugs or proteins.

12.4.3 Casting

Applying this process, a lipid or a lipid mixture is heated and then the drug or antigen is homogenously dispersed in the melt. The melt is subsequently cast into molds of different geometries and dimensions (Kreye et al. 2011a). For this technique, one needs to consider the thermal stress drugs or protein antigens will be exposed to, and the fact that the choice or composition of lipid/lipid blends impacts on drug release (Kreye et al. 2011b).

12.4.4 Extrusion

A simple extrusion process has been employed for collagen-based implants. For example, Higaki et al. (2001) used a lyophilized cake of collagen containing the antigens tetanus toxoid or diphtheria toxoid, and added a small amount of distilled water to achieve a swelling of the lyophilized matrix (Higaki et al. 2001). The spongy mass was then passed through a nozzle with an inner diameter of 1.7 mm and the resulting rod-shaped implants were air dried at 4 °C in a humid atmosphere for 24 h. Similarly, Yamagata et al. utilized a melted mixture of monoglycerides and lyophilized IFN-α which was loaded into a 14 G stainless steel needle and extruded to form a rod with a diameter of 1.2 mm (Yamagata et al. 2000).

12.4.4.1 Twin-Screw Extrusion

Twin-screw extrusion is a method that has been employed extensively for mixing and compounding of polymeric materials, not only in pharmaceutical processes but also in food processing. Two types of twin-screw extruders exist: corotating and counter-rotating screw extruders. They are often used to process heat- and shear-sensitive material. The extrusion process converts raw material into a product of uniform density and shape by pushing it through a die under controlled conditions. The lipid raw material can be fed into the extruder as a lipid mixture (premix), or alternatively some extruders offer the possibility to feed the single components into the extruder through one or two side stuffers. Extrusion offers the advantage of being a continuous process that can be scaled up according to production needs. The resulting extrudates often have a slightly larger diameter than the die, a phenomenon referred to as "die swell" where the sudden drop in pressure causes the polymer to show stress relaxation. In the pharmaceutical context, twin-screw extrusion has been used successfully for the preparation of implants from triglycerides and mixtures of triglycerides for the delivery of drugs and proteins (Reitz and Kleinebudde 2007; Schulze and Winter 2009; Sax et al. 2012a). Twin-screw extrudates have been prepared containing IFN-α-2a (Schulze and Winter 2009; Sax et al. 2012a) and lysozyme (Schwab et al. 2009; Sax and Winter 2012). However, this method has not been utilized for the preparation of implants for vaccine delivery where the implants have a much lower protein/peptide load (5 mg in drug delivery vs. 1–50 μg for vaccines). Studies on using twin-screw extrudates containing OVA as a model antigen are currently underway (Even et al. in preparation). One disadvantage of twin-screw extrusion is the temperature which needs to be utilized in order to extrude the lipids. The lipid mass must be converted into a semisolid mass that can be transported and mixed by the extrusion process. The temperature needed for this depends on the chosen lipid or lipid mixture. One needs to take into account that this elevated temperature may not only alter the lipid, but may also have an effect on the incorporated drug or protein.

12.5 Characterization

12.5.1 Surface Structure

To investigate the surface structure of the implants, the simplest method is to use an optical imaging system and acquire images, for example, at different time points of a release study. However, sometimes the magnification that can be achieved using optical microscopy may not be sufficient, meaning that further analysis involving the use of scanning electron microscopy is necessary. Here surface properties such as roughness, porosity, or pore formation during release can be investigated in more depth (Guse et al. 2006; Myschik et al. 2008a). Pore formation can also be analyzed after drying of the implants using mercury porosimetry where the application of a high pressure forces a non-wetting liquid, e.g., mercury, into the implant matrix. By doing so, the pore size can be determined based on the pressure needed to force the mercury into a pore against the force of the liquid's surface tension. Implants are usually dried and then placed in a vacuum (Mohl and Winter 2004). The degree of intactness of the implant surface as well as any observed pore formation will have a considerable effect on the release kinetics from implants, as will the occurrence of erosion or swelling. If erosion from the implant matrix occurs, the determination of the release profile becomes more challenging, as release is not independent on erosion. Some implant matrices exhibit swelling due to the choice of excipients. This is often the case if, for example, class II lipids such as phospholipids or monoglycerides are used (Witzleb et al. 2012; Small 1967).

12.5.2 *Distribution Within the Matrix*

Drug or antigen distribution within the implant matrix can be analyzed using techniques such as confocal microscopy. In order to do this, a fluorescently labeled drug has to be used and the concentration needs to be sufficiently high for signal detection. Depending on the implant preparation process, drug distribution can be strongly affected, as reported by Kreye et al. (2008). Directly compressed lipid implants generally display a continuous protein network in the matrix, whereas other preparation techniques may result in protein patches throughout the matrix.

12.5.3 *Stability of the Matrix*

An important aspect particularly for lipid-based implants is the stability of the lipid matrix. Differential scanning calorimetry can be employed to analyze the lipid matrix to detect the potential formation of unstable polymorphic forms of lipids, which may occur as a result of the extrusion procedure or impact of temperature (Schulze and Winter 2009; Pongjanyakul et al. 2004; Reitz and Kleinebudde 2007).

12.5.4 *Antigen Stability Within the Matrix*

Protein or antigen integrity within the implant matrix needs to be analyzed after implant formation. This can be done either in situ using spectroscopic techniques or following extraction of the protein or antigen from the solid matrix. The integrity of extracted protein can then be analyzed using a number of different techniques such as electrophoresis, chromatography, or in activity assays.

Fourier-transform infrared (FT-IR) spectroscopy can be used to analyze changes in the protein when present in the implant matrix, and attenuated Fourier-transform infrared (ATR-FITR) spectroscopy has been reported as a useful technique for locating drug either in the local surface layer or in deeper implant layers (Reitz and Kleinebudde 2007). For FT-IR measurements, usually the second derivative spectra is recorded and analyzed to observe conformational changes. Changes in the band intensity or a shift from α-helical to β-sheet or random coil-rich structures can be observed. However the low concentration of the antigen often makes the use of this technique challenging.

Electrophoresis and chromatography can be used to assess physical characteristics (such as size or charge) of the extracted protein. The extracted protein is commonly run on a polyacrylamide gel and protein integrity is analyzed by comparison to a standard (Mohl and Winter 2004; Herrmann et al. 2007a). Additionally occurring bands, or disappearance of bands may be a hint that protein integrity has been compromised. Size-exclusion chromatography can also be employed to determine the stability of the protein after extraction from the implant matrix. Here, the amount of monomeric protein in comparison to the presence of dimers, trimers, etc. can be analyzed. Again, the very low concentration of antigen in the implant matrix may pose a challenge in terms of antigen detection. Therefore, the use of fluorescence detection may be necessary. For proteins such as lysozyme, activity assays can be performed, and the amount of active drug released can be determined using *Micrococcus lysodeikticus* as a substrate (Pongjanyakul et al. 2004).

12.6 Drug Release Mechanisms

Drug release from a polymer or lipid matrix may occur as a result of diffusion through water-filled pores, diffusion through the polymer or lipid matrix, osmotic pressure, or erosion or degradation of the implant matrix. Sometimes processes such as diffusion and erosion may overlap, making drug release prediction more difficult. In addition, some excipients may lead to a swelling of the implant matrix, thereby increasing the diffusional barrier and hence also making prediction of drug release problematic. In some cases, it is the aim to achieve a longer-term zero-order kinetics release profile. However for vaccine delivery, it has been proposed that a bimodal or dual-pulsed release profile for antigens may be beneficial in order to induce protective immune responses (Wise et al. 1988). For this purpose, the use of pulsatile release implants may be feasible (Cardamone et al. 1997; Vogelhuber et al. 2001; Sanchez et al. 1996). An overview of different release profiles is given in Fig. 12.2.

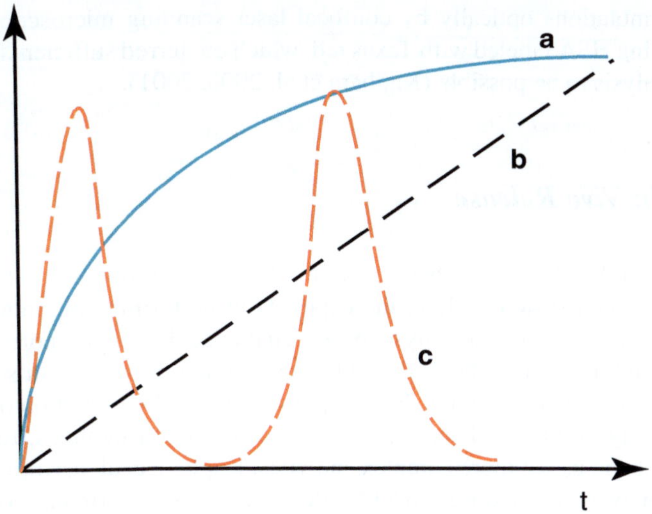

Fig. 12.2 In vitro release profiles: (**a**) first-order release profile with large amount of antigen released during the first time period, (**b**) zero-order release profile where the same amount of antigen is released per time period (idealized), (**c**) dual-pulsed release profile where a first pulse is released followed by a second pulse (idealized)

12.6.1 In Vitro Release

Determination of the in vitro release of proteins or antigens from implants is usually carried out in buffers such as phosphate-buffered saline (PBS), often at pH 7.4. Such a buffer is used to simulate isotonic conditions, and the temperature is kept to 37 °C in order to mimic physiological conditions. Sodium azide at a concentration of 0.01–0.05 % can be added as a preservative if studies are carried out over a period of several days, weeks, or even months. Sampling is usually carried out at predetermined time points. In some studies, a full exchange method is used, meaning that the release medium is completely removed and then replaced by fresh buffer. Other authors report a partial exchange of the buffer medium, which makes calculation of the final concentration of released drug/antigen more difficult. Even though most proteins and antigens are very soluble, one needs to take into account that sink-conditions should be maintained for the duration of a release study. Analysis of released drug may be carried out using UV-spectroscopy or fluorescence spectroscopy, or the protein concentration and monomer content may be assessed by size-exclusion chromatography (Lofthouse et al. 2001; Higaki et al. 2001; Schulze and Winter 2009). In addition, Kajihara et al. have observed drug release from silicone implant formulations optically by confocal laser scanning microscopy. This was achieved using HSA labeled with Texas red, which conferred sufficient fluorescence for such analysis to be possible (Kajihara et al. 2000, 2001).

12.6.2 *In Vivo Release*

In vivo release kinetic studies have been less frequently reported in literature with respect to implant systems. Here, the expenses in performing an animal study in addition to complicated data analysis pose a challenge. For small molecular weight drugs, in vivo kinetics may be analyzed by determining the plasma or serum values of the drug at certain time points. For antigens however, the formation of anti-drug-antibodies may eventually interfere with simple detection methods, requiring the development and use of more complex analytics. Kajihara et al. performed pharmacokinetic analysis in a mouse model by determining the maximum concentration (C_{max}), time of occurrence of (T_{max}), and the elimination rate (K_e) for IFN-silicone formulations (Kajihara et al. 2000).

12.7 Preclinical Studies

A number of preclinical studies on implants for vaccine delivery have been performed in different animal species, utilizing compressed, molten or extruded implants either from nonbiodegradable or biodegradable material. The most commonly used species include mice and sheep. Administration of the implant was either performed through a trochar or by minor surgery using a small incision. In the majority of studies, the antibody response was measured using an enzyme-linked immunoassay (ELISA). In some studies, histological investigation of the site of administration was also performed. An overview of these studies is given in Table 12.1.

Table 12.1 Overview of preclinical studies of implanted vaccine delivery systems in different animal species

Year	Author	Delivery system	Material	Antigen or drug	Investigations	Animals	Administration	Control	Assays	References
1976	Langer and Folkman	Polymer pellet	Ethylene-vinyl-acetate-copolymer	Soybean trypsin inhibitor	>100 days, zero-order kinetics	–	–	–	–	Langer and Folkman (1976)
			Molding procedure, "sandwich approach"	Lysozyme						
				Alkaline phosphatase catalase						
1979	Preis and Langer	Polymer pellet	Ethylene-vinyl-acetate-copolymer	Crystalline bovine albumin	Immunological in vivo study	C57Bl/6 mice	s.c. implanted through small incision	Polymer pellet without antigen	Antibody response induced by BSA	Preis and Langer (1979)
			Pellets coated with a thin film of pure polymer	Ribonuclease-A						
				Bovine γ-globulin						
1991	Khan et al.	Lipid pellet (5.5 mm)	Cholesterol or cholesterol with lecithin pellet	BSA	In vitro release, erosion	Quackenbush mice	s.c. implanted through small incision	Pellet without antigen	Antibody responses	Khan et al. (1991)
1993	Opdebeeck and Tucker	Lipid pellet (5.5 × 1.8 mm)	Compressed cholesterol pellet	BSA	In vivo release for at least 42 days	Quackenbush mice	s.c. insertion through small incision in the skin distal to the neck	Blank cholesterol pellets	Antibody response, Histology	Opdebeeck and Tucker (1993)
					Initial burst in the first 6 days, slowly thereafter					
1997	Cardamone et al.	Pulsatile drug delivery device (50 × 8 mm)	Device consisting of a plug, an active, a spacer, and a swelling agent	Tetanus toxoid	In vitro release methylene blue	BALB/c mice	s.c. implantation after incision	Two doses of commercial vaccine	Antibody response	Cardamone et al. (1997)
					Immunological in vivo study	Merino crossbred sheep				

(continued)

Table 12.1 (continued)

Year	Author	Delivery system	Material	Antigen or drug	Investigations	Animals	Administration	Control	Assays	References
1998	Walduck et al.	Lipid implant (5.5 × 1.8 mm)	(a) Cholesterol/lecithin	Recombinant *Dichelobacter nodosus* pili	In vitro release	Merino sheep	s.c. using a 25 mm 18G needle	Blank implants	Antibody response	Walduck et al. (1998)
			(b) Implants coated with cholesterol/lecithin	Adjuvant Quil-A	Within a few days				Tissue compatibility	
			(c) Enteric coated implants		Burst followed by trickle release when Quil-A was present in the implants					
			(d) Double-coated implants							
			Biodegradable							
1999	Ochiya et al.	Cylindrical collagen minipellet (0.6 mm diameter, 10 mm length)	Atelocollagen	plasmid DNA	In vitro release investigated over 10 days	ICR mice	i.m. using an injector equipped with an 18-gauge needle	Minipellet alone	Platelet count	Ochiya et al. (1999)
2000	Kajihara et al.	Silicone implant	Silicone matrix	INF/HAS lyophilized	In vitro	Athymic nude mice (nu/nu BALB/c background), 5 weeks old	s.c. injection on day 1	Tumor challenge	PK	Kajihara et al. (2000)
				Cured at RT for 3 days	Continuous release over 1 month		Blood samples: 0.5, 1,3, 10, 30, 60, 90		Tumor growth	
				Loading 5–30 %	Total amount released dependent on loading					
2001	Kajihara et al.	Silicone implant	Silicone implant	INF/HAS	In vitro	Athymic nude mice (BALB/c background)	s.c. injection		PK	Kajihara et al. (2001)
			Covered rod type	IFN/HAS/silicone inner part	Continuous release up to 100 days for HAS		Blood samples days 1, 4, 7, 14, 28			
				Additive free part: outer part						
				Cured for 3 days at RT						

2001	Higaki et al.	Collagen minipellet	Dried collagen rod, diameter 1 mm	Tetanus toxoid (TT), Diphtheria (DT) toxoid	In vitro release	BALB/c mice	s.c. injection	Alum-adsorbed TT or DT	Antibody response	Higaki et al. (2001)
					Cumulative release after 14 days ~35 %					
					In vivo immunological study					
2001	Lofthouse et al.	Collagen minipellet	Atellocollagen cylindrical solid dosage form	Avidin + IL-1β as adjuvant or clostridial vaccine as lyophilized powder	In vitro release first-order kinetic over 1 week	Female BALB/c mice and crossbred Merino sheep	s.c. injection	Avidin in alum or in saline	ELISA (antibodies)	Lofthouse et al. (2001)
			Degraded within 35 days of implantation				But: booster still required			
2001	Vogelhuber et al.	Embedded polymer implant	Polyanhydride cylinder embedded in PLGA or PLGA mantle	Pyranine	In vitro release and erosion	Immunodeficient NMRI (*nu/nu*) mice	s.c. administration in the left flank	–	Onset of Evan's blue release	Vogelhuber et al. (2001)
				Evan's blue	Pulsatile release depending on composition					
2002	Lofthouse et al.	Silicone implant	(a) Silicone implant	Avidin	In vitro release	12–24 months old merino sheep	Avidin 10–500 µg dose per sheep	Alum, 1 mL total volume, 18G needle	Antibody response	Lofthouse et al. (2002)
		(a) Matrix	(b) Covered rod type	*Clostridium tetani*	(a) First-order profile, 1 month		s.c. back of the neck			
		(b) Rod-shaped	Nonbiodegradable polymer	*Clostridium novyi* toxoid	(b) Zero-order profile, several months					
				Antigen + adjuvant lyophilized						
2002	Kemp et al.	Silicone implant	(a) Silicone implant	Avidin	In vitro release	Merino crossbred ewes and wethers	s.c. injection	Alzet™—mini-osmotic pump	Antibody responses	Kemp et al. (2002)
			(b) Co-extruded silicone implant with and without avidin	IL-1β	(a) First-order profile, 1 month					
					(b) Zero-order profile, several months					

(continued)

Table 12.1 (continued)

Year	Author	Delivery system	Material	Antigen or drug	Investigations	Animals	Administration	Control	Assays	References
2006	Guse et al.	Lipid pellet	Compressed triglycerides and blends with cholesterol and phospholipids	–	In vivo erosion and swelling	NMRI mice	s.c. implantation into the flank	–	Histology	Guse et al. (2006)
					In vivo biocompatibility					
2008	Myschik et al.	Lipid pellet	Compressed cholesterol and phospholipid	OVA	In vitro release	C57Bl/6 mice	s.c. implantation through trochar	Blank implant	Antibody response,	Myschik et al. (2008a, b, c, d)
				Adjuvant Quil-A or imiquimod or α-galactosylceramide analogue	In vivo immunological study				Cellular immune response	

12.8 Important Considerations

12.8.1 *Tissue Responses*

As a result of implantation, a tissue response may occur. This may be in the form of an encapsulation, or the development of tissue edema or inflammation. Encapsulation of implants in a membrane were reported by Opdebeeck and Tucker following administration of a BSA-loaded cholesterol pellet (Opdebeeck and Tucker 1993), and Khan et al. after the administration of cholesterol/lecithin implants (Khan et al. 1991). In addition, granuloma formation or increased vascularization may occur. Fibrous capsule formation was also reported by Sax et al. for implants prepared from a mixture of high and low melting point lipids (Sax et al. 2012b). Cardamone et al. observed some local foreign body responses to their implant, however, the authors concluded that this response may be beneficial for vaccination (Cardamone et al. 1997). Guse et al. observed foreign body reactions to implants made of phospholipids and glyceroltripalmitate (Guse et al. 2006).

12.8.2 *Biocompatibility*

For implant administration, the biocompatibility of the material is of major importance as the delivery system will remain at the site of implantation for several days, weeks, or even months. Good biocompatibility has been reported for glyceryl tripalmitate (Reithmeier et al. 2001), glyceryl palmitostearate (Gao et al. 1995), as well as implants made of triglycerides and cholesterol (Guse et al. 2006).

12.8.3 *Biodegradability Versus Non-Biodegradability*

One aspect that needs to be considered early on in implant formulation development is the decision as to whether biodegradable or nonbiodegradable systems are desired. Each system has its advantages and disadvantages. Biodegradable systems may degrade via bioerosion processes, meaning that over the time of incubation (or implantation), the diffusional barrier of the matrix becomes smaller. If, for example, the antigen is incorporated into a lipid matrix, this matrix may undergo enzymatic degradation via the action of lipases, as shown by Schwab et al. (2009, 2009, 2013). The occurrence and impact of lipase degradation depends on the type of lipid. Only in the case of defined triglycerides can degradation be investigated systematically. Short-chain triglycerides, for example, are degraded faster than long-chain triglycerides. Therefore, drug release can only be predicted for defined, synthetic triglyceride matrices.

In the case of polymers as matrix formers, three major approaches may be taken in order to facilitate degradation. If water-insoluble polymers are hydrolyzed, some can degrade into water-soluble fragments. This can also be achieved if functionalized groups within the polymer are affected by hydrolysis, ionization, or protonation, thereby converting the water-insoluble polymer into a soluble polymer. Thirdly, hydrophilic polymers can be converted into water-insoluble polymer networks through polymerization (Voigt 2000).

A disadvantage of implants that degrade over time is the loss of surface area due to the degradation process. This results in the observation of a decrease in drug release; a constant drug release profile can therefore not be achieved using these systems. Also, the time required for biodegradation depends heavily on the choice of lipid and the manufacturing technique (Schwab et al. 2009; Sax et al. 2012b). Sax et al. for example, observed that surprisingly, implants prepared by twin screw extrusion from a mixture of a high melting lipid (D118), a low melting lipid (H12 or E85), and a pore-forming agent (PEG 6000) showed biodegradability when tested in vivo in a rabbit model for over 6 months (Sax et al. 2012b). Compressed implants of D118 however stayed intact for more than 4 weeks (Schwab et al. 2008) and no macroscopic encapsulation was visible after surgical removal.

If the matrix is of a nonbiodegradable nature, later surgical removal of the implant matrix is in most circumstances inevitable. The nonbiodegradable material must have an excellent biocompatibility profile and must not show any adverse tissue reactions (Wise et al. 1988).

12.8.4 Immunogenicity

Last but not least, the immunogenicity of any antigen or vaccine must be preserved when incorporated into a controlled or sustained release implant. During sustained delivery, protein aggregation or the formation of protein particles may occur, thereby increasing the chance of generating immune responses. While this is an unwanted in the case of controlled delivery of therapeutic biopharmaceuticals (Jiskoot et al. 2012), it may prove advantageous and desirable in the controlled or sustained delivery of proteins for vaccine applications.

12.9 Summary

Certainly given the reports presented in literature, research on implants for controlled or sustained release of antigen is justified. However, a number of challenges as outlined above still need to be overcome in order to make such delivery systems applicable not only for veterinary vaccines but also for human applications.

References

Aguado MT, Lambert PH (1992) Controlled-release vaccines-biodegradable polylactide/polyglycolide (PL/PG) microspheres as antigen vehicles. Immunobiology 184(2–3):113–125. doi:10.1016/S0171-2985(11)80470-5

Appel B, Maschke A, Weiser B, Sarhan H, Englert C, Angele P, Blunk T, Göpferich A (2006) Lipidic implants for controlled release of bioactive insulin: effects on cartilage engineered in vitro. Int J Pharm 314(2):170–178. doi:10.1016/j.ijpharm.2005.11.049

Cardamone M, Lofthouse SA, Lucas JC, Lee RP, O'Donoghue M, Brandon MR (1997) In vitro testing of a pulsatile delivery system and its in vivo application for immunisation against tetanus toxoid. J Control Release 47(3):205–219. doi:10.1016/S0168-3659(97)01639-8

Demana PH, Davies NM, Hook S, Rades T (2005) Quil A—lipid powder formulations releasing ISCOMs and related colloidal structures upon hydration. J Control Release 103(1):45–59. doi:10.1016/j.jconrel.2004.11.027

Dixon FJ, Maurer PH (1955) Immunologic unresponsiveness induced by protein antigens. J Exp Med 101:237–245

Dresser DW (1962) Specific inhibition of antibody production. II. Paralysis induced in adult mice by small quantities of protein antigen. Immunology 5:378–388

Even MP, Young K, Winter G, Hook S, Engert J (2014) In vivo investigation of twin-screw extruded lipid implants for vaccine delivery. Eur J Pharm Biopharm 87(2):338–46. doi: 10.1016/j.ejpb.2014.02.014. Epub 2014 Mar 4.

Gao Z, Crowley WR, Shukla AJ, James JR, Reger JF (1995) Controlled release of contraceptive steroids from biodegradable and injection gel formulations: in vivo evaluation. Pharm Res 12:864–868

Guse C, Koennings S, Maschke A, Hacker M, Becker C, Schreiner S, Blunk T, Spruss T, Goepferich A (2006) Biocompatibility and erosion behavior of implants made of triglycerides and blends with cholesterol and phospholipids. Int J Pharm 314(2):153–160. doi:10.1016/j.ijpharm.2005.12.050

Herrmann S, Mohl S, Siepmann F, Siepmann J, Winter G (2007a) New insight into the role of polyethylene glycol acting as protein release modifier in lipidic implants. Pharm Res 24(8):1527–1537

Herrmann S, Winter G, Mohl S, Siepmann F, Siepmann J (2007b) Mechanisms controlling protein release from lipidic implants: effects of PEG addition. J Control Release 118(2):161–168. doi:10.1016/j.jconrel.2006.11.001

Higaki M, Azechi Y, Takase T, Igarashi R, Nagahara S, Sano A, Fujioka K, Nakagawa N, Aizawa C, Mizushima Y (2001) Collagen minipellet as a controlled release delivery system for tetanus and diphtheria toxoid. Vaccine 19(23–24):3091–3096. doi:10.1016/S0264-410X(01)00039-1

Jiskoot W, Randolph TW, Volkin DB, Middaugh CR, Schöneich C, Winter G, Friess W, Crommelin DJA, Carpenter JF (2012) Protein instability and immunogenicity: roadblocks to clinical application of injectable protein delivery systems for sustained release. J Pharm Sci 101(3):946–954. doi:10.1002/jps.23018

Kajihara M, Sugie T, Mizuno M, Tamura N, Sano A, Fujioka K, Kashiwazaki Y, Yamaoka T, Sugawara S, Urabe Y (2000) Development of new drug delivery system for protein drugs using silicone (I). J Control Release 66(1):49–61. doi:10.1016/S0168-3659(99)00257-6

Kajihara M, Sugie T, Hojo T, Maeda H, Sano A, Fujioka K, Sugawara S, Urabe Y (2001) Development of a new drug delivery system for protein drugs using silicone (II). J Control Release 73(2–3):279–291. doi:10.1016/S0168-3659(01)00302-9

Kemp JM, Kajihara M, Nagahara S, Sano A, Brandon M, Lofthouse S (2002) Continuous antigen delivery from controlled release implants induces significant and anamnestic immune responses. Vaccine 20(7–8):1089–1098. doi:10.1016/S0264-410X(01)00444-3

Khan MZI, Tucker IG, Opdebeeck JP (1991) Cholesterol and lecithin implants for sustained release of antigen: release and erosion in vitro, and antibody response in mice. Int J Pharm 76(1–2):161–170. doi:10.1016/0378-5173(91)90354-Q

Khan MZI, Tucker IG, Opdebeeck JP (1993) Evaluation of cholesterol-lecithin implants for sustained delivery of antigen: release in vivo and single-step immunisation of mice. Int J Pharm 90(3):255–262. doi:10.1016/0378-5173(93)90198-O

Koennings S, Garcion E, Faisant N, Menei P, Benoit JP, Goepferich A (2006) In vitro investigation of lipid implants as a controlled release system for interleukin-18. Int J Pharm 314(2):145–152. doi:10.1016/j.ijpharm.2005.08.031

Koennings S, Sapin A, Blunk T, Menei P, Goepferich A (2007) Towards controlled release of BDNF—Manufacturing strategies for protein-loaded lipid implants and biocompatibility evaluation in the brain. J Control Release 119(2):163–172. doi:10.1016/j.jconrel.2007.02.005

Kreye F, Siepmann F, Siepmann J (2008) Lipid implants as drug delivery systems. Expert Opin Drug Deliv 5(3):291–307. doi:10.1517/17425247.5.3.291

Kreye F, Siepmann F, Willart JF, Descamps M, Siepmann J (2011a) Drug release mechanisms of cast lipid implants. Eur J Pharm Biopharm 78(3):394–400. doi:10.1016/j.ejpb.2011.02.011

Kreye F, Siepmann F, Zimmer A, Willart JF, Descamps M, Siepmann J (2011b) Controlled release implants based on cast lipid blends. Eur J Pharm Sci 43(1–2):78–83. doi:10.1016/j.ejps.2011.03.013

Langer R, Folkman J (1976) Polymers for the sustained release of proteins and other macromolecules. Nature 263(5580):797–800

Lofthouse S, Nagahara S, Sedgmen B, Barcham G, Brandon M, Sano A (2001) The application of biodegradable collagen minipellets as vaccine delivery vehicles in mice and sheep. Vaccine 19(30):4318–4327. doi:10.1016/S0264-410X(01)00153-0

Lofthouse SA, Kajihara M, Nagahara S, Nash A, Barcham GJ, Sedgmen B, Brandon MR, Sano A (2002) Injectable silicone implants as vaccine delivery vehicles. Vaccine 20(13–14): 1725–1732. doi:10.1016/S0264-410X(02)00036-1

Mohl S, Winter G (2004) Continuous release of rh-interferon α-2a from triglyceride matrices. J Control Release 97(1):67–78. doi:10.1016/j.jconrel.2004.02.027

Myschik J, Eberhardt F, Rades T, Hook S (2008a) Immunostimulatory biodegradable implants containing the adjuvant Quil-A—Part I: Physicochemical characterisation. J Drug Target 16(3):213–223. doi:10.1080/10611860701848860

Myschik J, McBurney W, Hennessy T, Rades T, Hook S (2008b) Immunogenicity of lipid sustained release implants containing imiquimod, alpha-galactosylceramide, or Quil-A. Pharmazie 63(9):686–692

Myschik J, McBurney WT, Hennessy T, Phipps-Green A, Rades T, Hook S (2008c) Immunostimulatory biodegradable implants containing the adjuvant Quil-A—Part II: in vivo evaluation. J Drug Target 16(3):224–232. doi:10.1080/10611860701848886

Myschik J, McBurney WT, Rades T, Hook S (2008d) Immunostimulatory lipid implants containing Quil-A and DC-cholesterol. Int J Pharm 363(1–2):91–98. doi:10.1016/j.ijpharm.2008.07.014

O'Hagan DT, Jeffery H, Roberts MJ, McGee JP, Davis SS (1991) Controlled release microparticles for vaccine development. Vaccine 9(10):768–771

Ochiya T, Takahama Y, Nagahara S, Sumita Y, Hisada A, Itoh H, Nagai Y, Terada M (1999) New delivery system for plasmid DNA in vivo using atelocollagen as a carrier material: the Minipellet. Nat Med 5(6):707–710

Opdebeeck JP, Tucker IG (1993) A cholesterol implant used as a delivery system to immunize mice with bovine serum albumin. J Control Release 23(3):271–279. doi:10.1016/0168-3659(93)90008-S

Peralta O, Diaz S, Croxatto H (1995) Subdermal contraceptive implants. J Steroid Biochem Mol Biol 53(1–6):223–226. doi:10.1016/0960-0760(95)00051-Z

Pongjanyakul T, Medlicott NJ, Tucker IG (2004) Melted glyceryl palmitostearate (GPS) pellets for protein delivery. Int J Pharm 271(1–2):53–62. doi:10.1016/j.ijpharm.2003.10.017

Preis I, Langer RS (1979) A single-step immunization by sustained antigen release. J Immunol Methods 28(1–2):193–197

Ramsdell F, Fowlkes BJ (1992) Maintenance of in vivo tolerance by persistence of antigen. Science 257:1130–1134

Reithmeier H, Herrmann J, Göpferich A (2001) Lipid microparticles as a parenteral controlled release device for peptides. J Control Release 73(2–3):339–350. doi:10.1016/S0168-3659(01)00354-6

Reitz C, Kleinebudde P (2007) Solid lipid extrusion of sustained release dosage forms. Eur J Pharm Biopharm 67(2):440–448. doi:10.1016/j.ejpb.2007.03.008

Sanchez A, Gupta RK, Alonso MJ, Siber GR, Langer R (1996) Pulsed controlled-release system for potential use in vaccine delivery. J Pharm Sci 85(6):547–552. doi:10.1021/js960069y

Sax G, Winter G (2012) Mechanistic studies on the release of lysozyme from twin-screw extruded lipid implants. J Control Release 163(2):187–194. doi:10.1016/j.jconrel.2012.08.025

Sax G, Feil F, Schulze S, Jung C, Bräuchle C, Winter G (2012a) Release pathways of interferon α2a molecules from lipid twin screw extrudates revealed by single molecule fluorescence microscopy. J Control Release 162(2):295–302. doi:10.1016/j.jconrel.2012.07.014

Sax G, Kessler B, Wolf E, Winter G (2012b) In-vivo biodegradation of extruded lipid implants in rabbits. J Control Release 163(2):195–202. doi:10.1016/j.jconrel.2012.08.026

Schulze S, Winter G (2009) Lipid extrudates as novel sustained release systems for pharmaceutical proteins. J Control Release 134(3):177–185. doi:10.1016/j.jconrel.2008.11.026

Schwab M, Kessler B, Wolf E, Jordan G, Mohl S, Winter G (2008) Correlation of in vivo and in vitro release data for rh-INFα lipid implants. Eur J Pharm Biopharm 70(2):690–694. doi:10.1016/j.ejpb.2008.05.010

Schwab M, Sax G, Schulze S, Winter G (2009) Studies on the lipase induced degradation of lipid based drug delivery systems. J Control Release 140(1):27–33

Schwab M, McGoverin CM, Gordon KC, Winter G, Rades T, Myschik J, Strachan CJ (2013) Studies on the lipase-induced degradation of lipid-based drug delivery systems. Part II—Investigations on the mechanisms leading to collapse of the lipid structure. Eur J Pharm Biopharm 84(3):456–463. doi:10.1016/j.ejpb.2012.12.023

Small DM (1967) Phase equilibria and structure of dry and hydrated egg lecithin. J Lipid Res 8(6):551–557

Vogelhuber W, Rotunno P, Magni E, Gazzaniga A, Spruß T, Bernhardt G, Buschauer A, Göpferich A (2001) Programmable biodegradable implants. J Control Release 73(1):75–88. doi:10.1016/S0168-3659(01)00282-6

Voigt R (2000) Pharmazeutische Technologie: für Studium und Beruf, 9th edn. Deutscher Apotheker Verlag, Stuttgart, ISBN 3-7692-2649-6

Walduck AK, Opdebeeck JP, Benson HE, Prankerd R (1998) Biodegradable implants for the delivery of veterinary vaccines: design, manufacture and antibody responses in sheep. J Control Release 51(2–3):269–280. doi:10.1016/S0168-3659(97)00180-6

Wise DL, Trantolo DJ, Marino RT, Kitchell JP (1988) Opportunities and challenges in the design of implantable biodegradable polymeric systems for the delivery of antimicrobial agents and vaccines. Adv Drug Deliv Rev 1(3):269. doi:10.1016/0169-409X(88)90023-3

Witzleb R, Müllertz A, Kanikanti VR, Hamann HJ, Kleinebudde P (2012) Dissolution of solid lipid extrudates in biorelevant media. Int J Pharm 422(1–2):116–124. doi:10.1016/j.ijpharm.2011.10.037

Yamagata Y, Iga K, Ogawa Y (2000) Novel sustained-release dosage forms of proteins using polyglycerol esters of fatty acids. J Control Release 63(3):319–329. doi:10.1016/S0168-3659(99)00206-0

Chapter 13
Dendritic Cell-Based Vaccines

Olivier Gasser and Ian F. Hermans

13.1 Introduction

Early observations from half a century ago have implicated cell-mediated immunity in resistance and protection against cancer (Habel 1962). Decades later, the identification of the first tumor-associated antigens launched the era of cancer immunotherapy (Traversari et al. 1992; van der Bruggen et al. 1991), and there is now a large body of evidence that suggests that antigen-specific T-cells play a prominent role in the immunological control of cancer (Galon et al. 2012; Sato et al. 2005; Shankaran et al. 2001; Tosolini et al. 2011). Dendritic cells (DCs) are antigen-presenting cells (APCs) that stimulate proliferation of antigen-specific T-cells, and provide the key molecular signals required to drive optimal effector functions, including cytotoxicity (Steinman 1991). In terms of immunological potency, DCs are orders of magnitude better than any other professional APCs (Nussenzweig et al. 1980) and this heightened functionality has been shown to apply to cancer immunosurveillance (Petersen et al. 2010). Given this key role, optimizing DC function is essential to achieving potent immune responses, and it is for this reason that these cells are often the targets of the immunological adjuvants used in vaccination strategies (Gallucci et al. 1999; Steinman and Banchereau 2007; Coffman et al. 2010). With the advent of methodological approaches to culturing DCs in vitro, it became theoretically possible to precisely control aspects of DC activity, including the antigens and stimulation they receive, with the aim of maximizing their stimulatory function for antigen-specific T-cells; these cells could then be injected back into the patient to elicit desired responses. This approach to vaccination is attractive

O. Gasser (✉)
Malaghan Institute of Medical Research, PO Box 7060, Wellington 6242, New Zealand
e-mail: ogasser@malaghan.org.nz

I.F. Hermans
School of Biological Sciences, Victoria University Wellington, Wellington, New Zealand
e-mail: ihermans@malaghan.org.nz

C. Foged et al. (eds.), *Subunit Vaccine Delivery*, Advances in Delivery Science and Technology, DOI 10.1007/978-1-4939-1417-3_13

in the cancer setting, where DC function is often blunted or subverted by factors released by the tumor. Building on promising preliminary successes in animal models (Flamand et al. 1994; Inaba et al. 1990), the first attempt to use such DC-based vaccines therapeutically in humans was carried out in patients with B-cell lymphoma in the mid-1990s (Hsu et al. 1996). Several in vitro generated DC-based vaccine strategies have since been validated, culminating with the US-Food and Drug Administration (FDA)-approval of Sipuleucel-T for metastatic prostate cancer, a product composed of peripheral blood mononuclear cells differentiated with a proprietary fusion protein of granulocyte-macrophage colony stimulating factor (GM-CSF) and prostatic acid phosphatase (Small et al. 2006).

The immunotherapeutic use of DC vaccines in their various forms has been extensively documented, largely supporting their clinical safety (Aarntzen et al. 2008; Kalinski et al. 2013; Schuler 2010). However, a clear relationship between their immunological potency, as assessed by expansion of cancer-specific T-cells, and clinical impact has been hard to confirm. Among the disappointments, clinically, was a pioneering phase III clinical trial of DC vaccines in metastatic melanoma patients, which failed to demonstrate any efficacy as compared to standard of care alone (Schadendorf et al. 2006). A recent meta-analysis of 38 published DC-vaccine studies, including a total of 626 stage III-IV melanoma patients, concluded that less than a third of patients objectively respond to DC immunotherapy (including stable disease) (Engell-Noerregaard et al. 2009). An equally sobering fact is the modest 4 months median survival benefit from Sipuleucel-T-treatment in patients with metastatic castration-resistant prostate cancer, while the initial target of progression-free survival never reached statistical significance (Higano et al. 2009; Kantoff et al. 2010; Small et al. 2006). In this chapter, we aim at providing an extensive overview of the necessary steps to bring DC-based immunotherapy closer to wide, and justifiable, clinical use.

13.2 Challenges and Emerging Opportunities of DC-Based Immunotherapy

The in vitro generation of DCs offers a range of options including various culture conditions, sources of antigen and maturation protocols (Palucka and Banchereau 2012). Early attempts to isolate DCs directly from the peripheral circulation were successful but limited by a very low blood DC yield (Hsu et al. 1996). The generation of clinical grade DCs now typically involves the isolation of large numbers of monocytes from the peripheral circulation of patients, without the need for any pharmaceutical preconditioning, and in vitro culture for 5–7 days in the presence of GM-CSF and interleukin (IL)-4 (Romani et al. 1994; Sallusto and Lanzavecchia 1994). To induce maturation of immature DCs, a cytokine "cocktail" of IL-6, IL-1β, tumor necrosis factor (TNF-)α and prostaglandin (PG)E_2 represents the reference standard (Feuerstein et al. 2000; Schuler-Thurner et al. 2002). This particular combination of cytokines essentially represents a synthetic and standardized form of monocyte-conditioned medium (Jonuleit et al. 1997). Current standard operating procedures (SOPs) for the in vitro generation of clinical grade DCs are aimed at generating mature DCs with a stable phenotype. Validation studies are typically conducted to show that cells generated by the SOPs can efficiently present antigens to T-cells in vitro (Figdor et al. 2004).

In an effort to render DC-based immunotherapy more accessible in terms of costs and labor, attempts have been made to significantly shorten the ex vivo cell culture period. Mature DCs obtained after a mere 2-day culture period, the so-called "fastDCs," were shown, in seminal work conducted a decade ago, to be equivalent to standard mature DCs with regard to the essential capacity to induce antigen-specific T-cell responses (Dauer et al. 2003). The capacity of fastDCs to prime tumor-specific T-cells in particular was documented shortly after (Dauer et al. 2005). Herein we will use the term "fastDCs" broadly to encompass the various protocols described that last no more than 3 days. A few important implications of the shortening of the DC differentiation process have since been revealed. Importantly, fastDCs feature a lower spontaneous uptake of antigen and although amenable to genetic transduction, require proprietary electroporation protocols (Burdek et al. 2010; Kvistborg et al. 2009). They release less IL-12p70 (an important DC-derived soluble mediator of T-cell activation and polarization, as will be discussed in more detail later) in culture compared to DCs generated according to lengthier protocols (Kvistborg et al. 2009). By contrast, compared to standard DCs, fastDCs are characterized by higher yields, a higher CCL19-induced chemokinesis, better intracellular processing of antigen after it has been acquired, and a more effective priming of tumor-specific cytotoxic T-cells (CTLs) (Burdek et al. 2010; Dauer et al. 2005; Kvistborg et al. 2009). Based on preliminary recent observations, similarly functional fastDCs can be generated from unfractionated peripheral blood mononuclear cells (Kodama et al. 2013).

The qualitative enhancement of DC-vaccine products can occur on several levels, largely mirroring the essential signaling events that control effective T-cell activation (Arens and Schoenberger 2010). To yield maximal protection, in vitro generated DC-vaccines need to be able to efficiently present antigen for a prolonged duration sufficient to engage the T-cell receptor ("signal 1" for effective T-cell activation) and provide T-cells with appropriate co-stimulation through cell surface CD80/86 and CD70 (engaging T-cell-encoded CD28 and CD27 respectively; "signal" 2). Finally, the immunogenic cytokine IL-12p70 is thought to represent a crucial "third signal" for T-cell proliferation and function and should be ideally provided by DC-vaccines as well. In the section below we will discuss how these various requirements can be instilled into ex vivo generated DCs to yield highly effective vaccines.

13.2.1 Source of Antigen

Various DC antigen-loading strategies have been tested to date. Prominent among in vitro methodologies is the loading of DCs with tumor-derived peptides (Figdor et al. 2004), or cellular material from lysates or irradiated tissue. It is also possible to transduce DCs with autologous tumor-derived messenger (m)RNA or DNA

(Shurin et al. 2010) or to directly fuse DCs with autologous tumor cells (Lee 2011; Shu et al. 2007). The advantage of "whole cell" approaches, where cellular or genetic material is sourced from tumor cells, is that they are highly personalized, and will drive immunity towards patient-specific tumor-associated antigens (TAAs). A downside of such custom DC products is the level of difficultly, the variability between patients, the need for specialized equipment, and overall significant labor and costs. Most clinical trials to date have adopted a more off-the-shelf strategy, loading the DCs with common TAA(s), or peptides derived thereof. While accelerating the DC-manufacturing process, this selective strategy entails more restrictive patient selection criteria in terms of HLA-genotype and tumor phenotype, and is intrinsically biased. Indeed, in contrast to most pathogen-associated antigens that are reasonably well defined, the antigen-selection in cancer immunotherapy is far less established and still largely empirical (Buonaguro et al. 2011). In addition, while some TAAs such as cancer-testis antigens (CTAs) are shared among various cancer types, the restriction of other TAAs to specific tissues such as the mammary or prostatic epithelium prevents an excessively translational research approach (Buonaguro et al. 2011). Thus, while peptide-loaded DC-vaccines have enjoyed preferential attention, stringent comparative studies are needed to consolidate current antigen/peptide-selection criteria. Importantly, TAA selection and expression patterns significantly impact the quality of the ensuing immune response. As documented recently, sustained release of melanocyte-derived antigens promotes the induction of long-lived effector memory (EM) T-cells that are protective against melanoma (Byrne et al. 2011).

It is also important to note that mature DCs express the immunoproteasome, and therefore process and present antigen in a different manner than cells that have no professional antigen-presenting function (Van den Eynde and Morel 2001). The different proteolytic activities of the immunoproteasome compared to the ubiquitously expressed "constitutive" proteasome imply that DCs might prime T-cells that are unable to recognize antigenic peptides on target cells. DCs engineered to present antigen in the same form as encountered by T-cells in the periphery are therefore likely to provide better therapeutic benefit, as very recently described (Dannull et al. 2013).

13.2.2 Release of IL-12

A large body of work has focused on the capacity of DCs to produce bioactive IL-12p70, the third signal for optimal CTL expansion and acquisition of effector functions (Curtsinger et al. 1999; Trinchieri et al. 2003). Overall, current protocols do not seem to provide the necessary signals to optimally prime DCs for IL-12p70 production (Mailliard et al. 2004), a process that is known to rely on concerted activation of CD40 or toll-like receptors (TLR) and interferon (IFN)-γ-associated signaling pathways (Mosca et al. 2000; Snijders et al. 1998). CD40-mediated "licensing" of DCs to produce IL-12p70, and thus initiate productive CTL responses, has empirically been equated to CD4$^+$ T-cell "help." However it has recently become clear that the contribution of CD4$^+$ T-cells to a robust, memory-forming CTL immune response is not always restricted to CD40-signaling (Ballesteros-Tato et al. 2013) and that CD40-signaling itself can be mediated by other T-cell subsets such as CTLs themselves and type 1 (invariant) NKT (iNKT) cells (Bendelac et al. 2007; Frentsch et al. 2013).

Systemic administration of recombinant IL-12p70 is associated with dose-limiting toxicities (Atkins et al. 1997; Leonard et al. 1997) and thus cannot be incorporated into current vaccine designs. Emphasis has therefore been on changes to in vitro DC differentiation/maturation protocols that encourage IL-12p70 release (Mailliard et al. 2004). In this context, DCs matured in the presence of CD40 ligand and IFN-γ have shown promise in clinical application, with levels of DC-vaccine-derived IL-12p70 positively correlating with time to progression in a cohort of metastatic melanoma patients (Carreno et al. 2013). Another strategy has been to substitute IL-15 for IL-4 during in vitro DC differentiation, which enhances IL-12p70 secretion upon CD40-ligation (Anguille et al. 2009), and has other advantageous aspects such as shorter DC culture periods (2-3 days) and a superior capacity to prime CTLs (Anguille et al. 2009; Dubsky et al. 2007). Similar results were observed with "alpha-type-1-polarized" DCs (αDC1s), which are differentiated with IL-4 and GM-CSF but, in contrast to conventional monocyte-derived DCs, are matured with a combination of IFN-α, polyinosinic:polycytidylic acid (polyI:C), TNF-α, IL-1β, and IFN-γ. Improved functionality of αDC1 was observed both in vitro (Mailliard et al. 2004) and in vivo (Giermasz et al. 2009; Lee et al. 2008; Park et al. 2011; Wieckowski et al. 2011). Despite higher levels of IL-12p70 secretion, strong adherence to cell culture vessels was observed which unfortunately lowered yields; this might represent a significant impediment to the broad application of such type-1/IL-12p70-polarized DC-products (Arimoto-Miyamoto et al. 2010).

It is possible that type-1/IL-12p70-polarized programming of DCs can be promoted in vivo, thereby bypassing the challenges associated with strong adherence and low yields. To achieve this, DC-vaccination strategies must deliberately harness in vivo CD4$^+$ T-cell help. While short peptides with particular human leukocyte antigen (HLA)-restriction have been favored to date, synthetic long peptides (SLP), including both CD8$^+$ and CD4$^+$ T-cell epitopes, could address this problem. Importantly, simultaneous presentation of both CD4$^+$ and CD8$^+$ T-cell epitopes by vaccine DCs has been shown to act synergistically upon the induction of antitumor CTL responses (Tomita et al. 2013). The activation of CD4$^+$ T-cells would have the additional benefit of making use of their own antitumor activities, including local cytokine release, activation of local APCs, and direct cytotoxic activity towards tumor cells (Quezada et al. 2010).

Another strategy to provide in vivo IL-12p70-polarization of vaccine DCs is through interaction with innate-like T-cells. These are populations of T-cells with a restricted TCR repertoire that express markers and functions typical of Natural Killer (NK) cells. Included are iNKT cells, Vγ9Vδ2 T-cells, and mucosal-associated invariant T-cells (MAIT cells). The loading of DCs with the prototypical iNKT cell agonist α-galactosylceramide (α-GalCer) initiates in vivo crosstalk between these two cell types. The bipartite interaction induces CD40-ligand upregulation and IFN-γ-secretion by iNKT cells that in turn provide DCs with necessary signaling for IL-12p70 production (Kawano et al. 1997). Although iNKT cells have been successfully targeted in human clinical trials, a deliberate strategy to provoke iNKT cell crosstalk to DC vaccines presenting tumor antigens has yet to be assessed (Cerundolo et al. 2009; Hermans et al. 2003; Hunn and Hermans 2013). Interest is growing in incorporating agonists for Vγ9Vδ2 T-cells and MAIT cells in a similar manner. In this context, in vitro studies have shown improved T-cell responses to antigen-loaded DCs pulsed with bisphosphonates, as these compounds are known to stimulate Vγ9Vδ2 T-cells (Castella et al. 2011). Agonists for MAIT cells have only just been defined (Kjer-Nielsen et al. 2012), so there is still much to learn about exploiting crosstalk between these cells and DC vaccines. While iNKT cells are known to recognize glycolipids presented by the MHC class I-like molecule CD1d, there are other classes of T-cells that are restricted by the other CD1 molecules expressed in human that remain to be fully defined; some are likely to exert similar functionality (Vincent et al. 2002). Intriguingly, the in vivo interaction between vaccine DCs and innate-like T-cells may have the additional benefit of activating intrinsic anticancer activities in responding T-cells (Braza and Klein 2013; Song et al. 2009), resulting in an immunological "double-hit," as discussed above for $CD4^+$ T-cells.

Of note, the maximal IL-12p70 production by clinical grade DC-vaccines can vary massively between cancer patients (Carreno et al. 2013). Consistent with this finding, IL-12p70-secretion has been shown to be restricted to a specific fraction of in vitro cultured DCs that express high levels of CD1a. This cell subset represents 10-90% of any given vaccine product (Chang et al. 2000; Gogolak et al. 2007), thus possibly accounting for the large differences observed among patients. The lower capacity of IL-12p70 production by fastDCs (see above) is similarly associated with lower levels of CD1a expression (Kvistborg et al. 2009). The underlying mechanism(s) are as of yet poorly understood but will be instrumental for the validation of next-generation DC-manufacturing processes.

Importantly, the production of IL-12p70 is also a component of humoral immunity, both directly, by acting on naïve B-cells (Dubois et al. 1998), and indirectly, through the differentiation of follicular helper T-cells (Schmitt et al. 2009). As cancer cells are a known target for natural antibodies, providing the immunological environment for B-cell differentiation is likely to contribute to cancer-related immunosurveillance (Vollmers and Brandlein 2009). Furthermore, DCs that have been genetically engineered to produce bioactive IL-12 have been shown, in animal cancer models, to mediate their efficacy at least partly through the activation of NK-cells (Miller et al. 2003; Rodriguez-Calvillo et al. 2002; Tatsumi et al. 2007). The intratumoral injection of an autologous DC-vaccine harboring an adenoviral-encoded IL-12 gene has been evaluated in humans, inducing detectable peripheral blood NK-cell activation in 5 out of 17 patients (Mazzolini et al. 2005). More generally, it is well accepted that non-modified DC-vaccines, even in the absence of antigen, can induce NK-cell activation to a level that significantly contributes to antitumor activity (Lion et al. 2012). Preclinical observations have indeed demonstrated that

NK-cell activation is necessary, albeit not sufficient, to initiate T-cell mediated antitumor responses (Lion et al. 2012). Although the monitoring of NK-cell activity has been largely neglected in human DC-vaccine trials, some preliminary findings have been reported (Lion et al. 2012). Interestingly, the large majority of human DC-vaccine trials that recorded NK-cell data demonstrate some level of NK-cell activation, although these changes do not always seem to correlate with clinical outcome. Somewhat inconsistently, NK-cells, in the context of DC-vaccination studies, have been reported not to influence disease progression to any significant extent (Alfaro et al. 2011; Baek et al. 2011; Qi et al. 2012), to synergize with T cell-mediated immunity (Van Tendeloo et al. 2010), or to univariately predict disease outcome (Osada et al. 2006). The overwhelming number of variables that affect NK-cell recruitment and function in the context of DC vaccination, such as administration routes and regimen, and antigen choice, make it hard to draw unequivocal conclusions about their impact as of yet. More consistent monitoring of NK-cell parameters is needed to draw solid conclusions, but DC-vaccine improvement through NK-cell recruitment is likely to receive increasing attention (Schnurr et al. 2002; Tosi et al. 2004; Vujanovic et al. 2010).

13.2.3 Co-stimulation

Co-stimulation is a requirement for effective T-cell activation. The main co-stimulatory receptor for T-cells is CD28, which is constitutively expressed on the surface of naïve T-cells. Its two DC-encoded activatory ligands, CD80 and CD86 are rapidly upregulated upon DC maturation and licensing (Greenwald et al. 2005; Sharpe and Freeman 2002). Another group of important co-stimulatory molecules expressed by T-cells is the TNF receptor family, prominently represented by CD27. The monitoring of its DC-expressed ligand, CD70, is often overlooked, but is associated with improved survival of primed CTLs (Bowman et al. 1994; Dolfi and Katsikis 2007; Hendriks et al. 2003). Ideally, monitoring of its expression should be permanently integrated into DC-vaccine quality control (Arimoto-Miyamoto et al. 2010).

13.2.4 Combination with Check-Point Inhibitors

Infiltration of tumors by T-cells (Gooden et al. 2011), and having a favorable ratio of effector T-cells to regulatory T-cells (Tregs), has been associated with better disease prognosis (Gooden et al. 2011). However tumor-infiltrating lymphocytes (TIL) represent a largely "exhausted" T-cell population (Ahmadzadeh et al. 2009; Baitsch et al. 2011) reminiscent of similar populations found in human chronic infections (Day et al. 2006; Kim and Ahmed 2010). The specific triggering of inhibitory T-cell receptors by the tumor microenvironment—an "immune checkpoint" that halts T-cell activation—lays the basis for this dysfunction (Benencia et al. 2012; Condamine and Gabrilovich 2011; Teng et al. 2011). The blockade of such immune checkpoint(s) by monoclonal antibody-based therapy has been shown to efficiently antagonize the premature exhaustion of tumor-specific immune responses. Prominent among clinical targets are cytotoxic T-lymphocyte antigen (CTLA)-4 and programmed cell death (PD)-1 receptors. Bypass of both checkpoints (Freeman et al. 2000; Walunas et al. 1994) results in unbridled T-cell activity which is a potentially attractive prospect for cancer immunotherapy, if appropriately directed and controlled. The FDA has currently approved one such drug for cancer immunotherapy, ipilimumab, which blocks the interaction between the checkpoint molecule CTLA-4 and its ligands CD80 and CD86 (Hodi et al. 2010). Based on remarkable clinical results, an agent blocking interactions between another checkpoint molecule PD-1, and its ligands PDL1 and PDL2 has received breakthrough therapy designation from the FDA (Hamid et al. 2013). DC immunotherapy represents a prime candidate for combinatorial therapy with immune checkpoint regulators (Vanneman and Dranoff 2012). Importantly, as these "immunomodulatory" drugs are reliant on there being an immune response present in the first place, DC vaccines encoding whole tumor material may be particularly useful in driving broad immune responses that benefit from checkpoint blockade. Timely cycles of vaccination and checkpoint inhibition may prove to be very effective.

13.3 Conclusion

Over the past decades, the prospect of modulating the immune system to fight cancer has gone from being a possibility to a viable therapeutic option, particularly with the recent development of drugs that serve as checkpoint inhibitors. DC-based vaccines, while initially very exciting, have not yet progressed to practical therapies, although they have played an immense role in developing our understanding of tumor immunology and framing relevant questions regarding T-cell priming. Whether DC vaccines will yet find application in the clinic will depend on understanding and exploiting further intricacies of DC-mediated T-cell programming. Strategies that provide vaccine-DCs functional support in vivo, such as through engagement of $CD4^+$ T-cells, innate-like T-cells and NK-cells, are worthy of further examination. However, the improvements in efficacy will have to be significant to justify the complexity of vaccine manufacturing, although strategies to shorten and simplify the ex vivo generation of DC-vaccines are in development. With checkpoint inhibitors likely to become a common therapy in a range of cancers, it is important to evaluate the potential of combining simple DC vaccines with these new drugs, especially in patients who have not generated a spontaneous T-cell response to their tumors. Thus, despite early disappointments, DC-based immunotherapy may yet find a place in the fight against cancer.

References

Aarntzen EH, Figdor CG, Adema GJ, Punt CJ, de Vries IJ (2008) Dendritic cell vaccination and immune monitoring. Cancer Immunol Immunother 57(10):1559–1568. doi:10.1007/s00262-008-0553-y

Ahmadzadeh M, Johnson LA, Heemskerk B, Wunderlich JR, Dudley ME, White DE, Rosenberg SA (2009) Tumor antigen-specific CD8 T cells infiltrating the tumor express high levels of PD-1 and are functionally impaired. Blood 114(8):1537–1544. doi:10.1182/blood-2008-12-195792, blood-2008-12-195792 [pii]

Alfaro C, Perez-Gracia JL, Suarez N, Rodriguez J, Fernandez de Sanmamed M, Sangro B, Martin-Algarra S, Calvo A, Redrado M, Agliano A, Gonzalez A, Rodriguez I, Bolanos E, Hervas-Stubbs S, Perez-Calvo J, Benito A, Penuelas I, Vigil C, Richter J, Martinez-Forero I, Melero I (2011) Pilot clinical trial of type 1 dendritic cells loaded with autologous tumor lysates combined with GM-CSF, pegylated IFN, and cyclophosphamide for metastatic cancer patients. J Immunol 187(11):6130–6142. doi:10.4049/jimmunol.1102209, jimmunol.1102209 [pii]

Anguille S, Smits EL, Cools N, Goossens H, Berneman ZN, Van Tendeloo VF (2009) Short-term cultured, interleukin-15 differentiated dendritic cells have potent immunostimulatory properties. J Transl Med 7:109. doi:10.1186/1479-5876-7-109, 1479-5876-7-109 [pii]

Arens R, Schoenberger SP (2010) Plasticity in programming of effector and memory CD8 T-cell formation. Immunol Rev 235(1):190–205. doi:10.1111/j.0105-2896.2010.00899.x, IMR899 [pii]

Arimoto-Miyamoto K, Kadowaki N, Kitawaki T, Iwata S, Morimoto C, Uchiyama T (2010) Optimal stimulation for CD70 induction on human monocyte-derived dendritic cells and the importance of CD70 in naive CD4(+) T-cell differentiation. Immunology 130(1):137–149. doi:10.1111/j.1365-2567.2010.03220.x, IMM3220 [pii]

Atkins MB, Robertson MJ, Gordon M, Lotze MT, DeCoste M, DuBois JS, Ritz J, Sandler AB, Edington HD, Garzone PD, Mier JW, Canning CM, Battiato L, Tahara H, Sherman ML (1997) Phase I evaluation of intravenous recombinant human interleukin 12 in patients with advanced malignancies. Clin Cancer Res 3(3):409–417

Baek S, Kim CS, Kim SB, Kim YM, Kwon SW, Kim Y, Kim H, Lee H (2011) Combination therapy of renal cell carcinoma or breast cancer patients with dendritic cell vaccine and IL-2: results from a phase I/II trial. J Transl Med 9:178. doi:10.1186/1479-5876-9-178, 1479-5876-9-178 [pii]

Baitsch L, Baumgaertner P, Devevre E, Raghav SK, Legat A, Barba L, Wieckowski S, Bouzourene H, Deplancke B, Romero P, Rufer N, Speiser DE (2011) Exhaustion of tumor-specific CD8(+) T cells in metastases from melanoma patients. J Clin Invest 121(6):2350–2360. doi:10.1172/JCI46102, 46102 [pii]

Ballesteros-Tato A, Leon B, Lund FE, Randall TD (2013) CD4+ T helper cells use CD154-CD40 interactions to counteract T reg cell-mediated suppression of CD8+ T cell responses to influenza. J Exp Med 210(8):1591–1601. doi:10.1084/jem.20130097, jem.20130097 [pii]

Bendelac A, Savage PB, Teyton L (2007) The biology of NKT cells. Annu Rev Immunol 25:297–336. doi:10.1146/annurev.immunol.25.022106.141711

Benencia F, Sprague L, McGinty J, Pate M, Muccioli M (2012) Dendritic cells the tumor microenvironment and the challenges for an effective antitumor vaccination. J Biomed Biotechnol 2012:425476. doi:10.1155/2012/425476

Bowman MR, Crimmins MA, Yetz-Aldape J, Kriz R, Kelleher K, Herrmann S (1994) The cloning of CD70 and its identification as the ligand for CD27. J Immunol 152(4):1756–1761

Braza MS, Klein B (2013) Anti-tumour immunotherapy with Vgamma9Vdelta2 T lymphocytes: from the bench to the bedside. Br J Haematol 160(2):123–132. doi:10.1111/bjh.12090

Buonaguro L, Petrizzo A, Tornesello ML, Buonaguro FM (2011) Translating tumor antigens into cancer vaccines. Clin Vaccine Immunol 18(1):23–34. doi:10.1128/CVI.00286-10, CVI.00286-10 [pii]

Burdek M, Spranger S, Wilde S, Frankenberger B, Schendel DJ, Geiger C (2010) Three-day dendritic cells for vaccine development: antigen uptake, processing and presentation. J Transl Med 8:90. doi:10.1186/1479-5876-8-90, 1479-5876-8-90 [pii]

Byrne KT, Cote AL, Zhang P, Steinberg SM, Guo Y, Allie R, Zhang W, Ernstoff MS, Usherwood EJ, Turk MJ (2011) Autoimmune melanocyte destruction is required for robust CD8+ memory T cell responses to mouse melanoma. J Clin Invest 121(5):1797–1809. doi:10.1172/JCI44849, 44849 [pii]

Carreno BM, Becker-Hapak M, Huang A, Chan M, Alyasiry A, Lie WR, Aft RL, Cornelius LA, Trinkaus KM, Linette GP (2013) IL-12p70-producing patient DC vaccine elicits Tc1-polarized immunity. J Clin Invest 123(8):3383–3394. doi:10.1172/JCI68395, 68395 [pii]

Castella B, Riganti C, Fiore F, Pantaleoni F, Canepari ME, Peola S, Foglietta M, Palumbo A, Bosia A, Coscia M, Boccadoro M, Massaia M (2011) Immune modulation by zoledronic acid in human myeloma: an advantageous cross-talk between Vgamma9Vdelta2 T cells, alphabeta CD8+ T cells, regulatory T cells, and dendritic cells. J Immunol 187(4):1578–1590. doi:10.4049/jimmunol.1002514, jimmunol.1002514 [pii]

Cerundolo V, Silk JD, Masri SH, Salio M (2009) Harnessing invariant NKT cells in vaccination strategies. Nat Rev Immunol 9(1):28–38. doi:10.1038/nri2451, nri2451 [pii]

Chang CC, Wright A, Punnonen J (2000) Monocyte-derived CD1a+ and CD1a- dendritic cell subsets differ in their cytokine production profiles, susceptibilities to transfection, and capacities to direct Th cell differentiation. J Immunol 165(7):3584–3591

Coffman RL, Sher A, Seder RA (2010) Vaccine adjuvants: putting innate immunity to work. Immunity 33(4):492–503. doi:10.1016/j.immuni.2010.10.002, S1074-7613(10)00362-6 [pii]

Condamine T, Gabrilovich DI (2011) Molecular mechanisms regulating myeloid-derived suppressor cell differentiation and function. Trends Immunol 32(1):19–25. doi:10.1016/j.it.2010.10.002, S1471-4906(10)00149-3 [pii]

Curtsinger JM, Schmidt CS, Mondino A, Lins DC, Kedl RM, Jenkins MK, Mescher MF (1999) Inflammatory cytokines provide a third signal for activation of naive CD4+ and CD8+ T cells. J Immunol 162(6):3256–3262

Dannull J, Haley NR, Archer G, Nair S, Boczkowski D, Harper M, De Rosa N, Pickett N, Mosca PJ, Burchette J, Selim MA, Mitchell DA, Sampson J, Tyler DS, Pruitt SK (2013) Melanoma immunotherapy using mature DCs expressing the constitutive proteasome. J Clin Invest 123(7):3135–3145. doi:10.1172/JCI67544, 67544 [pii]

Dauer M, Obermaier B, Herten J, Haerle C, Pohl K, Rothenfusser S, Schnurr M, Endres S, Eigler A (2003) Mature dendritic cells derived from human monocytes within 48 hours: a novel strategy for dendritic cell differentiation from blood precursors. J Immunol 170(8):4069–4076

Dauer M, Schad K, Herten J, Junkmann J, Bauer C, Kiefl R, Endres S, Eigler A (2005) FastDC derived from human monocytes within 48 h effectively prime tumor antigen-specific cytotoxic T cells. J Immunol Methods 302(1–2):145–155. doi:10.1016/j.jim.2005.05.010, S0022-1759(05)00154-7 [pii]

Day CL, Kaufmann DE, Kiepiela P, Brown JA, Moodley ES, Reddy S, Mackey EW, Miller JD, Leslie AJ, DePierres C, Mncube Z, Duraiswamy J, Zhu B, Eichbaum Q, Altfeld M, Wherry EJ, Coovadia HM, Goulder PJ, Klenerman P, Ahmed R, Freeman GJ, Walker BD (2006) PD-1 expression on HIV-specific T cells is associated with T-cell exhaustion and disease progression. Nature 443(7109):350–354. doi:10.1038/nature05115, nature05115 [pii]

Dolfi DV, Katsikis PD (2007) CD28 and CD27 costimulation of CD8+ T cells: a story of survival. Adv Exp Med Biol 590:149–170. doi:10.1007/978-0-387-34814-8_11

Dubois B, Massacrier C, Vanbervliet B, Fayette J, Briere F, Banchereau J, Caux C (1998) Critical role of IL-12 in dendritic cell-induced differentiation of naive B lymphocytes. J Immunol 161(5):2223–2231

Dubsky P, Saito H, Leogier M, Dantin C, Connolly JE, Banchereau J, Palucka AK (2007) IL-15-induced human DC efficiently prime melanoma-specific naive CD8+ T cells to differentiate into CTL. Eur J Immunol 37(6):1678–1690. doi:10.1002/eji.200636329

Engell-Noerregaard L, Hansen TH, Andersen MH, Thor Straten P, Svane IM (2009) Review of clinical studies on dendritic cell-based vaccination of patients with malignant melanoma: assessment of correlation between clinical response and vaccine parameters. Cancer Immunol Immunother 58(1):1–14. doi:10.1007/s00262-008-0568-4

Feuerstein B, Berger TG, Maczek C, Roder C, Schreiner D, Hirsch U, Haendle I, Leisgang W, Glaser A, Kuss O, Diepgen TL, Schuler G, Schuler-Thurner B (2000) A method for the production of cryopreserved aliquots of antigen-preloaded, mature dendritic cells ready for clinical use. J Immunol Methods 245(1–2):15–29. doi:S0022-1759(00)00269-6 [pii]

Figdor CG, de Vries IJ, Lesterhuis WJ, Melief CJ (2004) Dendritic cell immunotherapy: mapping the way. Nat Med 10(5):475–480. doi:10.1038/nm1039, nm1039 [pii]

Flamand V, Sornasse T, Thielemans K, Demanet C, Bakkus M, Bazin H, Tielemans F, Leo O, Urbain J, Moser M (1994) Murine dendritic cells pulsed in vitro with tumor antigen induce tumor resistance in vivo. Eur J Immunol 24(3):605–610. doi:10.1002/eji.1830240317

Freeman GJ, Long AJ, Iwai Y, Bourque K, Chernova T, Nishimura H, Fitz LJ, Malenkovich N, Okazaki T, Byrne MC, Horton HF, Fouser L, Carter L, Ling V, Bowman MR, Carreno BM, Collins M, Wood CR, Honjo T (2000) Engagement of the PD-1 immunoinhibitory receptor by a novel B7 family member leads to negative regulation of lymphocyte activation. J Exp Med 192(7):1027–1034

Frentsch M, Stark R, Matzmohr N, Meier S, Durlanik S, Schulz AR, Stervbo U, Jurchott K, Gebhardt F, Heine G, Reuter MA, Betts MR, Busch D, Thiel A (2013) CD40L expression permits CD8+ T cells to execute immunologic helper functions. Blood 122(3):405–412. doi:10.1182/blood-2013-02-483586, blood-2013-02-483586 [pii]

Gallucci S, Lolkema M, Matzinger P (1999) Natural adjuvants: endogenous activators of dendritic cells. Nat Med 5(11):1249–1255. doi:10.1038/15200

Galon J, Pages F, Marincola FM, Thurin M, Trinchieri G, Fox BA, Gajewski TF, Ascierto PA (2012) The immune score as a new possible approach for the classification of cancer. J Transl Med 10:1. doi:10.1186/1479-5876-10-1, 1479-5876-10-1 [pii]

Giermasz AS, Urban JA, Nakamura Y, Watchmaker P, Cumberland RL, Gooding W, Kalinski P (2009) Type-1 polarized dendritic cells primed for high IL-12 production show enhanced activity as cancer vaccines. Cancer Immunol Immunother 58(8):1329–1336. doi:10.1007/s00262-008-0648-5

Gogolak P, Rethi B, Szatmari I, Lanyi A, Dezso B, Nagy L, Rajnavolgyi E (2007) Differentiation of CD1a- and CD1a+ monocyte-derived dendritic cells is biased by lipid environment and PPARgamma. Blood 109(2):643–652. doi:10.1182/blood-2006-04-016840, blood-2006-04-016840 [pii]

Gooden MJ, de Bock GH, Leffers N, Daemen T, Nijman HW (2011) The prognostic influence of tumour-infiltrating lymphocytes in cancer: a systematic review with meta-analysis. Br J Cancer 105(1):93–103. doi:10.1038/bjc.2011.189, bjc2011189 [pii]

Greenwald RJ, Freeman GJ, Sharpe AH (2005) The B7 family revisited. Annu Rev Immunol 23:515–548. doi:10.1146/annurev.immunol.23.021704.115611

Habel K (1962) Immunological determinants of polyoma virus oncogenesis. J Exp Med 115: 181–193

Hamid O, Robert C, Daud A, Hodi FS, Hwu WJ, Kefford R, Wolchok JD, Hersey P, Joseph RW, Weber JS, Dronca R, Gangadhar TC, Patnaik A, Zarour H, Joshua AM, Gergich K, Elassaiss-Schaap J, Algazi A, Mateus C, Boasberg P, Tumeh PC, Chmielowski B, Ebbinghaus SW, Li XN, Kang SP, Ribas A (2013) Safety and tumor responses with lambrolizumab (anti-PD-1) in melanoma. N Engl J Med 369(2):134–144. doi:10.1056/NEJMoa1305133

Hendriks J, Xiao Y, Borst J (2003) CD27 promotes survival of activated T cells and complements CD28 in generation and establishment of the effector T cell pool. J Exp Med 198(9):1369–1380. doi:10.1084/jem.20030916, jem.20030916 [pii]

Hermans IF, Silk JD, Gileadi U, Salio M, Mathew B, Ritter G, Schmidt R, Harris AL, Old L, Cerundolo V (2003) NKT cells enhance CD4+ and CD8+ T cell responses to soluble antigen in vivo through direct interaction with dendritic cells. J Immunol 171(10):5140–5147

Higano CS, Schellhammer PF, Small EJ, Burch PA, Nemunaitis J, Yuh L, Provost N, Frohlich MW (2009) Integrated data from 2 randomized, double-blind, placebo-controlled, phase 3 trials of active cellular immunotherapy with sipuleucel-T in advanced prostate cancer. Cancer 115(16):3670–3679. doi:10.1002/cncr.24429

Hodi FS, O'Day SJ, McDermott DF, Weber RW, Sosman JA, Haanen JB, Gonzalez R, Robert C, Schadendorf D, Hassel JC, Akerley W, van den Eertwegh AJ, Lutzky J, Lorigan P, Vaubel JM, Linette GP, Hogg D, Ottensmeier CH, Lebbe C, Peschel C, Quirt I, Clark JI, Wolchok JD, Weber JS, Tian J, Yellin MJ, Nichol GM, Hoos A, Urba WJ (2010) Improved survival with ipilimumab in patients with metastatic melanoma. N Engl J Med 363(8):711–723. doi:10.1056/NEJMoa1003466, NEJMoa1003466 [pii]

Hsu FJ, Benike C, Fagnoni F, Liles TM, Czerwinski D, Taidi B, Engleman EG, Levy R (1996) Vaccination of patients with B-cell lymphoma using autologous antigen-pulsed dendritic cells. Nat Med 2(1):52–58

Hunn MK, Hermans IF (2013) Exploiting invariant NKT cells to promote T-cell responses to cancer vaccines. Oncoimmunology 2(4):e23789. doi:10.4161/onci.23789, 2013ONCOIMM0019 [pii]

Inaba K, Metlay JP, Crowley MT, Steinman RM (1990) Dendritic cells pulsed with protein antigens in vitro can prime antigen-specific, MHC-restricted T cells in situ. J Exp Med 172(2):631–640

Jonuleit H, Kuhn U, Muller G, Steinbrink K, Paragnik L, Schmitt E, Knop J, Enk AH (1997) Proinflammatory cytokines and prostaglandins induce maturation of potent immunostimulatory dendritic cells under fetal calf serum-free conditions. Eur J Immunol 27(12):3135–3142. doi:10.1002/eji.1830271209

Kalinski P, Muthuswamy R, Urban J (2013) Dendritic cells in cancer immunotherapy: vaccines and combination immunotherapies. Expert Rev Vaccines 12(3):285–295. doi:10.1586/erv.13.22

Kantoff PW, Higano CS, Shore ND, Berger ER, Small EJ, Penson DF, Redfern CH, Ferrari AC, Dreicer R, Sims RB, Xu Y, Frohlich MW, Schellhammer PF (2010) Sipuleucel-T immunotherapy for castration-resistant prostate cancer. N Engl J Med 363(5):411–422. doi:10.1056/NEJMoa1001294

Kawano T, Cui J, Koezuka Y, Toura I, Kaneko Y, Motoki K, Ueno H, Nakagawa R, Sato H, Kondo E, Koseki H, Taniguchi M (1997) CD1d-restricted and TCR-mediated activation of valpha14 NKT cells by glycosylceramides. Science 278(5343):1626–1629

Kim PS, Ahmed R (2010) Features of responding T cells in cancer and chronic infection. Curr Opin Immunol 22(2):223–230

Kjer-Nielsen L, Patel O, Corbett AJ, Le Nours J, Meehan B, Liu L, Bhati M, Chen Z, Kostenko L, Reantragoon R, Williamson NA, Purcell AW, Dudek NL, McConville MJ, O'Hair RA, Khairallah GN, Godfrey DI, Fairlie DP, Rossjohn J, McCluskey J (2012) MR1 presents microbial vitamin B metabolites to MAIT cells. Nature 491(7426):717–723. doi:10.1038/nature11605, nature11605 [pii]

Kodama A, Tanaka R, Saito M, Ansari AA, Tanaka Y (2013) A novel and simple method for generation of human dendritic cells from unfractionated peripheral blood mononuclear cells within 2 days: its application for induction of HIV-1-reactive CD4 T cells in the hu-PBL SCID mice. Front Microbiol 4:292. doi:10.3389/fmicb.2013.00292

Kvistborg P, Boegh M, Pedersen AW, Claesson MH, Zocca MB (2009) Fast generation of dendritic cells. Cell Immunol 260(1):56–62. doi:10.1016/j.cellimm.2009.09.003, S0008-8749(09)00151-8 [pii]

Lee WT (2011) Dendritic cell-tumor cell fusion vaccines. Adv Exp Med Biol 713:177–186. doi:10.1007/978-94-007-0763-4_11

Lee JJ, Foon KA, Mailliard RB, Muthuswamy R, Kalinski P (2008) Type 1-polarized dendritic cells loaded with autologous tumor are a potent immunogen against chronic lymphocytic leukemia. J Leukoc Biol 84(1):319–325. doi:10.1189/jlb.1107737, jlb.1107737 [pii]

Leonard JP, Sherman ML, Fisher GL, Buchanan LJ, Larsen G, Atkins MB, Sosman JA, Dutcher JP, Vogelzang NJ, Ryan JL (1997) Effects of single-dose interleukin-12 exposure on interleukin-12-associated toxicity and interferon-gamma production. Blood 90(7):2541–2548

Lion E, Smits EL, Berneman ZN, Van Tendeloo VF (2012) NK cells: key to success of DC-based cancer vaccines? Oncologist 17(10):1256–1270. doi:10.1634/theoncologist.2011-0122, theoncologist.2011-0122 [pii]

Mailliard RB, Wankowicz-Kalinska A, Cai Q, Wesa A, Hilkens CM, Kapsenberg ML, Kirkwood JM, Storkus WJ, Kalinski P (2004) alpha-type-1 polarized dendritic cells: a novel immunization tool with optimized CTL-inducing activity. Cancer Res 64(17):5934–5937. doi:10.1158/0008-5472.CAN-04-1261, 64/17/5934 [pii]

Mazzolini G, Alfaro C, Sangro B, Feijoo E, Ruiz J, Benito A, Tirapu I, Arina A, Sola J, Herraiz M, Lucena F, Olague C, Subtil J, Quiroga J, Herrero I, Sadaba B, Bendandi M, Qian C, Prieto J, Melero I (2005) Intratumoral injection of dendritic cells engineered to secrete interleukin-12 by recombinant adenovirus in patients with metastatic gastrointestinal carcinomas. J Clin Oncol 23(5):999–1010. doi:10.1200/JCO.2005.00.463, JCO.2005.00.463 [pii]

Miller G, Lahrs S, Dematteo RP (2003) Overexpression of interleukin-12 enables dendritic cells to activate NK cells and confer systemic antitumor immunity. FASEB J 17(6):728–730. doi:10.1096/fj.02-0900fje, 02-0900fje [pii]

Mosca PJ, Hobeika AC, Clay TM, Nair SK, Thomas EK, Morse MA, Lyerly HK (2000) A subset of human monocyte-derived dendritic cells expresses high levels of interleukin-12 in response to combined CD40 ligand and interferon-gamma treatment. Blood 96(10):3499–3504

Nussenzweig MC, Steinman RM, Gutchinov B, Cohn ZA (1980) Dendritic cells are accessory cells for the development of anti-trinitrophenyl cytotoxic T lymphocytes. J Exp Med 152(4):1070–1084

Osada T, Clay T, Hobeika A, Lyerly HK, Morse MA (2006) NK cell activation by dendritic cell vaccine: a mechanism of action for clinical activity. Cancer Immunol Immunother 55(9):1122–1131. doi:10.1007/s00262-005-0089-3

Palucka K, Banchereau J (2012) Cancer immunotherapy via dendritic cells. Nat Rev Cancer 12(4):265–277. doi:10.1038/nrc3258, nrc3258 [pii]

Park MH, Yang DH, Kim MH, Jang JH, Jang YY, Lee YK, Jin CJ, Pham TN, Thi TA, Lim MS, Lee HJ, Hong CY, Yoon JH, Lee JJ (2011) Alpha-type 1 polarized dendritic cells loaded with apoptotic allogeneic breast cancer cells can induce potent cytotoxic T lymphocytes against breast cancer. Cancer Res Treat 43(1):56–66. doi:10.4143/crt.2011.43.1.56

Petersen TR, Dickgreber N, Hermans IF (2010) Tumor antigen presentation by dendritic cells. Crit Rev Immunol 30(4):345–386. doi:2a2a299c5ed1858d,124c256c468812db [pii]

Qi CJ, Ning YL, Han YS, Min HY, Ye H, Zhu YL, Qian KQ (2012) Autologous dendritic cell vaccine for estrogen receptor (ER)/progestin receptor (PR) double-negative breast cancer. Cancer Immunol Immunother 61(9):1415–1424. doi:10.1007/s00262-011-1192-2

Quezada SA, Simpson TR, Peggs KS, Merghoub T, Vider J, Fan X, Blasberg R, Yagita H, Muranski P, Antony PA, Restifo NP, Allison JP (2010) Tumor-reactive CD4(+) T cells develop cytotoxic activity and eradicate large established melanoma after transfer into lymphopenic hosts. J Exp Med 207(3):637–650. doi:10.1084/jem.20091918, jem.20091918 [pii]

Rodriguez-Calvillo M, Duarte M, Tirapu I, Berraondo P, Mazzolini G, Qian C, Prieto J, Melero I (2002) Upregulation of natural killer cells functions underlies the efficacy of intratumorally injected dendritic cells engineered to produce interleukin-12. Exp Hematol 30(3):195–204. doi:S0301472X01007925 [pii]

Romani N, Gruner S, Brang D, Kampgen E, Lenz A, Trockenbacher B, Konwalinka G, Fritsch PO, Steinman RM, Schuler G (1994) Proliferating dendritic cell progenitors in human blood. J Exp Med 180(1):83–93

Sallusto F, Lanzavecchia A (1994) Efficient presentation of soluble antigen by cultured human dendritic cells is maintained by granulocyte/macrophage colony-stimulating factor plus interleukin 4 and downregulated by tumor necrosis factor alpha. J Exp Med 179(4): 1109–1118

Sato E, Olson SH, Ahn J, Bundy B, Nishikawa H, Qian F, Jungbluth AA, Frosina D, Gnjatic S, Ambrosone C, Kepner J, Odunsi T, Ritter G, Lele S, Chen YT, Ohtani H, Old LJ, Odunsi K (2005) Intraepithelial CD8+ tumor-infiltrating lymphocytes and a high CD8+/regulatory T cell ratio are associated with favorable prognosis in ovarian cancer. Proc Natl Acad Sci U S A 102(51):18538–18543. doi:10.1073/pnas.0509182102, 0509182102 [pii]

Schadendorf D, Ugurel S, Schuler-Thurner B, Nestle FO, Enk A, Brocker EB, Grabbe S, Rittgen W, Edler L, Sucker A, Zimpfer-Rechner C, Berger T, Kamarashev J, Burg G, Jonuleit H, Tuttenberg A, Becker JC, Keikavoussi P, Kampgen E, Schuler G (2006) Dacarbazine (DTIC) versus vaccination with autologous peptide-pulsed dendritic cells (DC) in first-line treatment of patients with metastatic melanoma: a randomized phase III trial of the DC study group of the DeCOG. Ann Oncol 17(4):563–570. doi:10.1093/annonc/mdj138, mdj138 [pii]

Schmitt N, Morita R, Bourdery L, Bentebibel SE, Zurawski SM, Banchereau J, Ueno H (2009) Human dendritic cells induce the differentiation of interleukin-21-producing T follicular helper-like cells through interleukin-12. Immunity 31(1):158–169. doi:10.1016/j.immuni.2009.04.016, S1074-7613(09)00272-6 [pii]

Schnurr M, Scholz C, Rothenfusser S, Galambos P, Dauer M, Robe J, Endres S, Eigler A (2002) Apoptotic pancreatic tumor cells are superior to cell lysates in promoting cross-priming of cytotoxic T cells and activate NK and gammadelta T cells. Cancer Res 62(8):2347–2352

Schuler G (2010) Dendritic cells in cancer immunotherapy. Eur J Immunol 40(8):2123–2130. doi:10.1002/eji.201040630

Schuler-Thurner B, Schultz ES, Berger TG, Weinlich G, Ebner S, Woerl P, Bender A, Feuerstein B, Fritsch PO, Romani N, Schuler G (2002) Rapid induction of tumor-specific type 1 T helper cells in metastatic melanoma patients by vaccination with mature, cryopreserved, peptide-loaded monocyte-derived dendritic cells. J Exp Med 195(10):1279–1288

Shankaran V, Ikeda H, Bruce AT, White JM, Swanson PE, Old LJ, Schreiber RD (2001) IFNgamma and lymphocytes prevent primary tumour development and shape tumour immunogenicity. Nature 410(6832):1107–1111. doi:10.1038/35074122, 35074122 [pii]

Sharpe AH, Freeman GJ (2002) The B7-CD28 superfamily. Nat Rev Immunol 2(2):116–126. doi:10.1038/nri727

Shu S, Zheng R, Lee WT, Cohen PA (2007) Immunogenicity of dendritic-tumor fusion hybrids and their utility in cancer immunotherapy. Crit Rev Immunol 27(5):463–483, doi:451b19ed1fdac85b,17770a8f458fe738 [pii]

Shurin MR, Gregory M, Morris JC, Malyguine AM (2010) Genetically modified dendritic cells in cancer immunotherapy: a better tomorrow? Expert Opin Biol Ther 10(11):1539–1553. doi:10.1517/14712598.2010.526105

Small EJ, Schellhammer PF, Higano CS, Redfern CH, Nemunaitis JJ, Valone FH, Verjee SS, Jones LA, Hershberg RM (2006) Placebo-controlled phase III trial of immunologic therapy with sipuleucel-T (APC8015) in patients with metastatic, asymptomatic hormone refractory prostate cancer. J Clin Oncol 24(19):3089–3094. doi:10.1200/JCO.2005.04.5252, 24/19/3089 [pii]

Snijders A, Kalinski P, Hilkens CM, Kapsenberg ML (1998) High-level IL-12 production by human dendritic cells requires two signals. Int Immunol 10(11):1593–1598

Song L, Asgharzadeh S, Salo J, Engell K, Wu HW, Sposto R, Ara T, Silverman AM, DeClerck YA, Seeger RC, Metelitsa LS (2009) Valpha24-invariant NKT cells mediate antitumor activity via killing of tumor-associated macrophages. J Clin Invest 119(6):1524–1536. doi:10.1172/JCI37869, 37869 [pii]

Steinman RM (1991) The dendritic cell system and its role in immunogenicity. Annu Rev Immunol 9:271–296. doi:10.1146/annurev.iy.09.040191.001415

Steinman RM, Banchereau J (2007) Taking dendritic cells into medicine. Nature 449(7161):419–426. doi:10.1038/nature06175, nature06175 [pii]

Tatsumi T, Takehara T, Yamaguchi S, Sasakawa A, Miyagi T, Jinushi M, Sakamori R, Kohga K, Uemura A, Ohkawa K, Storkus WJ, Hayashi N (2007) Injection of IL-12 gene-transduced dendritic cells into mouse liver tumor lesions activates both innate and acquired immunity. Gene Ther 14(11):863–871. doi:10.1038/sj.gt.3302941, 3302941 [pii]

Teng MW, Ritchie DS, Neeson P, Smyth MJ (2011) Biology and clinical observations of regulatory T cells in cancer immunology. Curr Top Microbiol Immunol 344:61–95. doi:10.1007/82_2010_50

Tomita Y, Yuno A, Tsukamoto H, Senju S, Kuroda Y, Hirayama M, Irie A, Kawahara K, Yatsuda J, Hamada A, Jono H, Yoshida K, Tsunoda T, Kohrogi H, Yoshitake Y, Nakamura Y, Shinohara M, Nishimura Y (2013) Identification of promiscuous KIF20A long peptides bearing both CD4+ and CD8+ T-cell epitopes: KIF20A-specific CD4+ T-cell immunity in patients with malignant tumor. Clin Cancer Res 19(16):4508–4520. doi:10.1158/1078-0432.CCR-13-0197, 1078-0432.CCR-13-0197 [pii]

Tosi D, Valenti R, Cova A, Sovena G, Huber V, Pilla L, Arienti F, Belardelli F, Parmiani G, Rivoltini L (2004) Role of cross-talk between IFN-alpha-induced monocyte-derived dendritic cells and NK cells in priming CD8+ T cell responses against human tumor antigens. J Immunol 172(9):5363–5370

Tosolini M, Kirilovsky A, Mlecnik B, Fredriksen T, Mauger S, Bindea G, Berger A, Bruneval P, Fridman WH, Pages F, Galon J (2011) Clinical impact of different classes of infiltrating T cytotoxic and helper cells (Th1, th2, treg, th17) in patients with colorectal cancer. Cancer Res 71(4):1263–1271. doi:10.1158/0008-5472.CAN-10-2907, 0008-5472.CAN-10-2907 [pii]

Traversari C, van der Bruggen P, Luescher IF, Lurquin C, Chomez P, Van Pel A, De Plaen E, Amar-Costesec A, Boon T (1992) A nonapeptide encoded by human gene MAGE-1 is recognized on HLA-A1 by cytolytic T lymphocytes directed against tumor antigen MZ2-E. J Exp Med 176(5):1453–1457

Trinchieri G, Pflanz S, Kastelein RA (2003) The IL-12 family of heterodimeric cytokines: new players in the regulation of T cell responses. Immunity 19(5):641–644. doi:S1074761303002966 [pii]

Van den Eynde BJ, Morel S (2001) Differential processing of class-I-restricted epitopes by the standard proteasome and the immunoproteasome. Curr Opin Immunol 13(2):147–153. doi:S0952-7915(00)00197-7 [pii]

van der Bruggen P, Traversari C, Chomez P, Lurquin C, De Plaen E, Van den Eynde B, Knuth A, Boon T (1991) A gene encoding an antigen recognized by cytolytic T lymphocytes on a human melanoma. Science 254(5038):1643–1647

Van Tendeloo VF, Van de Velde A, Van Driessche A, Cools N, Anguille S, Ladell K, Gostick E, Vermeulen K, Pieters K, Nijs G, Stein B, Smits EL, Schroyens WA, Gadisseur AP, Vrelust I, Jorens PG, Goossens H, de Vries IJ, Price DA, Oji Y, Oka Y, Sugiyama H, Berneman ZN (2010) Induction of complete and molecular remissions in acute myeloid leukemia by Wilms' tumor 1 antigen-targeted dendritic cell vaccination. Proc Natl Acad Sci U S A 107(31):13824–13829. doi:10.1073/pnas.1008051107, 1008051107 [pii]

Vanneman M, Dranoff G (2012) Combining immunotherapy and targeted therapies in cancer treatment. Nat Rev Cancer 12(4):237–251. doi:10.1038/nrc3237, nrc3237 [pii]

Vincent MS, Leslie DS, Gumperz JE, Xiong X, Grant EP, Brenner MB (2002) CD1-dependent dendritic cell instruction. Nat Immunol 3(12):1163–1168. doi:10.1038/ni851, ni851 [pii]

Vollmers HP, Brandlein S (2009) Natural antibodies and cancer. Nat Biotechnol 25(5):294–298. doi:10.1016/j.nbt.2009.03.016, S1871-6784(09)00060-0 [pii]

Vujanovic L, Szymkowski DE, Alber S, Watkins SC, Vujanovic NL, Butterfield LH (2010) Virally infected and matured human dendritic cells activate natural killer cells via cooperative activity of plasma membrane-bound TNF and IL-15. Blood 116(4):575–583. doi:10.1182/blood-2009-08-240325, blood-2009-08-240325 [pii]

Walunas TL, Lenschow DJ, Bakker CY, Linsley PS, Freeman GJ, Green JM, Thompson CB, Bluestone JA (1994) CTLA-4 can function as a negative regulator of T cell activation. Immunity 1(5):405–413. doi:1074-7613(94)90071-X [pii]

Wieckowski E, Chatta GS, Mailliard RM, Gooding W, Palucka K, Banchereau J, Kalinski P (2011) Type-1 polarized dendritic cells loaded with apoptotic prostate cancer cells are potent inducers of CD8(+) T cells against prostate cancer cells and defined prostate cancer-specific epitopes. Prostate 71(2):125–133. doi:10.1002/pros.21228

Part III
Delivery Routes, Devices and Dosage Forms

Chapter 14
Parenteral Vaccine Administration: Tried and True

Pål Johansen and Thomas M. Kündig

14.1 Immunogenicity Is Defined by Antigen Location: A Roadmap for Immunogenicity

During lymphocyte development and differentiation, a variety of antigen-specific receptors on B- and T-cells are generated as a result of random rearrangement of genes encoding for B- and T-cell receptors. By consequence, the receptors can recognise a nearly infinite number of antigens derived from the environment. But on the other hand and in the case of autoimmunity, even endogenous or self-antigens can be recognised. Since the precursor frequency of a specific lymphocyte is less than one in a million, the probability that the lymphocyte will encounter the antigen in the periphery is rather low. Therefore, antigen recognition and induction of immune responses requires the transport of microbial or vaccine antigens from the port of entry into the parenchyma of draining lymph nodes where the antigens are presented to a large number of lymphocytes with a broad range of specificities. Today, with more exact knowledge of the signalling pathways leading to immune response induction, also thanks to technological inventions that improve and simplify assessment, this simplistic geographic concept of immunogenicity (Frey and Wenk 1957; MartIn-Fontecha et al. 2003) may look superficial and out-dated, but it remains a fact that the arrival of the antigen in a secondary lymphatic organ, e.g. lymph nodes, the spleen, and the gut- (GALT), bronchial- (BALT) and mucosal-associated lymphoid tissues (MALT), is the most undisputable requirement for the triggering or regulation of immune responses.

Draining antigens follow the one-way flow of the lymph. The lymph absorbs interstitial fluid from peripheral tissues and returns it back to blood after passing the

P. Johansen (✉) • T.M. Kündig
Department of Dermatology, University Hospital Zurich,
Gloriastrasse 31, 8091 Zurich, Switzerland
e-mail: pal.johansen@usz.ch; thomas.kuendig@usz.ch

C. Foged et al. (eds.), *Subunit Vaccine Delivery*, Advances in Delivery Science and Technology, DOI 10.1007/978-1-4939-1417-3_14

thoracic duct and after having passed through several lymph nodes (Randolph et al. 2005; von Andrian and Mempel 2003). This antigen drainage typically starts in the highly endocytic and permeable initial lymphatic vessels. In the skin, these 10–80 μm wide vessels form a network with channels lying approximately 100 μm apart just below the epidermis. The network has a structural function for lymphocytes and antigen-presenting cells (APCs), and it serves as a molecular sieve to allow entrance of only small molecules and particles (Roozendaal et al. 2008, 2009). Small antigens that are introduced in the upper layer of the skin will ultimately be captured by this conduit network before entering the more vascularised and anatomically deeper layer of the dermis. The cut-off for entering the vessels without the help of APCs is approx. 70 kDa (Pape et al. 2007). In the dermis, the lymph vessels assemble into larger ducts (100–220 μm), which further drain the lymph to collecting vessels in the subcutis. From here, the antigen-containing lymph is driven by peristalsis towards the next draining lymph node (Catron et al. 2004; Itano and Jenkins 2003), where it enters through the subcapsular sinuses. In the lymph node parenchyma, local APCs survey the lymph to acquire microbial or vaccine antigens (Bajenoff et al. 2003; Sixt et al. 2005).

If the antigens are not drained to the lymph node, they may arrive as cargo of professional APCs. This is normally the case for antigens larger than 70 kDa as well as for large particles and whole bacteria; smaller particles as well as viruses may passively drain to lymph nodes (Manolova et al. 2008). The APCs are majorly dendritic cells (DCs), Langerhans cells (LCs), macrophages and monocytes, and are situated in the body's periphery, typically in the skin and in the gut epithelia where they scavenge for infectious agents. In the skin, resident DCs are equipped with a wide array of pathogen or pattern recognition receptors such as Toll-like receptors (TLRs) that recognise bacteria and viruses, and C-type lectins such as Langerin, DC-SIGN, Dectin-1, and Dectin-2 that recognize carbohydrate moieties on pathogens. Upon encountering pathogens, antigens or other danger molecules, the APCs get activated and change their expression of surface adhesion and co-stimulatory molecules. This is part of the maturation and differentiation programme that serves two major tasks: firstly, the switching of cellular machinery from antigen uptake to antigen presentation, secondly, acquisition of a motile state that allows active and passive migration from the place of antigen encounter and uptake, to that of antigen presentation, which is within the T-cell zone of draining lymph nodes (Forster et al. 2012). Activation causes disruption of the E-cadherin-mediated adhesion to keratinocytes as well as upregulation of surface CCR7 levels (Jiang et al. 2007), which facilitate mobilisation of APCs. Activated LCs, but not dermal DCs (Kissenpfennig et al. 2005), also produce basement membrane-degrading enzymes such as matrix metalloproteinase (MMP)-2 and MMP-9 that allow the translocation of epidermal LCs to dermis (Ratzinger et al. 2002). This process further depends on the chemokine CXCL12 and its receptor CXCR4 on the LCs (Ouwehand et al. 2008). While these events of cell-surface adhesion and chemotaxis enable homing of APCs to secondary lymphatic tissues, the upregulation of co-stimulatory molecules such as CD80, CD86 and CD40 ensure effective presentation of antigen to T- and B-cells and thereby amplify the immune response (Cavanagh and Von Andrian 2002).

Hence, whether and how antigens find their way to the lymph nodes from the periphery highly depends on their size and on their nature, e.g. particulate or soluble. However, size alone is not a parameter that allows predicting the immunological fate of an antigen. On the one hand, large soluble antigens that remain in the periphery will hardly encounter specific lymphocytes and therefore remain immunologically ignored. On the other hand, small soluble antigens that drain through the lymph by their own means are often immunologically ignored due to lack of adjuvants, which typically activate the inflammasome (De Gregorio et al. 2009; Eisenbarth et al. 2008) or TLRs (Duthie et al. 2011) that would facilitate pro-inflammatory reactions and co-stimulatory signals. Moreover, the specific instructions made by APCs depend on the state of their maturation, which then programmes the type and strength of the immune response, e.g. antibody production by B-cells, cytotoxicity by $CD8^+$ T-cells, effector $CD4^+$ T-cell responses, or regulatory immune responses by $CD4^+$ T-regulatory cells. Due to their potency in stimulating and regulating immune responses, DCs are often subject to specific targeting (Birkholz et al. 2010; Kretz-Rommel et al. 2007; Tenbusch et al. 2012) or even used directly in autologous vaccines (Kantoff et al. 2010), as described in Chap. 13. In the latter case, DCs are isolated from human blood, loaded with antigen ex vivo and returned to the vaccinee/patient by injection. Sipuleucel-T (Provenge®) is a DC vaccine approved for treatment of prostate cancer (Kantoff et al. 2010).

14.2 Vaccine Development: From Live to Subunit Vaccines

When the father of vaccinology, Edward Jenner, in 1796 inoculated “a lad of the name of Phipps” with infectious material from infected cow utters (*vaccinia* cowpox virus) and then 2 months later challenged him with matter from a small pox (*variola* virus) pustule, he demonstrated that vaccination can prevent infections with the same or a related pathogen (1923). During the following 150–200 years, dozens of new vaccines were developed based on such live or attenuated pathogens, e.g. measles, mumps, rubella, yellow fever, polio, *M. bovis* Bacillus Calmette-Guérin (BCG), typhoid fever, cholera, pertussis, and influenza. These vaccines have been administered by almost all possible routes, subcutaneously, intradermally, intramuscularly, orally and nasally. Indeed, it appears to be an intrinsic property of live vaccines that they very efficiently drain to secondary lymphoid organs, independent of the site of inoculation. When certain routes are recommended, it is therefore often due to safety concerns, such as local toxicity. Subcutaneous delivery of BCG is for instance contraindicated in man, for which reason it is given intradermally. Also, the fear of faecal-oral transmissions of live oral polio virus vaccine (Sabin) is one reason why the subcutaneously given inactivated polio vaccine (Salk) is often preferred.

For reasons of safety and for reproducibility in the production and standardisation of vaccines, the last two decades have seen a preferential development of subunit vaccines, as new vaccines against new diseases or to replace already existing suboptimal live vaccines. However, the immunogenicity of subunit vaccines is typically poor. Therefore, the development of subunit antigens has been closely backed-up by the development of new and improved adjuvants, which may mimic some of the adjuvant properties of live vaccines (Duthie et al. 2011). The type and the nature of the adjuvants strongly affect the immunogenicity of a vaccine (Bachmann and Jennings 2010; Manolova et al. 2008). Other factors that influence vaccine properties are the antigen dose, as well as the frequency and kinetics in which the vaccine is administered (Johansen et al. 2008). These factors determine the type and strength of the local inflammation or innate immune responses, and this again orchestrates the adaptive immune responses. Moreover, the administration route of these new subunit vaccines has been shown to be more critical than was the case for their live predecessor. Hence, vaccine development has become a complex algorithm of antigen, adjuvant and route of administration. In the following, we will discuss some of the limitations and possibilities of parenteral administration of vaccines, including the classical subcutaneous, intradermal and the intramuscular routes, as well as systemic intravenous applications. In addition, we will discuss the potential of intralymphatic administration, a new method of delivering vaccines by direct injection into subcutaneous lymph nodes.

14.3 Routes of Administration

The route of vaccine administration unambiguously influences the outcome of immunisation with regard to the type and the strength of the stimulated immune responses. When the first anatomical hurdle is cleared, the composition of the underlying tissue decides the further fate of a vaccine, e.g. the lymph or blood perfusion of the tissue. The latter will determine whether the vaccine will spread or be cleared by the innate defence mechanisms, or if a vaccine is forwarded to the secondary lymphoid organs for elicitation of immune responses. Similarly, the route can be vital with respect to whether crucial vaccine properties such as safety, quality and efficacy will be successfully discovered or not. An inappropriate route of administration may render a potential vaccine ineffective (Cubas et al. 2009). On the other hand, to predict the efficacy of a new vaccine based on the known properties of a certain route is difficult. Table 14.1 lists a number of confounding factors that will influence the choice of route and consequently the efficacy and appropriateness of vaccination by that route.

In a simplistic view, a stronger immune response may be expected the closer a vaccine is administered to a lymph node or to an afferent lymphatic vessel, assuming epitope correctness and further appropriateness of the antigen and adjuvant. However, the above-described concept of lymphatic drainage may suggest that this is not necessarily the case as the upper layers of the skin are better equipped with lymphatic conduits and vessels as well as professional APCs than the lower skin layers. While microbes and intelligent antigen delivery systems may be tailored to find their way into the lymphatic organs, subunit vaccines would certainly benefit

Table 14.1 Factors that affect whether a certain route of vaccine administration is appropriate for a certain vaccine

Factor	Comment
Species	Anatomical and physiological differences
	Various expression of molecules (e.g. TLR molecules)
	Distribution of lymph nodes
	Thickness of skin
	Organisation of nasal cavity,
	Vascularisation of target organ
	Behaviour, life style
	Indoor or outdoor
	Nutritional habits, microflora and hygiene
Antigen	Virus
	Live, inactivated, recombinant
	Bacteria
	Live, inactivated, recombinant
	Cellular
	Dendritic cells
	Cell lysates, e.g. tumour
	Protein or peptide
	Molecular weight, hydrophobicity
	DNA or RNA
	Cytoplasmic, transmembrane or secreted expression
	Promoter type
Adjuvant	Soluble or particulate
	Depot
	Type of innate immune activation
	Inflammasome
	TLR stimulation
Disease	Infection
	Cytopathic or non-cytopathic microbes
	Intracellular or extracellular live cycles
	Replication speed
	Autoimmune disease
	Tumour
	Allergy
Wanted immune response	B-cells and antibodies
	Antibody isotype and subclass
	T-cell independent antibody responses (e.g. Alzheimer)
	T-cells
	Th1 or Th2 cells
	T-regulatory cells (e.g. allergy and autoimmunity)
	Cytotoxic $CD8^+$ T-cells
	Tolerance or anergy

from being administered into an anatomical site, which is rich in immune cells. The anatomical concept of vaccination also unveils different tissue-related bystander effects, which are related to disparities in the type, number and quality of cells available in a certain tissue. For instance, the most potent cells for the priming of immune responses, especially of $CD8^+$ T-cell responses, are DCs (Banchereau and Steinman 1998; Lee and Iwasaki 2007; Zitvogel 2002). DCs are positioned as sentinels in the periphery, where they frequently encounter foreign antigens or microbes, upon which they readily relocate to secondary lymphoid organs, particularly lymph nodes, to position themselves optimally for interactions with naïve or central memory T-cells. However, DCs in the skin, peritoneum, lung, muscles and secondary lymphatic organs have different potentials in priming immune responses and generating immunity (Everson et al. 1996). Moreover, DCs may receive highly variable tissue-dependent bystander signals from neighbouring cells for activation, migration and maturation (Rupec et al. 2010; Schroder et al. 2006). Within heterogeneous tissues, such as the skin, thorough knowledge of the structural differences between the epidermis, dermis and the subcutaneous layers are important for the decision whether to administer a vaccine to one of these sites (Azzi et al. 2005).

14.3.1 Epidermal Immunisation

The epidermis is the outer layer of the skin, and its thickness varies considerably across the body surface area. The human epidermis varies from typically 50–150 μm to more than 1,000 μm on the palms and the soles. By comparison, murine epidermis is typically less than 15 μm thick (Azzi et al. 2005). The top layer of the human epidermis, the *strateum corneum*, contains only dead and cornified keratinocytes that are constantly shed. In the deeper layers of the epidermis, the viability of the cells increases towards the basal membrane that separates the epidermis from the dermis. Of immunological importance is the high density of LCs present in the epidermis. LCs provide the frontline defence of the immune system. Similar to DCs, LCs are very efficient in recognising and taking up pathogens and antigens, and to migrate to lymph nodes for presentation of antigen. Their neighbouring keratinocytes, with which LCs interact via E-cadherins (Tang et al. 1993), can produce inflammatory cytokines such as tumour necrosis factor (TNF-)α, interleukin (IL)-1β and IL-18 (Cumberbatch et al. 2001; Trevejo et al. 2001; Wang et al. 1999) as well as chemotactic signals (Homey et al. 2006; Kimber et al. 2000; Stutte et al. 2010; Villablanca and Mora 2008; Xu et al. 2001) that efficiently activate LCs and epidermal macrophages to mature and to migrate. Importantly, the initial lymphatic vessels that capture antigens and pathogens for delivery to the lymph nodes are located below the basal membrane. Therefore, vaccines delivered into the epidermis have a very good chance of ending up in the lymph node parenchyma for stimulation of immune responses. While delivery of compounds to the epidermis is feasible by using transdermal patches and micro-needle arrays, as discussed in detail in Chap. 18 of this book, epidermal injection using conventional needles is not practical.

14.3.2 Intradermal Immunisation

Vaccines can also be administered into the dermis, which consists of a matrix of collagen, elastic tissue, reticular fibres, sweat glands, sebaceous glands and hair follicles. The dermis also varies in thickness depending on the location of the skin. In humans, the dermis is typically 3–10 mm on the back. In 13–15-week-old C57BL/6 mice, the dermis was on average 171 μm in females and 500 μm in males (Azzi et al. 2005). The only vaccine that is currently approved in the United States for administration by the intradermal route is a flu vaccine (Fluzone®). The same vaccine was originally licenced in Europe, Canada and Australia (Intanza®). The live viral rabies vaccine and *vaccinia* small pox vaccine were also given intradermally. The same holds for live *M. bovis*, the only vaccine against tuberculosis. BCG is still given in many part of the world, including many European countries. However, the methods of intradermal injection vary, and they typically require more training than the more conventional subcutaneous and intramuscular routes of vaccine administration.

Intradermal BCG injection is done by injecting only 0.05 mL using a special BCG syringe with a special BCG 26G needle. With the bevel of the needle facing up, the syringe and needle is laid almost flat along the skin, and the tip of the needle is inserted just under the skin, so that only the bevel and a little bit more disappears beneath the skin surface. If the vaccine is injected correctly, a clear, flat-topped swelling on the skin, like a mosquito bite, can be observed. If the BCG vaccine is injected too deep, i.e. into the subcutis, an abscess or enlarged glands may result (Centers for Disease Control and Prevention 2012).

An alternative way of intradermal vaccine administration has been done by scarification. This method was used in ancient India for variolation, i.e. smallpox vaccination using the *variola* virus. Pus or scab or the *variola* lesions of one patient was administered by scratching it on the forearm of the vaccinee. The method was brought to Europe from Turkey in the eighteenth century by Lady Mary Wortley Montagu and her family's physician Dr. Charles Maitland. Dr. Maitland "variolated" the Montagus' daughter as well as members of the royal family, though only after the forced smallpox inoculation of several condemned British prisoners. Also Edward Jenner used scarification when he later performed his famous *vaccinia* (cowpox) inoculations, and the method that was to be named vaccination by Louis Pasteur was extensively used for smallpox eradication, but provided variable and inefficient delivery into the skin.

Intradermal immunisation has more recently been shown to offer improved immunity with activation of innate immune responses in the skin (Pearton et al. 2010a, b), including protection against influenza and yellow fever (Roukens et al. 2012). Intradermal vaccination may also improve and simplify logistics of delivery, as compared to the most abundantly intramuscular immunisation, but the intradermal method in medicine is limited by the need for simple, reliable delivery methods (Hickling et al. 2011; Kenney et al. 2004; Lambert and Laurent 2008). Therefore, the current intradermal flu vaccines are provided with special applicators, the microinjection system Soluvia®, as to enable a safe and reproducible method for vaccine administration. Likewise, more experimental methods of intradermal vaccine administration using microneedle arrays are currently under development (Kim et al. 2012), as described in Chap. 18 of this book. Next to the relative difficulty in efficient intradermal vaccine administration, the very low number of intradermal vaccines might be due to the fact that they often contain adjuvants. Many adjuvants are contraindicated for intradermal applications due to the risk of local and long-lasting adverse reactions (Carlsson et al. 1996). Finally, studies on flu vaccines (Kenney et al. 2004), virosomal hepatitis A vaccines (Frosner et al. 2009) and rabies vaccines (World Health Organization 2000) in humans have demonstrated that vaccination costs can be significantly cut by changing to intradermal administrations just by reducing the immunising dose via intradermal administration. Intradermal administration with only 20 % of the recommended intramuscular dose of influenza or rabies vaccines elicited an immune response that was similar to or better than that elicited by intramuscular injection.

14.3.3 Subcutaneous Immunisation

Many vaccines, e.g. inactivated vaccine polio (Salk), measles, mumps, rubella, varicella, yellow fever, zoster, typhoid and Japanese encephalitis vaccines are administered just beneath the dermis into the subcutis. The subcutaneous tissue comprises a loose organisation of connective and adipose tissues in addition to blood vessels and nerve bundles. One great advantage of the subcutis is that it can take up larger vaccine volumes than the dermis. Its flexible structure causes less stimulation of pain and pressure noiceptors in the skin, for which reason the patients typically tolerate subcutaneous injections better than intradermal injections.

Due to the rather static properties of the connective and adipose tissue, the subcutis is particularly suited for vaccine depots, and the poorer vascularisation in the subcutis as compared to the dermis may result in slow mobilisation and processing of the vaccine. However, this longer persistence may also cause degradation of the vaccine before immunologically utilised (Zuckerman 2000). This is often a cause of vaccine failure (Poland et al. 1997) for example of hepatitis B, rabies and flu vaccines (Groswasser et al. 1997; Shaw et al. 1989). Due to the lower blood perfusion the subcutis is less efficient than the epidermis and the dermis in draining antigens via the conduits and lymphatic vessels, i.e. less of the administered vaccine is likely to reach the parenchyma of the lymph nodes. From an evolutionary viewpoint, the higher efficiency in recognising and draining antigens from the epidermis and the dermis makes perfect sense, because any immunological challenge is expected to originate from the exterior of the host organism and not from the interior. However, the lower draining efficiency in the subcutis can often be compensated for by the fact that the subcutis can take larger vaccine volumes (0.5–2 mL) or doses than the dermis (0.05–0.1 mL). The recommended site for subcutaneously administered vaccines are the thigh for infants and the upper triceps for those older than 1 year old, and a 23–25G needle is typically used (Centers for Disease Control and Prevention 2012).

The translation of data from immunological experiments on small rodents to humans is typically associated with many pitfalls. One difficulty in using small rodents is associated with the difference in the skin between species. Not only does the thickness of human and rodent skin vary, factors that will especially affect the quality of epidermal and intradermal vaccine administration, but also the anatomical build-up of the skin and especially its connection to neighbouring tissues is different in mice and men. While the subcutaneous tissue in humans is closely connected to its underlying tissues, such as bone and muscles, rodent skin is hardly attached to the underlying tissues. Such factors may very well influence the performance and developmental destiny of new vaccines. Hence, screening of vaccine candidates solely by the easy and convenient subcutaneous injection in mice may cause the rejection of vaccines with potential efficacy in other animal models, including humans.

14.3.4 Intramuscular Immunisation

Almost all inactivated vaccines are injected intramuscularly. Such vaccines are mostly based on protein and polysaccharide antigens [DTP, hepatitis A, hepatitis B, flu, haemophilius influenza type B (Hib), meningitis C, pneumococcus, and human papilloma virus (HPV)], and they often contain an adjuvant. For almost a century, salts of aluminium were the only adjuvants approved for human use (Chap. 3). The use of aluminium salts began in the 1930s, before regulatory guidelines became more stringent. More recently, approval has been obtained for MF59, an oil-in-water emulsion of squalene, polysorbate (Tween 80) and sorbitan trioleate (Span 85) (Chap. 4). MF59 is used as an adjuvant component of flu vaccines for elderly patients. Viral vaccines against hepatitis B as well as HPV have been approved with the adjuvant AS04, which is a combination of aluminium salts and monophosphoryl lipid A. However, because adjuvants can cause exaggerated local reactions, e.g. pain, swelling and erythema in the skin, the intramuscular route is generally recommended for adjuvanted vaccines (Centers for Disease Control and Prevention 2012). The site of intramuscular vaccine administration is typically the *vastus lateralis* muscle (anterolateral thigh) and the deltoid muscle (upper arm). Injection at these sites reduces the risk of injecting into larger neuronal or vascular structures, because such tissues are void in the mentioned sites. A 22–25G needle is used.

If the administration of several vaccines is planned for a single doctor visit, it is recommended that each vaccine preparation is administered to a different site (Centers for Disease Control and Prevention 2012). If several vaccines are injected in a single limb of infants or young children, it is recommended to use the thigh because of the greater muscle mass. The injections should be separated by at least 2–3 cm as to be able to distinguish any local reaction from one vaccine to that of another. Vaccines that frequently produce local reaction should be split on different limbs if possible.

14.3.5 *Intravenous Immunisation*

Systemic administration of vaccines has been used widely in small animals, e.g. intravenous and intraperitoneal vaccine administration in mice, but not in humans. However, and as indicated above, during pre-clinical development of vaccines, other routes are often used to test immunogenicity and efficacy of the vaccine—often only for the sake of convenience. For instance, live attenuated sporozoites *Plasmodium falciparum* has long been an attractive approach to vaccinate against malaria (Mons 1991), because such sporozoites administered by mosquito bites are the only immunogens that have clearly been demonstrated to induce protection in humans (Roestenberg et al. 2009, 2011). During the pre-clinical testing of a candidate vaccine in mice, intravenous administration was compared and found superior to subcutaneous and intradermal vaccine administrations (Epstein et al. 2011). However, for the first clinical testing in 80 volunteers, intradermal and subcutaneous routes of administration were applied, but failed reaching the primary outcome, for which reason the authors have concluded that vaccine should be given intravenously in man (Epstein et al. 2011).

The last decade has also seen the development of several autologous vaccines based on DCs (Chap. 13). Such vaccines have especially found resonance in cancer immunotherapy. Briefly, DCs are prepared by purification and culturing of the patient's own blood. The cells are then cultured and pulsed with tumour antigens and then returned to the patient by injection. The route of administration of autologous DC vaccines has been intradermal, but mostly intravenous. Indeed, the only FDA-approved autologous DC vaccine, sipuleucel-T, which is indicated for the treatment of asymptomatic or minimally symptomatic metastatic castrate resistant prostate cancer, is given by intravenous injection (Kantoff et al. 2010).

14.3.6 *Intralymphatic Vaccination: As Good as It Gets*

When the first anatomical hurdle is cleared, the composition of the underlying tissue decides the further fate of a vaccine. The environment, including the tissue perfusion will determine whether the vaccine will be spread or cleared by the innate defence mechanisms, or if a vaccine is carried forward to the secondary lymphoid organs for induction of immune responses. Consequently, the chosen route can be crucial with respect to whether vaccine properties such as safety, quality and efficacy will be successfully discovered or not, and the incorrect choice may render a possible vaccine ineffective.

Independent on where the vaccine is administered, the goal of vaccination is to bring the vaccine to a lymph node, the major site for adaptive immune stimulation. For that simple reason, direct lymph node injection represents a short cut and the strongest possible method for stimulation of immune responses. In many ways, lymph nodes function as in vivo petri dishes where APCs and lymphocytes are in close proximity and where the antigen concentration can be kept high, situations that are highly beneficial for effective APC activation and antigen presentation. Although direct intralymphatic injection is not a generally recommended method of vaccination, it has been employed in a number of pre-clinical investigations as well as in human clinical trials of cancer and allergen immunotherapy (Johansen et al. 2012).

In mice, but not in humans, the procedure is invasive. In mice, a 5–10 mm incision is typically made in the inguinal region of anesthetised animals (Johansen et al. 2005). This enables the localisation of the inguinal lymph node, which is then immobilised by help of forceps. Using a syringe with a 28–30G short hypodermic needle and a short bevel, 10–20 μL of the vaccine is injected under visual control, e.g. the swelling of the lymph node. During the injection, the bevel should be facing up. Finally, the incision is closed with a single stitch using surgical sutures. A video article describes the procedure of intralymphatic injections in mice in detail (Johansen and Kündig 2014).

In humans, the procedure is guided by ultrasound, which is used to image both the lymph node and the inserted needle. In contrast to in mice, there are typically several subcutaneous lymph nodes in the inguinal region of humans. The one chosen for injection is typically among the group localised in the outer upper quadrant of the groin area, because these lymph nodes are superficially localised and have a slower flow-through of lymph than the bigger ones draining the lymph from the leg which are also located deeper down in the dermis. This will facilitate longer retention times of the vaccine, which again is beneficial for effective stimulation of immune responses (Huppa et al. 2003). A 0.5–1 mL syringe equipped with a 25G hypodermic needle is used for injection, and the injection volume is typically 100 μL. Figure 14.1 illustrates the procedure.

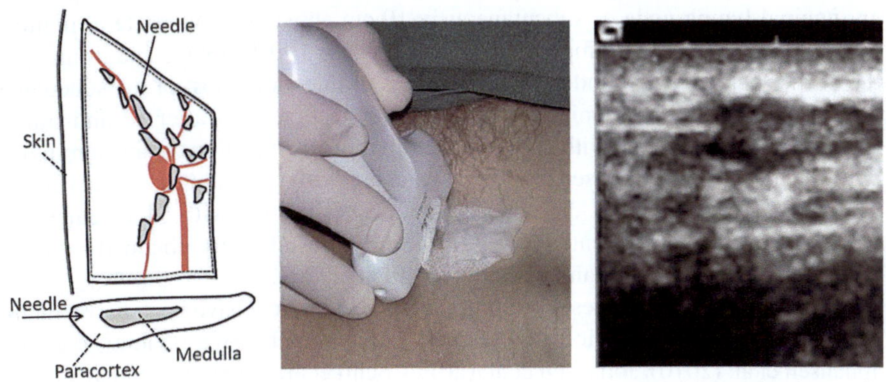

Illustration showing chains of subcutaneous LNs along blood vessels and the LN with paracortex and medulla.

Intralymphatic vaccination is typically guided by ultrasound (here, inguinal LN).

Ultrasound image showing needle (straight whitish line) inserted into the paracortex of the LN (peanut-shaped hypoechoic area).

Fig. 14.1 Intralymphatic vaccine administration is done by locating subcutaneous lymph nodes (LNs) with the aid of ultrasound. The hypodermal needle is inserted into the paracortex of the LN, and the injection can be controlled by observation immediate LN swelling in the ultrasound image. This figure is reprinted, with minor modifications, with kind permission from Springer (Kündig et al. 2012)

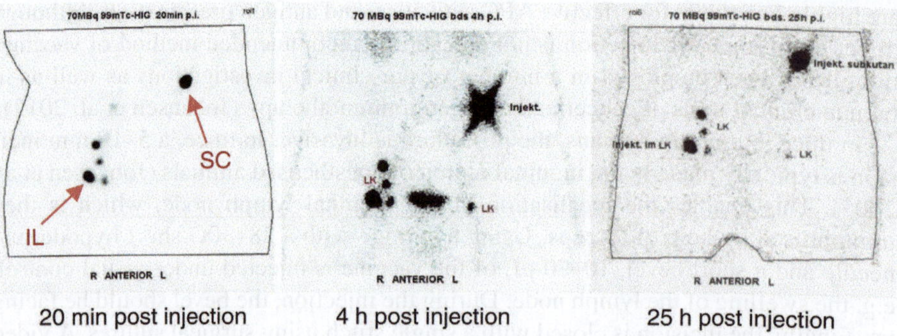

Fig. 14.2 Biodistribution of 99mTc-labelled human IgG after intralymphatic (IL) and subcutaneous (SC) injection in man. The tracing of radioactive substance was made by gamma imaging. The *arrows* indicate were the injection was made: directly into the right lymph node or subcutaneously 10 cm above the contralateral left lymph node. This figure is reprinted, with minor modifications, with kind permission from Springer (Kündig et al. 2012)

Only a small fraction of a vaccine injected into a peripheral tissue reaches a lymph node. This is one of the reasons why so far intralymphatic vaccination or immunotherapy has been superior to any other peripheral route of administration with regard to the lymphatic system. In mice, and using a radioactively labelled antigen, 100-fold more antigen was found in the lymph nodes upon direct injection as compared with subcutaneous injection in the same anatomical region (Martinez-Gomez et al. 2009). Comparable observations were made in humans (Senti et al. 2011) as illustrated in Fig. 14.2. Proteins were labelled with 99mTc and injected in an inguinal lymph node or subcutaneously 10 cm above the contralateral inguinal lymph nodes. While the lymph node injection cause 100 % uptake and draining to the deeper pelvic lymph nodes within minutes of injection, most of the subcutaneously injected protein remained at the site of injection 24 h after administration (Senti et al. 2011). In line with this, three intralymphatic injections of an aluminium-adjuvanted and protein-based hay fever vaccine within 2 months stimulated the same level of protection as did conventional desensitisation with 54 subcutaneous injections over 3 years (Senti et al. 2008). Similar results were obtained for treatment of patients with cat-dander allergy (Senti et al. 2012).

A large number of studies, both pre-clinical and clinical have been based on the intralymphatic administration of vaccines. For comprehensive reviews, see refs. Johansen et al. (2010), Johansen et al. (2012), Senti et al. (2009), Senti et al. (2011), and Table 14.2 lists a summary of investigations where vaccine antigens were administered by intralymphatic injection. A large variety of vaccines have been tested in animals for the applicability of intralymphatic vaccination. In mice, peptides, proteins, DNA, mRNA as well as DC-based vaccine were investigated in various models. Direct administration of major histocompatibility complex (MHC) class I binding peptide vaccines into lymph nodes showed strongly enhanced $CD8^+$ T-cell responses that were protective against a viral challenge and tumour growth in mice (Johansen et al. 2005). Similarly, intranodal immunisation also dramatically

Table 14.2 An overview of investigations describing intralymphatic immunisation and immunotherapy in experimental models and in man

Specie	Comment	References
Mouse	Soluble, particulate and bacterial vaccines. Intralymphatic vaccination always superior to intramuscular, subcutaneous and/or intradermal vaccination	Johansen et al. (2005); Johansen et al. (2010); Maloy et al. (2001); Martinez-Gomez et al. (2009); Mohanan et al. (2010); Waeckerle-Men et al. (2013a)
Cat	Protein-based HIV (FIV) vaccine	Finerty et al. (2001)
Dog	Cellular- and protein-based cancer therapy studies	Juillard and Boyer (1977); Juillard et al. (1979); Juillard et al. (1976); Juillard et al. (1977)
Non-human primate	Protein-based HIV vaccines with ISCOMs or with aluminium adjuvant	Koopman et al. (2007); Lehner et al. (1994); Lehner et al. (1998); Lu et al. (1998)
Human	Cancer immunotherapy with DCs	Bedrosian et al. (2003); Lesimple et al. (2003), Lesterhuis et al. (2011), Schwaab et al. (2009)
	Cancer immunotherapy with tumour cells	Juillard et al. (1978); Lacour et al. (1992); Moy et al. (1985); Williams et al. (1992); Wiseman et al. (1986); Wiseman et al. (1989)
	Cancer immunotherapy with pDNA/peptide	Ribas et al. (2011); Weber et al. (2008); Weber et al. (2011)
	Cancer immunotherapy with recomb. virus	Adamina et al. (2010); Brown et al. (2003); Spaner et al. (2006)
	Cancer adjuvant therapy with BCG	Kirkwood et al. (1980); Kirkwood et al. (1982)
	Allergy immunotherapy with protein and aluminium	Senti et al. (2012); Senti et al. (2008) Hylander et al. (2013)

enhanced the efficacy of plasmid DNA (pDNA) vaccination in mice (Heinzerling et al. 2006; Maloy et al. 2001) as well as RNA vaccination (Kreiter et al. 2010, 2011). By changing the route of pDNA delivery from subcutaneous, intradermal or the most abundantly used intramuscularly route of administration to the intralymphatic route, the dose of pDNA necessary to stimulate cytotoxic T-cell responses could be reduced by several orders of magnitude, e.g. 0.2 μg pDNA by intralymphatic injection was more effective than intramuscular injection of 200 μg of the same pDNA (Maloy et al. 2001).

Intranodal immunisation with proteins for induction of antibodies was performed surprisingly early. At a time when it was difficult to purify large quantities of proteins, researchers were looking for a more efficient route of immunisation (Sigel et al. 1983). By intralymphatic immunisation, only nanograms of protein were required to elicit sufficiently strong immune responses (Nilsson et al. 1987) Likewise, targeted lymph node administration is also extensively documented to be the most efficient way to immunise macaques against SIV (Kawabata et al. 1998; Lehner et al. 1998; Lu et al. 1998). Similar results were obtained in macaques vaccinated with other proteins (Bogers et al. 2004a, b; Kawabata et al. 1998; Klavinskis et al. 1996; Lehner et al. 1994, 2000). Also DCs (Johansen et al. 2008; Waeckerle-Men et al. 2013b) and bacteria (Waeckerle-Men et al. 2013a) have been used as vehicles for antigen in intralymphatic vaccination in mice for stimulation of $CD4^+$ and $CD8^+$ T-cell responses against tumours and tuberculosis.

Several clinical trials with intranodal therapy have confirmed these initial pre-clinical studies. Most of these studies have been performed in cancer patients usually as autologous vaccination using antigen-pulsed DCs (Barth et al. 2010; Czerniecki et al. 2007; Fadul et al. 2011; Lesimple et al. 2006; Lesterhuis et al. 2011; Schwaab et al. 2009; Yi et al. 2010). While some clinical trials using intranodal therapy with DCs have suggested enhanced immune responses (Bedrosian et al. 2003; Lesimple et al. 2003), other trials failed to demonstrate an advantage of intranodal over intradermal DC delivery (Brown et al. 2003; Fong et al. 2001). We also found that non-professional APCs, such as a fibro-sarcoma cell line efficiently induced antigen-specific $CD8^+$ T-cell responses in lymph nodes via direct antigen presentation on MHC class I molecules present on the fibro-sarcoma (Kündig et al. 1995; Ochsenbein et al. 2001). Intranodal therapy with tumour cells has been tried in both human cancer patients and dogs with indication of success (Juillard and Boyer 1977; Juillard et al. 1976, 1977, 1978, 1979).

In other cancer trials, the intranodal vaccines were based on pDNA (Weber et al. 2008) or pDNA prime and peptide boost (Ribas et al. 2011; Weber et al. 2011). Intranodal injections of vaccines based on viral vectors have been studied in melanoma patients, who responded with strong cytotoxic and other immunological T-cell responses as well as some clinical benefits (Adamina et al. 2010; Spaner et al. 2006).

Finally, three recent trials on patients with hay fever revealed that allergen-specific immunotherapy in this patient group can be strongly improved by changing from the conventional subcutaneous route to the novel intralymphatic route of injection. While subcutaneous immunotherapy (SCIT) typically requires more than 50 injections over at least 3 years, successful intralymphatic immunotherapy (ILIT) was achieved by only three injections with 1 month interval (Hylander et al. 2013; Senti et al. 2008, 2012). Moreover, SCIT is frequently associated with local and systemic adverse events (Windom and Lockey 2008). Because the number of injections as well as the injected dose is much lower in ILIT than in SCIT, and because the lymph nodes do not contain mast cells, there is also a safety benefit with ILIT. Finally, the short treatment and the few side effects have strongly improved the patient compliance of ILIT as compared to SCIT (Senti et al. 2008). Hence, it is expected that ILIT will become a real alternative to SCIT as well as the sublingual immunotherapy (SLIT), which is approved and practiced in Europe (Dretzke et al. 2013).

14.4 Comparative Studies of Administration Routes

Compared to the rather weak immunogenicity of modern sub-unit vaccines, the historical and still abundant live or attenuated microbial childhood vaccines have been very effective in protecting vaccinees from infectious diseases (Chap. 2), for

Table 14.3 Comparative studies on the effect of administration routes on vaccine performance in animals

Animal model	Vaccine formulation	References
Mouse	Virus or bacteria	Brockstedt et al. (1999); Delagrave et al. (2012); Eo et al. (2001); Goetsch et al. (2001); Goetsch et al. (2000); Nnalue and Stocker (1989); Waeckerle-Men et al. (2013a)
	Synthetic particles	Baldwin et al. (2009); Carcaboso et al. (2004); Cubas et al. (2009); Mohanan et al. (2010); Slutter et al. (2011); Stertman et al. (2004)
	Dendritic cells	Eggert et al. (1999); Malowany et al. (2006); Okada et al. (2001)
	Protein with alum or other adjuvant	Johansen et al. (2005); Koutsonanos et al. (2012); Lobaina et al. (2010); Martinez-Gomez et al. (2009)
	Plasmid DNA	Boutennoune et al. (2012); Hartikka et al. (2012); Lai et al. (2009); Maloy et al. (2001); Morel et al. (2004); Tu et al. (2007); Yoshida et al. (2000)
Bovine and sheep	Virus, bacteria, blood lysate or alum-adsorbed protein	Gramzinski et al. (1998); Samina et al. (1998); Tomita et al. (1998); Woolums et al. (2013)
Non-human primates	Protein or pDNA with ISCOMs, liposomes or alum	Gramzinski et al. (1998); Koopman et al. (2007); Lehner et al. (1998); Lu et al. (1998)

which reason an increase in efficacy was seldom pursued. For weakly immunogenic sub-unit vaccines however, there may be much to gain in optimising not only the antigen and the adjuvant, but also the route of administration. Therefore, several studies have been conducted in a variety of animal models as to test how the route influences the immunological performance as well as the safety of new vaccines or model vaccines. Pre-clinical works in rodents often tested and evaluated several parenteral routes side-by-side. In larger animals, in non-human primates and in man, such studies have often been performed to compare invasive and non-invasive methods, such as an injection method against a mucosal vaccination method, e.g. nasal, oral and vaginal. However, few studies are available that compare the real efficacy of various invasive methods of vaccination. Table 14.3 lists a number of comparative studies done in animals, and Table 14.4 summarises a representative, but not complete selection of such studies in man. The overview is limited to the comparison of two or more invasive or injectable vaccination methods only. Comparisons between one injectable method and one or more mucosal vaccination methods as to demonstrate efficacy of the latter are ignored in the current review.

When reviewing the data from studies in animals, it is unfortunately not the fact that the different administration routes produce a clear fingerprint with respect to the type and the strength of immune responses they stimulate. A rather stronger effect on the stimulated immune responses is found with respect to the species, the type and the dose and the dosing frequency of antigen, the adjuvant, as well as the type and the severity of pathogenicity when the vaccine efficacy is tested in a

Table 14.4 Comparative clinical studies on the effect of administration routes on vaccine performance

Formulation	Vaccine	References
Protein, polysaccharide with alum	DT, DTP, DTP-Hib, MenC, Antrax	Carlsson et al. (1999); Mark et al. (1999); Pittman (2002); Pittman et al. (2002); Ruben et al. (2001)
Protein ± MF59	Flu	Van Damme et al. (2010)
DNA	HIV	Bansal et al. (2008)
DCs	Melanoma	Lesterhuis et al. (2011)
Adenovirus	HIV	Koblin et al. (2011)
Live/attenuated virus	HepA	Fisch et al. (1996); Reynolds et al. (2006)
Bacteria	BCG	Davids et al. (2006)

challenge model of the disease. Consequently, the most optimal route for a given vaccine must be investigated individually. In each case, one also has to consider the safety of vaccine administration as one route may be safe for one vaccine and one specie, but not for another. Also, the global compliance or vaccination coverage may be favoured by a certain administration route, e.g. non-invasive, and especially oral methods of vaccination are favoured by many vaccinees or patients and enable vaccine experts and health authorities or institutions such as WHO to reach out to a greater number of persons.

One often met dogma in vaccinology is that the induction of mucosal immunity with IgA switch for antibodies must be obtained by mucosal vaccination, and vice a versa, the systemic immunity with serum IgG antibodies must be raised by parenteral vaccination. However, numerous studies have shown that this ruling is not that stringent. Many parenteral vaccines can produce protective mucosal immunity (Delagrave et al. 2012; Opriessnig et al. 2011). The properties of mucosal vaccine are described in detail in the Chaps. 6 (oral), 15 (nasal), 16 (pulmonary) and 17 (vaginal) of this book.

14.5 Conclusion

In conclusion, parenteral routes of vaccine administration have been and will maintain perhaps the most important way of immunisation or vaccination. While some of the potential parenteral routes can easily be excluded for a new vaccine based on safety issues, the prediction of the most effective routes with regard to vaccine efficacy is difficult, if not impossible. However, minor, moderate and even large differences in the efficacy can be found for certain vaccines, but these differences are not independent on the type of vaccine formulation tested. Hence, vaccine development is a complex algorithm of antigen, adjuvant, disease and route of administration. Similarities with previous studies with similar vaccines may be sought and found, and this review should provide a tool to start searching in the ocean of publications on vaccine administration.

References

Adamina M, Rosenthal R, Weber WP, Frey DM, Viehl CT, Bolli M, Huegli RW, Jacob AL, Heberer M, Oertli D, Marti W, Spagnoli GC, Zajac P (2010) Intranodal immunization with a vaccinia virus encoding multiple antigenic epitopes and costimulatory molecules in metastatic melanoma. Mol Ther 18:651–659

Azzi L, El-Alfy M, Martel C, Labrie F (2005) Gender differences in mouse skin morphology and specific effects of sex steroids and dehydroepiandrosterone. J Invest Dermatol 124:22–27

Bachmann MF, Jennings GT (2010) Vaccine delivery: a matter of size, geometry, kinetics and molecular patterns. Nat Rev Immunol 10:787–796

Bajenoff M, Granjeaud S, Guerder S (2003) The strategy of T cell antigen-presenting cell encounter in antigen-draining lymph nodes revealed by imaging of initial T cell activation. J Exp Med 198:715–724

Baldwin SL, Bertholet S, Kahn M, Zharkikh I, Ireton GC, Vedvick TS, Reed SG, Coler RN (2009) Intradermal immunization improves protective efficacy of a novel TB vaccine candidate. Vaccine 27:3063–3071

Banchereau J, Steinman RM (1998) Dendritic cells and the control of immunity. Nature 392:245–252

Bansal A, Jackson B, West K, Wang S, Lu S, Kennedy JS, Goepfert PA (2008) Multifunctional T-cell characteristics induced by a polyvalent DNA prime/protein boost human immunodeficiency virus type 1 vaccine regimen given to healthy adults are dependent on the route and dose of administration. J Virol 82:6458–6469

Barth RJ Jr, Fisher DA, Wallace PK, Channon JY, Noelle RJ, Gui J, Ernstoff MS (2010) A randomized trial of ex vivo CD40L activation of a dendritic cell vaccine in colorectal cancer patients: tumor-specific immune responses are associated with improved survival. Clin Cancer Res 16:5548–5556

Bedrosian I, Mick R, Xu S, Nisenbaum H, Faries M, Zhang P, Cohen PA, Koski G, Czerniecki BJ (2003) Intranodal administration of peptide-pulsed mature dendritic cell vaccines results in superior CD8+ T-cell function in melanoma patients. J Clin Oncol 21:3826–3835

Birkholz K, Schwenkert M, Kellner C, Gross S, Fey G, Schuler-Thurner B, Schuler G, Schaft N, Dorrie J (2010) Targeting of DEC-205 on human dendritic cells results in efficient MHC class II-restricted antigen presentation. Blood 116:2277–2285

Bogers WM, Bergmeier LA, Ma J, Oostermeijer H, Wang Y, Kelly CG, Ten Haaft P, Singh M, Heeney JL, Lehner T (2004a) A novel HIV-CCR5 receptor vaccine strategy in the control of mucosal SIV/HIV infection. AIDS 18:25–36

Bogers WM, Bergmeier LA, Oostermeijer H, ten Haaft P, Wang Y, Kelly CG, Singh M, Heeney JL, Lehner T (2004b) CCR5 targeted SIV vaccination strategy preventing or inhibiting SIV infection. Vaccine 22:2974–2984

Boutennoune H, Qaqish A, Al-Aghbar M, Abdel-Hafez S, Al-Qaoud K (2012) Induction of T helper 1 response by immunization of BALB/c mice with the gene encoding the second subunit of Echinococcus granulosus antigen B (EgAgB8/2). Parasite 19:183–188

Brockstedt DG, Podsakoff GM, Fong L, Kurtzman G, Mueller-Ruchholtz W, Engleman EG (1999) Induction of immunity to antigens expressed by recombinant adeno-associated virus depends on the route of administration. Clin Immunol 92:67–75

Brown K, Gao W, Alber S, Trichel A, Murphey-Corb M, Watkins SC, Gambotto A, Barratt-Boyes SM (2003) Adenovirus-transduced dendritic cells injected into skin or lymph node prime potent simian immunodeficiency virus-specific T cell immunity in monkeys. J Immunol 171:6875–6882

Carcaboso AM, Hernandez RM, Igartua M, Rosas JE, Patarroyo ME, Pedraz JL (2004) Enhancing immunogenicity and reducing dose of microparticulated synthetic vaccines: single intradermal administration. Pharm Res 21:121–126

Carlsson U, Brudin L, Eliasson I, Hansson BG (1996) Hepatitis A vaccination by intracutaneous low dose administration: a less expensive alternative. Scand J Infect Dis 28:435–438

Carlsson RM, Claesson BA, Kayhty H, Selstam U, Iwarson S (1999) Studies on a Hib-tetanus toxoid conjugate vaccine: effects of co-administered tetanus toxoid vaccine, of administration route and of combined administration with an inactivated polio vaccine. Vaccine 18:468–478

Catron DM, Itano AA, Pape KA, Mueller DL, Jenkins MK (2004) Visualizing the first 50 hr of the primary immune response to a soluble antigen. Immunity 21:341–347

Cavanagh LL, Von Andrian UH (2002) Travellers in many guises: the origins and destinations of dendritic cells. Immunol Cell Biol 80:448–462

Centers for Disease Control and Prevention (2012) Appendix D: vaccine administration. In: Atkinson W, Wolfe C, Hamborsky J (eds) Epidemiology and prevention of vaccine-preventable diseases. The pink book. Public Health Foundation, Washington, DC

Cubas R, Zhang S, Kwon S, Sevick-Muraca EM, Li M, Chen C, Yao Q (2009) Virus-like particle (VLP) lymphatic trafficking and immune response generation after immunization by different routes. J Immunother 32:118–128

Cumberbatch M, Dearman RJ, Antonopoulos C, Groves RW, Kimber I (2001) Interleukin (IL)-18 induces Langerhans cell migration by a tumour necrosis factor-alpha- and IL-1beta-dependent mechanism. Immunology 102:323–330

Czerniecki BJ, Koski GK, Koldovsky U, Xu S, Cohen PA, Mick R, Nisenbaum H, Pasha T, Xu M, Fox KR, Weinstein S, Orel SG, Vonderheide R, Coukos G, DeMichele A, Araujo L, Spitz FR, Rosen M, Levine BL, June C, Zhang PJ (2007) Targeting HER-2/neu in early breast cancer development using dendritic cells with staged interleukin-12 burst secretion. Cancer Res 67:1842–1852

Davids V, Hanekom WA, Mansoor N, Gamieldien H, Gelderbloem SJ, Hawkridge A, Hussey GD, Hughes EJ, Soler J, Murray RA, Ress SR, Kaplan G (2006) The effect of bacille Calmette-Guerin vaccine strain and route of administration on induced immune responses in vaccinated infants. J Infect Dis 193:531–536

De Gregorio E, D'Oro U, Wack A (2009) Immunology of TLR-independent vaccine adjuvants. Curr Opin Immunol 21:339–345

Delagrave S, Hernandez H, Zhou CH, Hamberger JF, Mundle ST, Catalan J, Baloglu S, Anderson SF, DiNapoli JM, Londono-Hayes P, Parrington M, Almond J, Kleanthous H (2012) Immunogenicity and efficacy of intramuscular replication-defective and subunit vaccines against herpes simplex virus type 2 in the mouse genital model. PLoS One 7(10):e46714

Dretzke J, Meadows A, Novielli N, Huissoon A, Fry-Smith A, Meads C (2013) Subcutaneous and sublingual immunotherapy for seasonal allergic rhinitis: a systematic review and indirect comparison. J Allergy Clin Immunol 131:1361–1366

Duthie MS, Windish HP, Fox CB, Reed SG (2011) Use of defined TLR ligands as adjuvants within human vaccines. Immunol Rev 239:178–196

Eggert AA, Schreurs MW, Boerman OC, Oyen WJ, de Boer AJ, Punt CJ, Figdor CG, Adema GJ (1999) Biodistribution and vaccine efficiency of murine dendritic cells are dependent on the route of administration. Cancer Res 59:3340–3345

Eisenbarth SC, Colegio OR, O'Connor W, Sutterwala FS, Flavell RA (2008) Crucial role for the Nalp3 inflammasome in the immunostimulatory properties of aluminium adjuvants. Nature 453:1122–1126

Eo SK, Gierynska M, Kamar AA, Rouse BT (2001) Prime-boost immunization with DNA vaccine: mucosal route of administration changes the rules. J Immunol 166:5473–5479

Epstein JE, Tewari K, Lyke KE, Sim BK, Billingsley PF, Laurens MB, Gunasekera A, Chakravarty S, James ER, Sedegah M, Richman A, Velmurugan S, Reyes S, Li M, Tucker K, Ahumada A, Ruben AJ, Li T, Stafford R, Eappen AG, Tamminga C, Bennett JW, Ockenhouse CF, Murphy JR, Komisar J, Thomas N, Loyevsky M, Birkett A, Plowe CV, Loucq C, Edelman R, Richie TL, Seder RA, Hoffman SL (2011) Live attenuated malaria vaccine designed to protect through hepatic CD8(+) T cell immunity. Science 334:475–480

Everson MP, McDuffie DS, Lemak DG, Koopman WJ, McGhee JR, Beagley KW (1996) Dendritic cells from different tissues induce production of different T cell cytokine profiles. J Leukoc Biol 59:494–498

Fadul CE, Fisher JL, Hampton TH, Lallana EC, Li Z, Gui J, Szczepiorkowski ZM, Tosteson TD, Rhodes CH, Wishart HA, Lewis LD, Ernstoff MS (2011) Immune response in patients with newly diagnosed glioblastoma multiforme treated with intranodal autologous tumor lysate-dendritic cell vaccination after radiation chemotherapy. J Immunother 34:382–389

Finerty S, Stokes CR, Gruffydd-Jones TJ, Hillman TJ, Barr FJ, Harbour DA (2001) Targeted lymph node immunization can protect cats from a mucosal challenge with feline immunodeficiency virus. Vaccine 20:49–58

Fisch A, Cadilhac P, Vidor E, Prazuck T, Dublanchet A, Lafaix C (1996) Immunogenicity and safety of a new inactivated hepatitis A vaccine: a clinical trial with comparison of administration route. Vaccine 14:1132–1136

Fong L, Brockstedt D, Benike C, Wu L, Engleman EG (2001) Dendritic cells injected via different routes induce immunity in cancer patients. J Immunol 166:4254–4259

Forster R, Braun A, Worbs T (2012) Lymph node homing of T cells and dendritic cells via afferent lymphatics. Trends Immunol 33:271–280

Frey JR, Wenk P (1957) Experimental studies on the pathogenesis of contact eczema in the guinea-pig. Int Arch Allergy Appl Immunol 11:81–100

Frosner G, Steffen R, Herzog C (2009) Virosomal hepatitis a vaccine: comparing intradermal and subcutaneous with intramuscular administration. J Travel Med 16:413–419

Goetsch L, Plotnicky-Gilquin H, Champion T, Beck A, Corvaia N, Stahl S, Bonnefoy JY, Nguyen TN, Power UF (2000) Influence of administration dose and route on the immunogenicity and protective efficacy of BBG2Na, a recombinant respiratory syncytial virus subunit vaccine candidate. Vaccine 18:2735–2742

Goetsch L, Plotnicky-Gilquin H, Aubry JP, De-Lys P, Haeuw JF, Bonnefoy JY, Nguyen NT, Corvaia N, Velin D (2001) BBG2Na an RSV subunit vaccine candidate intramuscularly injected to human confers protection against viral challenge after nasal immunization in mice. Vaccine 19:4036–4042

Gramzinski RA, Millan CL, Obaldia N, Hoffman SL, Davis HL (1998) Immune response to a hepatitis B DNA vaccine in Aotus monkeys: a comparison of vaccine formulation, route, and method of administration. Mol Med 4:109–118

Groswasser J, Kahn A, Bouche B, Hanquinet S, Perlmuter N, Hessel L (1997) Needle length and injection technique for efficient intramuscular vaccine delivery in infants and children evaluated through an ultrasonographic determination of subcutaneous and muscle layer thickness. Pediatrics 100:400–403

Hale-White W (1923) The Jenner celebrations. Br Med J 1:203–206

Hartikka J, Bozoukova V, Morrow J, Rusalov D, Shlapobersky M, Wei Q, Boutsaboualoy S, Ye M, Wloch MK, Doukas J, Sullivan S, Rolland A, Smith LR (2012) Preclinical evaluation of the immunogenicity and safety of plasmid DNA-based prophylactic vaccines for human cytomegalovirus. Hum Vaccin Immunother 8:1595–1606

Heinzerling L, Basch V, Maloy K, Johansen P, Senti G, Wuthrich B, Storni T, Kundig TM (2006) Critical role for DNA vaccination frequency in induction of antigen-specific cytotoxic responses. Vaccine 24:1389–1394

Hickling JK, Jones KR, Friede M, Zehrung D, Chen D, Kristensen D (2011) Intradermal delivery of vaccines: potential benefits and current challenges. Bull World Health Organ 89:221–226

Homey B, Steinhoff M, Ruzicka T, Leung DY (2006) Cytokines and chemokines orchestrate atopic skin inflammation. J Allergy Clin Immunol 118:178–189

Huppa JB, Gleimer M, Sumen C, Davis MM (2003) Continuous T cell receptor signaling required for synapse maintenance and full effector potential. Nat Immunol 4:749–755

Hylander T, Latif L, Petersson-Westin U, Cardell LO (2013) Intralymphatic allergen-specific immunotherapy: an effective and safe alternative treatment route for pollen-induced allergic rhinitis. J Allergy Clin Immun 131:412–420

Itano AA, Jenkins MK (2003) Antigen presentation to naive CD4 T cells in the lymph node. Nat Immunol 4:733–739

Jiang A, Bloom O, Ono S, Cui W, Unternaehrer J, Jiang S, Whitney JA, Connolly J, Banchereau J, Mellman I (2007) Disruption of E-cadherin-mediated adhesion induces a functionally distinct pathway of dendritic cell maturation. Immunity 27:610–624

Johansen P, Kündig TM (2014). Intralymphatic immunotherapy and vaccination in mice. J Vis Exp (84):e51031

Johansen P, Haffner AC, Koch F, Zepter K, Erdmann I, Maloy K, Simard JJ, Storni T, Senti G, Bot A, Wuthrich B, Kundig TM (2005) Direct intralymphatic injection of peptide vaccines enhances immunogenicity. Eur J Immunol 35:568–574

Johansen P, Storni T, Rettig L, Qiu Z, Der-Sarkissian A, Smith KA, Manolova V, Lang KS, Senti G, Mullhaupt B, Gerlach T, Speck RF, Bot A, Kundig TM (2008) Antigen kinetics determines immune reactivity. Proc Natl Acad Sci U S A 105:5189–5194

Johansen P, Mohanan D, Martinez-Gomez JM, Kundig TM, Gander B (2010) Lympho-geographical concepts in vaccine delivery. J Control Release 148:56–62

Johansen P, von Moos S, Mohanan D, Kundig TM, Senti G (2012) New routes for allergen immunotherapy. Hum Vaccin Immunother 8:1525–1533

Juillard GJ, Boyer PJ (1977) Intralymphatic immunization: current status. Eur J Cancer 13: 439–440

Juillard GJ, Boyer PJ, Snow HD (1976) Intralymphatic infusion of autochthonous tumor cells in canine lymphoma. Int J Radiat Oncol Biol Phys 1:497–503

Juillard GJ, Boyer PJ, Yamashiro CH, Snow HD, Weisenburger TH, McCarthy T, Miller RJ (1977) Regional intralymphatic infusion (ILI) of irradiated tumor cells with evidence of distant effects. Cancer 39:126–130

Juillard GJ, Boyer PJ, Yamashiro CH (1978) A phase I study of active specific intralymphatic immunotherapy (ASILI). Cancer 41:2215–2225

Juillard GJ, Boyer PJ, Niewisch H, Hom M (1979) Distribution and consequences of cell suspensions following intralymphatic infusion. Bull Cancer 66:217–228

Kantoff PW, Higano CS, Shore ND, Berger ER, Small EJ, Penson DF, Redfern CH, Ferrari AC, Dreicer R, Sims RB, Xu Y, Frohlich MW, Schellhammer PF, Ahmed T, Amin A, Arseneau J, Barth N, Bernstein G, Bracken B, Burch P, Caggiano V, Chin J, Chodak G, Chu F, Corman J, Curti B, Dawson N, Deeken JF, Dubernet T, Fishman M, Flanigan R, Gailani F, Garbo L, Gardner T, Gelmann E, George D, Godfrey T, Gomella L, Guerra M, Hall S, Hanson J, Israeli R, Jancis E, Jewett MAS, Kassabian V, Katz J, Klotz L, Koeneman K, Koh H, Kratzke R, Lance R, Lech J, Leichman L, Lemon R, Liang J, Libertino J, Lilly M, Malik I, Martin SE, McCaffrey J, McLeod D, McNeel D, Miles B, Murdock M, Nabhan C, Nemunaitis J, Notter D, Pantuck A, Perrotte P, Pessis D, Petrylak D, Polikoff J, Pommerville P, Ramanathan S, Rarick M, Richards J, Rifkin R, Rohatgi N, Rosenbluth R, Santucci R, Sayegh A, Seigne J, Shapira I, Shedhadeh N, Shepherd D, Sridhar S, Stephenson R, Teigland C, Thaker N, Vacirca J, Villa L, Vogelzang N, Wertheim M, Wolff JH, Wurzel R, Yang C, Young J, Investigators IS (2010) Sipuleucel-T immunotherapy for castration-resistant prostate cancer. N Engl J Med 363:411–422

Kawabata S, Miller CJ, Lehner T, Fujihashi K, Kubota M, McGhee JR, Imaoka K, Hioi T, Kiyono H (1998) Induction of Th2 cytokine expression for p27-specific IgA B-cell responses after targeted lymph node immunization with simian immunodeficiency virus in rhesus macaques. J Infect Dis 177:26–33

Kenney RT, Frech SA, Muenz LR, Villar CP, Glenn GM (2004) Dose sparing with intradermal injection of influenza vaccine. N Engl J Med 351:2295–2301

Kim YC, Jarrahian C, Zehrung D, Mitragotri S, Prausnitz MR (2012) Delivery systems for intradermal vaccination. Curr Top Microbiol Immunol 351:77–112

Kimber I, Cumberbatch M, Dearman RJ, Bhushan M, Griffiths CE (2000) Cytokines and chemokines in the initiation and regulation of epidermal Langerhans cell mobilization. Br J Dermatol 142:401–412

Kirkwood JM, Ariyan S, Mitchell M, Nordlund JJ, Lerner A, Papac RJ, Forget B, Knobf M (1980) Intralymphatic (Il) and local-regional (Lr) adjuvant immunotherapy with mercaptoethanol-extracted residue (Mer) of Bacille Calmette-Guerin (Bcg) in high-risk melanoma (M). Proc Am Assoc Canc Res 21:240–240

Kirkwood JM, Ariyan S, Nordlund JJ, Forget BM (1982) Granulomatous lymphangitis—a complication of intralymphatic immunotherapy with methanol extraction residue of Bcg (Mer). Cancer 50:1299–1303

Kissenpfennig A, Henri S, Dubois B, Laplace-Builhe C, Perrin P, Romani N, Tripp CH, Douillard P, Leserman L, Kaiserlian D, Saeland S, Davoust J, Malissen B (2005) Dynamics and function of Langerhans cells in vivo: dermal dendritic cells colonize lymph node areas distinct from slower migrating Langerhans cells. Immunity 22:643–654

Klavinskis LS, Bergmeier LA, Gao L, Mitchell E, Ward RG, Layton G, Brookes R, Meyers NJ, Lehner T (1996) Mucosal or targeted lymph node immunization of macaques with a particulate SIVp27 protein elicits virus-specific CTL in the genito-rectal mucosa and draining lymph nodes. J Immunol 157:2521–2527

Koblin BA, Casapia M, Morgan C, Qin L, Wang ZM, Defawe OD, Baden L, Goepfert P, Tomaras GD, Montefiori DC, McElrath MJ, Saavedra L, Lau CY, Graham BS, Network NHVT (2011) Safety and immunogenicity of an HIV adenoviral vector boost after DNA plasmid vaccine prime by route of administration: a randomized clinical trial. PLoS One 6:e24517

Koopman G, Bogers WM, van Gils M, Koornstra W, Barnett S, Morein B, Lehner T, Heeney JL (2007) Comparison of intranasal with targeted lymph node immunization using PR8-Flu ISCOM adjuvanted HIV antigens in macaques. J Med Virol 79:474–482

Koutsonanos DG, Vassilieva EV, Stavropoulou A, Zarnitsyn VG, Esser ES, Taherbhai MT, Prausnitz MR, Compans RW, Skountzou I (2012) Delivery of subunit influenza vaccine to skin with microneedles improves immunogenicity and long-lived protection. Sci Rep 2:357

Kreiter S, Selmi A, Diken M, Koslowski M, Britten CM, Huber C, Tureci O, Sahin U (2010) Intranodal vaccination with naked antigen-encoding RNA elicits potent prophylactic and therapeutic antitumoral immunity. Cancer Res 70:9031–9040

Kreiter S, Diken M, Selmi A, Diekmann J, Attig S, Husemann Y, Koslowski M, Huber C, Tureci O, Sahin U (2011) FLT3 ligand enhances the cancer therapeutic potency of naked RNA vaccines. Cancer Res 71:6132–6142

Kretz-Rommel A, Qin F, Dakappagari N, Torensma R, Faas S, Wu D, Bowdish KS (2007) In vivo targeting of antigens to human dendritic cells through DC-SIGN elicits stimulatory immune responses and inhibits tumor growth in grafted mouse models. J Immunother 30:715–726

Kündig TM, Bachmann MF, DiPaolo C, Simard JJ, Battegay M, Lother H, Gessner A, Kuhlcke K, Ohashi PS, Hengartner H et al (1995) Fibroblasts as efficient antigen-presenting cells in lymphoid organs. Science 268:1343–1347

Kündig TM, Bot A, Senti G (2012) Intralymphatic vaccination. In: Thalhamer J, Weiss R, Scheiblhofer S (eds) Gene vaccines. Springer, Wien, pp 205–222

Lacour JP, Caldani C, Thyss A, Schneider M, Ortonne JP (1992) Vitiligo-like depigmentation and morpheas after specific intralymphatic immunotherapy for malignant-melanoma. Dermatology 184:283–285

Lai MD, Yen MC, Lin CM, Tu CF, Wang CC, Lin PS, Yang HJ, Lin CC (2009) The effects of DNA formulation and administration route on cancer therapeutic efficacy with xenogenic EGFR DNA vaccine in a lung cancer animal model. Genet Vaccines Ther 7:2

Lambert PH, Laurent PE (2008) Intradermal vaccine delivery: will new delivery systems transform vaccine administration? Vaccine 26:3197–3208

Lee HK, Iwasaki A (2007) Innate control of adaptive immunity: dendritic cells and beyond. Semin Immunol 19:48–55

Lehner T, Bergmeier LA, Tao L, Panagiotidi C, Klavinskis LS, Hussain L, Ward RG, Meyers N, Adams SE, Gearing AJ et al (1994) Targeted lymph node immunization with simian immunodeficiency virus p27 antigen to elicit genital, rectal, and urinary immune responses in nonhuman primates. J Immunol 153:1858–1868

Lehner T, Wang Y, Cranage M, Bergmeier LA, Mitchell E, Tao L, Hall G, Dennis M, Cook N, Jones I, Doyle C (1998) Protective mucosal immunity elicited by targeted lymph node immunization with a subunit SIV envelope and core vaccine in macaques. Dev Biol Stand 92:225–235

Lehner T, Mitchell E, Bergmeier L, Singh M, Spallek R, Cranage M, Hall G, Dennis M, Villinger F, Wang Y (2000) The role of gammadelta T cells in generating antiviral factors and beta-chemokines in protection against mucosal simian immunodeficiency virus infection. Eur J Immunol 30:2245–2256

Lesimple T, Moisan A, Carsin A, Ollivier I, Mousseau M, Meunier B, Leberre C, Collet B, Quillien V, Drenou B, Lefeuvre-Plesse C, Chevrant-Breton J, Toujas L (2003) Injection by various routes of melanoma antigen-associated macrophages: biodistribution and clinical effects. Cancer Immunol Immunother 52:438–444

Lesimple T, Neidhard EM, Vignard V, Lefeuvre C, Adamski H, Labarriere N, Carsin A, Monnier D, Collet B, Clapisson G, Birebent B, Philip I, Toujas L, Chokri M, Quillien V (2006) Immunologic and clinical effects of injecting mature peptide-loaded dendritic cells by intralymphatic and intranodal routes in metastatic melanoma patients. Clin Cancer Res 12:7380–7388

Lesterhuis WJ, de Vries IJ, Schreibelt G, Lambeck AJ, Aarntzen EH, Jacobs JF, Scharenborg NM, van de Rakt MW, de Boer AJ, Croockewit S, van Rossum MM, Mus R, Oyen WJ, Boerman OC, Lucas S, Adema GJ, Punt CJ, Figdor CG (2011) Route of administration modulates the induction of dendritic cell vaccine-induced antigen-specific T cells in advanced melanoma patients. Clin Cancer Res 17:5725–5735

Lobaina Y, Trujillo H, Garcia D, Gambe A, Chacon Y, Blanco A, Aguilar JC (2010) The effect of the parenteral route of administration on the immune response to simultaneous nasal and parenteral immunizations using a new HBV therapeutic vaccine candidate. Viral Immunol 23:521–529

Lu X, Kiyono H, Lu D, Kawabata S, Torten J, Srinivasan S, Dailey PJ, McGhee JR, Lehner T, Miller CJ (1998) Targeted lymph-node immunization with whole inactivated simian immunodeficiency virus (SIV) or envelope and core subunit antigen vaccines does not reliably protect rhesus macaques from vaginal challenge with SIVmac251. AIDS 12:1–10

Malowany JI, McCormick S, Santosuosso M, Zhang X, Aoki N, Ngai P, Wang J, Leitch J, Bramson J, Wan Y, Xing Z (2006) Development of cell-based tuberculosis vaccines: genetically modified dendritic cell vaccine is a much more potent activator of CD4 and CD8 T cells than peptide- or protein-loaded counterparts. Mol Ther 13:766–775

Maloy KJ, Erdmann I, Basch V, Sierro S, Kramps TA, Zinkernagel RM, Oehen S, Kundig TM (2001) Intralymphatic immunization enhances DNA vaccination. Proc Natl Acad Sci U S A 98:3299–3303

Manolova V, Flace A, Bauer M, Schwarz K, Saudan P, Bachmann MF (2008) Nanoparticles target distinct dendritic cell populations according to their size. Eur J Immunol 38:1404–1413

Mark A, Carlsson RM, Granstrom M (1999) Subcutaneous versus intramuscular injection for booster DT vaccination of adolescents. Vaccine 17:2067–2072

Martinez-Gomez JM, Johansen P, Erdmann I, Senti G, Crameri R, Kundig TM (2009) Intralymphatic injections as a new administration route for allergen-specific immunotherapy. Int Arch Allergy Immunol 150:59–65

MartIn-Fontecha A, Sebastiani S, Hopken UE, Uguccioni M, Lipp M, Lanzavecchia A, Sallusto F (2003) Regulation of dendritic cell migration to the draining lymph node: impact on T lymphocyte traffic and priming. J Exp Med 198:615–621

Mohanan D, Slutter B, Henriksen-Lacey M, Jiskoot W, Bouwstra JA, Perrie Y, Kundig TM, Gander B, Johansen P (2010) Administration routes affect the quality of immune responses: a cross-sectional evaluation of particulate antigen-delivery systems. J Control Release 147:342–349

Mons B (1991) A live attenuated malaria vaccine: science or fiction? Acta Leiden 60:181–190

Morel PA, Falkner D, Plowey J, Larregina AT, Falo LD (2004) DNA immunisation: altering the cellular localisation of expressed protein and the immunisation route allows manipulation of the immune response. Vaccine 22:447–456

Moy PM, Golub SH, Calkins E, Morton DL (1985) Effects of intralymphatic immunotherapy on natural-killer activity in malignant-melanoma patients. J Surg Oncol 29:112–117

Nilsson BO, Svalander PC, Larsson A (1987) Immunization of mice and rabbits by intrasplenic deposition of nanogram quantities of protein attached to Sepharose beads or nitrocellulose paper strips. J Immunol Methods 99:67–75

Nnalue NA, Stocker BA (1989) Vaccination route, infectivity and thioglycollate broth administration: effects on live vaccine efficacy of auxotrophic derivatives of Salmonella choleraesuis. Microb Pathog 7:299–310

Ochsenbein AF, Sierro S, Odermatt B, Pericin M, Karrer U, Hermans J, Hemmi S, Hengartner H, Zinkernagel RM (2001) Roles of tumour localization, second signals and cross priming in cytotoxic T-cell induction. Nature 411:1058–1064

Okada N, Tsujino M, Hagiwara Y, Tada A, Tamura Y, Mori K, Saito T, Nakagawa S, Mayumi T, Fujita T, Yamamoto A (2001) Administration route-dependent vaccine efficiency of murine dendritic cells pulsed with antigens. Br J Cancer 84:1564–1570

Opriessnig T, Gomes-Neto JC, Hemann M, Shen HG, Beach NM, Huang YW, Halbur PG, Meng XJ (2011) An experimental live chimeric porcine circovirus 1-2a vaccine decreases porcine circovirus 2b viremia when administered intramuscularly or orally in a porcine circovirus 2b and porcine reproductive and respiratory syndrome virus dual-challenge model. Microbiol Immunol 55:863–873

Ouwehand K, Santegoets SJ, Bruynzeel DP, Scheper RJ, de Gruijl TD, Gibbs S (2008) CXCL12 is essential for migration of activated Langerhans cells from epidermis to dermis. Eur J Immunol 38:3050–3059

Pape KA, Catron DM, Itano AA, Jenkins MK (2007) The humoral immune response is initiated in lymph nodes by B cells that acquire soluble antigen directly in the follicles. Immunity 26:491–502

Pearton M, Kang SM, Song JM, Anstey AV, Ivory M, Compans RW, Birchall JC (2010a) Changes in human Langerhans cells following intradermal injection of influenza virus-like particle vaccines. PLoS One 5:e12410

Pearton M, Kang SM, Song JM, Kim YC, Quan FS, Anstey A, Ivory M, Prausnitz MR, Compans RW, Birchall JC (2010b) Influenza virus-like particles coated onto microneedles can elicit stimulatory effects on Langerhans cells in human skin. Vaccine 28:6104–6113

Pittman PR (2002) Aluminum-containing vaccine associated adverse events: role of route of administration and gender. Vaccine 20(Suppl 3):S48–S50

Pittman PR, Kim-Ahn G, Pifat DY, Coonan K, Gibbs P, Little S, Pace-Templeton JG, Myers R, Parker GW, Friedlander AM (2002) Anthrax vaccine: immunogenicity and safety of a dose-reduction, route-change comparison study in humans. Vaccine 20:1412–1420

Poland GA, Borrud A, Jacobson RM, McDermott K, Wollan PC, Brakke D, Charboneau JW (1997) Determination of deltoid fat pad thickness. Implications for needle length in adult immunization. JAMA 277:1709–1711

Randolph GJ, Angeli V, Swartz MA (2005) Dendritic-cell trafficking to lymph nodes through lymphatic vessels. Nat Rev Immunol 5:617–628

Ratzinger G, Stoitzner P, Ebner S, Lutz MB, Layton GT, Rainer C, Senior RM, Shipley JM, Fritsch P, Schuler G, Romani N (2002) Matrix metalloproteinases 9 and 2 are necessary for the migration of Langerhans cells and dermal dendritic cells from human and murine skin. J Immunol 168:4361–4371

Reynolds MG, Yorita KL, Kuehnert MJ, Davidson WB, Huhn GD, Holman RC, Damon IK (2006) Clinical manifestations of human monkeypox influenced by route of infection. J Infect Dis 194:773–780

Ribas A, Weber JS, Chmielowski B, Comin-Anduix B, Lu D, Douek M, Ragavendra N, Raman S, Seja E, Rosario D, Miles S, Diamond DC, Qiu Z, Obrocea M, Bot A (2011) Intra-lymph node prime-boost vaccination against Melan A and tyrosinase for the treatment of metastatic melanoma: results of a phase 1 clinical trial. Clin Cancer Res 17:2987–2996

Roestenberg M, McCall M, Hopman J, Wiersma J, Luty AJ, van Gemert GJ, van de Vegte-Bolmer M, van Schaijk B, Teelen K, Arens T, Spaarman L, de Mast Q, Roeffen W, Snounou G, Renia L, van der Ven A, Hermsen CC, Sauerwein R (2009) Protection against a malaria challenge by sporozoite inoculation. N Engl J Med 361:468–477

Roestenberg M, Teirlinck AC, McCall MB, Teelen K, Makamdop KN, Wiersma J, Arens T, Beckers P, van Gemert G, van de Vegte-Bolmer M, van der Ven AJ, Luty AJ, Hermsen CC, Sauerwein RW (2011) Long-term protection against malaria after experimental sporozoite inoculation: an open-label follow-up study. Lancet 377:1770–1776

Roozendaal R, Mebius RE, Kraal G (2008) The conduit system of the lymph node. Int Immunol 20:1483–1487

Roozendaal R, Mempel TR, Pitcher LA, Gonzalez SF, Verschoor A, Mebius RE, von Andrian UH, Carroll MC (2009) Conduits mediate transport of low-molecular-weight antigen to lymph node follicles. Immunity 30:264–276

Roukens AH, Gelinck LB, Visser LG (2012) Intradermal vaccination to protect against yellow fever and influenza. Curr Top Microbiol Immunol 351:159–179

Ruben FL, Froeschle JE, Meschievitz C, Chen K, George J, Reeves-Hoche MK, Pietrobon P, Bybel M, Livingood WC, Woodhouse L (2001) Choosing a route of administration for quadrivalent meningococcal polysaccharide vaccine: intramuscular versus subcutaneous. Clin Infect Dis 32:170–172

Rupec RA, Boneberger S, Ruzicka T (2010) What is really in control of skin immunity: lymphocytes, dendritic cells, or keratinocytes? Facts and controversies. Clin Dermatol 28:62–66

Samina I, Zakay-Rones Z, Weller JI, Peleg BA (1998) Host factors affecting the homologous and heterologous immune response of cattle to FMDV: genetic background, age, virus strains and route of administration. Vaccine 16:335–339

Schroder JM, Reich K, Kabashima K, Liu FT, Romani N, Metz M, Kerstan A, Lee PH, Loser K, Schon MP, Maurer M, Stoitzner P, Beissert S, Tokura Y, Gallo RL (2006) Who is really in control of skin immunity under physiological circumstances—lymphocytes, dendritic cells or keratinocytes? Exp Dermatol 15:913–929

Schwaab T, Schwarzer A, Wolf B, Crocenzi TS, Seigne JD, Crosby NA, Cole BF, Fisher JL, Uhlenhake JC, Mellinger D, Foster C, Szczepiorkowski ZM, Webber SM, Schned AR, Harris RD, Barth RJ Jr, Heaney JA, Noelle RJ, Ernstoff MS (2009) Clinical and immunologic effects of intranodal autologous tumor lysate-dendritic cell vaccine with Aldesleukin (Interleukin 2) and IFN-{alpha}2a therapy in metastatic renal cell carcinoma patients. Clin Cancer Res 15:4986–4992

Senti G, Prinz Vavricka BM, Erdmann I, Diaz MI, Markus R, McCormack SJ, Simard JJ, Wuthrich B, Crameri R, Graf N, Johansen P, Kundig TM (2008) Intralymphatic allergen administration renders specific immunotherapy faster and safer: a randomized controlled trial. Proc Natl Acad Sci U S A 105:17908–17912

Senti G, Johansen P, Kundig TM (2009) Intralymphatic immunotherapy. Curr Opin Allergy Clin Immunol 9:537–543

Senti G, Johansen P, Kundig TM (2011) Intralymphatic immunotherapy: from the rationale to human applications. Curr Top Microbiol Immunol 352:71–84

Senti G, Crameri R, Kuster D, Johansen P, Martinez-Gomez JM, Graf N, Steiner M, Hothorn LA, Gronlund H, Tivig C, Zaleska A, Soyer O, van Hage M, Akdis CA, Akdis M, Rose H, Kundig TM (2012) Intralymphatic immunotherapy for cat allergy induces tolerance after only 3 injections. J Allergy Clin Immunol 129:1290–1296

Shaw FE Jr, Guess HA, Roets JM, Mohr FE, Coleman PJ, Mandel EJ, Roehm RR Jr, Talley WS, Hadler SC (1989) Effect of anatomic injection site, age and smoking on the immune response to hepatitis B vaccination. Vaccine 7:425–430

Sigel MB, Sinha YN, VanderLaan WP (1983) Production of antibodies by inoculation into lymph nodes. Methods Enzymol 93:3–12

Sixt M, Kanazawa N, Selg M, Samson T, Roos G, Reinhardt DP, Pabst R, Lutz MB, Sorokin L (2005) The conduit system transports soluble antigens from the afferent lymph to resident dendritic cells in the T cell area of the lymph node. Immunity 22:19–29

Slutter B, Bal SM, Ding Z, Jiskoot W, Bouwstra JA (2011) Adjuvant effect of cationic liposomes and CpG depends on administration route. J Control Release 154:123–130

Spaner DE, Astsaturov I, Vogel T, Petrella T, Elias I, Burdett-Radoux S, Verma S, Iscoe N, Hamilton P, Berinstein NL (2006) Enhanced viral and tumor immunity with intranodal injection of canary pox viruses expressing the melanoma antigen, gp100. Cancer 106:890–899

Stertman L, Strindelius L, Sjoholm I (2004) Starch microparticles as an adjuvant in immunisation: effect of route of administration on the immune response in mice. Vaccine 22:2863–2872

Stutte S, Quast T, Gerbitzki N, Savinko T, Novak N, Reifenberger J, Homey B, Kolanus W, Alenius H, Forster I (2010) Requirement of CCL17 for CCR7- and CXCR4-dependent migration of cutaneous dendritic cells. Proc Natl Acad Sci U S A 107:8736–8741

Tang A, Amagai M, Granger LG, Stanley JR, Udey MC (1993) Adhesion of epidermal Langerhans cells to keratinocytes mediated by E-cadherin. Nature 361:82–85

Tenbusch M, Ignatius R, Nchinda G, Trumpfheller C, Salazar AM, Topfer K, Sauermann U, Wagner R, Hannaman D, Tenner-Racz K, Racz P, Stahl-Hennig C, Uberla K (2012) Immunogenicity of DNA vaccines encoding simian immunodeficiency virus antigen targeted to dendritic cells in rhesus macaques. PLoS One 7:e39038

Tomita GM, Nickerson SC, Owens WE, Wren B (1998) Influence of route of vaccine administration against experimental intramammary infection caused by Escherichia coli. J Dairy Sci 81:2159–2164

Trevejo JM, Marino MW, Philpott N, Josien R, Richards EC, Elkon KB, Falck-Pedersen E (2001) TNF-alpha -dependent maturation of local dendritic cells is critical for activating the adaptive immune response to virus infection. Proc Natl Acad Sci U S A 98:12162–12167

Tu CF, Lin CC, Chen MC, Ko TM, Lin CM, Wang YC, Lai MD (2007) Autologous neu DNA vaccine can be as effective as xenogenic neu DNA vaccine by altering administration route. Vaccine 25:719–728

Van Damme P, Arnou R, Kafeja F, Fiquet A, Richard P, Thomas S, Meghlaoui G, Samson SI, Ledesma E (2010) Evaluation of non-inferiority of intradermal versus adjuvanted seasonal influenza vaccine using two serological techniques: a randomised comparative study. BMC Infect Dis 10:134

Villablanca EJ, Mora JR (2008) A two-step model for Langerhans cell migration to skin-draining LN. Eur J Immunol 38:2975–2980

von Andrian UH, Mempel TR (2003) Homing and cellular traffic in lymph nodes. Nat Rev Immunol 3:867–878

Waeckerle-Men Y, Bruffaerts N, Liang Y, Jurion F, Sander P, Kundig TM, Huygen K, Johansen P (2013a) Lymph node targeting of BCG vaccines amplifies CD4 and CD8 T-cell responses and protection against Mycobacterium tuberculosis. Vaccine 31:1057–1064

Waeckerle-Men Y, Mauracher A, Hakerud M, Mohanan D, Kundig TM, Hogset A, Johansen P (2013b) Photochemical targeting of antigens to the cytosol for stimulation of MHC class-I-restricted T-cell responses. Eur J Pharm Biopharm 85:34–41

Wang B, Amerio P, Sauder DN (1999) Role of cytokines in epidermal Langerhans cell migration. J Leukoc Biol 66:33–39

Weber J, Boswell W, Smith J, Hersh E, Snively J, Diaz M, Miles S, Liu X, Obrocea M, Qiu Z, Bot A (2008) Phase 1 trial of intranodal injection of a Melan-A/MART-1 DNA plasmid vaccine in patients with stage IV melanoma. J Immunother 31:215–223

Weber JS, Vogelzang NJ, Ernstoff MS, Goodman OB, Cranmer LD, Marshall JL, Miles S, Rosario D, Diamond DC, Qiu Z, Obrocea M, Bot A (2011) A phase 1 study of a vaccine targeting preferentially expressed antigen in melanoma and prostate-specific membrane antigen in patients with advanced solid tumors. J Immunother 34:556–567

Williams TW, Yanagimoto JM, Mazumder A, Wiseman CL (1992) Interleukin-2 increases the antibody-response in patients receiving autologous intralymphatic tumor-cell vaccine immunotherapy. Mol Biother 4:66–69

Windom HH, Lockey RF (2008) An update on the safety of specific immunotherapy. Curr Opin Allergy Clin Immunol 8:571–576

Wiseman C, Rao VS, Bakke A, Kerman R, Bouzaglou A, Presant C, Mckenna RJ, Smith JD, Sakulsky SB (1986) Increased T-helper lymphocytes following active specific intralymphatic immunotherapy of cancer. J Biol Response Mod 5:490–497

Wiseman CL, Rao VS, Kennedy PS, Presant CA, Smith JD, Mckenna RJ (1989) Clinical-responses with active specific intralymphatic immunotherapy for cancer—a phase I-Ii trial. West J Med 151:283–288

Woolums AR, Berghaus RD, Berghaus LJ, Ellis RW, Pence ME, Saliki JT, Hurley KA, Galland KL, Burdett WW, Nordstrom ST, Hurley DJ (2013) Effect of calf age and administration route of initial multivalent modified-live virus vaccine on humoral and cell-mediated immune responses following subsequent administration of a booster vaccination at weaning in beef calves. Am J Vet Res 74:343–354

World Health Organization (2000) Intradermal application of rabies vaccine. Report of a WHO Consultation. Communicable Disease Surveillance and Control WHO/CDS/CSR/APH/2000.5

Xu H, Guan H, Zu G, Bullard D, Hanson J, Slater M, Elmets CA (2001) The role of ICAM-1 molecule in the migration of Langerhans cells in the skin and regional lymph node. Eur J Immunol 31:3085–3093

Yi Q, Szmania S, Freeman J, Qian J, Rosen NA, Viswamitra S, Cottler-Fox M, Barlogie B, Tricot G, van Rhee F (2010) Optimizing dendritic cell-based immunotherapy in multiple myeloma: intranodal injections of idiotype-pulsed CD40 ligand-matured vaccines led to induction of type-1 and cytotoxic T-cell immune responses in patients. Br J Haematol 150:554–564

Yoshida A, Nagata T, Uchijima M, Higashi T, Koide Y (2000) Advantage of gene gun-mediated over intramuscular inoculation of plasmid DNA vaccine in reproducible induction of specific immune responses. Vaccine 18:1725–1729

Zitvogel L (2002) Dendritic and natural killer cells cooperate in the control/switch of innate immunity. J Exp Med 195:F9–F14

Zuckerman JN (2000) The importance of injecting vaccines into muscle. Different patients need different needle sizes. BMJ 321:1237–1238

Chapter 15
Nasal Administration of Vaccines

Regina Scherließ

15.1 Anatomy and Physiology of the Nasal Cavity

Nasal administration of vaccines makes use of the inner nasal surface to induce an immune response. The human adult nose has an inner surface area of about 160 cm^2, separated into two sides being separated by the nasal septum and can be divided in different sections (Jones 2001). The anterior region of the nose is represented by the nostrils (the "outer part of the nose"); their inner surface is covered by squamous epithelium similar to normal skin (Harkema et al. 2006) and also bears nasal hairs filtering the inspired air. The nasal valve separates this part from the nasal cavity. The nasal valve is the narrowest part of the nose with a free diameter of as less as 0.25 mm (Swift 1981). The nasal turbinates can be differentiated in the lower, middle and upper turbinates (where the olfactory region is located) (Jones 2001). Small holes allow entrance to the nasal sinuses. The turbinates of both nasal passages are reunited in the nasopharynx, which is the posterior part of the nose. As the olfactory bulb in the upper turbinates has a direct connection to the central nervous system (CNS), this route is believed to mediate nose-to-brain transport. With respect to vaccines, this could be dangerous as pathogens or other immunogenic compounds such as antigens or adjuvants may cause side effects in the CNS (Neutra and Kozlowski 2006; Lycke 2012).

The nasal epithelium is a ciliated respiratory epithelium covered with a mucus film (Mygind and Dahl 1998). The ciliary beat moves the mucus lining towards the posterior part of the nose, clearing the nose from particles being deposited on the

R. Scherließ (✉)
Department of Pharmaceutics and Biopharmaceutics, Kiel University,
Grasweg 9a, 24118 Kiel, Germany
e-mail: rscherliess@pharmazie.uni-kiel.de

C. Foged et al. (eds.), *Subunit Vaccine Delivery*, Advances in Delivery Science and Technology, DOI 10.1007/978-1-4939-1417-3_15

mucus. Nasal clearance takes 12–15 min in healthy adults, but may be prolonged significantly if ciliary function is impaired (Turker et al. 2004). Nasal clearance is an important challenge for nasally administered vaccines, as the formulations only have a short period of time for interaction with the epithelium and the immunocompetent cells. Residence time, however, can be extended significantly by the use of mucoadhesives (Jabbal-Gill 2010).

It has to be mentioned that the nose is a highly patient-individual organ which may differ in size, air passages and obstruction due to an asymmetric nasal septum, nasal polyps and the nasal cycle, which is present in 80 % of the population and is characterised by alternate swelling of one side of the nose and preferential breezing through the other side (Hanif et al. 2000). This cycle changes every 3–5 h throughout the day. With a pH of about 6.4 (Washington et al. 2000) and low enzymatic activity compared to the gastro-intestinal tract, the nasal mucosa offers a good target for subunit vaccines.

The physiological function of the nose is to pre-warm and humidify the air upon nasal inhalation (Harkema et al. 2006). The nose is designed to filter, heat and humidify the inspired air before entering to the lungs (Mygind and Dahl 1998), hence, it has certain particle collection capabilities. It has been stated that in vivo, about 80 % of particles of a size up to 12.5 µm and 100 % of particles being larger than 50 µm are retained in the nose (Jones 2001), whereas about 50 % of small particles in the range of 2–4 µm will pass to the lower airways (Chap. 16). Particles are not deposited in certain regions of the nose, but may be found distributed throughout the nasal cavity as the cut-off between different regions of the nose is not as sharp as in the lungs.

As the respiratory tract and especially the nose as part of the upper respiratory tract is one of the main entry ports for pathogens, it is well equipped with immunocompetent cells which are the target for nasal vaccines. The lymphoid tissue of the nose (nose-associated lymphoid tissue, NALT) is located in the nasopharynx (Davis 2001). It is part of the common mucosal immune system and as such, may induce an immune response in distant mucosal sites such as the urogenital tract, in addition to a local immune response in the respiratory tract and systemic immune reactions (Neutra and Kozlowski 2006). Whereas the NALT is located in confined structures, dendritic cells (DCs) as antigen-presenting cells (APCs) are present throughout the complete epithelium.

Intranasal vaccination stimulates an immune response in these tissues and has been shown to be especially effective for live-attenuated vaccines such as the intranasal influenza vaccine FluMist (MedImmune) or Fluenz (AstraZeneca), respectively, and can be a promising strategy for vaccination against sexually transmitted diseases due to the induction of an immune response in the uro-genital tract upon nasal vaccination (Lycke 2012). Furthermore, nasal vaccination can be an attractive alternative for those infections where the respiratory mucosa is the normal entry route for the pathogen, such as pertussis (Haneberg and Holst 2002).

15.2 Formulation Requirements for Nasal Administration

To induce an immune response in the nose, the antigen has to be taken up by APCs in a particulate form (Pavot et al. 2012). Hence, the antigen needs to be formulated with a particulate vaccine carrier. This is easy if the antigen itself is particulate such as an attenuated pathogen, but is more complicated for subunit vaccines. A detailed description of particulate vaccine preparations, which may comprise liposomes, immunostimulating complexes, emulsion droplets and polymeric particles, can be found in Chaps. 2–10 of this book and in literature (Alpar et al. 2005; Kaye et al. 2009; De Temmerman et al. 2011; Csaba et al. 2009). The antigen carrier should also protect the antigen from instability and degradation and may further be functionalised to guide uptake, release and adjuvant effect (Nicolas et al. 2012).

For effective uptake of particulate antigen carriers via DCs, the size of the particles should be in the nanometer range above 50–100 nm (De Temmerman et al. 2011), as very small nanoparticles are not processed locally, but are directly drained to the lymph nodes similar to soluble antigens. It could also be shown that there is a certain size-dependent efficacy in terms of immune response, as smaller particles of about 200 nm provoke a higher immune response than larger particles of 700 nm (Li et al. 2011). Other studies found a particle size of 200–300 nm to be optimal for DC uptake. Nonetheless, nanoparticles may only carry a small antigen load. This can be increased largely with microparticulate carriers, which are not very well taken up by DCs, but can be taken up by M-cells in the NALT up to a size of 10 μm (Tafaghodi et al. 2004).

All primary antigen carriers are too small to be delivered directly to the nasal cavity, they would mostly get inhaled to the lung; hence, they need to be processed further to a formulation which can be deposited in the nose. Due to their large surface area, dry nanoparticles tend to form agglomerates, whereas nanoparticles in suspension would be delivered within larger spray droplets. Spray droplet size will mainly be defined by the spray nozzle of the device as well as by further parameters like viscosity and surface tension of the dispersion medium. For nasal spray products, FDA guidelines require most of the spray droplets to be larger than 10 μm to ensure nasal deposition without a major postnasal fraction which would get inhaled to the lung (FDA 2002). For nasal dry powders, Hickey et al. propose to use particles larger than 50 μm to ensure predominant nasal deposition (Garmise and Hickey 2009).

15.3 Formulation Aspects

Unlike parenteral vaccine formulations, which need to be injectable liquids, nasal formulations comprise solutions or suspensions administered by nasal sprays, pressurised nasal sprays or nebulisers as well as dry powder formulations which can be administered by passive or active dry powder dispensers. Examples of these devices are given below. Liquid preparations face a high risk of instability over storage due to high molecular mobility and with this, increased likelihood of chemical reactions and physical instability (Hasija et al. 2013). This is the reason why many liquid vaccine preparations need to be stored and transported under refrigeration. For a liquid antigen preparation, stability can be optimised via appropriate selection of pH buffering salts, and often amino acids are also used for stabilisation (Brandau et al. 2003; Taneja and Ahmad 1994). The aim is to ensure and maintain optimal antigen hydration without physical instability or chemical degradation. Hence, osmolarity may also play a critical role. For the nasal mucosa, osmolarity and pH of the administered liquid (solution or suspension) are also important parameters, as preparations, which deviate largely from physiological conditions may cause irritancy. In addition, it has been shown that buffer ionic strength may influence the uptake of nanoparticles into M-cells (Rajapaksa et al. 2010).

Stability, especially thermal stability, can be increased largely if the antigen can be stabilised and dried. Liquid preparations are often freeze-dried to enhance storage stability, with this minimising molecular mobility and hence, risk of intermolecular reactions, and need to be redispersed in buffer directly prior to administration. Here, the antigen needs to be stable in the liquid pre-step and has to be stabilised during freezing (cryoprotection) and the subsequent drying step (lyoprotection). During freezing the molecules need to be protected from harmful effects of the forming ice crystals and a shift in pH, which may easily occur due to the formation of saturated solutions differing in salt composition from the original buffer during freezing. Afterwards, the molecule needs to be stabilised from dehydration during the removal of water. This can be achieved by an exchange of water with other hydrophilic molecules which may replace it as hydrogen bond forming partner. Another possibility is the formation of a sugar glass matrix, which has been shown to stabilise vaccine preparations (Amorij et al. 2007). This principle can also be used in other drying techniques such as spray-freeze drying or spray drying (Chap. 16).

Formulations, which are administered as dry powders, face the same problem as intermediate formulations in the dried state: the antigen and its carrier system need to be stabilised during drying. Furthermore, the dried formulations need to have a particle size allowing nasal deposition, should have good dispersion characteristics, and low agglomeration and adhesive tendencies to allow powder handling, packaging and efficient release from the device. Particle size can be controlled by the parameters of the drying procedure. Here, processes resulting in a dispersible dry powder in one step (such as spray drying) are favoured to freeze drying, where the freeze dried cake might need to undergo a milling step to obtain the desired particle size (Garmise et al. 2006). Cohesive and adhesive behaviour are in parts determined by particle size: the larger the particles, the better their flow characteristics and the lower their agglomeration. Powder characteristics can further be controlled by the use of dispersion modifiers, which either cover the surface of the microparticles resulting in reduced hygroscopicity and surface energy (Minne et al. 2008; Raula et al. 2010; Weiler et al. 2010) or which form separate particles in the dry powder (Westmeier and Steckel 2008) increasing the dispersion capability.

Stability of the antigen may also be achieved by its particulate carrier. Drying nanoparticles without further bulking excipients normally leads to highly aggregated particles of undefined size, which are difficult to redisperse, and a very low yield. Therefore, further excipients can be added which serve as matrix, embedding and stabilising the individual nanoparticles and increasing redispersibility upon matrix dissolution (Trows and Scherließ 2012a). The matrix component should consist of a material which is capable of quickly releasing the particulate vaccine carrier upon dispersion in media or deposition in the nasal cavity, as only the nanoparticles will be taken up. Normally this is secured by the use of water soluble carbohydrate matrices, which dissolve in the aqueous mucus.

If a dry formulation is directly administered to the nose, particles may cause physical irritancy depending on their size and concentration. Furthermore, all water soluble components start dissolving in the nasal mucus. This may result in a concentrated solution of high osmolarity, which can also cause irritancy and increased ingression of water to dilute the substance causing a running nose.

Finally, the nose is a highly sensitive organ for olfaction. Therefore, formulation smell is an important factor for patient compliance as well as taste, because all formulations will be cleared to the pharynx and will also be tasteable on the tongue.

In order to increase nasal retention time and with this the time for interaction between the formulation and the nasal mucosa to allow uptake of particulate antigen preparations, mucoadhesive substances (hydrophilic polymers such as chitosan, hydroxypropyl methylcellulose or carbomer) can be used.

Apart from the variability in formulations, another advantage of nasal administration is the difference in microbiological requirements as a nasal formulation does not need to be sterile. Further additives may comprise preservatives, which are mandatory for liquid multidose devices to ensure microbiological stability, and adjuvanting substances. Preservatives in nasal formulations are under controversial discussion especially in chronic use, as they may have an effect on ciliary function (Marple et al. 2004; Merkus et al. 2001).

The choice of an effective and non-toxic adjuvant for nasal vaccination is a challenging task. Especially, it must be tested, whether vaccine components can enter the CNS and cause safety problems. It has been shown by molecular imaging for a botulism vaccine in monkeys that the antigen did not enter the CNS upon nasal administration (Yuki et al. 2010), but the reports on Bell's palsy following a nasal administration of an influenza vaccine adjuvanted with the heat labile *E. coli* enterotoxin (LT) are allocated to translocation of the adjuvant component to the CNS, which led to withdrawal of the vaccine from the market (Perrie et al. 2008). Adjuvants with good efficacy on mucosal routes comprise lipopolysaccharide (LPS)-protein-complexes such as the cholera enterotoxin (CT) and the heat-labile enterotoxin (LT) from *E. coli*, monophosphoryl lipid A (MPL), muramyl dipeptide, oligonucleotids (CpG), saponins like QuilA (e.g. in ISCOMs), nonionic block polymers (Poloxamers), dehydroepiandosterone (DHEA) and cytokines (interleukin (IL)-1, IL-12) (Baudner and Del Giudice 2010; Chadwick et al. 2009; Holmgren et al. 2003; Ribeiro and Schijns 2010; Lawson et al. 2011). Antigen-adjuvant interactions are discussed in more detail in Chaps. 20 and 21. It has to be noted, that enterotoxins such as the heat-labile toxin from *E. coli* or the cholera toxin, which have proven to be effective mucosal adjuvants, may not be used in the nose due to their possible neurotoxic effects (Lycke 2012) as mentioned earlier. Non-toxic mutants or derivatives such as MPL can be a feasible alternative, but the efficacy for subunit vaccine preparations may be questionable.

In the best case, a nasal vaccine only needs to be given once, possibly with one or two boost doses. Here, the preparation should be packed as a single dose, which would be protected by a closed container prior to application. Thus, no further doses need to be taken from the same container, with this minimising the risk of microbial contamination and hence avoiding the use of preservatives. This is especially true for dry powder formulations, which are less susceptible to microbial growth due to their dry nature. The application scheme can be different in the case of therapeutic vaccines, which could be needed in shorter time intervals or even every day. Here, a multidose device is of advantage. Again, a dry powder formulation would bear the least risk of microbial growth and contamination during storage and use.

15.4 Formulation Approaches for Nasal Subunit Vaccines

15.4.1 Mucoadhesive Gels

A straight forward approach is to disperse the antigen, or the particulate antigen carrier, within a liquid of increased viscosity to enhance residence time on the nasal mucosa (Chap. 11). This has been tested for several formulations. Nochi et al. prepared a nanogel of a cationic type of cholesteryl-group-bearing pullulan comprising a subunit antigen from *Clostridium botulinum*, which was administered intranasally to mice (Nochi et al. 2010). They found a strong antigen-specific humoral immune response, which could be repeated with a tetanus subunit antigen indicating that the nanogel may be used as a "universal" delivery system for vaccines. Furthermore, they found no sign for translocation of the antigen to the brain or accumulation in the olfactory bulb. For administration purposes the gel should only be slightly viscous. This drawback can be overcome if a thermosensitive gel formulation is used, with low viscosity at lower temperatures, but increased viscosity upon warming on the nasal mucosa (Wu et al. 2012; Gordon et al. 2010).

Another possibility is to use a dry powder formulation with an excipient which instantly gels upon contact with water, so that the particles are glued on the mucosa and the nasal retention time of the formulation is increased. This is the principle behind ChiSys®, a proprietary nasal delivery technology based on chitosan glutamate (Illum et al. 2002; Koch 2002; Illum 2012). Utilising this technology, a spray dried nasal vaccine comprising virus-like particles (VLPs) with a Norovirus antigen is under development and has been shown to be effective in rabbits (Vodak et al. 2012). A similar system is GelVac® (Sullivan et al. 2009; Garmise and Hickey 2009).

15.4.2 Polymeric Particles

Chitosan is a polymer which is often used for the formulation of antigens for nasal vaccination. Formulation can be as simple as utilising a spray drying step of a chitosan solution comprising the antigen, resulting in microparticles (Scherließ and Trows 2011b; Trows and Scherließ 2012b; Westmeier 2010), but normally aims at the formation of nanoparticles as antigen carriers for particulate uptake using various techniques. Ionic gelation is one method to produce nanoparticles by dissolving chitosan in acidic media to obtain positively charged groups which can be complexed by negatively charged counterparts such as tripolyphosphate (Amidi et al. 2007) or bile salts (Scherließ and Buske 2012) resulting in gel-like nanoparticles. If the antigenic protein is introduced during the process, it will also interact with the charged groups and with this is incorporated in the nanoparticle. Chitosan micro- and nanoparticles have been evaluated as nasal vaccine delivery systems comprising a range of antigens such as diphtheria toxoid (van der Lubben et al. 2003) or cross-reactive material from diphtheria toxin (McNeela et al. 2001), pertussis (Jabbal-Gill et al. 1998), influenza subunit antigen (Amidi et al. 2007) and model antigens such as ovalbumin (Boonyo et al. 2007; Gordon et al. 2008; Bal et al. 2012). As chitosan has permeation enhancing effects and mucoadhesive properties in addition to an adjuvant activity when administered together with an antigen, chitosan-containing vaccine delivery systems are very promising for nasal delivery (Illum et al. 2001). Hence, chitosan can also be used as coating for other polymeric particles (Jaganathan and Vyas 2006). Here, it has been shown that poly-lactic-co-glycolic acid (PLGA) microparticles prepared by the double emulsion method loaded with the Hepatitis B surface antigen (HBsAg) and being coated with chitosan rendering them cationic could provoke a higher local as well as systemic immune response in mice upon nasal administration than unmodified particles. This was mainly attributed to enhanced residence time on the mucosa by the authors.

Cationic particles have also been shown to be beneficial in other nasal delivery studies using the HBsAg (Debin et al. 2002). Vice versa, other polymers have also been used to coat chitosan particles. In one study, alginate-coated particles also comprising HBsAg were evaluated for their ability to induce an immune response upon nasal administration. The alginate coating primarily served as protection of the antigen on the surface of chitosan particles and was shown to result in an antigen-specific humoral immune response, whereas the uncoated particles did not (Borges et al. 2008).

Other polymeric particles under investigation for nasal subunit vaccine delivery are particles made from PLGA, which have been shown to be effective upon nasal administration, but result in a milder immune response compared to chitosan-containing particles (Slütter et al. 2010a) or more hydrophobic polycaprolactone particles (Singh et al. 2006). Further modification with poly ethylene glycol (PEG) or other surface-active substances such as poloxamer or polyvinylalcohol can enhance the immunogenicity of PLGA particles (Csaba et al. 2009).

A wide variety of other particulate systems is under investigation for mucosal vaccination and may also be feasible for nasal delivery; a summary can be found in Chadwick et al. (2010).

15.4.3 Conjugates/Complexes

With the intention to decrease the size of a vaccine carrier system while maintaining the advantages over a soluble small antigen, Slütter et al. evaluated the use of nano-conjugates to induce an immune response upon nasal administration (Slütter et al. 2010b). For this they covalently linked the model antigen ovalbumin to trimethyl-chitosan (TMC) and compared antibody response of this system to ovalbumin administered by TMC nanoparticles. Nasal uptake and immune response (sIgA and IgG) were greatly enhanced by the nanoconjugate system, which led the authors to the conclusion that efficient co-delivery of antigen and adjuvant might be more important for immunogenicity than the particulate form.

Another complex formulation is the Shigella vaccine Invaplex 50. It is based on a macromolecular complex of serotype-specific LPS and different Shigella integrins in aqueous dispersion. Preclinical studies proved safety, immunogenicity and efficacy (Tribble et al. 2010) and initial clinical trials in human volunteers using nasal administration by pipette or by the Dolphin sprayer confirmed this (Riddle et al. 2011). Interestingly, the integrin complex may also be used as an effective nasal adjuvant which can be combined with other antigens (Kaminski et al. 2006).

15.4.4 Lipid Systems

Differently charged liposomes made with addition of diverse mucoadhesive agents such as carbomer, chitosan or hyaluronic acid and with tetanus toxoid as antigen were tested for nasal vaccination in mice. It was shown that after the first immunisation only negatively charged liposomes and liposomes containing chitosan enhanced the antibody response compared to free antigen (Alpar et al. 2005). Liposomes may also be functionalised by integration of other substances to increase uptake or immune response (Kojima et al. 2008; Arigita et al. 2003; White et al. 2006).

A variation of liposomes—virosomes—has also been tested for nasal vaccination using haemagglutinin and neuraminidase as the antigenic subunits of the influenza virus. The study showed that nasal administration to ferrets resulted in almost total prevention of virus shedding in contrast to parenteral vaccination (Lambkin et al. 2004).

To increase the stability of a lipid-based delivery system, lipid microparticles can be used instead. Due to their solid nature, they are less prone to physical instabilities. In a study using HBsAg in lipid microparticles made of soy lecithin with or without addition of stearylamine (SA) it was shown that the formulation was taken up upon nasal administration and resulted in an immune response, which was more pronounced in the case of cationic particles comprising SA (Saraf et al. 2006).

15.4.5 Dry Powder Formulations

All particulate systems can also be transferred to dry powder formulations having an increased stability profile. In order to formulate a dry powder which fits the particle size requirements for nasal deposition, different approaches can be chosen. Although the primary particle size of a spray dried powder might be too small for nasal deposition, deposition characteristics strongly depend on agglomeration of the powder and dispersion efficiency of the used device. With this, it is possible to deliver a spray dried powder of low micron sized particles to the nasal cavity without the risk of a high postnasal fraction (Trows and Scherließ 2012a). If the particles are well dispersible resulting in a large postnasal fraction, a carrier-based formulation can help to deposit the majority of the formulation in the nasal cavity (Westmeier 2010; Scherließ 2011a; Buske and Scherließ 2012). Another approach is to spray-coat the suspension comprising the nanoparticulate antigen on a larger carrier particle (Scherließ et al. 2013).

Dry powder formulations of subunit vaccines, which were investigated in vivo following intranasal administration, comprise alginate microspheres with tetanus toxoid (Tafaghodi and Rastegar 2010), a spray-freeze dried powder of trehalose with Anthrax recombinant protective antigen (Wang et al. 2012) and the aforementioned Norovirus vaccine (Vodak et al., 2012).

These dry powder studies were performed in rabbits, which might be due to the sometimes reported difficulties to administer dry powder formulations to small rodents (Illum et al. 2001). Nonetheless, most of the cited studies are performed in small rodents such as mice, mostly Balb/c or C57Bl/6 strains, or rats. These are standardised laboratory animals with a well-defined immunology, which are easy and cheap to experiment with. Nonetheless, it has to be kept in mind, that a rodent's nose physiology is far from that of a human (Harkema et al. 2006). An animal model, which resembles the human respiratory tract more closely, is the guinea pig, which also shows a lot of similarities in terms of hormone and immune system (Hanif and Garcia-Contreras 2012), whereas closest would be macaques or other non-human or human primates. Another important point to keep in mind is whether the chosen animal model comprises the relevant targets in terms of response to the antigen. Finally, the way of administration is an important point. Most studies are performed by dispersing the respective vaccine system in a liquid, often PBS, and administering this to the nose by using a pipette or a syringe. The used volume is up to 100 μL, which will flood a mouse's nose and probably drains down to the lung resulting in a more systemic immune response. Furthermore, the immune response may differ depending on whether the animal was dosed under anaesthesia or when fully awake resulting in a higher systemic immune response upon anaesthesia due to increased drainage to the lung (Janakova et al. 2002). This needs to be considered if the ratio of local vs. systemic response plays a role in the study. Formulations can also be administered to animals intranasally using powder formulations. This can be easily performed in larger animals as devices can be used, which are equally feasible for human use (Huang et al. 2004), but it is also possible in small animals such as mice e.g. by utilising the PennCentury Powder Insufflator, which is normally in use for intratracheal administration (Duret et al. 2012).

15.5 Nasal Administration Devices for Humans

Nasal administration systems comprise devices for liquid formulations such as nasal drops, liquid or suspension sprays, nasal nebulisation as well as dry powder systems. The dispersion can be performed by active dispersion, where the device is used to actively disperse the formulation to droplets or particles by a defined energy created by the patient during actuation or from an external source. Here, patient inhalation or exhalation is not needed to release the powder from the device, but the patient only needs to actuate the system. Other systems rely on the patient's inhalation/exhalation airflow or other mechanisms relying on patient capabilities such as pushing a syringe plunger or a belly-like device. Performance of these systems may vary from patient to patient and hence, are regarded as less suitable. Whereas liquid systems may not be affected largely by the inhalation manoeuvre (Guo et al. 2005), dry powder systems relying on patient inhalation or exhalation can vary in dispersion efficiency, delivered dose and nasal deposition depending on the patient's inhalation speed and duration. As it is far more difficult to teach patients to breathe in a controlled manner than to teach them to hold breath during administration of a nasal product, it is believed that active dispersion during breath hold will lead to the most reproducible results.

Aqueous liquid formulations are mostly delivered to the nose by using normal spray pumps or nasal drop systems as commonly known from nasal decongestants. Whereas drops are appropriate for infants, children and adults should rather use nasal sprays as the formulation can be distributed more evenly in the nasal cavity.

Nasal sprays are mechanically actuated pump systems which actively disperse a single dose of the formulation to a spray. These systems normally comprise a liquid reservoir, a volumetric dose metering chamber, the actuation spring and a nozzle dispersing the formulation (Bommer 2006). As nasal sprays usually are multidose devices, contamination must be prevented. This is performed by the use of special seals or antimicrobial materials such as a silver-coated spring. To match the device to the respective formulation and its viscosity and concentration, adaptations of the dosing volume and nozzle characteristics can be made influencing device performance. Modern trends in nasal spray devices are side actuation and the addition of dose indicators.

Nasal sprays may also be single-dose or bi-dose systems (Fig. 15.1a), which are more appropriate for discontinuous use such as in migraine or vaccination (Marx et al. 2010). A very simple, but effective device is the VaxINator (Teleflex Medical, LMA Atomisation, Co. Westmeath, Ireland; Fig. 15.1b). It consists of a cone-shaped nozzle as nasal adapter being connected to a syringe (Wolfe and Denton 2012). When pushing the plunger, the liquid from the syringe is dispersed into fine droplets. Nasal sprays can also be delivered by using pressurised metered dose inhalers (Fig. 15.1c) similar to oral inhalation products (Righton and Harrison 2013). The use of these multidose systems overcome some drawbacks of nasal spray pumps such as dripping of the formulation to the pharynx and with this an unpleasant aftertaste as the dispersed volume is usually smaller and droplets are

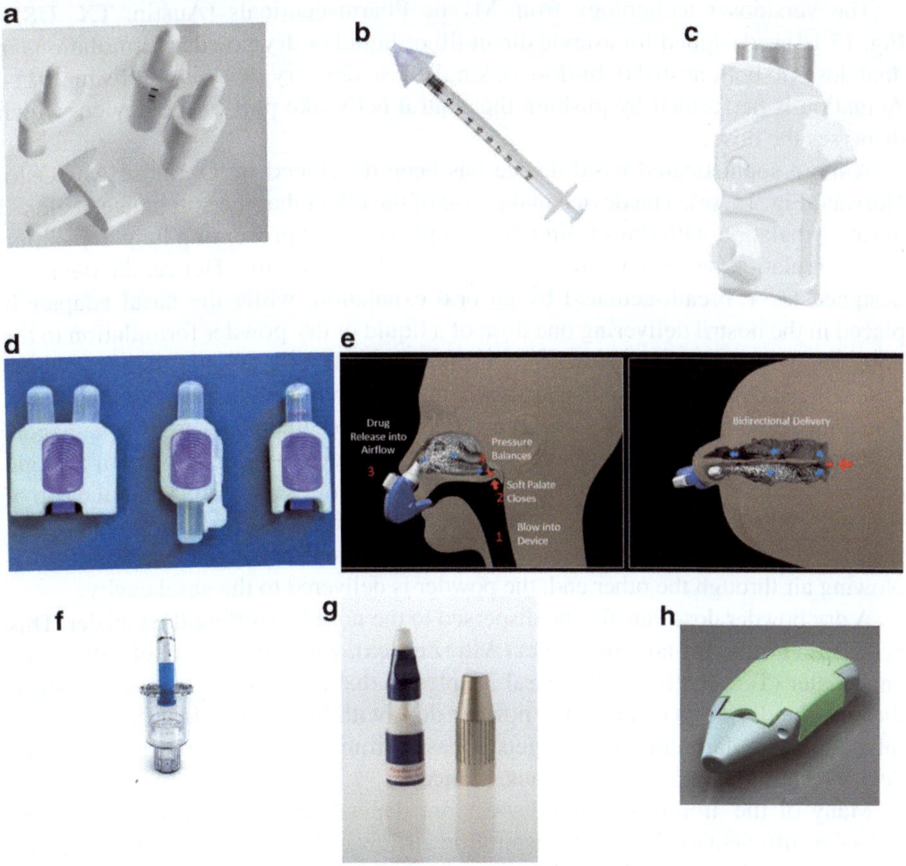

Fig. 15.1 Nasal devices for the delivery of liquid and dry powder formulations. (**a**) Passive unit-dose and bi-dose nasal delivery devices (picture: courtesy of Aptar Pharma), (**b**) VaxINator (Courtesy of Teleflex Medical Europe Limited. Unauthorized use prohibited), (**C**) nasal pMDI (Copyright© 2013, 3 M. All rights reserved), (**d**) three VersiDoser® Nasal Delivery Devices (Copyright© 2010–2013 Mystic Pharmaceuticals, Inc.), (**e**) Breath Powered™ Bi-Directional™ delivery (courtesy of Optinose), (**f**) Powder UDS device (picture: courtesy of Aptar Pharma), (**g**) PowderJet (courtesy of RPC Formatec), (**h**) Naltos (© Alchemy Pharmatech Limited, UK, www.alchemypharmatech.com)

finer than of a nasal pump spray. In this way, the dose gets distributed as a fine spray in the nasal cavity.

Kurve Technology (Lynnwood, WA, USA) has developed a special nasal electronic atomizer, the ViaNase, which is reported to be capable of delivering a wide range of formulations such as solutions and suspensions, vaccines, peptides, proteins and monoclonal antibody formulations to the nose with minimal lung or oral deposition, making use of their "Controlled Particle Dispersion" (CPD) technology which relies on a vertical flow for dispersion (Giroux et al. 2005).

The Versidoser technology from Mystic Pharmaceuticals (Austin, TX, USA; Fig. 15.1d) is designed for aseptic direct fill of liquid or dry powder formulations in dual dose (to both nostrils), bi-dose or single dose delivery devices (Sullivan 2011). Actuation is performed by pushing the central belly-like part of the device, which disperses the dose.

A more sophisticated nasal device has been developed by Optinose AS (Oslo, Norway; Fig. 15.1e). The device makes use of the effect that the soft palate is closed during exhalation with this closing the nasopharynx and preventing lung deposition of a formulation being delivered to the nose at the same time. Hence, the device is designed to be breath-actuated by an oral exhalation, while the nasal adapter is placed in the nostril delivering one dose of a liquid or dry powder formulation to the nose (Djupesland 2005). This principle is called "bi-directional" delivery as the dose will get deposited in both nostrils which are interconnected by the nasopharynx. The same principle is used by the DirectHaler Nasal (DirectHaler AS, Copenhagen, Denmark; now Trimel Pharmaceuticals) which is made for nasal delivery of a single dry powder dose (Keldmann 2005, 2006). The powder dose is protected in the cap of the straw-like disposable device and gets released when the foil is removed. Afterwards the cap is removed, one end of the straw is placed in the nostril and when blowing air through the other end, the powder is delivered to the nasal cavity.

A dry powder dose can also be dispersed to the nose by sniffing the powder. This is the case for the Rhinocort product (AstraZeneca), which makes use of a dry powder inhaler (Turbohaler) with a nasal adapter so that the patient can nasally inhale through the device to disperse the powder dose with his air flow instead of an oral inhalation. Similarly, the powder gets released from the Aptar bi-dose dry powder device (Aptar Pharma, Louveciennes, France).

Many of the aforementioned devices are dependent on patient capabilities to induce a sufficient expiration or inspiration airflow or to squeeze a belly in the correct manner. User-dependency, however, is always associated with higher performance variability. This can be overcome by the use of user-independent, active devices such as the Powder UDS system (Aptar Pharma, Louveciennes, France, Fig. 15.1f) (Marx et al. 2011). Here, the single powder dose is placed in a central container. Upon actuation of the device, a defined volume of air is compressed and at the point of release entrains the powder delivering it to the nose. A similar principle is used by the PowderJet (RPC Formatec, Mellrichstadt, Germany; Fig. 15.1g), which is an active multidose dry powder device (Scherließ and Trows 2011a). In this device, a volumetric dosing cavity is pushed into the central powder reservoir upon actuation, while a defined volume of air is compressed in the lower part of the device, which results in dispersion of the dose at the point of release. Also for dry powder dispersion, a propellant may be used as it is the case for the Naltos device (alchemy pharmatech, Daresbury, UK; Fig. 15.1h). Here, the powder dose and the propellant (HFA 134a) are stored in two separate chambers. Upon actuation of the device, these chambers are connected and the propellant disperses the powder (Harrison 2013).

15.6 Physicochemical Characterisation of Nasal Vaccine Formulations

Apart from the immunological effect of a nasal subunit vaccine, a range of physico-chemical parameters are of importance for characterising a nasal subunit vaccine formulation (Fig. 15.2). Firstly, an essential parameter to characterise is the size of the nanoparticulate vaccine carrier, which should fit the requirements for uptake to immunocompetent APCs. This can be done by means of dynamic light scattering (DLS); visual confirmation might be performed by scanning electron microscopy (SEM) or transmission electron microscopy (TEM). Here, attention should be given to the sample preparation not to alter the particle characteristics largely (e.g. by the drying step or evacuation). The size of the particulate vaccine carrier should not only be looked at directly after preparation, but also after drying and redispersion and during storage to ensure stability. As the formulation will be in contact with the nasal fluid and might redisperse therein in the case of dry powder formulations, effect of changes in the dispersion media on vaccine carrier size should also be taken into account. Aggregation upon contact with the nasal fluid can lead to a significant increase in particle size and the vaccine carrier might not be taken up any more.

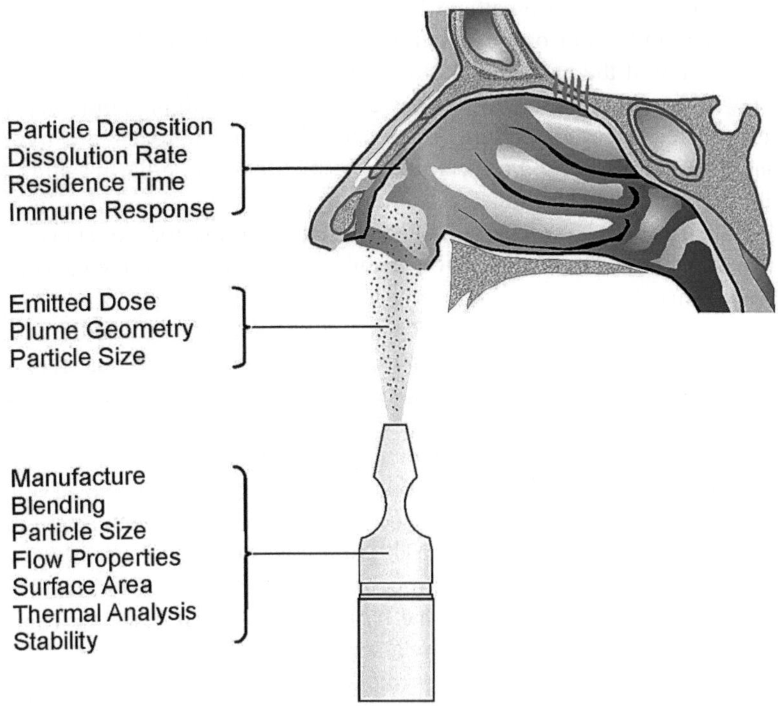

Fig. 15.2 Important parameters for characterising a nasal subunit vaccine formulation, adapted from Garmise and Hickey (2009)

Surface characteristics such as charge might be important parameters depending on the formulation and can be looked at by measuring for example the zeta-potential of the particles, but these may also change largely with a change in environment.

Similar to oral inhalation products, nasal product performance is strongly dependent on the dispersion device being used with the formulation. Hence, formulation development and characterisation should be performed in combination with the respective administration system. Particle or droplet size upon dispersion is an important determinant for nasal deposition and location of deposition. Particle size distribution is usually analysed by means of laser diffraction using a nasal spray actuator, where the device can be fitted to the system and can be actuated making use of the dispersion mechanism (Kippax and Fracassi 2003). Dispersion differs from device to device and this can alter particle size of the dispersed formulation. Actuation can be standardised either by constant actuation force or by constant actuation distance. Nasal deposition can be estimated from these particle size data taking into account a cutoff size for nasal deposition.

More accurate nasal deposition studies can be performed utilising a nasal cast model representing a human nose of the respective patient group (adults, children) (Scherließ 2011b; Hughes et al. 2008; Newman et al. 2004; Schönbrodt et al. 2010; Kelly et al. 2004). As nasal administration can be performed without or during inspiration, different modes can be tested with these casts either simulating inspiration airflow (normal nasal breathing for an adult is estimated to be 15 L/min) or simulating breath hold or oral exhalation leading to closure of the soft palate and with this closure of the nasopharynx towards the lower airways resulting in a lack of airflow in the nasal passage. This may alter deposition significantly and should be taken into account when defining user instructions. If a more detailed look at the post-nasal fraction is needed to determine the fraction of fine particles in the formulation, which might get inhaled to the lung, it is recommended to use of an impactor such as the Next Generation Pharmaceutical Impactor (NGI) in combination with a cast model or an expansion chamber for collection of the nasal fraction (Scherließ 2010).

As a vaccine preparation for nasal delivery should interact with APCs within the nose, nasal retention time can be an important parameter to test (Soane et al. 2001). Nonetheless, this is difficult to test in vitro. Here, viscosity may serve as a surrogate and can be tested using artificial nasal fluid and performing viscosity measurements (Callens et al. 2003). Another method, more resembling natural conditions, is to use freshly excised pig nasal mucosa and measure strip-off forces of a defined specimen covered with the formulation or the respective mucoadhesive agent used in the formulation from the mucosa (Zscherpe 2009) or to look at adhesion of microparticles of the respective polymer on ex vivo mucosa (Lehr et al. 1992). In vivo it can be measured by radioactive labelling of the formulation of interest, so that deposition and clearance can be followed by gamma-scintigraphy (Tafaghodi et al. 2004).

Usually, a particulate nasal vaccine formulation should not release the antigen into the nasal fluid upon nasal deposition, but should first release the antigen after uptake, because soluble antigens on mucosal surfaces or flooding with large amounts of antigen might result in tolerance (Neutra and Kozlowski 2006). Hence, formulations should be designed for delayed release or release following a certain trigger such as change of temperature, pH or enzymatic conditions. Once taken up, the particulate vaccine carrier should easily release the antigen to allow rapid recognition and processing. Testing release triggers and release rate can hence be an interesting parameter to analyse during formulation characterisation.

Finally, registration authorities (EMA, FDA) like to see data for spray pattern and plume geometry for nasal spray formulations (Suman 2009), but these are measures for batch-to-batch quality rather than being linked to any performance characteristic in terms of action.

15.7 Outlook

Until now, there are no subunit vaccines for nasal administration on the market. This may be due to the low immunogenicity compared to attenuated pathogens, which goes along with increased safety, but complicates induction of a sound immune response. This must be mediated with adjuvants, as it is done for parenteral administration of subunit vaccines. Here, the difficulty is to find a powerful, non-toxic and effective nasal mucosal adjuvant (De Magistris 2006). Nonetheless, a range of products is under investigation and extensive research is done in the area of nasal subunit vaccines targeting different antigens and making use of various formulation approaches, so that nasal administration of subunit vaccines will be a viable option in the near future.

References

Alpar HO, Somavarapu S, Atuah KN, Bramwell VW (2005) Biodegradable mucoadhesive particulates for nasal and pulmonary antigen and DNA delivery. Adv Drug Deliv Rev 57:411–430

Amidi M, Romeijn SG, Verhoef JC, Junginger HE, Bungener L, Huckriede A, Crommelin D, Jiskoot W (2007) N-Trimethyl chitosan (TMC) nanoparticles loaded with influenza subunit antigen for intranasal vaccination: biological properties and immunogenicity in a mouse model. Vaccine 25:144–153

Amorij J-P, Meulenaar J, Hinrichs WLJ, Stegmann T, Huckriede A, Coenen F, Frijlink HW (2007) Rational design of an influenza subunit vaccine powder with sugar glass technology: preventing conformational changes of haemagglutinin during freezing and freeze-drying. Vaccine 25:6447–6457

Arigita C, Bevaart L, Everse LA, Koning GA, Hennink WE, Crommelin D, van de Winkel JG, van Vugt MJ, Kersten GF, Jiskoot W (2003) Liposomal meningococcal B vaccination: role of dendritic cell targeting in the development of a protective immune response. Infect Immun 71(9):5210–5218

Bal SM, Slütter B, Verheul R, Bouwstra JA, Jiskoot W (2012) Adjuvanted, antigen loaded N-trimethyl chitosan nanoparticles for nasal and intradermal vaccination: adjuvant- and site-dependent immunogenicity in mice. Eur J Pharm Sci 45:475–481

Baudner BC, Del Giudice G (2010) Determining the activity of mucosal adjuvants. In: Davies G (ed) Vaccine adjuvants. Springer Science+Business Media, LCC, New York, pp 261–285

Bommer R (2006) Drug delivery: nasal route. In: Swabrick J (ed) Encyclopedia of pharmaceutical technology, vol 3. Informa Healthcare, New York, pp 1201–1208
Boonyo W, Junginger HE, Waranuch N, Polnok A, Pitaksuteepong T (2007) Chitosan and trimethyl chitosan chloride (TMC) as adjuvants for inducing immune responses to ovalbumin in mice following nasal administration. J Control Release 121:168–175
Borges O, Cordeiro-da-Silva A, Tavares J, Santarem N, de Sousa A, Borchard G, Junginger HE (2008) Immune response by nasal delivery of hepatitis B surface antigen and codelivery of a CpG ODN in alginate coated chitosan nanoparticles. Eur J Pharm Biopharm 69(2):405–416
Brandau DT, Jones LS, Wiethoff CM, Rexroad J, Middaugh CR (2003) Thermal stability of vaccines. J Pharm Sci 92(2):218–231
Buske S, Scherließ R (2012) Dispersion behaviour of dry powder nanoparticle-in-microparticle formulations for nasal vaccine delivery. Paper presented at the 8th world meeting on pharmaceutics, biopharmaceutics and pharmaceutical technology, Istanbul
Callens C, Ceulemans J, Ludwig A, Foreman P, Remon JP (2003) Rheological study on mucoadhesivity of some nasal powder formulations. Eur J Pharm Biopharm 55:323–328
Chadwick S, Kriegel C, Amiji MM (2009) Delivery strategies to enhance mucosal vaccination. Expert Opin Biol Ther 9(4):427–440
Chadwick S, Kriegel C, Amiji MM (2010) Nanotechnology solutions for mucosal immunization. Adv Drug Deliv Rev 62:394–407
Csaba N, Garcia-Fuentes M, Alonso MJ (2009) Nanoparticles for nasal vaccination. Adv Drug Deliv Rev 61:140–157
Davis SS (2001) Nasal vaccines. Adv Drug Deliv Rev 51:21–42
De Magistris MT (2006) Mucosal delivery of vaccine antigens and its advantages in pediatrics. Adv Drug Deliv Rev 58:52–67
De Temmerman M-L, Rejman J, Demeester J, Irvine DJ, Gander B, De Smedt SC (2011) Particulate vaccines: on the quest for optimal delivery and immune response. Drug Discov Today 16(13/14):569–582
Debin A, Kravtzoff R, Santiago JV, Cazales L, Sperandio S, Melber K, Janowicz Z, Betbeder D, Moynier M (2002) Intranasal immunization with recombinant antigens associated with new cationic particles induces strong mucosal as well as systemic antibody and CTL responses. Vaccine 20(21–22):2752–2763
Djupesland PG (2005) Breath-actuated bi-directional delivery sets the nasal market on a new course. OnDrugDelivery (3rd issue):20–23
Duret C, Wauthoz N, Merlos R, Goole J, Maris C, Roland I, Sebti T, Vanderbist F, Amighi K (2012) In vitro and in vivo evaluation of a dry powder endotracheal insufflator device for use in dose-dependent preclinical studies in mice. Eur J Pharm Biopharm 81:627–634
FDA (2002) Guidance for industry: nasal spray and inhalation solution, suspension, and spray drug products—chemistry, maufacturing, and controls documentation. FDA, Rockville
Garmise RJ, Hickey A (2009) Dry powder nasal vaccines as an alternative to needle-based delivery. Crit Rev Ther Drug Carrier Syst 26(1):1–27
Garmise RJ, Mar K, Crowder TM, Hwang CR, Ferriter M, Huang J, Mikszta JA, Sullivan VJ, Hickey AJ (2006) Formulation of a dry powder influenza vaccine for nasal delivery. AAPS PharmSciTech 7(1):E1–E7
Giroux M, Hwang P, Prasad A (2005) Controlled particle dispersion™: applying vortical flow to optimize nasal drug deposition. Drug Delivery Technol 5(3):44–49
Gordon S, Saupe A, McBurney WT, Rades T, Hook S (2008) Comparison of chitosan nanoparticles and chitosan hydrogels for vaccine delivery. J Pharm Pharmacol 60:1591–1600
Gordon S, Teichmann E, Young K, Finnie K, Rades T, Hook S (2010) In vitro and in vivo investigation of thermosensitive chitosan hydrogels containing silica nanoparticles for vaccine delivery. Eur J Pharm Sci 41:360–368
Guo Y, Laube B, Dalby R (2005) The effect of formulation variables and breathing patterns on the site of nasal deposition in an anatomically correct model. Pharm Res 22:1871–1878
Haneberg B, Holst J (2002) Can nonliving nasal vaccines be made to work? Expert Rev Vaccines 1(2):227–232

Hanif J, Jawad SSM, Eccles R (2000) The nasal cycle in health and disease. Clin Otolaryngol 25:461–467

Hanif SNM, Garcia-Contreras L (2012) Pharmaceutical aerosols for the treatment and prevention of tuberculosis. Front Cell Infect Microbiol 2(Article 118):1–11

Harkema JR, Carey SA, Wagner JG (2006) The nose revisited: a brief review of the comparative structure, function, and toxicologic pathology of the nasal epithelium. Toxicol Pathol 24:252–269

Harrison I (2013) Nasal dry powder delivery—device development & optimization. Paper presented at the Nasal and Buccal Drug Delivery Conference, London

Hasija M, Li L, Rahman N, Ausar SF (2013) Forced degradation studies: an essential tool for the formulation development of vaccines. Vaccine 3:11–33

Holmgren J, Czerkinsky C, Eriksson K, Mharandi A (2003) Mucosal immunisation and adjuvants: a brief overview of recent advances and challenges. Vaccine 21:S2/89–S82/95

Huang J, Garmise RJ, Crowder TM, Mar K, Hwang CR, Hickey AJ, Miksztaa JA, Sullivan VJ (2004) A novel dry powder influenza vaccine and intranasal delivery technology: induction of systemic and mucosal immune responses in rats. Vaccine 23:794–801

Hughes R, Watterson J, Dickens C, Ward D, Banaszek A (2008) Development of a nasal cast model to test medicinal nasal devices. Proc Inst Mech Eng H 222(7):1013–1022

Illum L (2012) Nasal drug delivery—recent developments and future prospects. J Control Release 161:254–263

Illum L, Iqbal K, Dodane V (2002) Chitosan technology to enhance the effectiveness of nasal drug delivery. Drug Development Delivery 2(2)

Illum L, Jabbal-Gill I, Hinchcliffe M, Fisher AN, Davis SS (2001) Chitosan as a novel nasal delivery system for vaccines. Adv Drug Deliv Rev 51:81–96

Jabbal-Gill I (2010) Nasal vaccine innovation. J Drug Target 18(10):771–786

Jabbal-Gill I, Fisher AN, Rappuolit R, Davis SS, Illum L (1998) Stimulation of mucosal and systemic antibody responses against Bordetella pertussis filamentous haemagglutinin and recombinant pertussis toxin after nasal administration with chitosan in mice. Vaccine 16(20): 2039–2046

Jaganathan KS, Vyas SP (2006) Strong systemic and mucosal immune responses to surface-modified PLGA microspheres containing recombinant hepatitis B antigen administered intranasally. Vaccine 24(19):4201–4211

Janakova L, Bakke H, Haugen IL, Berstad AKH, Høiby EA, Aaberge IS, Haneberg B (2002) Influence of intravenous anesthesia on mucosal and systemic antibody responses to nasal vaccines. Infect Immun 70(10):5479–5484

Jones N (2001) The nose and paranasal sinuses physiology and anatomy. Adv Drug Deliv Rev 51:5–19

Kaminski RF, Turbyfill KR, Oaks EV (2006) Mucosal adjuvant properties of the Shigella invasin complex. Infect Immun 74:2856–2866

Kaye RS, Purewal TS, Alpar OH (2009) Development and testing of particulate formulations for the nasal delivery of antibodies. J Control Release 135:127–135

Keldmann T (2005) Advanced simplification of nasal delivery technology: anatomy+innovative device=added value opportunity. OnDrugDelivery (3rd issue):4–7

Keldmann T (2006) Simplicity wins—from product conceptualisation to drug delivered. Drug Delivery Report (Spring/Summer):49–52

Kelly JT, Asgharian B, Kimbell JS, Wong BA (2004) Particle deposition in human nasal airway replicas manufactured by different methods. part I: inertial regime particles. Aerosol Sci Tech 38(11):1063–1071

Kippax P, Fracassi J (2003) Particle size characterisation in nasal sprays and aerosols. LabPlus Int (Feb/March)

Koch M (2002) Drug delivery via the nose. Drug Delivery & Formulation:90–94

Kojima N, Biao L, Nakayama T, Ishii M, Ikehara Y, Tsujimura K (2008) Oligomannose-coated liposomes as a therapeutic antigen-delivery and an adjuvant vehicle for induction of in vivo tumor immunity. J Control Release 129:26–32

Lambkin R, Oxford JS, Bossuyt S, Mann A, Metcalfe IC, Herzog C, Viret J-F, Glück R (2004) Strong local and systemic protective immunity induced in the ferret model by an intranasal virosome-formulated influenza subunit vaccine. Vaccine 22:4390–4396

Lawson LB, Norton EB, Clements JD (2011) Defending the mucosa: adjuvant and carrier formulations for mucosal immunity. Curr Opin Immunol 23(3):414–420

Lehr C-M, Bouwstra JA, Schacht EH, Junginger HE (1992) In vitro evaluation of mucoadhesive properties of chitosan and some other natural polymers. Int J Pharm 78:43–48

Li X, Sloat BR, Yanasarn N, Cui Z (2011) Relationship between the size of nanoparticles and their adjuvant activity: data from a study with an improved experimental design. Eur J Pharm Biopharm 78:107–116

Lycke N (2012) Recent progress in mucosal vaccine development: potential and limitations. Nat Rev Immunol 12:592–605

Marple B, Roland P, Benninger M (2004) Safety review of benzalkonium chloride used as a preservative in intranasal solutions: an overview of conflicting data and opinions. Otolaryngol Head Neck Surg 130:131–141

Marx D, Leitz M, Fagot C (2011) Do we need new devices for intranasal vaccination? Drug Dev Delivery 11(3):54–59

Marx D, Leitz M, Pfitzer K (2010) Intranasal vaccination. Inhalation 4(3):8–11

McNeela EA, O'Connor D, Jabbal-Gill I, Illum L, Davis SS, Pizza M, Peppoloni S, Rappuoli R, Mills KHG (2001) A mucosal vaccine against diphtheria: formulation of cross reacting material (CRM197) of diphtheria toxin with chitosan enhances local and systemic antibody and Th2 responses following nasal delivery. Vaccine 19:1188–1198

Merkus P, Romeijn SG, Verhoef JC, Merkus FWHM, Schouwenburg PF (2001) Classification of cilio-inhibiting effects of nasal drugs. Laryngoscope 111:595–602

Minne A, Boireau H, Horta MJ, Vanbever R (2008) Optimization of the aerosolization properties of an inhalation dry powder based on selection of excipients. Eur J Pharm Biopharm 70(3):839–844

Mygind N, Dahl R (1998) Anatomy, physiology and function of the nasal cavities in health and disease. Adv Drug Deliv Rev 29:3–12

Neutra MR, Kozlowski PA (2006) Mucosal vaccines: the promise and the challenge. Nat Rev Immunol 6:148–158

Newman SP, Pitcairn GP, Dalby RN (2004) Drug delivery to the nasal cavity: in vitro and in vivo assessment. Crit Rev Ther Drug Carrier Syst 21:21–66

Nicolas J, Mura S, Brambilla D, Mackiewicz N, Couvreur P (2012) Design, functionalization strategies and biomedical applications of targeted biodegradable/biocompatible polymer-based nanocarriers for drug delivery. Chem Soc Rev 42:1147–1235. doi:10.1039/c2cs35265f

Nochi T, Yuki Y, Takahashi H, Sawada S-I, Mejima M, Kohda T, Harada N, Kong IG, Sato A, Kataoka N, Tokuhara D, Kurokawa S, Takahashi Y, Tsukada H, Kozaki S, Akiyoshi K, Kiyono H (2010) Nanogel antigenic protein-delivery system for adjuvant-free intranasal vaccines. Nat Mater 9:572–578

Pavot V, Rochereau N, Genin C, Verrier B, Paul S (2012) New insights in mucosal vaccine development. Vaccine 30:142–154

Perrie Y, Mohammed AR, Kirby DJ, McNeil SE, Bramwell VW (2008) Vaccine adjuvant systems: enhancing the efficacy of sub-unit protein antigens. Int J Pharm 364:272–280

Rajapaksa TE, Bennett KM, Hamer M, Lytle C, Rodgers VGJ, Lo DD (2010) Intranasal M cell uptake of nanoparticles is independently influenced by targeting ligands and buffer ionic strength. J Biol Chem 285(31):23739–23746

Raula J, Thielmann F, Naderi M, Lehto V-P, Kauppinen EI (2010) Investigations on particle surface characteristics vs. dispersion behaviour of l-leucine coated carrier-free inhalable powders. Int J Pharm 385:79–85

Riddle, M.S., Robert W. Kaminski, Carlos Williams, Chad Porter, Shahida Baqar, Alexis Kordis, Theron Gilliland, Joyce Lapa, Melissa Coughlin, Chris Soltis, Erica Jones, Jackie Saunders, Paul B. Keiser, Ryan T. Ranallo, Robert Gormley, Michael Nelson, K. Ross Turbyfill, David

Tribble and Edwin V. Oaks (2011) Safety and immunogenicity of an intranasal Shigella flexneri 2a Invaplex 50 vaccine. Vaccine 29:7009–7019
Ribeiro CMS, Schijns VEJC (2010) Immunology of vaccine adjuvants. In: Davies G (ed) Vaccine adjuvants. Methods in molecular biology, vol 626. Springer Science+Business Media LCC, New York, pp 1–14
Righton L, Harrison L (2013) Moving toward patient-preferred nasal drug delivery systems. OnDrugDelivery (April):4–7
Saraf S, Mishra D, Asthana A, Jain R, Singh S, Jain NK (2006) Lipid microparticles for mucosal immunization against hepatitis B. Vaccine 24(1):45–56
Scherließ R (2010) Comparison of in vitro methods to determine nasal versus lung deposition of a protein formulation. Paper presented at the DDL 21, Edinburgh
Scherließ R (2011a) In vitro evaluation of dry powder nasal deposition from a single use nasal device. Paper presented at the DDL 22, Edinburgh
Scherließ R (2011b) In vitro particle deposition in the nasal cavity. Inhalation 5(3):14–18
Scherließ R, Buske S (2012) Dry powder nanoparticulate formulations for mucosal vaccination. In: Tiddy G, Tan RB (eds) Nanoformulation. RSC, Singapore, pp 104–112
Scherließ R, Trows S (2011a) Characterization of nasal deposition and post-nasal fraction of a model vaccine protein formulation. In: Dalby RN (ed) Respiratory drug delivery Europe 2011, vol 2. Davies Healthcare International, Paris, pp 321–325
Scherließ R, Trows S (2011b) Novel formulation concept for particulate uptake of vaccines via the nasal associated lymphoid tissue. Procedia Vaccinol 4:113–119
Scherließ R, Trows S, Buske S (2013) Overcoming challenges in development of nasal vaccines through intelligent particle engineering. Inhalation 6:8–14
Schönbrodt T, Egen M, Heyder K, Kohler D, Kranz Y, Müller C, Schiewe J (2010) Method development for deposition studies in a nasal cast. In: Dalby RN (ed) Respiratory drug delivery 2010, vol 2. Davies Healthcare International, Orlando, pp 445–449
Singh J, Pandit S, Bramwell VW, Alpar HO (2006) Diphtheria toxoid loaded poly-(E-caprolactone) nanoparticles as mucosal vaccine delivery systems. Methods 38:96–105
Slütter B, Bal S, Keijzer C, Mallants R, Hagenaars N, Que I, Kaijzel E, van Eden W, Augustijns P, Löwik C, Bouwstra J, Broere F, Jiskoot W (2010a) Nasal vaccination with N-trimethyl chitosan and PLGA based nanoparticles: nanoparticle characteristics determine quality and strength of the antibody response in mice against the encapsulated antigen. Vaccine 28:6282–6291
Slütter B, Bal SM, Que I, Kaijzel E, Löwik C, Bouwstra J, Jiskoot W (2010b) Antigen-adjuvant nanoconjugates for nasal vaccination: an improvement over the use of nanoparticles? Mol Pharm 7(6):2207–2215
Soane RJ, Hinchcliffe M, Davis SS, Illum L (2001) Clearance characteristics of chitosan based formulations in the sheep nasal cavity. Int J Pharm 217(1–2):183–191
Sullivan T (2011) Novel advances in rapid manufacturing, development, and delivery of intranasally administered drugs and biologics. OnDrugDelivery April:31–34
Sullivan T, Taraporewala I, Zielinski W (2009) Innovations in intranasal vaccine delivery technology. BioPharm International Supplements (October):1–5
Suman J (2009) In vitro nasal spray characterisation. Inhalation (June) 1–4
Swift DL (1981) Aerosol deposition and clearance in the human upper airways. Ann Biomed Eng 9:593–604
Tafaghodi M, Rastegar S (2010) Preparations and in vitro study of dry powder microspheres for nasal immunization. J Drug Target 18(3):235–242
Tafaghodi M, Tabassi SAS, Jaafaria M-R, Zakavid SR, Momen-Nejad M (2004) Evaluation of the clearance characteristics of various microspheres in the human nose by gamma-scintigraphy. Int J Pharm 280(1–2):125–135
Taneja S, Ahmad F (1994) Increased thermal stability of proteins in the presence of amino acids. Biochem J 303:147–153
Tribble D, Kaminski R, Cantrell J, Nelson M, Porter C, Baqar S, Williams C, Arora R, Saunders J, Ananthakrishnan M, Sanders J, Zaucha G, Turbyfill R, Oaks E (2010) Safety and immunoge-

nicity of a Shigella flexneri 2a Invaplex 50 intranasal vaccine in adult volunteers. Vaccine 28:6076–6085

Trows S, Scherließ R (2012a) Preparation and characterization of dry powder agarose nano-in-microparticles for nasal vaccination. In: Dalby RN (ed) Respiratory drug delivery 2012, vol 2. Davies Healthcare International, Phoenix, pp. 491–496

Trows S, Scherließ R (2012b) Stability of antigen loaded dry powder microparticles for nasal vaccination. Paper presented at the 8th world meeting on pharmaceutics, biopharmaceutics and pharmaceutical technology, Istanbul

Turker S, Onur E, Ozer Y (2004) Nasal route and drug delivery systems. Pharm World Sci 26(3):137–142

van der Lubben IM, Kersten G, Fretz MM, Beuvery C, Verhoef JC, Junginger HE (2003) Chitosan microparticles for mucosal vaccination against diphtheria: oral and nasal efficacy studies in mice. Vaccine 21:1400–1408

Vodak DT, Dobry DE, Falk R, Dubose D, Kuehl PJ, Chand R, Foubert TR, Steadman BL (2012) Enhancing the immune response to intranasal vaccination with norovirus VLPs—size matters. Paper presented at the DDL 23, Edinburgh

Wang SH, Kirwan SM, Abraham SN, Staats HF, Hickey AJ (2012) Stable dry powder formulation for nasal delivery of anthrax vaccine. J Pharm Sci 101(1):31–47

Washington N, Steele RJC, Jackson SJ, Bush D, Mason J, Gill DA, Pitt K, Rawlins DA (2000) Determination of baseline human nasal pH and the effect of intranasally administered buffers. Int J Pharm 198:139–146

Weiler C, Egen M, Trunk M, Langguth P (2010) Force control and powder dispersibility of spray dried particles for inhalation. J Pharm Sci 99(1):303–316

Westmeier R (2010) Novel formulation concept for nasal deposition of dry powder vaccines. In: Dalby RN (ed) Respiratory drug delivery 2010, vol 2. Davies Healthcare International, Orlando, pp 523–527

Westmeier R, Steckel H (2008) In-situ fine particle excipient as dispersion modifier for a dry powder inhalation product. Paper presented at the DDL 19, Edinburgh

White KL, Rades T, Furneaux RH, Tyer PC, Hook S (2006) Mannosylated liposomes as antigen delivery vehicles for targeting to dendritic cells. J Pharm Pharmacol 58(6):729–737

Wolfe T, Denton M (2012) Intranasal vaccine delivery—a promising future. OnDrugDelivery:10–12

Wu Y, Wu S, Hou L, Wei W, Zhou M, Su Z, Wu J, Chen W, Ma G (2012) Novel thermal-sensitive hydrogel enhances both humoral and cell-mediated immune responses by intranasal vaccine delivery. Eur J Pharm Biopharm 81:486–497

Yuki Y, Nochi T, Harada N, Katakai Y, Shibata H, Mejima M, Kohda T, Tokuhara D, Kurokawa S, Takahashi Y, Ono F, Kozaki S, Terao K, Tsukada H, Kiyono H (2010) In vivo molecular imaging analysis of a nasal vaccine that induces protective immunity against Botulism in nonhuman primates. J Immunol 185:5436–5443

Zscherpe J (2009) Nasale Pulverformulierungen zur systemischen Wirkstoffapplikation. Kiel University, Kiel

Chapter 16
Pulmonary Administration of Subunit Vaccines

Wouter F. Tonnis, Anke L.W. Huckriede, Wouter L.J. Hinrichs, and Henderik W. Frijlink

16.1 Introduction

Traditionally, vaccines are administered by intramuscular or subcutaneous injection. The administration time by needle is short and the dosing is very precise. Although this percutaneous route of administration has been used for many years, injection of vaccines has some major disadvantages. First, due to the re-use of needles and accidental needle stick injuries, transmission of HIV and hepatitis B virus can increase. Second, the administration of vaccines by needle is depending on trained healthcare workers. Third, usage of needles leads to needle waste, which is expensive to discard. In the fourth place, due to needle phobia some people might not get vaccinated at all. A final disadvantage of administration by needle is the limited stimulations of mucosal immunity.

Needle-free administration of vaccines can overcome these disadvantages. The lungs are an interesting site for administration of vaccines since they provide an enormous interface between the outside world and the body. Due to this large interface the respiratory tract is a preferred port of entry for many pathogens. By applying vaccines to this site, pulmonary vaccination follows the natural route

W.F. Tonnis (✉) • H.W. Frijlink • W.L.J. Hinrichs
Department of Pharmaceutical Technology and Biopharmacy, University of Groningen, Groningen, The Netherlands
e-mail: w.f.tonnis@rug.nl; h.w.frijlink@rug.nl; w.l.j.hinrichss@rug.nl

A.L.W. Huckriede
Molecular Virology Section, Department of Medical Microbiology, University of Groningen, Groningen, The Netherlands

University Medical Centre Groningen, Groningen, The Netherlands
e-mail: a.l.w.huckriede@umcg.nl

C. Foged et al. (eds.), *Subunit Vaccine Delivery*, Advances in Delivery Science and Technology, DOI 10.1007/978-1-4939-1417-3_16

of infection. For pathogens that are not transmitted via the respiratory tract, pulmonary administration has still the advantage of being a needle-free way of vaccine administration.

In the past, pulmonary vaccination with non-subunit vaccines against measles, influenza, and tuberculosis has been proven to be effective in humans (Rosenthal et al. 1968; Waldman et al. 1969; Khanum et al. 1987; Dilraj et al. 2000; Bennett et al. 2002; Low et al. 2008). However, little is known about the pulmonary administration of subunit vaccines in clinical studies. In this chapter the immune system of the lungs, the devices, and dosage forms, the target populations and the safety of pulmonary administration of subunit vaccines will be described.

16.2 The Lung Immune System

Anatomically, the respiratory tract can be divided into an upper part, comprising the nose, the oral cavity and the throat, and a lower part comprising the trachea and the lungs, the two parts being separated by the glottis (Sato and Kiyono 2012). The upper respiratory tract is directly exposed to the incoming air and is generally colonized by (commensal) microorganisms. The lower respiratory tract was long considered to be sterile in healthy individuals. However, molecular techniques, being much more sensitive than traditional microbiological techniques, provided evidence that the same microorganisms found in the upper respiratory tract are also present in the lungs, albeit in much lower numbers (Charlson et al. 2011). Maintenance of the immunological homeostasis of the respiratory tract is a major challenge and involves a range of physical, chemical, and immunological mechanisms. Profound knowledge of these mechanisms is important for exploitation of the respiratory tract for vaccination.

16.2.1 Innate Immune Mechanisms

The conducting airways in the different parts of the respiratory tract are lined by ciliated epithelial cells, which are covered by a thick layer of mucus (Martin and Frevert 2005). Invading microorganisms as well as other particles are trapped in the mucus and are transported with the mucus to the mouth (mucociliary clearance) where they are expelled or swallowed. The mucus also contains a range of antimicrobial substances such as lysozyme, lactoferrin, defensins, and surfactants (Martin and Frevert 2005).

The next line of defence against microorganisms invading the airways is formed by cellular components, epithelial cells, and alveolar macrophages in particular (Sato and Kiyono 2012). The epithelial cells of the airways express a set of pattern-recognition receptors, including the Toll-like receptors (TLR), with which they can sense the presence of invaders (Martin and Frevert 2005). In response they can produce defensins as well as a range of proinflammatory chemokines and cytokines, thus recruiting neutrophils and natural killer (NK) cells. Alveolar macrophages form by far the largest population of leukocytes in the lungs (Guilliams et al. 2013). They phagocytize very actively all sorts of particles that reach the alveolar space. Yet, in the non-infected host they are of an anti-inflammatory phenotype and usually repress the induction of immune responses to the phagocytized compounds in order to keep the homeostasis in the lungs.

16.2.2 Adaptive Immune Mechanisms

The adaptive immune system of the respiratory tract comprises overall antibodies of diverse subclasses, T-helper (Th) cells ($CD4^+$) and cytotoxic T lymphocytes (CTL, $CD8^+$). Immunoglobulin A (IgA) is the most prominent and a particularly important subtype of antibodies in the respiratory tract (Renegar et al. 2004). IgA produced by B-cells in the submucosa can be transported to the airway lumen where it is released as secretory IgA (SIgA) (Sato and Kiyono 2012). There is evidence that IgA can neutralize microorganisms present in the lamina propria, inside epithelial cells as well as at the luminal side of the respiratory epithelium (Sato and Kiyono 2012). Experiments with a murine influenza infection model revealed that IgA alone can protect against nasal infection with influenza virus (Renegar and Small 1991). Furthermore, IgA but not immunoglobulin G (IgG) can neutralize influenza virus in the nasal cavity and prevent initial infection (Renegar et al. 2004). In contrast, IgA is dispensable for clearance of pulmonary influenza infection (Mbawuike et al. 1999) although it probably contributes to protection (Onodera et al. 2012). Furthermore, influenza-specific IgA can provide some cross-protection against heterologous influenza strains [recently reviewed in Van Riet et al. (2012)]. In contrast to IgG, IgA does not fix complement and thus does not trigger inflammatory responses.

IgG is also highly important in protection of the respiratory tract from and clearance of respiratory infections, particularly in the lower respiratory tract (Renegar et al. 2004). This is illustrated by the fact that parenteral vaccines, e.g., influenza vaccines, inducing systemic IgG, but not mucosal IgA, are highly effective in preventing infection (Clements and Murphy 1986).

Antibodies in the respiratory tract are produced by local B-cells present in the submucosa of the linings of the airways and in the interstitial space of the lungs (Sato and Kiyono 2012). These B-cells have undergone differentiation in local germinal centres (Onodera et al. 2012). Alternatively, antibodies produced by long-lived plasma cells in the bone marrow can reach the lower airways by transudation from serum. It is becoming increasingly clear that memory B-cells located in the lungs can rapidly differentiate into IgA- and IgG-producing plasma cells and play an important role in protection from infection (Onodera et al. 2012).

The role of local T-cells in respiratory infections has been extensively reviewed recently by Kohlmeier and Woodland (2009). In the airways of the naïve host T-cells are present in low numbers. Upon primary infection, T-cells start to appear in the airways and lung parenchyma after about 6–7 days, peak about 10 days post infection, and then undergo a phase of contraction. Both Th-cells and CTLs are contributing to the clearance of infections and protection. The type of T-cells activated and the pattern of chemokines/cytokines they produce depends on the infectious agent. Studies in a murine influenza infection model showed that during the first months after infection, substantial numbers of influenza-specific $CD4^+$ and $CD8^+$ T-cells remain present in the lungs and can effectively control reinfection. The T-cell numbers in the respiratory tract wane with time but a limited number of memory T-cells stays indefinitely. These can respond very quickly upon reinfection by producing cytokines and limit virus replication in the early phase of the infection. Later, the lung-resident T-cells are replaced by memory T-cells recruited from the blood and from the local lymph nodes (Kohlmeier and Woodland 2009).

An increasing body of evidence indicates that the local presence of memory B- and T-cells is very important for optimal protection against respiratory infections (Joo et al. 2008; Teijaro et al. 2011). The induction of these cells, which probably will require local or at least mucosal presence of antigen, should therefore be a primary aim of vaccination.

16.2.3 *Uptake and Processing of Antigen in the Respiratory Tract*

An effective immune response in the respiratory tract requires homing of B- and T-cells to this site. Only when activated by specialized dendritic cells (DCs) coming from a mucosal site to a nearby lymph node will the B- and T-cells be imprinted to express mucosal homing receptors (Lycke 2012). Originally it was thought that mucosally primed lymphocytes could recirculate through all mucosal tissues. Yet, currently it is known that efficient induction of immune responses in the respiratory tract requires immunization via the intranasal, sublingual, or pulmonary route (Lycke 2012).

Immune-inductive sites in the mucosal tissues are called mucosa-associated lymphoid tissue (MALT). In the upper respiratory system, MALT is found in the nasal-associated lymphoid tissue (NALT) of mice and the adenoid glands and tonsils, which together form Waldeyer's Ring in humans. The lower respiratory tract of specific pathogen-free mice and young children does not contain organized lymphoid tissue (Moyron-Quiroz et al. 2004). However, bronchus-associated lymphoid tissue (BALT) is induced upon respiratory infection and is referred to as iBALT (Moyron-Quiroz et al. 2004).

Presentation of antigen by activated DCs to T-cells is an essential step in the initiation of immune responses. Diverse populations of DCs are present throughout the respiratory tract in the submucosal tissue or in the airway or alveolar lumen [see Guilliams et al. (2013) for a recent review]. A specialized DC population, delineated as CD11b^{low}CD103^{+} DC, can acquire antigen directly from the airway lumen via protrusions which extend through the epithelial cell layer. Alternatively, antigen is taken up by the so-called M-cells and transported to underlying DCs which are mostly of a CD11b^{high}CD103^{-} phenotype. M-cells are found in the NALT and BALT but possibly also outside organized lymphoid tissue (Sato and Kiyono 2012). There is convincing evidence that T-cell induction can take place directly in the iBALT (Moyron-Quiroz et al. 2004). Yet, the more common mechanism is that upon antigen uptake the DCs migrate to draining lymph nodes (Moyron-Quiroz et al. 2004). Only if properly activated these DCs can mature and are then capable of activating naïve T-cells, either directly or through antigen transfer to lymph node-resident DCs.

While DCs and M-cells are important in antigen sampling, the vast majority of particles reaching the lungs will be taken up by alveolar macrophages (Blank et al. 2013). Macrophages are by far the most abundant cell type in the alveolar space and are very active in phagocytosis. They have an important function in dampening immune responses in the lungs. Accordingly, they are poorly migratory and do not contribute decisively to the induction of adaptive immune responses (Braciale et al. 2012). Tissue-resident macrophages also actively take up antigen and transport it to local lymph nodes but do probably not participate in the activation of naïve T-cells.

16.2.4 *Implications for Pulmonary Vaccine Design*

With respect to optimization of pulmonary vaccines, four important conclusions can be drawn from the above: (1) The induction of iBALT by pulmonary vaccines might be advantageous; (2) vaccine antigen uptake by macrophages should as far as possible be avoided; (3) instead antigen should be designed for optimal uptake by DCs; and (4) the vaccine needs to be capable of properly activating DCs. So far, very few studies have been published that investigate the significance of iBALT formation and targeting of vaccines to specific cell populations which could provide a rational basis for pulmonary vaccine design.

One aspect of optimization that has been studied is the place in the respiratory tract where the antigen should be delivered. The two studies available on this issue both underline that delivery to the deep lungs results in better immune responses than delivery to the upper parts of the airways (Minne et al. 2007; Todoroff et al. 2013). These studies discuss a prolonged residence time of the antigens as a possible reason for the improved immune responses. No attempt was made to investigate whether the site of delivery had an effect on the specific type of DC taking up and presenting the antigen.

Another aspect, which has been studied recently, is the effect of particle size on antigen uptake by different cell populations in the lungs (Blank et al. 2013). This study showed that in all parts of the lung investigated (trachea, bronchoalveolar fluid (BALF), lung parenchyma) particles predominantly ended up in macrophages. In the BALF and lung parenchyma, CD11^{low} DCs were more active than CD11b^{high} DCs in particle uptake, while in the trachea both DC populations contributed equally. Furthermore, smaller particles (20–100 nm) were more efficiently taken up than large ones (1,000 nm).

While these studies give first indications of how to optimize pulmonary vaccines, they also emphasize that much has to be learned from a detailed and systematic analysis of the fate of vaccines in the respiratory tract and its relation to the immune response induced.

16.3 Delivery to the Lungs

To successfully deliver subunit vaccines to the lungs, an aerosol of powder particles or droplets containing the vaccine in the correct size range should be inhaled by the vaccinee via the mouth. When the powder particles or droplets are too small, they will be exhaled (Byron 1986). On the other hand, when the particles or droplets are too large, they will not pass the upper part of the trachea, simply because larger particles cannot take the sharp bend at the back of the throat. Particles with an aerodynamic size between 1 and 5 μm are considered to be ideal for pulmonary administration (Patton and Byron 2007). When the inhalation maneuver is performed correctly, the majority of the particles in this size range will be deposited in the lungs.

Until today, it is not clear which part of the lungs should be targeted for the optimal immune response to the inhaled vaccine, although it seems that, as mentioned before, deep lung delivery yields the highest immune response (Minne et al. 2007; Todoroff et al. 2013). If peripheral delivery to the alveoli is aimed for, the median particle size should be closer to 1 μm than to 5 μm. If, however, the upper part of the lungs should be targeted, the median particle size should be closer to 5 μm than to 1 μm.

16.4 Devices and Dosage Forms for Mass Vaccination Programs

The lungs have been studied for many years as delivery site for drugs, especially for local therapy with corticosteroids and bronchodilators for patients suffering from asthma and chronic obstructive pulmonary disease (COPD). Many devices have been developed for this patient population. These inhalation devices can be divided into three major categories: Nebulizers, pressurized metered dose inhalers (pMDIs), and dry powder inhalers (DPIs). Within each category, a distinction can be made between single-use and multi-use devices. For the administration of vaccines, single-use devices are preferred, since administration of vaccines is done only once per vaccinee or in some cases once per year (influenza). Multi-use devices could in theory be used for mass vaccination programs, but would need thorough cleaning after each usage to prevent transmission of diseases from one person to another. Multi-use devices can also be used with a disposable user part that could prevent contamination of the device by the vaccinee. The three types of devices are discussed in the next paragraphs.

16.4.1 *Nebulizers*

Nebulizers generate an aerosol of an aqueous solution or suspension of the active compound and can be divided in ultrasonic, jet and vibrating mesh nebulizers (Le Brun et al. 2000). Ultrasonic nebulizers use high-frequency waves to disperse the solution into small droplets. These waves are generated by applying high-frequency pulses from an oscillating piezo element to the aqueous solution or suspension. The droplet size is dependent on the frequency of the pulses. Jet nebulizers use a two-fluid nozzle to produce an aerosol. Air from a compressor or pressurized air (from the mains) passes the two channels by which droplets are formed. To narrow down the size distribution, a baffle is placed above the nozzle, which collects the large particles that subsequently flow back into the fluid reservoir. The aerosol generated by both ultrasonic and jet nebulizers is inhaled by tidal breathing. With every inhalation only a small part of the dose is deposited in the lungs, depending on the output rate of the nebulizer. Therefore, it can last up to 10–15 min before the whole dose is inhaled. For young children both types of nebulizers can be equipped with a facemask that covers the nose and the mouth.

In the more recently developed vibrating mesh nebulizers, the piezo technology is combined with a perforated membrane (mesh) which is in contact with the drug solution (Kesser and Geller 2009). Two different technologies are available; those in which the oscillation is applied to the membrane itself and those in which the oscillation comes from a piston that vibrates in the liquid reservoir (Kesser and Geller 2009). Vibrating mesh nebulizers deliver more condensed aerosols than jet nebulizers, which increases the output rate and reduces the administration time.

Since most nebulizers are re-usable devices, they need to be cleaned and disinfected after each usage. This makes nebulizer less suitable for mass vaccination programs. Some single-use (disposable) jet nebulizers are available, but may not be as effective as re-usable jet nebulizer in terms of output rate and droplet size distribution (Vecellio et al. 2011). Furthermore, these devices still require clean pressurized air, which makes them less portable.

In most studies on pulmonary administration of measles vaccine in Mexican children, a jet nebulizer has been used, the so-called "Classical Mexican Device" (Fig. 16.1). In these studies, the lyophilized measles vaccine is reconstituted and placed on crushed ice to limit microbial growth (Bennett et al. 2002). A compressor that is connected to a power supply is used to generate the aerosol, which is subsequently inhaled by the children through a facemask. A disposable paper cone is used to prevent contamination of the mask by the children. Even though this device has been used for immunizing large groups of children, the device is not portable and needs an external power source. Therefore, the World Health Organization (WHO) is searching for a substitute nebulizer in the context of the "Measles Aerosol Project" (Laube 2005). However, it is at present not known which nebulizer will be or has been chosen by the WHO (Henao-Restrepo et al. 2010).

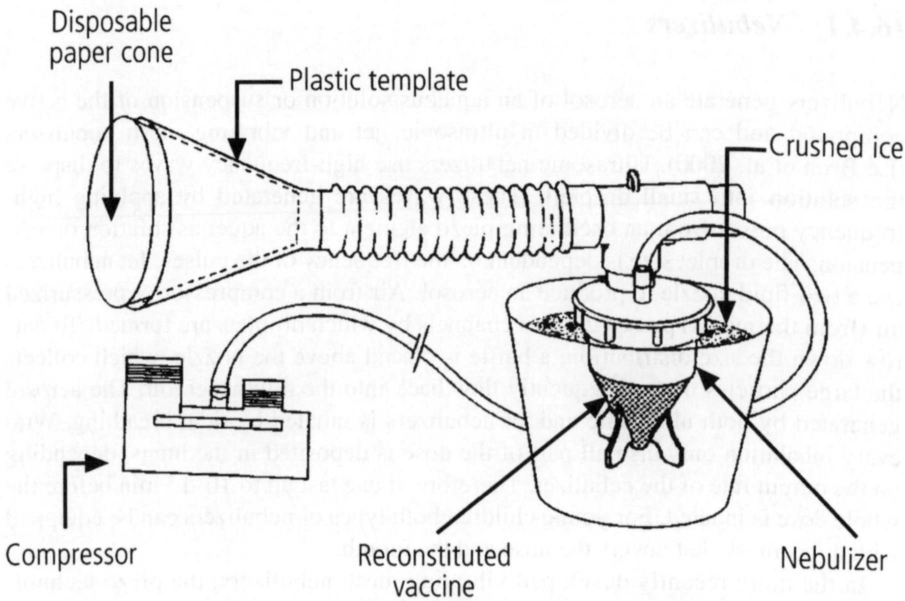

Fig. 16.1 Components of the "Classical Mexican Device." Air from the compressor is fed through the vaccine solution or suspension which is subsequently aerosolized. The aerosol is inhaled by the vaccinee through the disposable paper cone [reprinted from Bennett et al. (2002) with permission from the World Health Organization]

16.4.1.1 Vaccine Formulations for Nebulization

Developing aqueous vaccine formulations for nebulization is relatively straight forward, as compared to other formulations for pulmonary administration. The few requirements for aqueous formulations for inhalation as described by the European Pharmacopoeia are (1) a pH between 3 and 8.5, and (2) preferably the solution should be sterile and isotonic, although the latter two are not strict requirements (European Pharmacopoeia 2008). Preservatives can be added to maintain sterility during storage.

Parameters like the viscosity of the aqueous solution or suspension, the jet pressure, and the breathing pattern of the vaccinee influence the final droplet size coming out of the nebulizer (McCallion et al. 1995; Brand et al. 2000; de Boer et al. 2003). When developing formulations for nebulizers, these influences should be simulated in vitro to make sure droplets within the correct size range are administered and deposited properly in the lungs (Lexmond et al. 2013).

In clinical research, two aqueous subunit vaccine formulations suitable for administration by using a nebulizer have been developed, one against the human papilloma virus (HPV16) and one against *Streptococcus pneumoniae* (23-PPV) (Nardelli-Haefliger et al. 2005; Menzel et al. 2005). The formulations consisted solely of the antigen suspended in saline. Pulmonary administration of the HPV vaccine resulted in both a systemic and a mucosal immune response in healthy adult female (Nardelli-Haefliger et al. 2005). The 23-PPV vaccine induced serum IgG antibodies after pulmonary administration, although the level of antibodies was not as high as after intramuscular administration (Menzel et al. 2005).

16.4.1.2 Application in Vaccine Delivery

Nebulization is used in the majority of clinical trials on pulmonary administration of subunit vaccines as well as non-subunit vaccines (Tonnis et al. 2013). The main reason for this is probably the relatively simple formulation work that is required, as well as the availability of nebulizers to be used for almost any aqueous product (in contrast to pMDIs and DPIs for which no of-the-shelf device exists that can be used for a standard formulation). Although nebulization is used in many studies, there are a couple of disadvantages that limit the use of this type of device in mass vaccination programs. First of all, the deposition in the lungs is low due to the tidal breathing. With every exhalation, the fraction of the aerosol that is still in the dead volume (trachea and bronchi) is exhaled. Therefore, the administration time is long, usually around 10–15 min, and a large amount of the vaccine formulation is needed leading to increased costs per dosage. Furthermore, classic nebulizers (both jet and ultrasonic) have relatively large residual volumes (several milliliters), which further increases the costs per dosage. Since vaccines are often expensive to prepare, this, in addition to the relatively high price of most nebulizers, makes nebulization a costly method to administer vaccines to the lungs. Another disadvantage is that the classical nebulizers are dependent on a power source in order to operate, which is not always available. The newer mesh nebulizers may overcome some of these disadvantages. However, these devices are often equipped with chip technology to adjust the nebulization procedure to the drug solution and to the breathing maneuver of the patient (adaptive aerosol delivery) or to monitor patient adherence and compliance. This makes such devices expensive and therefore, their use is mainly confined to therapies against relatively rare diseases like cystic fibrosis.

Another concern about nebulization of vaccines is the chemical and physical stability of the vaccine during nebulization. Shear stresses can severely damage the antigen leading to administration of degraded products (Khatri et al. 2001). Furthermore, during ultrasonic nebulization the temperature of the solution can increase (Phipps et al. 1990), which may lead to heat-induced degradation of the antigen.

16.4.2 Pressurized Metered Dose Inhalers

A pressurized metered dose inhaler (pMDI) consists of a canister closed off by a metering valve, an actuator, and a mouthpiece. The canister contains the vaccine, either dissolved or suspended in a propellant under pressure (Smyth 2005). By pressing the canister down, a fixed amount of formulation is released. The rapid expansion of the propellant disperses the vaccine formulation into small, inhalable particles. pMDIs are multidose devices by virtue of their design.

The aerosol is released from the pMDI at high velocity (5 m/s) (Hochrainer and Hölz 2005). Therefore, the release of the dose and the inhalation should be at the exact same moment. When the inhalation is too late, the whole dose is deposited in the back of the mouth. To properly administer the dose to the lungs, good hand-lung coordination is required. This coordination can be taught, but takes time to learn. For everyday administration of drugs to the lungs, this might be worth the investment, but for a one-time administration of vaccines this might require too much time. To improve the lung deposition in children, the dose can be dispersed into a spacer or valved holding chamber (VHC), from which it can be slowly inhaled in multiple inhalations. The final lung deposition, however, is quite low due to losses in the spacer/VHC resulting from impaction and sedimentation of the aerosol. Furthermore, electrostatic charging may lead to attraction of the aerosol particles to the wall of the spacer/VHC, also resulting in improper dosing. On the other hand, an advantage of using a VHC is that the multidose pMDI cannot be contaminated by the vaccinee, because exhalations are diverted from the chamber by the valve. For mass vaccination programs, a cheap and disposable VHC is preferred. Another option would be a VHC with a cheap and disposable mouthpiece (including the valve).

16.4.2.1 Vaccine Formulations

Formulations for pMDIs consist of the antigen dissolved or dispersed in a propellant [nowadays only hydrofluoroalkane (HFA) propellants are allowed for use]. When the antigen is suspended in the propellant, dose fluctuations due to sedimentation or floating of the antigen may occur. Therefore, the canister should be shaken properly before use and the suspension should be physically stable. Another option is to dissolve the antigen in the propellant by using surfactants or cosolvents (e.g., ethanol) (Smyth 2005).

The main challenge in developing vaccine formulations for pMDIs is protein integrity and stability in the propellant. The hydrophobic nature of the propellant may induce denaturation of protein-based antigens (Shoyele and Slowey 2006). So far only one study on pulmonary vaccine administration by using a pMDI has been published (Brown et al. 1997), although not on subunit vaccines. In this study, the killed whole bacterium *Streptococcus suis* was dissolved in a propellant (dimethylether, liquefied by high pressure) using a surfactant for pulmonary administration to swine with the help of a nose-cone. Only 7–12 % of the aerosol particles had an aerodynamic diameter in the correct size range. Moreover, the antigenicity was partially lost after aerosolization.

16.4.2.2 Application

Hardly any studies have been performed using pMDIs as device for the delivery of vaccines to the lungs. Even though pMDIs are cheap and portable, the difficulties of developing stable vaccine formulations make them less suitable for this application.

Additionally, the administration is inefficient in case of improper hand–lung coordination. The efficiency can be improved by using a spacer/VHC, although losses in this accessory device occur as well. The use of an entirely disposable VHC or one with a disposable mouthpiece would allow for the use of multidose pMDIs for mass vaccination programs when the antigen is found to be stable in the propellant.

16.4.3 Dry Powder Inhalers

Many different dry powder inhalers (DPIs) are commercially available, which can be classified into devices for single use and devices for multiple uses. Multi-use devices can be further classified into single-dose, multiple unit-dose, and multidose devices. Multiple unit-dose devices have an interchangeable part holding multiple doses (e.g., a disk with eight doses). Multidose devices have a compartment that contains multiple doses of the powder formulation. Both multiple unit-dose and multidose devices are extensively used in the treatment of chronic lung diseases that require daily administration (Laube et al. 2011). Vaccination however, requires no more than a few administrations in a lifetime or annually in the case of influenza vaccination. Therefore, multi unit-dose and multidose devices are considered to be less suitable for pulmonary vaccination. On the contrary, multiuse single-dose devices are loaded with one dose at a time, for example with a capsule. These devices are more suitable, although the use of one device by more than one person is not ideal due to the risk of transmission of diseases from one vaccinee to another. To decrease the risk of transmission, the device should have a disposable user interface, preferably with a one direction valve, which prevents that one can exhale into the device.

Single-use, thus disposable devices would be ideal for pulmonary administration of vaccines, since there is no risk of transmission of diseases. These disposable devices should be cheap to produce in order to keep the price per dose low.

Dry powder vaccine formulations (see Sect. 16.4.3.1) consist of particles in the appropriate size range. These small particles attract each other by van der Waals forces resulting in the formation of agglomerates, which are too large to be deposited in the lungs. Therefore, the DPI should have a dispersion mechanism to break up these agglomerates and deliver the powder particles of their primary particle size. Several dispersion mechanisms have been developed, which are either based on shear forces between the powder and the device, on drag and lift forces exerted by the moving air on the powder, or on impaction forces generated by particle–device or particle–particle collisions (Frijlink and De Boer 2004; Islam and Cleary 2012). The forces exerted on the agglomerates are generated by the inhaled airstream. Therefore, the powder is delivered from the DPI only when the vaccinee is inhaling through the device. Proper hand–lung coordination, as encountered with pMDIs, is therefore not important when using DPIs.

16.4.3.1 Vaccine Formulations

Antigens are produced as aqueous solutions. Obviously, when the antigen is envisaged to be administered as a powder, the aqueous antigen solution should be dried. Dry powder particles with the appropriate particle size distribution can be produced either by one-step or by two-step processes (Amorij et al. 2008). One-step processes for the production of powder particles are (1) spray drying, (2) spray-freeze drying, or (3) supercritical fluid drying. Two-step processes consist of a combination of a drying technique (e.g., freeze drying, vacuum drying, foam drying) and a particle size reduction method, e.g., milling. However, one-step processes are preferred because two-step processes can lead to contamination and increased production costs.

During each of the mentioned drying processes the antigen is exposed to harsh conditions, e.g., to shear, heat, and dehydration stresses during spray drying, shear, freezing, and dehydration stresses during spray-freeze drying and shear and dehydration stresses during supercritical fluid drying. Because of their proteinaceous nature, subunit vaccines are usually fragile molecules which can easily deteriorate when exposed to these harsh conditions. Consequently, stabilizing excipients should be used to maintain the immunogenicity of the antigen during drying. Often used excipients that stabilize proteinaceous drug substances are sugars. If dried properly, the sugar forms a glass which also stabilizes the antigen during subsequent storage. It has even been shown that influenza subunit vaccine spray dried or spray-freeze dried in the presence of inulin is stabilized to such an extent that it does not require refrigerated storage and transport, the so-called cold chain (Saluja et al. 2010). Therefore, to increase the shelf-life and to avoid dependence on the expensive cold-chain, also antigens intended to be administered as a liquid formulation (e.g., for nebulization), might be dried in the presence of sugar after production and reconstituted before use.

Various theories have been described to explain the protective action of sugars (Grasmeijer et al. 2013). The two main ones are (Chang and Pikal 2009):

- The water replacement theory: In an aqueous solution, the antigen forms many hydrogen bonds with the surrounding water molecules. During drying the water molecules are gradually replaced by sugar molecules. The hydroxyl groups of the sugar form hydrogen bonds with the antigen comparable with water by which the structural integrity of the antigen is maintained.
- The vitrification theory: During drying the sugar turns into its glassy state thereby incorporating the antigen. Because the molecular mobility of the sugar molecules in the glassy state is very low, the mobility of the molecules of the antigen is very low as well. Because most degradation reactions require molecular mobility, the degradation rate of the antigen incorporated in the matrix of sugar glass will decrease.

The glassy state of the sugar is essential for its protective action. Above the so-called glass transition temperature (T_g), sugar glasses turn into the rubbery state which is undesirable for two reasons. First, the molecular mobility of the sugar and thereby the incorporated drug strongly increases. Second, in the rubbery state, sooner or later, crystallization of the sugar will occur. By this phase separation, the hydrogen bonds between the sugar and the antigen are broken by which the protective action of the sugar is lost. Moreover, the crystallization process itself may cause severe damage to the antigen. Therefore, a sugar with a high T_g is preferred. However, the T_g not only depends on the type of sugar but also on the water content after drying. Water acts as a plasticizer that strongly decreases the T_g. Due to the hygroscopic nature of sugar glasses, the T_g may drop below ambient temperature when the glass is exposed to air of relatively high humidity. Because subunit vaccines contain amino groups, the applied sugar should not contain reducing groups to prevent Maillard browning. Often used sugars for the stabilization of proteinaceous drug substances, such as subunit vaccines, are the disaccharides, trehalose and sucrose (Wang 2000; Amorij et al. 2008). However, also the oligosaccharide inulin has been shown to be an excellent stabilizer (Saluja et al. 2010).

The formulation used in a DPI should have a primary aerodynamic particle size distribution that is optimal for inhalation (1–5 μm). Usually, particle size distributions are measured by laser diffraction. However, with laser diffraction the geometric particle size is measured and not the aerodynamic particle size. The aerodynamic particle size can be calculated from the geometric particle size using the equation:

$$d = d_{ae} = \sqrt{\frac{\rho}{\rho_0 \chi}}$$

where d_{ae} is the aerodynamic diameter, d_e the geometric particle size, ρ_p the density of the particles (g/cm^3), ρ_0 the unit density (1 g/cm^3), and χ the dynamic shape factor (one for a sphere).

More information on the aerodynamic particle size distribution can be obtained by cascade impactor analysis through which the so-called fine particle fraction (FPF) can be determined. The FPF is defined as the volume fraction of powder particles with an aerodynamic particle size smaller than 5 μm.

Generally, two types of powder formulations can be distinguished: Drug-only agglomerates and carrier-based formulations (Fig. 16.2). Drug-only agglomerates consist of primary powder particles containing the antigen, which form agglomerates. During inhalation the agglomerates are broken up into the primary particles (which should thus be 1–5 μm). Carrier-based formulations are blends of the micron-sized formulated antigen powder particles and a coarse carrier (usually lactose). These formulations are prepared by blending agglomerates of the primary formulation with the carrier. During blending the agglomerates are broken up and the particles are dispersed over the surface of the coarse carrier. During inhalation, the primary particles should detach from the carrier. In general, carrier-based formulations have a better flowability and are physically more stable than drug-only agglomerates.

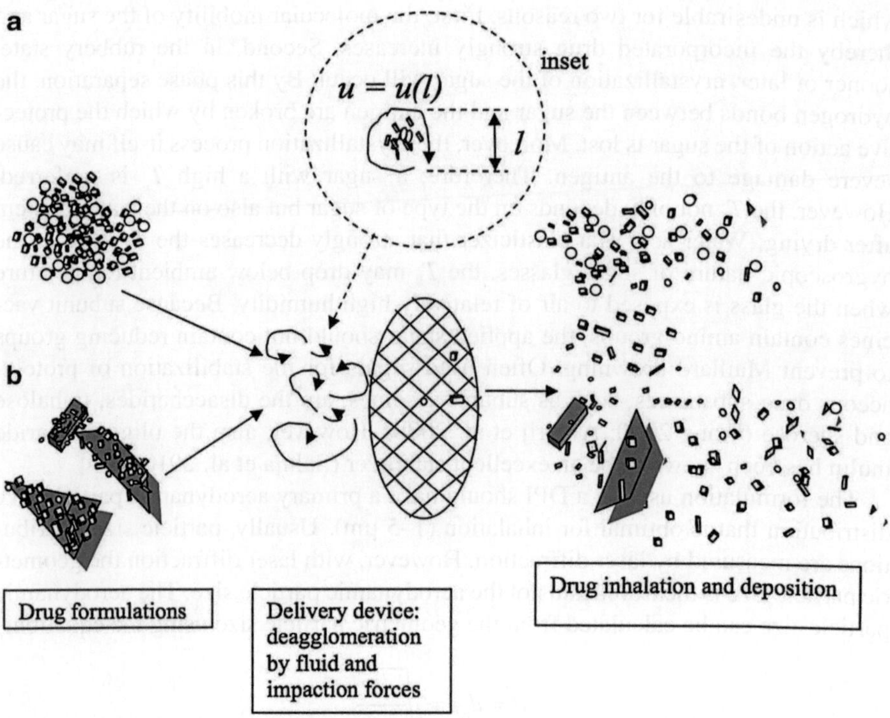

Fig. 16.2 Two types of powder formulations as described by Islam and Cleary (2012). (**a**) Drug-only agglomerates; (**b**) carrier-based formulation. The dispersion mechanism in the delivery device breaks up the formulation into primary particles upon inhalation [reprinted from Islam and Cleary (2012), with permission from Elsevier].

16.4.3.2 Application

Until today, no clinical studies have been published on pulmonary vaccination using a dry powder inhaler. Recently, a clinical trial was finalized, in which two different DPIs [the Puffhaler® and Solovent™ (Fig. 16.3)] were investigated for the administration of a measles vaccine, but the results have not been published yet. Both devices are re-usable devices but have a disposable user interface. The powder is dispersed into a disposable reservoir or spacer by applying pressure on a blister or capsule using a squeeze bulb or a syringe, respectively. Next, the reservoir is detached and the dose is subsequently inhaled by the vaccinee. Both devices have been tested in macaques with drug-only agglomerates of a myo-inositol-based powder formulation produced by supercritical fluid drying (Lin et al. 2011). The results showed that both devices performed equally well in inducing an immune response.

Although no clinical trials have been published on DPI vaccine formulations, a lot of preclinical research has been performed on this type of formulation (Table 16.1). In many of these studies, a Dry Powder Insufflator™ was used to

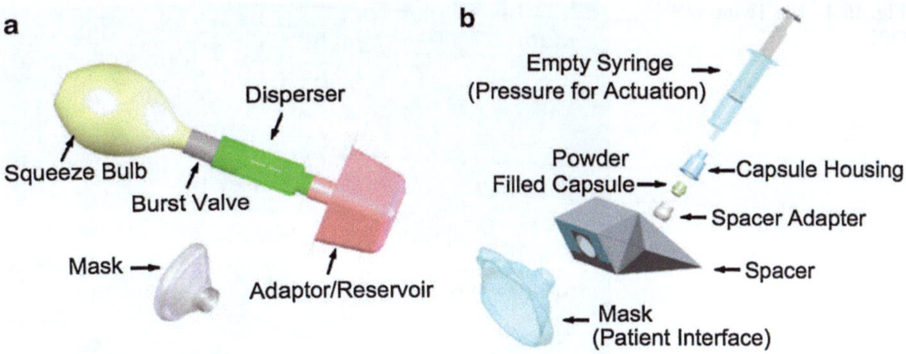

Fig. 16.3 Two DPIs: (**a**) PuffHaler®; (**b**) Solovent™ [reprinted from (Lin et al. 2011), with permission from PNAS]

Table 16.1 Powder formulations of subunit vaccines tested in preclinical research using the Dry Powder Insufflator™

Antigen	Formulation	Production method	Primary particle size (μm)	Fine particle fraction (<5 μm) from insufflator	Animal	Ref
Diphtheria toxoid	Antigen encapsulated in chitosan and dextran microparticles	SCF	1–7[a]	55 %[c]	Guinea pigs	Amidi et al. (2007)
Hepatitis B surface antigen	Microparticles of leucine and antigen encapsulated in PLGA/PEG nanoparticles	SD	7[a]	50 %[3] (MMAD =4.8 μm)	Guinea pigs	Muttil et al. (2010) 2010
Influenza subunit vaccine	Antigen encapsulated in lipid microparticles	SD	1–5[b]	NA	Rats	Smith et al. (2003)
Influenza subunit vaccine	Antigen encapsulated in inulin	SFD	4–24[a]	38 %[c]	Mice	Amorij et al. (2007)
Influenza subunit vaccine	Antigen encapsulated in inulin	SD SFD	1–6[a] 4–24[a]	37 %[c,d] 23 %[c,d]	Mice	Saluja et al. (2010)
Antigen 85B (tuberculosis vaccine)	Antigen encapsulated in PLGA	SD	NA	69 %[c]	Guinea pigs	Lu et al. (2010)

MMAD mass mean aerodynamic diameter, NA not available, *PEG* polyethylene glycol, *PLGA* poly(lactic-co-glycolic acid), *SCF* supercritical fluid drying, *SD* spray drying, *SFD*: spray-freeze drying

[a]Determined by laser diffraction analysis

[b]Determined by scanning electron microscopy

[c]Determined by cascade impactor analysis

[d]Fine particle fraction from Twincer inhaler, not from insufflator

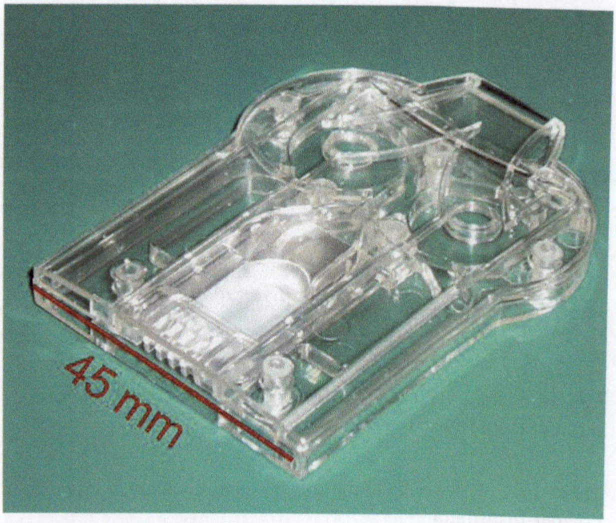

Fig. 16.4 The Twincer™ DPI

administer the powder to the lungs of small animals. Various subunit vaccine powder formulations have been produced and tested using this device (Table 16.1). Saluja et al. (2010) compared spray drying with spray-freeze drying of the influenza subunit vaccine in the presence of inulin. It was found that both powder formulations induced an immune response after pulmonary administration, which was significantly higher than after intramuscular administration of the vaccine. Furthermore, in vitro characterization showed that both formulations could be dispersed by using the Twincer™ (Fig. 16.4) in a particle size suitable for pulmonary administration. The Twincer™ is a disposable DPI, which consists of three plastic plates and an aluminum blister, and is therefore cheap to manufacture (Friebel and Steckel 2010).

16.5 Target Populations

Proper deposition of a vaccine in the lungs is mainly dependent on three factors: (1) The formulation, (2) the device, and (3) the performance of the vaccinee. Not every device is suitable for all age groups. Whether or not an inhalation device is suitable for a target group for vaccination depends on (1) the inspiratory flow rate that they can generate, (2) their lung volume, (3) their emotional state, as well as (4) their understanding of and ability to master the inhalation maneuver. Infants are an important target group for administration of vaccines. This group is not capable of performing a controlled inhalation maneuver simply because infants do not comprehend how to do it. Furthermore, infants are nose breathers only. When a nebulizer with face mask or a DPI with a spacer attached to a facemask (e.g., Puffhaler or Solovent) is used for this group, the antigen will most likely be deposited in the nasal cavity instead of in the lungs. A risk of nasal administration is that the antigen might be transported to the central nervous system of the infant via the olfactory bulb, which directly connects the nose with the brain (Illum 2000).

With increasing age, children become increasingly more capable of understanding how to perform the desired inhalation maneuver. In general, from the age of 6 or 7 years onwards, DPIs can be used. However, many DPIs need 1.5 L of air or more to disperse and release the total dose, for which the lung volume of young children may be insufficient (Koopman et al. 2011). For this age group, a DPI is desirable that releases its entire dose within the first 0.5 to 1.0 L of air.

From adolescence onwards, any type of device is generally suitable in terms of the understanding and physical capacities of the vaccinee. Elderly on the contrary, who receive the annual influenza vaccine and might need revaccination at some point, may have physical constraints that would limit the use of DPIs. Table 16.2 shows different age groups and pulmonary delivery devices that would be suitable.

Table 16.2 Pulmonary delivery device options specified per age group

	Age (years)	Physical limitation	Ability to master inhalation maneuver	Suitable pulmonary delivery devices
Infants	0–1	Small lung volume, nose breathing	None	None
Toddlers	1–3	Small lung volume	None to poor	Nebulizer, pMDI-VHC or active DPI in combination with facemask
Preschool children	3–6	Small lung volume	Poor	Nebulizer, pMDI-VHC or active DPI (in combination with facemask)
				Possibly DPI
School children	6–12	(Small lung volume)	(Mostly) good	Nebulizer
				pMDI-VHC
				DPI
Adolescents	12–18	None	Good	Nebulizer
				pMDI-VHC
				DPI
Adults	18–65	None	Good	Nebulizer
				pMDI-VHC
				DPI
Elderly	>65	Reduced lung capacity, possibly impaired fine motor skills	Possibly impaired	Nebulizer
				pMDI with VHC (Active) DPI

pMDI-VHC pMDI equipped with a valved holding chamber, *active DPI* aerosol is formed by the device itself, independent of the patient

16.6 Delivery Systems for Pulmonary Administration of Vaccines

In the past decade, various new delivery systems for pulmonary delivery of subunit vaccines have been developed (Table 16.3). Antigens for subunit vaccines are often poorly immunogenic, especially in the dissolved state. When formulated with immune potentiators (TLR ligands) or formulated as for example polymeric microparticles, nanoparticle conjugates, liposomes or incorporated in the immunostimulating complex matrix (ISCOMATRIX) their immunogenicity can be increased. Although all formulations listed in Table 16.3 are dry products, they are all suspended in an aqueous solution prior to administration in animals. In theory, these formulations could be administered as a powder using a DPI.

Thomas et al. found that the immune response to the hepatitis B vaccine formulated as poly(lactic-co-glycolic acid) (PLGA) microparticles is highly dependent on the size and the charge of the microparticles (Thomas et al. 2009, 2010). It was found that PLGA particles with a low positive charge induced a greater immune response after pulmonary administration than PLGA microparticles with a negative surface charge. Pulmonary administration of PLGA particles with a mass mean aerodynamic diameter (MMAD) of 1 μm as well as PLGA particles with a MMAD 4.5 μm induced an immune response that was comparable to pulmonary administration of a hepatitis B vaccine formulation containing the antigen in saline. Interestingly, pulmonary administration of PLGA particles with a MMAD of 2.5 μm led to a three- to fourfold increase in the immune response, as compared to pulmonary administration of the plain formulation. The difference in immune response is explained by the authors as the difference in geometrical particle size. Due to the low density of the particles, the aerodynamic diameter of all formulation is in the correct size range (MMAD of 1, 2.5, and 4.5 μm, respectively) but the geometrical particle size is much larger (mean values of 3, 5, and 12 μm, respectively).

Table 16.3 New delivery systems for pulmonary vaccination

Pathogen or toxin/Antigen	Delivery system	Animal	Ref
Cytomegalovirus/Glycoprotein B	ISCOMATRIX	Sheep	Vujanic et al. (2010)
Hepatitis B virus/Hepatitis B surface antigen	PLGA microparticles	Rats	Thomas et al. (2009, 2010)
Hepatitis B virus/Hepatitis B surface antigen	PLGA nanoparticles	Rats	Thomas et al. (2011)
Influenza virus/Influenza subunit vaccine	ISCOMATRIX	Sheep	Wee et al. (2008)
Mycobacterium tuberculosis/ Antigen 85B	Conjugate of antigen and nanoparticles	Mice	Ballester et al. (2011)
Ricin Ricin toxoid	Liposome	Rats	Griffiths et al. (1997)
Yersinia pestis/F1 and V subunit vaccine	PLGA microparticles	Mice	Eyles et al. (2000)

All of these particulate formulations are administered as suspensions
ISCOMATRIX immunostimulating complex matrix, *PLA* polylactic acid, *PLGA* poly(lactic-co-glycolic acid)

The geometrical particle size determines whether the particle is internalized by antigen-presenting cells (APC). It was found that particles with a geometrical size above 10 μm were not phagocytized by alveolar macrophages, while 4–5 μm particles were. Therefore, a geometrical particle size above 10 μm is simply too big to allow for uptake by APC, leading to a lower immune response after pulmonary administration. When developing non-dissolving delivery systems like PLGA microparticles, it is important that not only the aerodynamic diameter is taken into account but also the geometrical diameter of the particles.

16.7 Safety

One of the main concerns for every new dosage form is its safety profile. Vaccines are given to large groups of healthy people, in whom a lower risk-benefit ratio is acceptable than for new drugs for diseased people. In some cases, vaccination can lead to an allergic reaction to the vaccine or to a component of the formulation (e.g., egg protein-containing vaccines). Side effects might be minor if this happens at the site of injection, but might be far more invasive when this is manifested in the lungs.

Groups that need extra attention concerning safety are immunocompromised patients (e.g., HIV-infected patients) and patients suffering from pulmonary diseases such as COPD and asthma. Pulmonary administration of a vaccine might lead to bronchoconstriction or exacerbations in asthmatic patients. In a study by Minne et al. (2008), an ovalbumin-sensitized murine model of asthma was used to determine whether pulmonary administration of a split influenza virus vaccine would cause exacerbations. Results showed that there was no development of airway hyperreactivity, no significant increase in allergen-specific IgE, and no increase in inflammatory cells in the bronchoalveolar lavage (BAL).

In another safety study, the measles vaccine was pulmonary administered to immunocompetent and immunosuppressed macaques (de Swart et al. 2006). The immune system of the second group of animals was suppressed by a combination of cyclosporine A and prednisolone. Although the number of animals was low, results showed that there was no safety hazard for pulmonary administration of the vaccine both to healthy and immunosuppressed animals.

In a study by Audouy et al. (2011) the safety profile of a powder formulation and a liquid formulation of an influenza vaccine after pulmonary administration in mice was researched. It was found that pulmonary administration led to a temporary increase (up to 72 h) of neutrophils and a decrease in the number of macrophages in the lungs. Histopathological examination of the lungs showed no changes compared to the control group. It was concluded that the reaction to the administration was short and mild, but not absent.

Until today, safety studies on pulmonary vaccination are scarce. The studies that have been published show that in those cases pulmonary vaccination was safe. At this stage, large clinical trials are required to assess the short- and long-term safety profiles of pulmonary administrated vaccines. In these studies, not only healthy volunteers should be included but in particular groups that might be at risk.

16.8 Perspectives for Future Developments

As has been shown in many studies, pulmonary administration of vaccines can be a promising alternative to administration by using needles. Although a lot of research has been performed, not one single product has been licensed for vaccination via the lungs yet. To license a product for pulmonary vaccination, the combination of formulation and inhalation device should be licensed as a whole. For this reason, formulation development and development of a suitable device should go hand in hand. Next to the combined licensure, developing a powder formulation when there is no device available for both the formulation and the target group is rather useless.

In all published clinical trials until today, nebulization was used as inhalation technique and proven to be effective in terms of inducing an immune response. However, the long administration time, the portability of the device, and the microbial and chemical stability of the aqueous solutions remain an issue. Mesh nebulizers might be an alternative to the traditional jet nebulizer, but these devices are re-usable devices and therefore need proper cleaning after each administration, which makes them less suitable for mass vaccination programs. The first choice of device would be a cheap and disposable dry powder inhaler or a dry powder inhaler with a disposable user interface. The development and production of powder formulations for pulmonary administration is more difficult than liquid or semi-liquid formulations, but this is outweighed by the improved stability of the antigen in the dry state. The next step in development should therefore be clinical trials in which DPIs are tested.

A concern regarding the use of DPI in mass vaccination programs is the understanding and capability of the vaccinee. A large target group for vaccination is small children who might not be able to use a DPI properly. Especially for deep lung delivery the inhalation maneuver should be performed properly. The alveoli has been suggested to be the optimal part of the lungs for vaccine deposition, (Minne et al. 2007; Todoroff et al. 2013). However, delivery to the upper part of the lungs has also been shown to elicit an immune response (Sievers et al. 2012). Therefore, more research is needed to elucidate the influence of the site of deposition of the antigen on the immune response. If deep lung targeting is not essential for inducing an immune response, concerns about proper inhalation technique are less important, which would facilitate the application of DPIs especially in small children.

References

Amidi M, Pellikaan HC, Hirschberg H et al (2007) Diphtheria toxoid-containing microparticulate powder formulations for pulmonary vaccination: preparation, characterization and evaluation in guinea pigs. Vaccine 25:6818–6829

Amorij J-P, Saluja V, Petersen AH et al (2007) Pulmonary delivery of an inulin-stabilized influenza subunit vaccine prepared by spray-freeze drying induces systemic, mucosal humoral as well as cell-mediated immune responses in BALB/c mice. Vaccine 25:8707–8717

Amorij J-P, Huckriede A, Wilschut J et al (2008) Development of stable influenza vaccine powder formulations: challenges and possibilities. Pharm Res 25:1256–1273

Audouy SAL, van der Schaaf G, Hinrichs WLJ et al (2011) Development of a dried influenza whole inactivated virus vaccine for pulmonary immunization. Vaccine 29:4345–4352

Ballester M, Nembrini C, Dhar N et al (2011) Nanoparticle conjugation and pulmonary delivery enhance the protective efficacy of Ag85B and CpG against tuberculosis. Vaccine 29:6959–6966

Bennett JV, Fernandez de Castro J, Valdespino-Gomez JL et al (2002) Aerosolized measles and measles-rubella vaccines induce better measles antibody booster responses than injected vaccines: randomized trials in Mexican schoolchildren. Bull World Health Organ 80:806–812

Blank F, Stumbles PA, Seydoux E et al (2013) Size-dependent uptake of particles by pulmonary antigen-presenting cell populations and trafficking to regional lymph nodes. Am J Respir Cell Mol Biol 49:67–77

Braciale TJ, Sun J, Kim TS (2012) Regulating the adaptive immune response to respiratory virus infection. Nat Rev Immunol 12:295–305

Brand P, Friemel I, Meyer T et al (2000) Total deposition of therapeutic particles during spontaneous and controlled inhalations. J Pharm Sci 89:724–731

Brown A, George D, Matteson D (1997) Vaccinator device for delivering propellant-driven aerosols of Streptococcus suis bacterin into the respiratory tracts of swine. Vaccine 15:1165–1173

Byron P (1986) Prediction of drug residence times in regions of the human respiratory tract following aerosol inhalation. J Pharm Sci 75:433–438

Chang L, Pikal M (2009) Mechanisms of protein stabilization in the solid state. J Pharm Sci 98:2886–2908

Charlson ES, Bittinger K, Haas AR et al (2011) Topographical continuity of bacterial populations in the healthy human respiratory tract. Am J Respir Crit Care Med 184:957–963

Clements ML, Murphy BR (1986) Development and persistence of local and systemic antibody responses in adults given live attenuated or inactivated influenza A virus vaccine. J Clin Microbiol 23:66–72

De Boer AH, Hagedoorn P, Frijlink HW (2003) The choice of a compressor for the aerosolisation of tobramycin (TOBI®) with the PARI LC PLUS® reusable nebuliser. Int J Pharm 268:59–69

De Swart RL, Kuiken T, Fernandez-de Castro J et al (2006) Aerosol measles vaccination in macaques: preclinical studies of immune responses and safety. Vaccine 24:6424–6436

Dilraj A, Cutts FT, de Castro JF et al (2000) Response to different measles vaccine strains given by aerosol and subcutaneous routes to schoolchildren: a randomised trial. Lancet 355:798–803

European Pharmacopoeia (2008) Preparations or inhalation (Ph Eur monograph 0671). In: European pharmacopoeia, 6th edn. Council of Europe, Strasbourg, pp 740–742

Eyles JE, Williamson ED, Spiers ID et al (2000) Protection studies following bronchopulmonary and intramuscular immunisation with yersinia pestis F1 and V subunit vaccines coencapsulated in biodegradable microspheres: a comparison of efficacy. Vaccine 18:3266–3271

Friebel C, Steckel H (2010) Single-use disposable dry powder inhalers for pulmonary drug delivery. Expert Opin Drug Deliv 7:1359–1372

Frijlink HW, De Boer AH (2004) Dry powder inhalers for pulmonary drug delivery. Expert Opin Drug Deliv 1:67–86

Grasmeijer N, Stankovic M, De Waard H et al (2013) Unraveling protein stabilization mechanisms: vitrification and water replacement in a glass transition temperature controlled system. Biochim Biophys Acta 1834:763–769

Griffiths GD, Bailey SC, Hambrook JL et al (1997) Liposomally-encapsulated ricin toxoid vaccine delivered intratracheally elicits a good immune response and protects against a lethal pulmonary dose of ricin toxin. Vaccine 15:1933–1939

Guilliams M, Lambrecht BN, Hammad H (2013) Division of labor between lung dendritic cells and macrophages in the defense against pulmonary infections. Mucosal Immunol 6:464–473

Henao-Restrepo AM, Greco M, Laurie X et al (2010) Measles aerosol vaccine project. Proc Vaccinol 2:147–150

Hochrainer D, Hölz H (2005) Comparison of the aerosol velocity and spray duration of Respimat® Soft Mist™ Inhaler and pressurized metered dose inhalers. J Aerosol Med 18:273–282
Illum L (2000) Transport of drugs from the nasal cavity to the central nervous system. Eur J Pharm Sci 11:1–18
Islam N, Cleary MJ (2012) Developing an efficient and reliable dry powder inhaler for pulmonary drug delivery—a review for multidisciplinary researchers. Med Eng Phys 34:409–427
Joo HM, He Y, Sangster MY (2008) Broad dispersion and lung localization of virus-specific memory B cells induced by influenza pneumonia. Proc Natl Acad Sci U S A 105:3485–3490
Kesser KC, Geller DE (2009) New aerosol delivery devices for cystic fibrosis. Respir Care 54:754–767
Khanum S, Uddin N, Garelick H et al (1987) Comparison of Edmonston-Zagreb and Schwarz strains of measles vaccine given by aerosol or subcutaneous injection. Lancet 150–153
Khatri L, Taylor KM, Craig DQ et al (2001) An assessment of jet and ultrasonic nebulisers for the delivery of lactate dehydrogenase solutions. Int J Pharm 227:121–131
Kohlmeier JE, Woodland DL (2009) Immunity to respiratory viruses. Annu Rev Immunol 27:61–82
Koopman M, Zanen P, Kruitwagen CLJJ et al (2011) Reference values for paediatric pulmonary function testing: The Utrecht dataset. Respir Med 105:15–23
Laube BL (2005) The expanding role of aerosols in systemic drug delivery, gene therapy, and vaccination. Respir Care 50:1161–1176
Laube BL, Janssens HM, de Jongh FHC et al (2011) What the pulmonary specialist should know about the new inhalation therapies. Eur Respir J 37:1308–1331
Le Brun PP, De Boer AH, Heijerman HG et al (2000) A review of the technical aspects of drug nebulization. Pharm World Sci 22:75–81
Lexmond AJ, Hagedoorn P, Frijlink HW et al (2013) Challenging the two-minute tidal breathing challenge test. J Aerosol Med Pulm Drug Deliv 26:1–7
Lin W-H, Griffin DE, Rota PA et al (2011) Successful respiratory immunization with dry powder live-attenuated measles virus vaccine in rhesus macaques. Proc Natl Acad Sci U S A 108: 2987–2992
Low N, Kraemer S, Schneider M et al (2008) Immunogenicity and safety of aerosolized measles vaccine: systematic review and meta-analysis. Vaccine 26:383–398
Lu D, Garcia-Contreras L, Muttil P et al (2010) Pulmonary immunization using antigen 85-B polymeric microparticles to boost tuberculosis immunity. AAPS J 12:338–347
Lycke N (2012) Recent progress in mucosal vaccine development: potential and limitations. Nat Rev Immunol 12:592–605
Martin TR, Frevert CW (2005) Innate immunity in the lungs. Proc Am Thorac Soc 2:403–411
Mbawuike IN, Pacheco S, Acuna CL et al (1999) Mucosal immunity to influenza without IgA: an IgA knockout mouse model. J Immunol 162:2530–2537
McCallion ONM, Taylor KMG, Thomas M et al (1995) Nebulization of fluids of different physicochemical properties with air-jet and ultrasonic nebulizers. Pharm Res 12:1682–1688
Menzel M, Muellinger B, Weber N et al (2005) Inhalative vaccination with pneumococcal polysaccharide in healthy volunteers. Vaccine 23:5113–5119
Minne A, Louahed J, Mehauden S et al (2007) The delivery site of a monovalent influenza vaccine within the respiratory tract impacts on the immune response. Immunology 122:316–325
Minne A, Huaux F, Jaworska J et al (2008) Safety evaluation of pulmonary influenza vaccination in healthy and "asthmatic" mice. Vaccine 26:2360–2368
Moyron-Quiroz JE, Rangel-Moreno J, Kusser K et al (2004) Role of inducible bronchus associated lymphoid tissue (iBALT) in respiratory immunity. Nat Med 10:927–934
Muttil P, Prego C, Garcia-Contreras L et al (2010) Immunization of guinea pigs with novel hepatitis B antigen as nanoparticle aggregate powders administered by the pulmonary route. AAPS J 12:330–337
Nardelli-Haefliger D, Lurati F, Wirthner D et al (2005) Immune responses induced by lower airway mucosal immunisation with a human papillomavirus type 16 virus-like particle vaccine. Vaccine 23:3634–3641

Onodera T, Takahashi Y, Yokoi Y et al (2012) Memory B cells in the lung participate in protective humoral immune responses to pulmonary influenza virus reinfection. Proc Natl Acad Sci U S A 109:2485–2490

Patton JS, Byron PR (2007) Inhaling medicines: delivering drugs to the body through the lungs. Nat Rev Drug Discov 6:67–74

Phipps PR, Pharm B, Gonda I (1990) Droplets produced by medical nebulizers. Some factors affecting their size and solute concentration. Chest 97:1327–1332

Renegar KB, Small PA (1991) Immunoglobulin A mediation of murine nasal anti-influenza virus immunity. J Virol 65:2146–2148

Renegar KB, Small PA, Boykins LG et al (2004) Role of IgA versus IgG in the control of influenza viral infection in the murine respiratory tract. J Immunol 173:1978–1986

Rosenthal SR, McEnery JT, Raisys N (1968) Aerogenic BCG vaccination against tuberculosis in animal and human subjects. J Asthma 5:309–323

Saluja V, Amorij J-P, Kapteyn JC et al (2010) A comparison between spray drying and spray freeze drying to produce an influenza subunit vaccine powder for inhalation. J Control Release 144:127–133

Sato S, Kiyono H (2012) The mucosal immune system of the respiratory tract. Curr Opin Virol 2:225–232

Shoyele SA, Slowey A (2006) Prospects of formulating proteins/peptides as aerosols for pulmonary drug delivery. Int J Pharm 314:1–8

Sievers RE, Cape SP, McAdams DH et al (2012) Optimizing the effectiveness of live-attenuated measles vaccine aerosols : effects of different delivery modes. In: Respiratory drug delivery conference, Phoenix, 2012

Smith DJ, Bot S, Dellamary L et al (2003) Evaluation of novel aerosol formulations designed for mucosal vaccination against influenza virus. Vaccine 21:2805–2812

Smyth H (2005) Propellant-driven metered-dose inhalers for pulmonary drug delivery. Expert Opin Drug Deliv 2:53–74

Teijaro JR, Turner D, Pham Q et al (2011) Cutting edge: tissue-retentive lung memory CD4 T cells mediate optimal protection to respiratory virus infection. J Immunol 187:5510–5514

Thomas C, Gupta V, Ahsan F (2009) Influence of surface charge of PLGA particles of recombinant hepatitis B surface antigen in enhancing systemic and mucosal immune responses. Int J Pharm 379:41–50

Thomas C, Gupta V, Ahsan F (2010) Particle size influences the immune response produced by hepatitis B vaccine formulated in inhalable particles. Pharm Res 27:905–919

Thomas C, Rawat A, Hope-Weeks L et al (2011) Aerosolized PLA and PLGA nanoparticles enhance humoral, mucosal and cytokine responses to hepatitis B vaccine. Mol Pharm 8: 405–415

Todoroff J, Ucakar B, Inglese M et al (2013) Targeting the deep lungs, Poloxamer 407 and a CpG oligonucleotide optimize immune responses to Mycobacterium tuberculosis antigen 85A following pulmonary delivery. Eur J Pharm Biopharm 84:40–48

Tonnis WF, Lexmond AJ, Frijlink HW et al (2013) Devices and formulations for pulmonary vaccination. Expert Opin Drug Deliv 10:1383–1397

Van Riet E, Ainai A, Suzuki T et al (2012) Mucosal IgA responses in influenza virus infections; thoughts for vaccine design. Vaccine 30:5893–5900

Vecellio L, Abdelrahim ME, Montharu J et al (2011) Disposable versus reusable jet nebulizers for cystic fibrosis treatment with tobramycin. J Cyst Fibros 10:86–92

Vujanic A, Wee JLK, Snibson KJ et al (2010) Combined mucosal and systemic immunity following pulmonary delivery of ISCOMATRIX adjuvanted recombinant antigens. Vaccine 28:2593–2597

Waldman RH, Mann JJ, Small PA (1969) Immunization against influenza: prevention of illness in man by aerosolized inactivated vaccine. JAMA 207:520–524

Wang W (2000) Lyophilization and development of solid protein pharmaceuticals. Int J Pharm 203:1–60

Wee JLK, Scheerlinck J-PY, Snibson KJ et al (2008) Pulmonary delivery of ISCOMATRIX influenza vaccine induces both systemic and mucosal immunity with antigen dose sparing. Mucosal Immunol 1:489–496

Chapter 17
Vaginal Delivery of Subunit Vaccines

Deborah Lowry

17.1 Vaginal Drug Delivery

Vaginal drug administration has a long history dating back to the Middle Ages. This route of administration has long been employed for postmenopausal delivery of hormones, and vaginal administration of recreational drugs such as cocaine has been reported in the medical literature. Clinicians are already using drugs routinely in the vagina; bromocriptine is used vaginally in the treatment of hyperprolactinemia in women who suffer from nausea and vomiting with oral administration. A study of vaginal delivery of misoprostol compared with oral administration for the induction of labour showed the vaginal route to be superior (Zieman and Fong 1997), while a further study demonstrated that vaginally administered indomethacin was more effective than oral indomethacin for halting preterm labour (Abramov et al. 2000).

17.2 The Human Vagina Anatomy and Physiology

The vagina is the terminal portion of the female genital tract. It is a fibromuscular tube lined by non-keratinized stratified epithelium which extends from the vestibule to the uterus. The bladder and urethra lie anteriorly; the rectum and anal canal lie posteriorly. The vagina ascends in a shallow S-shaped curve at an angle of over 90° to the uterine axis, but which varies with the contents of the bladder and rectum (Gray and Williams 1995). It extends from the lower part of the uterine cervix to the

D. Lowry (✉)
Aston Pharmacy School, Aston University, Aston Triangle, Birmingham, UK
e-mail: d.lowry@aston.ac.uk

C. Foged et al. (eds.), *Subunit Vaccine Delivery*, Advances in Delivery Science and Technology, DOI 10.1007/978-1-4939-1417-3_17

external part of the vulva known as the labia minora. The vault of the vagina is divided into four areas relative to the cervix. These are the posterior fornix, which is capacious, the anterior fornix, which is shallow and two lateral fornices. The inner surface of the walls of the vagina, are ordinarily in contact with the other, its lumen forming an H-shaped cleft. The anterior wall of the vagina averages 6–7.5 cm in length; the posterior is slightly longer about 9 cm; its width increases as it ascends (Gray and Williams 1995).

In addition to the barrier properties offered by the vaginal epithelial tissue, the cervical mucus, vaginal secretions and local bacterial flora also help to protect the vagina against infection. The stratified squamous epithelium sheds constantly making it difficult for organisms to invade or access the basement membrane/capillary bed (Alexander et al. 2004).

17.2.1 Mucosa

The mucosa of the vagina adheres firmly to the muscular layer. It has numerous transverse folds or rugae of about 2–5 mm in thickness and is lined with non-keratinized stratified squamous epithelium. The cellular structure of the vaginal mucosal epithelium consists of five distinct cytological layers: the basal, parabasal, intermediate, transitional and superficial layers. The basal cells, typically cuboidal in shape and characterised by the presence of microvilli on the surface of the cell membrane, are responsible for the continuous production of squamous cells. Parabasal cells are polygonal in shape and differ slightly from the basal cells in having a substantially greater amount of surface microvilli (Tubin and Novak 1956). These cells store considerable amounts of glycogen. The vaginal wall consists of an inner mucosal layer, an intermediate muscular layer and an outer adventitial layer.

On average, the human vaginal epithelium consists of 32 ± 5 layers in thick regions and 19 ± 5 layers in thin regions (Thompson and Van der Bijl 2001). Beneath the squamous epithelial cells lies the lamina propria, which exhibits two distinct regions. The outer region immediately below the epithelium is highly cellular loose connective tissue. The deeper region, adjacent to the muscular layer, is more dense and the cells here are joined by junctional complexes. This layer may be considered a submucosa. The epithelium must therefore receive nourishment from the underlying connective tissue, which has larger intercellular spaces that can accommodate blood vessels and nerves. Numerous elastic fibres are present immediately below the epithelium and some of the fibres extend into the muscular layer. Many lymphocytes and leukocytes are found in the lamina propria and many migrate into the epithelium. The number of lymphocytes and leukocytes in the mucosa and vaginal lumen dramatically increases around the time of menstrual flow (Ross and Lee 1995).

17.2.2 Arterial Supply and Venous Drainage

Arteries, blood vessels and lymphatic vessels are abundant in the walls of the vagina. The main blood supply to the vagina originates from the internal iliac artery; the uterine artery and ovarian artery supply the anterior and posterior walls of the vagina and the lower urethra with oxygen and nourishment. The veins are continuous with the course of the arteries and form a plexus which communicates with the vesicle, uterine and rectal plexuses and with the veins of the vulva. These drain into the internal iliac vein.

17.2.3 Immune Cells

The lymphatic drainage of the vagina is distributed between the left and right sides of the pelvis. Generally the upper third of the vagina drains into the external iliac nodes, the middle third drains into the common and internal iliac nodes and the lower third drains into the common iliac, superficial vaginal and perirectal nodes. Protective immunity is provided by both the cellular and humoral systems. Langerhans cells can be found with dendritic extensions exposed to the lumen of the vaginal epithelium, thus possibly serving as guardians of the local immune system. These cells can pass antigens to dendritic cells (DCs) that migrate to the lymph nodes where they activate B-cells and $CD4^+$ T-cells. Activated B lymphocytes return to the subepithelium where they become IgA-secreting cells. The IgA is taken up by the epithelial cells and dimerized prior to release into the lumen. Cervical mucus contains both IgG and IgM as well as IgA antibodies (McGhee et al. 1992). An antigenic challenge at the epithelial surface is afforded by intraepithelial T lymphocytes, DCs and a subepithelial population of B lymphocytes that synthesize IgA locally (Alexander et al. 2004).

17.2.4 Vaginal pH

The vaginal pH varies throughout life, being approximately 6 at birth, rising to near neutrality within 1 month and staying at this value until puberty (Lang 1995). In a healthy mature woman, the vaginal fluid has a typical pH of between 3.5 and 5 thereby creating an inhospitable environment for the growth of most endogenous pathogenic bacteria. Desquamated cells have a secondary use to provide a source of intracellular glycogen that can be converted to lactic acid by the lactobacilli that proliferate near the epithelium. The pH value is maintained by *Lactobacillus acidophilus* which produces lactic acid from the glycogen contained in the sloughed mature cells of the vaginal mucosa (Graves et al. 1980; Croughan and

Behbehani 1988; Pettit et al. 1999; Aroutcheva and Simoes 2001). Some species of *Lactobacillus* suppress the growth of other endogenous bacteria in the vagina through the production of organic acids such as lactic acid, hydrogen peroxidase and bacteriocins or lactocins (Aroutcheva and Simoes 2001; Mardh 1991; Antonio et al. 1999). Vaginal secretions contain a mixture of aerobic and anaerobic bacterial flora, at an average concentration of 10 million/mL in healthy women of reproductive age (Roy et al. 1994).

17.2.5 Vaginal Fluid

The vaginal fluid consists primarily of transudate which passes through the vaginal wall from the blood vessels. It is mixed with vulval secretions from sebaceous and sweat glands, with minor contributions from Bartholin's and Skene's glands (Burgos and Roig de Varnas-Linares 1978; Deschpande 1992). The fluid then becomes contaminated with cervical mucus and sloughed cells from the vaginal epithelia. Endometrial and oviduct fluids may also contribute to its chemical composition. Vaginal fluids may also contain various proteinase inhibitors, proteins, carbohydrates, lactic acid, acetic acid, glycerol, urea, glycogen, glucose, hydroxy-ketones and aromatic compounds (Owen and Katz 1999). The chief component of vaginal fluid is cervical mucus, produced by glandular units within the cervical canal. Studies suggest that approximately 6 g of vaginal fluid is produced daily with approximately 0.5–0.75 g present at any one time in the vagina. Human cervical mucus contains proteins in two basic forms, soluble and mucin. Mucin plays an important role in the physicochemical characteristics of cervical mucus, including viscosity and surface tension (Wang and Lee 2002). The epithelium of the cervix and its glands secrete mucin which has a defence function in the uterus and vagina. Mucin keeps pathogens out, maintains moisture and acts as a lubricant.

Vaginal hormonal levels are affected during hormonal changes introduced by pregnancy and delivery. Medication has been found to have a direct influence on the vaginal microflora, bacterial adherence to the vaginal epithelium and local vaginal antibodies. A significant factor in the pathogenesis of any infection appears to be the ability of bacteria to adhere to the epithelial surface. The degree of adherence is governed by the nature of the bacterial and vaginal cells themselves, for example the hydrophobic/hydrophilic nature of bacterial cell membranes and surface epithelial antibodies.

17.3 Advantages of Vaginal Drug Delivery

The main advantages of vaginal drug delivery are summarized in Table 17.1. The vagina has a rich system of defences and a dynamic microbiology, as well as a rich vascular plexus that makes it ideal for absorbing drugs. Vaginal administration enables the use of prolonged dosing regimes, lower daily doses and continuous

Table 17.1 Advantages of vaginal drug delivery

Advantages of vaginal drug delivery
Large surface area (rugae)
Dense vascular network
Avoids first-pass effect
Local delivery minimizes side effects
Ease of administration

release of medication. The presence of vaginal rugae increases the surface area of the vagina, helping to retain a medication that is placed in the vagina and enhancing drug absorption (Nelson 2005).

The vaginal histology is defined by a stratified squamous epithelium; the upper vagina is a mucous membrane that is highly vascular and contains elastic fibres, as well as dense connective tissue and muscle. Arteries and veins form a plexus around the vagina, and venous drainage of the vagina does not immediately pass through the liver. These features make the vagina effective in the uptake of many drugs and stable for long-term, continuous drug delivery systems such as a vaginal ring (Nelson 2005).

The vaginal defence system is analogous to those in the gastrointestinal (GI) and respiratory tracts. Langerhans cells (specialized DCs) scavenge bacterial antigens on the vaginal surface and present the antigens to T- and B-cells, which eventually results in cell-mediated immunity and in the secretion of vagina-specific antibodies. In addition, continuous epithelial sloughing creates a physical barrier to bacterial invasion.

One of the major advantages of vaginal administration over oral administration is that drugs avoid GI absorption and the hepatic first-pass effect. Blood leaving the vagina enters the peripheral circulation via a rich venous plexus which empties primarily into the internal iliac veins and ultimately into the vena cava, thus initially bypassing the portal circulation (Richardson and Illum 1992). GI blood drains into the portal vein and is passed directly through the liver before reaching the general circulation and target tissues. Absorption from the GI tract can be unpredictable and may be compromised by vomiting, drug–drug interference or decreased intestinal absorption capacity. Also the GI lumen and the liver are sites of elimination for many compounds (Wu and Benet 1995). Avoidance of the hepatic first-pass effect is particularly advantageous for compounds that undergo a high degree of hepatic metabolism. Therefore the potential benefits include lower dosing and lower systemic exposure plus lower incidences of side effects while achieving the same pharmacodynamic effect (Alexander et al. 2004).

Hormones placed in the vagina are absorbed as through other mucosae. Absorption is dependent upon transport in blood and/or lymph (Einer-Jensen and Kotwica 1993). Lymph vessels are found that run from the cranial part of the vagina toward those originating from the uterine cervix, both ending into the hypogastric lymph glands (Williams and Warwick 1989). The lymphatic system of the upper part of the vagina being in direct communication with those of the uterus may represent a potential route of substances applied into the vagina (Cicinelli and deZiegler 1999).

17.4 Factors Affecting the Vaginal Absorption of Drugs

Drugs administered vaginally can be transported across the vagina membrane by a number of different mechanisms. These are (1) the intracellular route by diffusion through the cell due to a concentration gradient, (2) by a vesicular or receptor-mediated transport mechanism and (3) by the intercellular route where there is diffusion between the cells through the tight junctions (Richardson and Illum 1992).

17.4.1 Physiological Factors

Cyclic changes in the thickness and porosity of the vaginal epithelium may affect the vaginal absorption of drugs. It is thought that the increased vascularity of the vagina could contribute to the improved absorption (Richardson and Illum 1992). It is thought that the volume, viscosity and pH of vaginal fluids may also affect the absorption of drugs. The presence of a film of moisture on the vaginal epithelium is an advantage as the drug must be in solution before it is absorbed. However, the presence of thick cervical mucus may present a barrier to drug absorption, and a large volume of vaginal secretions may cause the removal of a vaginal dosage form, thereby reducing absorption. The pH can also affect vaginal absorption, as most drugs are weak electrolytes and it is expected that they will be absorbed more readily when unionized. The pathways for drug diffusion across vaginal epithelium are essentially similar to other epithelial tissues (Richardson and Illum 1992) and are well represented by the 'fluid mosaic model' as a lipid continuum interspersed with aqueous pores, the latter forming an aqueous 'shunt' route (Singer and Nicolson 1972). The lipid continuum predominates in vaginal drug absorption (Woolfson 2003).

17.4.2 Physiochemical Properties

Vaginal absorption of drugs may also be affected by the physicochemical properties of the drug itself such as molecular weight, lipophilicity, ionization, surface charge and chemical nature (Hussain and Ahsan 2005). Vaginal permeability is much greater to lipophilic steroids such as progesterone and estrone than to hydrophilic steroids such as hydrocortisone and testosterone (Brannon-Peppas 1992). It is generally accepted that low molecular weight lipophilic drugs are absorbed to a higher extent than the high molecular weight hydrophilic drugs. However, it has been suggested that higher molecular weight drugs are absorbed more in the vagina than for other mucosal surfaces (Sanders and Matthews 1990). Drugs intended for vaginal delivery should show some degree of solubility in water as vaginal fluid contains a large amount of water. A proposed physical model for the uptake of drugs across the vaginal epithelium suggests that the epithelium could be regarded as an aqueous diffusion layer in series with a membrane consisting of aqueous pores and lipoidal pathways (Hwang and Owada 1976). The external cell layers and the basal cell layers of the vagina retain most of the enzyme activity (Woolfson and Malcolm 2000). Among the enzymes present proteases are likely to be the prominent barrier for the absorption of intact peptide and protein drugs and vaccines into the systemic circulation (Lee 1988).

For drugs with a high vaginal membrane permeability coefficient, absorption is mainly controlled by the permeability across the hydrodynamic diffusion layer formed by vaginal fluid sandwiched between the vaginal epithelial membrane and the delivery device (Chien 1982). For drugs with low vaginal membrane permeability, vaginal absorption is mainly controlled by the permeability across the vaginal epithelium (Chien 1982). For systemic drug delivery to occur, the penetrating substance must have sufficient lipophilicity to diffuse through the lipid continuum of the membrane, but also require some degree of aqueous solubility to ensure dissolution in vaginal fluid (Woolfson et al. 2003).

17.5 Vaginal Immunity

Developments in controlled release methods for drug delivery have extended the means by which vaccine antigens may be delivered with microspheres, implants and pumps being available. These systems introduce prolonged antigen delivery profiles as an alternative delivery technique, in contrast to bolus administration by injection or short-term delayed release resulting from injection with adjuvants such as alum or emulsions (Lofthouse 2002).

There is great interest and demand for the development of vaccines that target mucosal sites of infection so as to prevent virus entry and/or the establishment and dissemination of infection (Azizi et al. 2010; Chen and Cerutti 2010). Numerous strategies have been proposed to reduce or prevent sexually acquired human immunodeficiency virus (HIV) infection. An effective HIV vaccine would be the most obvious solution, but despite extensive research, an effective HIV vaccine for human use has not been successfully developed. Mucosal immunization with an HIV-1 vaccine aims at inducing a specific humoral response either systemically or locally as well as inducing cellular immunity (Letvin 1998; Mascola et al. 2000). It has been shown that for HIV-1, the envelope spike is the only viral target available for neutralizing antibodies (Zhou et al. 2007) resulting in the majority of vaccine candidates being targeted to the envelope glycoproteins of the virus. Many vaginal vaccination approaches involve antigen in a buffer solution (Bernstein 2000; Kwant and Rosenthal 2004), which have the potential for leakage, rapid enzymatic degradation of the antigen and inadequate exposure of antigen to the mucosal lymphoid tissue possibly resulting in ineffective uptake of the antigen by the vaginal mucosa (O'Hagan 1992; Park et al. 2003).

17.5.1 Structure of HIV

HIV is spherical in shape with a diameter of approximately 100 nm and containing an electron-dense core surrounded by a lipid envelope derived from the host cell membrane (Fig. 17.1). The outer core of the virus, known as the viral envelope, is composed of a phospholipid bilayer derived from the membrane of the host cell from which a newly formed virus particle buds. Embedded in this envelope are proteins from the host cell which often form spikes. The virus core contains several core proteins, two strands of genomic RNA and the enzyme reverse transcriptase. The major HIV proteins associated with the envelope are gp120 and gp41. Gp160 is a glycoprotein that is cleaved into gp120 and gp41. These function as the viral attachment proteins, binding to the CD4 receptors on certain cells and facilitating entry into the cells. Gp120 is a glycoprotein which is heavily glycosylated and is bound to the outside of the membrane and it is noncovalently attached to gp41, the transmembrane protein spanning the bilayer.

A cellular protease cleaves gp160 to generate gp41 and gp120. The gp41 moiety contains the transmembrane domain of *Env*, while gp120 is located on the surface of the infected cell and of the virion through noncovalent interactions with gp41. *Env* exists as a trimer on the surface of infected cells and virions (Bernstein et al. 1995). T- and B-cells may be attracted to mucosal sites by cytokines such as tumour necrosis factor (TNF)-α and interferon (INF)-γ which are secreted by DCs in the vagina and macrophages in the vaginal subepithelium (Lehner et al. 1991, 1994; Lehner 2003). Macrophages, DCs, and natural killer (NK) cells also secrete chemokines such as RANTES, macrophage inflammatory protein (MIP)-1α, and MIP-1β (CCL 5, 3, and 4) which can bind to M-tropic-CCR5 or T-tropic-HIV-1-CXCR4

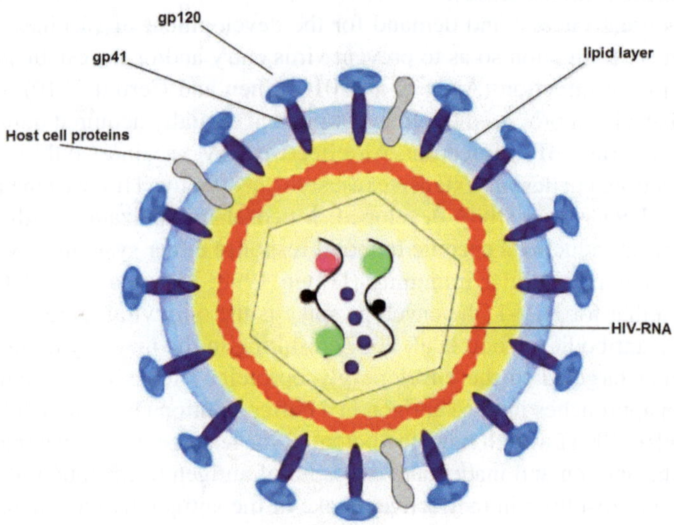

Fig. 17.1 Structure of the human immunodeficiency virus (HIV)

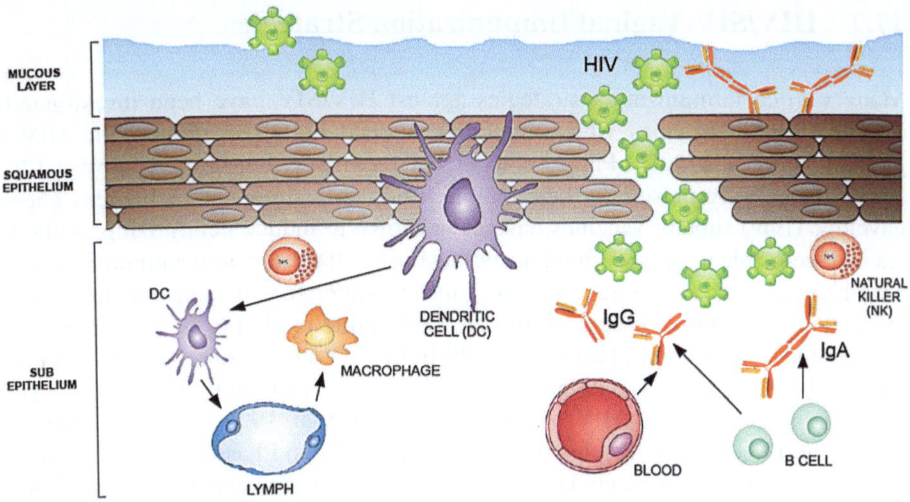

Fig. 17.2 Vaginal immunity and entry of HIV. IgA, produced by mucosal B cells, is present in vaginal secretions which intercept antigens. Mucosal IgG is derived from the blood by diffusion from local fenestrated capillaries. Pathogens may be captured by dendritic cells (DCs) and macrophages which are carried to draining nodes. The current consensus is that at least two main routes are available for HIV to cross the vaginal epithelium, transepithelial migration of infected Langerhans cells and penetration of the virus through damaged epithelial tissue

co-receptors and accordingly mediate HIV infection in vitro. Some studies have also shown that high amounts of these chemokines may down-regulate the cell-surface expression of the CCR5 receptor (Cocchi et al. 1995). NK cells are able to destroy HIV-infected cells directly or through antibody-dependent cellular toxicity (ADCC) (Chung et al. 2009). These findings support the view that innate immunity may control HIV-1 replication. Figure 17.2 presents the mechanisms of vaginal immunity and shows how HIV can penetrate through the vaginal mucosal.

17.6 Vaccines for the Female Genital Tract

The female genital tract is considered a component of the common mucosal immune system (McDermott and Bienenstock 1979). Mucosal immune responses in the genital tract can be induced by the administration of antigen to distal or local mucosal surfaces (Sato and Igarashi 1990). Studies comparing immunization at the female genital tract by delivering plasmid DNA intranasally, intrarectally and vaginally demonstrate that vaginal immunization induces better mucosal immunity (Livingston and Lu 1998). Professional antigen-presenting cells (DCs, Langerhans cells), T-cells and B-cells populate the cervix and vagina of the human and murine female genital tract, indicating the potential for production of mucosal immunity at the genital tract by the local application of plasmid DNA (Parr and Parr 1991). Ease of access to the vaginal surface also makes the local immunization practically possible.

17.7 HIV/SIV Vaginal Immunization Strategies

Many vaginal immunization strategies against HIV/SIV have been investigated. Protein subunit vaccines were initially developed based on monomeric HIV-1 gp160 (Zhang et al. 2004; Phogat et al. 2008), these were later followed by gp120. The antigens were present in their soluble form using alum as an adjuvant. These envelope (Env) subunit vaccines have been shown to induce neutralizing antibodies and were able to protect chimpanzees against a challenge with a homologous of near-homologous HIV-1 strain but not against a challenge with a distant virus strain (Berman et al. 1990, 1996; Girard et al. 1991; Fultz et al. 1992). Similar results were found with vaccines based on gp120 in the SIV/macaque model. A trial carried out between 1988 and 1996 on a candidate Env vaccine found no induction of neutralizing antibodies to HIV-1 isolates (McElrath et al. 1996; Burton and Moore 1998). Modified Env molecules were developed and two Phase II clinical trials of a gp120 subunit vaccine found that neither study showed a statistically significant reduction of HIV infection (Graham et al. 1998; Gilbert et al. 2010).

Various formulations for vaginal delivery have been investigated for HIV vaccines. These are listed in Table 17.2 and summarized below. A tampon containing 200 μg of ADP 740-8 resulted in enhanced levels in the rat of vaginal IgA and IgG. ADP 740-8 is a peptide consisting of residues of 102–121 HIV-1 gp120 in lysophosphatidyl glycerol (LPG) (O'Hagan 1992). Several strategies have been investigated in mice. A phosphate-buffered saline (PBS) solution containing 20 μg pcMN160 induced the production of vaginal immunoglobulins (IgG > IgA) which specifically bind to HIV-1 *env* and neutralize cell-free HIV-1 infectivity in vitro (Wang et al. 1997). VC1/CT, a vaccine candidate, was administered as a wax cylinder with 15, 75 and 150 μg of VC1 and 10 μg of CT at 2-week intervals. A strong antigen-specific IgA response was found in the vaginal wash resulting in a higher response after the second and third immunizations. It was also established that the vaginal route was better than the oral route (Kato et al. 2000). In the monkey model, a suspension containing 200 μL of 10^{11} cfu of *S. gordoni* (expresses the V3 domain of HIV-1 gp120) applied for 1 min on days 0, 54 and 64 induced vaginal IgA, serum IgG and T-cell-mediated immune responses (Di Fabio and Medaglini 1998).

Various systems have been investigated to provide sustained and controlled release of vaccines vaginally (Pavelic et al. 2001, 2005). Liposomes have been found to be compatible with agents that increase the viscosity of delivery vehicles

Table 17.2 HIV vaccine strategies

System	Formulation	Animal/human	Results
Tampon	200 μg ADP 740-8	Rat	Vaginal IgA and IgG
PBS	20 μg pcMN160	Mouse	Vaginal IgG and IgA (IgG > IgA)
Wax	15, 75, 150 μg VC1 10 μg CT	Mouse	Strong IgA response
Suspension	10^{11} cfu *S. gordoni*	Monkey	Vaginal IgA, serum IgG, T-cells

such as cellulose derivatives and acrylic acid polymers (Foldvari 1996; Skalko et al.. 1998). Liposomes can be incorporated into hydrogels which provide good mucoadhesive properties (Woolfson et al. 2003; Owen et al. 2000). A further formulation of lyophilized liposome hydroxyethyl cellulose (HEC) rods was evaluated by Gupta et al. (2012). The rods are designed to revert to a gel following vaginal application. The liposomes were found to exhibit good encapsulation efficiency and mucoadhesive properties offering a potential dosage form.

17.8 Non-HIV Immunization Strategies

A number of non-HIV immunization strategies have also been reported. These are summarized in Table 17.3. Several strategies have been carried out in the mouse vagina model. One dose of a PBS solution containing 10^8 cfu of *S. gordonii* resulted in an induction of HPV-specific vaginal IgA and serum IgG (Medaglini and Rush 1997). Vagina ferritin releasing poly(ethylene-co-vinyl acetate-co-carbon monoxide) (PEVAc) vaginal rings were applied every 10 days or ferritin soaked tampons were applied daily for 10 days. The single vaginal ring produced higher titres than the multiple tampons (Wyatt and Sodroski 1998). A solution containing 20 μL of 6.0×10^6 PFU/mL of attenuated herpes simplex virus 2 (HSV-2) resulted in an increase in the vaginal production of IgG. Vaginal immunization protected the vagina against a challenge with wild-type HSV-2 (Parr and Parr 1999). In the guinea-pig model, a PBS solution containing a gH-deleted disabled infectious single cycle (DISC) HSV-1 vaccine was applied either intranasally, orally or vaginally in two doses 18 days apart. It caused a reduction in the primary disease symptoms in all the vaccination regimes. However the intranasal immunization route resulted in greater levels than the vaginal route, and the vaginal immunization route resulted in greater levels than the oral immunization route. The vaginal immunization gave the biggest reduction in the vaginal virus titres (McClean and Reid 1995). A poly(ethylene-co-vinyl) acetate matrix was compared to a PBS solution containing HSV-2 gp B (rgB)/CpG oligodeoxynucleotide. A total loading of 45 μg was in the matrix and was used for 16 weeks. There was an induction of specific IgA in the vaginal secretions after 2 weeks and this was maintained until 11 weeks. The DNA in the PBS did

Table 17.3 Non-HIV vaccine strategies

System	Formulation	Animal/human	Results
PBS	*S. gordoni*	Mouse	HPV-specific vaginal IgA and IgG
Solution	HSV-2	Mouse	Increased vaginal IgG
PBS	HSV-1	Guinea-pig	Reduction in vaginal titres
Vinyl acetate matrix	HSV-2	Guinea-pig	Induction of vaginal IgA
Gel	HPV	Guinea-pig	Induction vaginal IgA
Gel/solution	CTB	Human	Serum IgA and IgG
Gel	CTB	Human	Detectable IgA and IgG

not induce detectable IgA levels and there was no serum IgA detected (Sher and Fisch 2002). A thermoreversible gel system containing 100 μg of HPV-16 L1 protein or 100 μg of DNA plasmid was administered in 10 μL of gel. An initial dose was given followed by a boost 21 days later. The use of the mucoadhesive, thermoreversible systems significantly enhanced the vaginal IgA titres compared to the buffer solutions (Han and Kim 2006). Several strategies have also been applied to the human vagina. A 1 mL solution containing inactivated Salk polio vaccine was administered into the human vagina or uterus on three successive days. This resulted in an induction of IgA and IgG in the vaginal secretions which were detectable between 1 and 20 weeks post vaccination (Ogra and Ogra 1973). A solution containing 1 mg of cholera toxin B (CTB) was applied three times at 14-day intervals and produced an increased IgA in the genital tract secretions (Kozlowski and Cu-Uvin 1999). A 2 mL polysaccharide gel/ PBS solution containing 1.0 mg CTB was applied and was boosted after 2 weeks. It was compared with nasal administration. The vaginal and nasal administration induced the IgA and IgG in the serum (Johansson et al. 2001).

A clinical trial conducted by Wassen et al. (2006) compared local vaginal and oral vaccination of a whole cell/B subunit (CTB) oral cholera vaccine. Two groups of unimmunized volunteers received either an oral vaccine or a local vaccination applied vaginally. The CTB vaccine was mixed with a gel (eldexomer) for vaginal application. Three doses of vaccine were administered at 2-week intervals. The first dose was given on day 10 of the menstrual cycle. Before the first dose and at 8–10 days following the final vaccination peripheral blood and cervical mucus was sampled. Detectable IgA and IgG anti-CTB antibodies were found in 6 out of the 7 vaginally immunized volunteers compared to 3 out of 7 for the oral vaccine. Serum anti-CTB IgG increased in orally vaccinated volunteers with 4 out of 7 exhibiting specific IgA serum titres. Only 3 of out 7 vaginally vaccinated volunteers showed an increase in serum IgG and IgA.

The major factor which impedes absorption at mucosal sites is the low and incomplete transport across the epithelial barrier. A transient and reversible opening of the tight junctions between the epithelium cells by safe penetration enhancers would allow for the permeation of non-absorbable drugs across the epithelial barrier and subsequent uptake of the drugs into the systemic blood circulation (Junginger and Verhoef 1998). For vaccine delivery, the lymphoid tissue should be targeted. Access to mucosal lymphoid tissue is provided by antigen-sampling cells. These M cells are located in between the epithelial cells and take up antigens and microparticles smaller than 10 μm (Wang and Xiang 1998). However, the use of penetration enhancers for the administration of HIV vaccines would be a very risky strategy since the tissue disruption associated would also facilitate uptake of the virus, and provide access to a much higher population of immune cells contained within the underlying tissue. Another possibility is to sustain the release of antigen at the mucosal surface through increased residence time of the antigen delivery system. In an attempt to overcome these problems, the aim is to develop a non-invasive, female controlled, antigen-delivery modality with enhanced retention and immunogenicity in which prolonged duration of delivery may sustain protective mucosal immune responses.

17.9 Conclusion

The large surface area of the vagina allows for efficient absorption of drugs and vaccines. Studies have found local and systemic production of IgA and IgG after delivery of subunit vaccines for HIV and HSV. The studies detailed above have shown that the vagina has potential for the delivery of subunit vaccines. Further research must be explored to enhance understanding of the immune response found in the vagina.

Acknowledgments The author gratefully acknowledges Damien Lowry for providing Fig. 17.2.

References

Abramov Y, Nadjari M, Weinstein D, Ben-Shachar I, Plotkin V, Ezra Y (2000) Indomethacin for oreterm labor: a randomized comparison of vaginal and rectal-oral routes. Obstet Gynecol 95(4):482–486

Alexander NJ, Baker E, Kaptein M, Karck U, Miller L, Zampaglione E (2004) Why consider vaginal drug administration? Fertil Steril 82(1):1–12

Antonio MA, Hawes SE, Hillier SL (1999) The identification of vaginal Lactobacillus species and the demographic and microbiologic characteristics of women colonized by these species. J Infect Dis 180(6):1950–1956

Aroutcheva A, Simoes JA (2001) The inhibitory effect of clindamycin on Lactobacillus in vitro. Infect Dis Obstet Gynecol 9(4):239–244

Azizi A, Ghunaim H, Diaz-Mitoma F, Mestecky J (2010) Mucosal HIV vaccines: a holy grail or a dud? Vaccine 28:4015–4026

Berman PW, Gregory TJ, Riddle L, Nakamura GR, Champe MA, Porter JP (1990) Protection of chimpanzees from infection by HIV-1 after vaccination with recombinant glycoprotein gp120 but not gp160. Nature 345:622–625

Berman PW, Murthy KK, Wrin T, Vennari JC, Cobb EK, Eastman DJ (1996) Protection of MN-rgp120-immunized chimpanzees from heterologous infection with a primary isolate of human immunodeficiency virus type 1. J Infect Dis 173:52–59

Bernstein DI (2000) Effect of route of vaccination with vaccinia virus expressing HSV-2 glycoprotein D on protection from genital HSV-2 infection. Vaccine 18:1351–1358

Bernstein HS, Tucker SP, Kar SR, McPherson SA, McPherson DT, Dubay JW, Lebowitz J, Compans RW, Hunter E (1995) Oligomerization of the hydrophobic heptad repeat of gp41. J Virol 69:2745–2750

Brannon-Peppas L (1992) Novel vaginal drug release applications. Adv Drug Deliv Rev 11:169–177

Burgos MH, Roig de Varnas-Linares CE (1978) Ultrastructure of the vaginal mucosa. In: Hafez ESE, Evans TN (eds) The human vagina. Elsevier/North Holland Biomedical Press, Amsterdam

Burton DR, Moore JP (1998) Why do we not have an HIV vaccine and how can we make one? Nat Med 4:495–498

Chen K, Cerutti A (2010) Vaccination strategies to promote mucosal antibody responses. Immunity 33:479–491

Chien Y (1982) Intravaginal controlled release drug administration Novel drug delivery systems. Marcel Dekker, New York

Chung AW, Rollman E, Center RJ, Kent SJ, Stratov I (2009) Rapid degranulation of NK cells following activation by HIV-specific antibodies. J Immunol 182(2):1202–1210

Cicinelli E, deZiegler D (1999) Transvaginal progesterone: evidence for a new functional portal system flowing from the vagina to the uterus. Hum Reprod Update 5:365–372
Cocchi F, DeVico AL, Garzino-DemoA ASK, Gallo RC, Lusso P (1995) Identification of RANTES MIP-1 alpha, and MIP-1 beta as the major HIV-suppressive factors produced by CD8+ T cells. Science 27(5243):1811–1815
Croughan WS, Behbehani AM (1988) Comparative study of inactivation of herpes simplex virus types 1 and 2 by commonly used antiseptic agents. J Clin Microbiol 26(2):213–215
Deschpande A (1992) Intravaginal drug delivery. Drug Dev Ind Pharm 18:1225–1279
Di Fabio S, Medaglini D (1998) Vaginal immunization of Cynomolgus monkeys with Streptococcus gordonii expressing HIV-1 and HPV 16 antigens. Vaccine 16(5):485–492
Einer-Jensen N, Kotwica J (1993) Rapid absorption and local redistribution of progesterone after vaginal application in gifts. Acta Vet Scand 34(1):1–7
Foldvari M (1996) Effect of vehicle on topical liposomal drug delivery: petrolatum bases. J Microencapsul 13:589–600
Fultz PN, Nara P, Barre-Sinoussi F, Chaput A, Greenberg ML, Muchmore E (1992) Vaccine protection of chimpanzees against challenge with HIV-1-infected peripheral blood mononuclear cells. Science 256:1687–1690
Gilbert P, Wang M, Wrin T, Petropoulos C, Gurwith M, Sinangil F (2010) Magnitude and breadth of a nonprotective neutralizing antibody response in an efficacy trial of a candidate HIV-1 gp120 vaccine. J Infect Dis 202:595–605
Girard M, Kieny MP, Pinter A, Barre-Sinoussi F, Nara P, Kolbe H (1991) Immunization of chimpanzees confers protection against challenge with human immunodeficiency virus. Proc Natl Acad Sci U S A 88:542–546
Graham BS, McElrath MJ, Connor RI, Schwartz DH, Gorse GJ, Keefer MC (1998) Analysis of intercurrent human immunodeficiency virus type 1 infections in phase I and II trials of candidate AIDS vaccines. AIDS vaccine evaluation group, and the correlates of HIV immune protection group. J Infect Dis 177:310–319
Graves S, Gotv T et al (1980) Effect of pH on the motility and virulence of Treponema pallidum (Nichols) and Treponema paraluis-cuniculi in vitro under anaerobic conditions. Br J Vener Dis 56(4):269
Gray H, Williams P (1995) Grays anatomy the anatomical basis of medicine and surgery. Churchill Livingstone, London
Gupta PN, Pattani A, Curran R, Kett VL, Andrews GP, Morrow RJ, Woolfson AD, Malcolm RK (2012) Development of liposome gel based formulations for intravaginal delivery of the recombinant HIV-1 envelope protein CN54gp140. Eur J Pharm Sci 46:315–322
Han IK, Kim YB (2006) Thermosensitive and mucoadhesive delivery systems of mucosal vaccines. Methods 38(2):106–111
Hussain A, Ahsan F (2005) The vagina as a route for systemic drug delivery. J Control Release 103(2):301–313
Hwang S, Owada E (1976) Systems approach to vaginal delivery of drugs. II. In situ vaginal absorption of unbranched alipathatic alcohols. J Pharm Sci 65:1574–1578
Johansson EL, Wassen L et al (2001) Nasal and vaginal vaccinations have differential effects on antibody responses in vaginal and cervical secretions in human. Infect Immun 69(12): 7481–7486
Junginger H, Verhoef J (1998) Macromolecules as safe penetration enhancers for hydrophilic drugs—a fiction? Pharm Sci Technol Today 1(9):370–376
Kato H, Bukawa H et al (2000) Rectal and vaginal immunization with a macromolecular multicomponent peptide vaccine candidate for HIV-1 infection induces HIV-specific preventive immune responses. Vaccine 18(13):1151–1160
Kozlowski PA, Cu-Uvin S (1999) Mucosal vaccination strategies for women. J Infect Dis 179(S3):S493–S498
Kwant A, Rosenthal KL (2004) Intravaginal immunization with viral subunit protein plus CpG oligodeoxynucleotides induces protective immunity against HSV-2. Vaccine 22(23–24): 3098–3104

Lang W (1995) Vaginal acidity and pH: a review. Obstet Gynecol Surv 10:546–560
Lee VHL (1988) Enzymatic barriers to peptide and protein absorption. Crit Rev Ther Drug Carrier Syst 5:69–98
Lehner T (2003) Innate and adaptive mucosal immunity in protection against HIV Infection. Vaccine 21(S2):S68–S76
Lehner T, Hussain L, Wilson J, Chapman M (1991) Mucosal transmission of HIV. Nature 353(6346):709
Lehner T, Tao L, Panagiotidi C, Klavinskis LS, Brookes R, Hussain L (1994) Mucosal model of genital immunization in male rhesus macaques with a recombinant simian immunodeficiency virus p27 antigen. J Virol 68(3):1624–1632
Letvin NL (1998) Progress in the development of an HIV-1 vaccine. Science 280:1875–1880
Livingston J, Lu S (1998) Immunization of the female genital tract with a DNA-based vaccine. Infect Immun 66(1):322–329
Lofthouse S (2002) Immunological aspects of controlled antigen delivery. Adv Drug Deliv Rev 54(6):863–870
Mardh P (1991) The vaginal ecosystem. Am J Obstet Gynecol 165(4):1163–1168
Mascola JR, Frankel SS, Broliden K (2000) HIV-1 entry at the mucosal surface: role of antibodies in protection. Science 14:S167–S174
McClean HL, Reid M (1995) Healthy alliances? Other sexual health services and their views of genitourinary medicine. Genitourin Med 71(6):396–399
McDermott MR, Bienenstock J (1979) Evidence for a common mucosal immunologic system. I. Migration of B immunoblasts into intestinal, respiratory, and genital tissues. J Immunol 122(5):1892–1898
McElrath MJ, Corey L, Greenberg PD, Matthews TJ, Montefiori DC, Rowen L (1996) Human immunodeficiency virus type 1 infection despite prior immunization with a recombinant envelope vaccine regimen. Proc Natl Acad Sci U S A 93:3972–3977
McGhee JR, Yamamoto M et al (1992) Isotype of anti-SIV responses in infected rhesus macaques and in animals immunized by mucosal routes. AIDS Res Hum Retroviruses 8(8):1389
Medaglini D, Rush CM (1997) Commensal bacteria as vectors for mucosal vaccines against sexually transmitted diseases: vaginal colonization with recombinant streptococci induces local and systemic antibodies in mice. Vaccine 15(12–13):1330–1337
Nelson AL (2005) The vagina: new options for the administration of medications. http://www.medscape.com
O'Hagan D (1992) Oral delivery of vaccines. Formulation and clinical pharmacokinetic considerations. Clin Pharmacokinet 22(1):1–10
Ogra PL, Ogra SS (1973) Local antibody response to poliovaccine in the human female genital tract. J Immunol 110(5):1307–1311
Owen DH, Katz DF (1999) A vaginal fluid stimulant. Contraception 59:91–95
Owen DH, Peters JJ, Katz DF (2000) Rheological properties of contraceptive gels. Contraception 62:321–326
Park J, Oh Y, Kang M, Kim C (2003) Enhanced mucosal and systemic immune responses following intravaginal immunization with human papillomavirus 16 L1 virus-like particle vaccine in thermosensitive mucoadhesive delivery systems. J Med Virol 70(4):633–641
Parr MB, Parr EL (1991) Langerhans cells and T lymphocyte subsets in the murine vagina and cervix. Biol Reprod 44(3):491–498
Parr EL, Parr MB (1999) Immune responses and protection against vaginal infection after nasal or vaginal immunization with attenuated herpes simplex virus type-2. Immunology 98(4): 639–645
Pavelic Z, Skalko-Basnet N, Schubert R (2001) Liposomal gels for vaginal drug delivery. Int J Pharm 219:139–149
Pavelic Z, Skalko-Basnet N, Filipovic-Grcic J, Anita Martinac A, Jalsenjak I (2005) Development and in vitro evaluation of a liposomal vaginal delivery system for acyclovir. J Control Release 106:34–43

Pettit RK, McAllister SC et al (1999) Response of gonococcal clinical isolates to acidic conditions. J Med Microbiol 48(2):149–156
Phogat S, Svehia K, Tang M, Spadaccini A, Muller J, Mascola J, Berkower I, Wyatt R (2008) Analysis of the human immunodeficiency virus type 1 gp41 membrane proximal external arrayed on hepatitis B surface antigen particles. Virology 373(1):72–84
Richardson J, Illum L (1992) Routes of delivery case studies: the vaginal route of peptide and protein drug delivery. Adv Drug Deliv Rev 8:341–366
Ross RA, Lee MT (1995) Effect of Candida albicans infection and clotrimazole treatment on vaginal microflora in vitro. Obstet Gynaecol 86:925–930
Roy S, Sharma M et al (1994) A quantitative microbiological study of bacterial vaginosis. Indian J Med Res 100:172–176
Sanders J, Matthews H (1990) Vaginal absorption of polyvinyl alcohol in Fischer 344 rats. Hum Exp Toxicol 9:71–77
Sato A, Igarashi H (1990) Identification and localization of vpr gene product of human immunodeficiency virus type 1. Virus Genes 4:303–312
Sher G, Fisch D (2002) Effect of vaginal sildenafil on the outcome of in vitro fertilization (IVF) after multiple IVF failures attributed to poor endometrial development. Fertil Steril 78(5):1073–1076
Singer SJ, Nicolson GL (1972) The fluid mosaic model of the structure of cell membranes. Science 175:720–722
Skalko N, Cajkovac M, Jalsenjak I (1998) Liposomes with metronidazole for topical use: the choice of preparation method and vehicle. J Liposome Res 8:283–293
Thompson I, Van der Bijl P (2001) A comparative light-microscopic, electron-microscopic and chemical study of human vaginal and buccal epithelium. Arch Oral Biol 46(12):1091–1098
Tubin I, Novak J (1956) Integrated gynecology. McGraw-Hill, New York
Wang B, Dang K, Agadjanyan MG, Srikantans V, Li F, Ugen KE, Boyer J, Merva M, Williams WV, Weiner DB (1997) Mucosal immunization with a DNA vaccine induces immune responses against HIV-1 at a mucosal site. Vaccine 15(8):821–825
Wang Y, Lee CH (2002) Characterization of a female controlled drug delivery system for microbicides. Contraception 66(4):281–287
Wang Y, Xiang Z (1998) Effect of passive immunization or maternally transferred immunity on the antibody response to a genetic vaccine to rabies virus. J Virol 72:1790–1796
Wassen L, Jertborn M (2006) Influence of exogenous reproductive hormones on specific antibody production in genital secretions after vaginal vaccination with recombinant chlorea toxin B subunit in humans. Clinical and Vaccine Immunology 13(2):202–207
Williams P, Warwick R (1989) Gray's anatomy. Churchill Livingstone, Edinburgh
Woolfson A (2003) Intravaginal drug delivery technologies. In: Rathbone MJ, Hadgraft J, Roberts MS (eds) Modified release drug delivery technology. Marcel Dekker, New York, pp 759–774
Woolfson A, Malcolm R (2000) Drug delivery by the intravaginal route. Crit Rev Ther Drug Carrier Syst 17:509–555
Woolfson AD, Malcolm RK, Gallagher RJ (2003) Design of a silicone reservoir intravaginal ring for the delivery of oxybutynin. J Contr Rel 91(3): 465–476
Wu CY, Benet LZ (1995) Differentiation of absorption and first-pass gut and hepatic metabolism in humans: studies with cyclosporine. Clin Pharmacol Ther 58(5):492–497
Wyatt R, Sodroski J (1998) The HIV-1 envelope glycoproteins: fusogens, antigens, and immunogens. Science 19:1884–1888
Zhang R, Li M, Chen C, Yao Q (2004) SHIV virus-like particles bind and activate human dendritic cells. Vaccine 23(2): 139–147
Zhou Q, Hidajat R, Peng B, Venzon D, Aldrich MK, Richardson E (2007) Comparative evaluation of oral and intranasal priming with replication-competent adenovirus 5 host range mutant (Ad5hr)-simian immunodeficiency virus (SIV) recombinant vaccines on immunogenicity and protective efficacy against SIV(mac251). Vaccine 25:8021–8035
Zieman M, Fong SK (1997) Absorption kinetics of misoprostol with oral or vaginal administration. Obstet Gynecol 90(1):88–92

Chapter 18
Transcutaneous Immunization

Teerawan Rattanapak, Camilla Foged, and Sarah Hook

18.1 Introduction

Vaccination is regarded as the most cost-effective approach for controlling infectious disease (Coudeville et al. 2005). Currently, the main routes of delivery for vaccines are via either the oral or parenteral routes. Delivery by injection has many drawbacks; for example, vaccinators require injection training and there is a risk of needle-borne diseases associated with improper disposal of needles (Miller and Pisani 1999; Aylward et al. 1995). As a consequence, needle-free immunization has been investigated and developed for the safety of the vaccinator, patient and community. Additionally, it is likely that compliance will be improved by decreasing or eliminating injection site pain (Brown et al. 2006). The non-invasive vaccination routes include oral, buccal, nasal, pulmonary, vaginal and topical routes. In this chapter vaccination via the skin, transcutaneous immunization will be reviewed.

18.2 The Skin

The skin provides the first barrier of protection against the invasion of pathogens into the body. The skin is composed of two main layers, the dermis and the epidermis, which are separated by the epidermal-dermal junction (Fig. 18.1). The dermis

T. Rattanapak • S. Hook (✉)
School of Pharmacy, University of Otago, P.O. Box 56, Dunedin 9054, New Zealand
e-mail: ratte848@student.otago.ac.nz; sarah.hook@otago.ac.nz

C. Foged
Department of Pharmacy, Faculty of Health and Medical Sciences, University of Copenhagen, Universitetsparken 2, 2100 Copenhagen Ø, Denmark
e-mail: camilla.foged@sund.ku.dk

C. Foged et al. (eds.), *Subunit Vaccine Delivery*, Advances in Delivery Science and Technology, DOI 10.1007/978-1-4939-1417-3_18

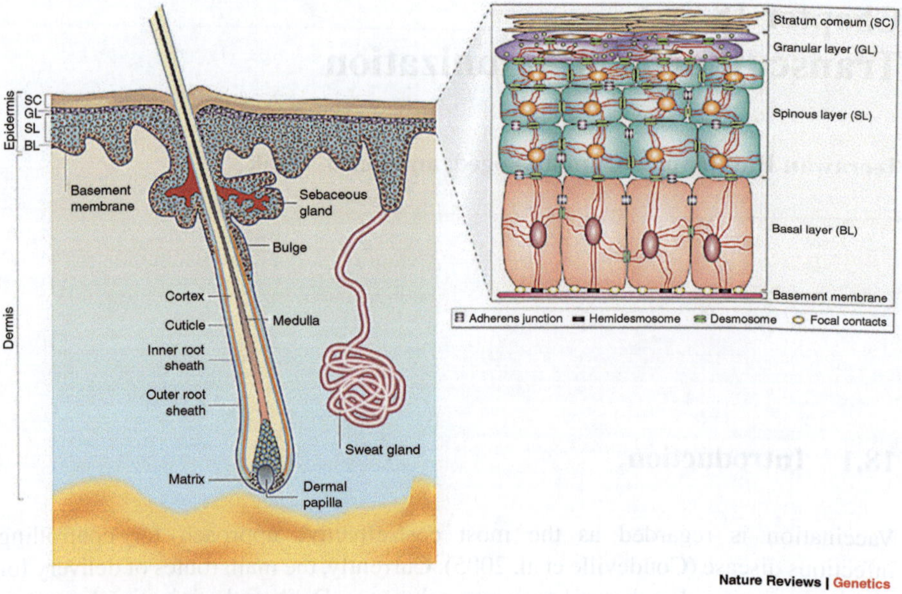

Fig. 18.1 The structure of human skin. The epidermis and dermis are separated by the basement membrane. The epidermis (*inset*) is composed of the stratum basale, the stratum spinosum, the stratum granulosum and the stratum corneum. Figure taken with permission from Fuchs and Raghavan (2002)

is made up of connective tissue, collagen, glycosaminoglycans and elastin. The dermis is a highly vascularized layer and provides the avascular upper layer, the epidermis, with nutrients. The epidermis is the most superficial layer of the skin consisting of keratinocytes and has a thickness of approximately 50–200 μm depending on the body region (Lambert and Laurent 2008).

The epidermis is divided into four different layers (Fig. 18.1). The stratum basale is a single layer of columnar basal cells which remain attached to the basement membrane. The cells begin to flatten and elongate in the stratum spinosum and the cells have lost their nuclei in the stratum granulosum. The stratum granulosum produces and organizes keratin proteins and water-proofing lipids. The stratum corneum (SC) is primarily composed of corneocytes (~90 %), which are flattened, dead, keratin-filled cells. These cells are surrounded by a cell envelope consisting of an inner layer of cross-linked proteins (cornified envelope proteins) and an outer layer of covalently bound lipid envelope (Menon 2002; Bouwstra et al. 2003; Proksch and Jensen 2008).

The SC resembles a brick wall. The corneocytes serve as the bricks and extracellular lipids as the mortar (Michaels et al. 1975). The densely packed and highly conformationally ordered arrangement of the SC results in low diffusion of drugs into skin. Thus, diffusion into the SC can be described as the rate limiting step and the main obstacle to transdermal delivery (Barry 2001).

18.3 Immune Surveillance in the Skin

Different skin layers contain different types of immune cell. $CD8^+$ T-cells and Langerhans cells (LCs) are found in the epidermis (Krueger and Stingl 1989), while the dermis contains various immune cells including macrophages, mast cells, dermal dendritic cells (DDCs), $CD4^+$ T-cells, $\gamma\delta$ T-cells and natural killer T (NK T) cells (Nestle et al. 2009). The two key antigen-presenting cell (APC) subsets in the skin are the LCs and DDCs. These skin APCs possess the ability to take up and process antigen, migrate to draining lymph nodes and to present processed antigen to naïve T-cells (Glenn et al. 2003).

18.3.1 Langerhans Cells

LCs were first discovered by Paul Langerhans in 1868. LCs reside in the epidermis, where approximately 1,000 LCs are present per mm^2 (Flacher et al. 2010) of skin, equating to about 3–5 % of the total epidermal cells (Merad et al. 2008). LCs are surrounded by keratinocytes and the dendrites branching out from the LCs extend between individual keratinocytes (Pearton et al. 2010). Once activated, LCs disengage from the surrounding keratinocytes and migrate across the epidermal/dermal junction to the local draining lymph node (Dearman et al. 2004). LCs can be identified by their unique physical characteristics (presence of many dendrites), their location, the presence of Birbeck granules and high levels of expression of the C-type lectin langerin (CD207) (Valladeau et al. 2000).

LCs have been speculated to be the first APCs involved in capturing antigens delivered by transcutaneous immunization (TCI) due to their location in the epidermis. Kubo and colleagues (2009) found that LCs can extend their dendrites through tight junctions (TJ) and take up antigens via the dendrite tip. Romani et al. (2010) have postulated that the role of LCs in TCI will be dependent on several factors, such as the vaccination area, the amount of vaccine applied and the type of antigen and adjuvant used. For example, LCs were found to express toll-like receptor (TLR) 2, TLR4, and TLR9 but lack TLR7 (Mitsui et al. 2004). Hence, the type of adjuvant used in TCI should be taken into consideration when designing vaccine formulations to activate LCs. In addition, the site of vaccination has been shown to impact on LC activation. Wang et al. (2008) found LC activation was observed at the flank area but was absent in the ear. They suggested this was due to the SC in the flank area being much thicker than in the ear resulting in the vaccine accumulating in the upper skin layer leading to more opportunities for LCs to take up vaccine. More recently there have been conflicting reports on the role of LCs in stimulating effector immune responses and they have been reported to have an immunoregulatory function. In mice specifically depleted of LCs, contact hypersensitivity (CHS) responses were significantly augmented (Bobr et al. 2010). However in mice deficient in $CD207^+$ DDCs there was no difference in the CHS response (Honda et al. 2010). It can thus be concluded that LCs suppressed antigen-specific CHS responses (Bobr et al. 2010).

18.3.2 Dermal Dendritic Cells

The role of DDCs in TCI has been less studied due to their location in the dermis and the idea that therefore antigen uptake by DDCs would occur only rarely. However, recent evidence suggests that DDCs play a vital role in antigen-specific immune responses in the skin. Bursch et al. (2007b) found that LCs were not activated after epicutaneous immunization with a combination of peptide vaccine and adjuvant whereas DDCs migrated and accumulated in the dermis beneath the immunized area. In addition, surface expression of maturation makers was increased and DDCs migrated to draining lymph nodes stimulating T-cell proliferation.

DDCs reside in the dermis and are mostly found adjacent to the epidermal-dermis junction. Some DDCs cluster around hair follicles which has been suggested to facilitate contact with antigens that penetrate via hair follicles (Bursch et al. 2007a). Skin DDCs can be categorized into two subsets based on the expression of CD207. The main population of DCs in the dermis are the $CD207^-$ DDCs (82.1 %) (Henri et al. 2010). Although LCs and $CD207^+$ DDCs both express CD207 and are possibly derived from the same monocyte precursor, they do not have the same function (Ginhoux et al. 2006). Several studies have shown that efficient cross-presentation (Igyártó Botond et al. 2011) and activation of $CD8^+$ T-cells requires priming by $CD207^+$ DDCs (Elnekave et al. 2010; Henri et al. 2010; Stoecklinger et al. 2011). Stoecklinger et al. (2011) reported that following gene gun immunization with plasmid DNA $CD207^+$ DDCs were critical for the activation and functional differentiation of $CD8^+$ T-cells, but not for $CD4^+$ T-cell activation. In addition, the function of $CD207^+$ DDCs was specifically influenced by the nature of the antigen with protein vaccines being unable to stimulate protective immune responses. In the same study, they also reported that $CD207^-$ DDCs biased towards $CD4^+$ T-cell stimulation.

18.4 Immune Modulators for Transcutaneous Immunization

Most transcutaneous vaccines use proteins or peptide antigens and an issue with these are that they are either poorly immunogenic or non-immunogenic. Therefore, potent substances known as adjuvants are required to be delivered with the antigens to improve the immune response. Adjuvants enhance the immune response to vaccine antigens by several different means. For example, adjuvants are capable of increasing the immunogenicity of weak antigens and also of improving the speed and duration of the resulting immune response (Singh and O'Hagan 2003). Additionally, the utilization of adjuvants might decrease in the amount of antigen required to induce immunity, thus reducing costs and helping to overcome antigen competition in combination vaccines (O'Hagan et al. 2001).

Adjuvants play a critical role in TCI. The most common adjuvants used for TCI are cholera toxin (CT) and heat-labile enterotoxin (LT) (O'Hagan et al. 2001; Glenn et al. 1999). Numerous studies have demonstrated that these mucosal adjuvants can enhance immune responses without toxicity after topical application (Scharton-Kersten et al. 1999; Chen et al. 2002; Eyles et al. 2004; Skountzou et al. 2006). Recently, bacterial lipopolysaccharide (LPS) has become an attractive adjuvant for TCI. According to Kahlon and Dutz (2003), LPS and its derivatives can activate TLR4 expressed by LCs and DCs. Additionally, Quil A (QA) has been incorporated into TCI formulations to enhance skin penetration and immune responses (Madsen et al. 2009). Combining adjuvants that act through different pathways can be used to further optimize immune responses (Garçon et al. 2007).

18.5 Transcutaneous Delivery Strategies

There are three possible pathways for compounds to penetrate into skin; the intracellular, the intercellular and the appendageal routes. The intracellular pathway is where the compound penetrates through the cells deeper into skin. The compounds that preferentially take this route are small hydrophilic molecules (Sznitowska et al. 1998). The intercellular pathway is where the compounds can penetrate into the skin through the extracellular lipids, fatty acids and cellular fluids, located between cells. Most of the compounds that preferentially use this pathway are lipophilic. The last pathway is the appendageal pathway which utilizes the sweat glands and hair follicles (Bolzinger et al. 2012). This route is of interest for nanoparticle delivery into the skin as the appendages can also act as a depot for particles from which drug can be slowly released (Liu et al. 2011; Morgen et al. 2011; Patzelt et al. 2011). Despite drug delivery via hair follicles being an effective delivery route, it cannot be a major route due to the fact that the number of pores in skin is only 0.1 % of the entire surface (Otberg et al. 2004). For larger molecules such as peptide and proteins, transcutaneous delivery is a challenge as even if minimal CD4 and CD8 peptides are used, they are still in excess of 500 Da and will therefore not be able to penetrate into the skin according to the "500 Dalton rule" which states that molecules with a molecular weight above 500 Da cannot cross the skin barrier (Bos and Meinardi 2000). Moreover, peptides and proteins are mostly hydrophilic compounds and according to Fick's law of diffusion (equation shown below) penetration of these large hydrophilic molecules without utilization of a skin penetration enhancer is not possible.

$$J = \frac{DK\Delta c}{h}$$

where J is the flux per unit area and per unit time, D is the diffusion coefficient, K is the skin-vehicle partition coefficient, Δc is the concentration difference across the skin and h is the length of the diffusion path.

The major obstacle for TCI is therefore penetration of the vaccine antigen (peptide, protein and DNA) through the densely packed and highly conformationally ordered corneocytes of the SC. As a result, diffusion through the SC can be described as the rate limiting step for TCI (Michaels et al. 1975). Several approaches have been investigated to enhance skin penetration. These include both chemical and physical methods that, in general, work by temporarily reducing or disrupting the skin barrier and/or by providing a mechanism for actively driving the vaccine into the skin. These methods do not need to be utilized in isolation and there may be advantages or synergies to using combined approaches, for example Rattanapak et al. reported that using a physical penetration enhancer (microneedles) in combination with a lipid-based colloidal system (cubosomes) improved vaccine retention in the skin (Rattanapak et al. 2013).

18.5.1 Chemical Penetration Enhancers

The main mechanism for enhanced penetration by chemical enhancers is through the removal of the barrier provided by the SC. This occurs through a disordering of the intercellular lipid structure of the SC and through interactions with keratin. In addition, chemical enhancers increase the partitioning of drugs resulting in an increased diffusion rate (Hadgraft and Walters 1992; Parhi et al. 2012). The most commonly used penetration enhancers are alcohols (Morimoto et al. 2002), propylene glycol (Díez-Sales et al. 2005) and surfactants such as polysorbate (Akhtar 2011).

Ethanol is widely used as a solvent because it can increase the solubility of active ingredients in formulations. Ethanol is also well known as a potent skin penetration enhancer. Many studies have shown that ethanol can significantly increase drug permeation through the skin (Obata et al. 1993; Morimoto et al. 2002; Kobayashi et al. 1994). Ethanol enhances skin permeation and penetration by decreasing skin polarity (Kobayashi et al. 1994) and solubilizing the lipid components of the SC (Kai et al. 1990). Due to the concentration-dependent effect of ethanol on skin permeation enhancement, ethanol has been described by Heard et al. (2006) as having a so-called "pull" or "drag" effect.

Propylene glycol (PG) is regularly used in the cosmetics industry as a penetration enhancer. The skin penetration enhancement is due to hydrogen bonding with keratin (Takeuchi et al. 1992) and interactions with the polar head groups of the lipid bilayers (Bouwstra et al. 1991). Consequently, the structure of the SC is disordered and drug penetration into the skin is increased. Díez-Sales et al. (2005) reported the enhancing effect of PG on acyclovir penetration through human epidermis. Adding 50 % PG to a carbopol gel formulation increased drug permeation as compared to the unmodified gel (Díez-Sales et al. 2005).

Surfactants can be anionic, cationic or non-ionic. Cationic surfactants have the most potential to enhance skin penetration due to electrostatic interactions with negatively charged fatty acids in the SC (Lampe et al. 1983). However, the efficiency of the surfactant action is directly proportional to the amount of skin irritation induced. Thus, non-ionic surfactants are extensively incorporated into topical formulations due to their non-toxic properties. A mechanism for skin penetration enhancement by non-ionic surfactants proposed by Nokhodchi et al. (2003) is that the surfactant molecules may fluidize SC intercellular lipids and also bind to the keratin, leading to disordering of the densely packed SC. Tween 80 is a commonly used non-ionic surfactant in topical formulations. The structure of tween 80 with its ethylene oxide and long hydrocarbon chain is relevant to the surfactant role. The lipophilic part modifies the intercellular lipid lamellae in the SC and the hydrophilic part disrupts protein domains of the corneocytes (Shokri et al. 2001). Akhtar (2011) reported that tween 80 increased the permeation of ascorbic acid through a hairless rabbit skin with an enhancement ratio of 5.07 in relation to the control formulation.

18.5.2 Lipid-Based Colloidal Systems

One of the most controversial methods for enhancing drug penetration into skin is the utilization of lipid vesicles. Around 30 years ago, vesicles were introduced for topical drug delivery by Mezei and Gulasekharam (1980). These authors suggested that intact liposomes were able to penetrate into skin. This investigation brought about numerous studies on vesicles for skin delivery (Fang et al. 2008a; Deshmukh et al. 2008; Lopes et al. 2007).

18.5.2.1 Liposomes

Liposomes are spherical phospholipid vesicles (see Chap. 5). They self-assemble spontaneously into bilayered structures containing an inner aqueous cavity (Castro and Ferreira 2008). Liposomes can be classified into three categories according to vesicle size and the number of lipid bilayers (Torchilin 1996). Vesicles with sizes in the range of 500–5,000 nm with several lipid bilayers are categorized as multilamellar vesicles (MLVs). Large unilamellar vesicles (LUVs) are liposomes with a single lipid bilayer with sizes in the range of 200 to 800 nm. Vesicles with a size of about 100 nm and a single lipid bilayer are referred to as small unilamellar liposomes (SUVs). Multilamellar liposomes can be reduced to LUVs or SUVs by extrusion through stacks of filters.

Phospholipids are biocompatible and biodegradable and these properties make liposomes a safe system, able to be used in the pharmaceutical field (Cosco et al. 2008). Liposomes can prevent the degradation of antigens resulting in prolonged primary activation of T-cells in vivo (Combadiere and Mahe 2008). Many studies have investigated the ability of liposomes to act as an immunological adjuvant and delivery system for subunit vaccines (Davidsen et al. 2005; Brunel et al. 1999; Holten-Andersen et al. 2004).

The ability of liposomes to increase transdermal drug delivery (as compared with non-vesicle formulations such as aqueous solutions, hydro-gels and creams) has been proposed to be due to the ability of vesicular systems to enhance drug penetration (Betz et al. 2005), improve pharmacological properties (Sharma et al. 1994), control drug release (Fang et al. 2004) and serve as a photoprotection system for drugs (Arsic and Vuleta 1999). The penetration enhancement mechanism of liposomes is thought to be through disruption of the stratum corneum. Liposomes remain on the exterior of the skin, mixing with and fluidizing skin lipids thus disordering and loosening the SC resulting in improved drug penetration (El Maghraby et al. 2008). Because of their rigid membranes liposomes do not appear to be able to utilize the intercellular mechanism of penetration leading to the development of elastic vesicles such as transfersomes and ethosomes to improve skin penetration.

18.5.2.2 Transfersomes

Transfersomes, a more recent class of modified liposomes, was first reported by Cevc and Blume (1992) and are variously described as deformable, highly deformable, elastic or ultra-flexible liposomes or vesicles (Benson 2006). They are claimed to improve in vitro transdermal delivery of a variety of drugs. The deformability possessed by transfersomes is the outcome of incorporation of an edge activator within the phospholipid bilayers and this improves elasticity by means of lipid bilayer destabilization (Dubey et al. 2007). Edge activators commonly used are single chain surfactants such as sodium cholate (Boinpally et al. 2003) and tween 80 (Akhtar 2011). Transfersomes are claimed to be able to squeeze through conduits one-tenth the diameter of the vesicles, allowing them to spontaneously penetrate the stratum corneum (Cevc 1996). Moreover, Cevc and Blume (1992) reported that the driving force for penetration into the skin was the osmotic gradient. The osmotic gradient is caused by the difference in water content between the relatively dehydrated skin surface (varying from 15 to 20 % water in the SC) and the hydrated viable epidermis (approximately 70 % water). Aqueous lipid colloidal dispersions applied to the skin are subject to evaporation and this provides the impetus for the lipid system to follow the natural water gradient across the epidermis. Therefore, assuming this proposed mechanism is correct, transfersomes should not be applied under occluded conditions since this would decrease the osmotic effect (Cevc et al. 2002). Interestingly, transfersomes have been found to enhance skin permeation under occlusive condition in vitro whereas the opposite trend was observed when transfersomes were applied in vivo. It was suggested that the difference between in vitro and in vivo occurred because simple diffusion of free drug was a major pathway for permeation in vitro while the osmotic effect of vesicles was the major pathway in vivo (Cevc et al. 2008).

Topical delivery of peptides and proteins by transfersomes has been extensively investigated (Mishra et al. 2006). Transfersomes have been reported to improve vaccine entrapment efficiency, skin retention and penetration across the SC as compared to traditional vesicles (Paul et al. 1998). More robust immune responses were induced by antigen-loaded transfersomes compared with those induced by antigen-loaded liposomes and vaccine solutions (Li et al. 2011). Mishra (2010) reported that hepatitis B surface antigen (HBsAg)-loaded transfersomes triggered improved antigen-specific systemic and mucosal responses against HBsAg in vivo as compared to other formulations including a physical mixture of transfersomes and HBsAg, HBsAg solution and intramuscularly administered alum-adsorbed HBsAg.

18.5.2.3 Ethosomes

Ethosomes have also shown potential for TCI. The efficacy and safety of ethosomal formulations has been convincingly demonstrated as compared to other transcutaneous carriers such as gels (Ainbinder and Touitou 2005), patches (Touitou et al. 2001) and conventional liposomes (Fang et al. 2008b). Ethosomes were first developed by Touitou and colleagues (2000). They are vesicles composed of phospholipid hydrated in water with a high ethanol concentration (up to 45 %). Some of the physical characteristics of ethosomes are their softness, flexibility and deformability. An additional characteristic of ethosomes is their multilayered structure, which is expected to increase drug entrapment, resulting in improved therapeutic efficacy. Furthermore, ethosomes have a negatively charged surface, due to the presence of high amounts of ethanol, which is one factor implicated in their ability to increase the permeation of drugs through the skin (Verma and Pathak 2010). According to Ogiso et al. (2001), the penetration rate of melatonin entrapped in negatively charged liposomes across the skin was higher than that of positively charged liposomes.

Touitou et al. (2000) developed a model to describe how ethosomes facilitate penetration. They proposed that free ethanol disrupts the SC by interacting with the polar head group region of lipid molecules. This interaction with free ethanol causes the structure of the SC to become loosely disordered, increasing fluidity and membrane permeability. Then ethosomal vesicles, which are flexible and deformable, easily penetrate through the disordered SC into deeper layers of the skin. Free drug in the ethosomal system can also penetrate into the skin via the loosened SC. An additional proposed mechanism is the fusion of ethosomes with skin lipids, resulting in drug release from the vesicles. Dayan and Touitou (2000) reported that ethosomes significantly increased the depth of penetration of a fluorescent probe (D-289) into skin as compared to classic liposomes. Moreover, the transcutaneous delivery of ammonium glycyrrhizinate in ethosomes was able to improve the anti-inflammatory effect of this drug as compared to ethanolic or aqueous solutions (Paolino et al. 2005). The immune enhancing abilities of ethosomes have also been reported. Mishra et al. (2010) reported enhanced antigen uptake by human DCs incubated with HBsAg-loaded ethosomes and the subsequent triggering of an efficient Th1-type immune response. However, it must be noted that the presence of ethanol as a component of ethosomes increased cell apoptosis. As regards safety, organic solvents are not necessary for the production of ethosomes whereas liposomes or transfersomes require organic solvents for dissolving the lipid phase, which may be a problem if these formulations contain residual solvents.

18.5.2.4 Cubosomes

Cubosomes are colloidal dispersions of the bicontinuous cubic liquid crystalline phase (see Chap. 7) and they possess the same microstructure as the parent cubic phase. Cubosomes have a significantly larger surface area and a lower viscosity than the bulk cubic phase. The low aqueous solubility of cubic phase-forming lipids allows cubosomes to exist at almost any dilution level, as opposed to most liquid crystalline systems that convert into micelles at higher dilutions. Thus, cubosomes can be easily incorporated into product formulations.

Variable entrapment and release of active pharmaceutical ingredient (API) from cubosomes has been reported and it has been suggested that this is due to the size of the API and any interactions occurring between the API and the cubosomes. Boyd (Boyd 2003) reported that release from bulk cubic phases was driven by simple diffusion resulting in the burst release of a small lipophilic drug. However, drug release from cubosomes is possibly influenced by the molecular weight of the drug. Rizwan et al. (2009) reported high entrapment and retarded release of the model protein ovalbumin (MW ~45,000 Da) from cubosomes. These particles have also been reported to act as an effective vaccine delivery system with increased interferon (IFN)-γ production in animals vaccinated subcutaneously with cubosomes containing ovalbumin and QA as compared to control groups (Gordon et al. 2012).

Cubosomes have been utilized as transdermal drug carriers. The penetration of hinokitiol, a hair growth promotion agent, was increased upon formulation into cubosomes (Kwon and Kim 2010). It has been reported that the penetration enhancing effect of cubosomes is due to the lipids of the particles forming a mixture with the lipids of the SC, which is facilitated by their similar cubic phase structure (Norlen and Al-Amoudi 2004; Esposito et al. 2005). Bender et al. (2008) visualized skin penetration of a fluorescence hydrophilic model drug formulated in cubic phase monoolein using two-photon microscopy and found high fluorescence intensity in micro-fissures and in a three-dimensional network of thin threads in the skin.

18.5.3 Other Delivery Systems

In addition to the lipid-based delivery systems, polymer-based delivery systems and virus-like particles (VLPs) have been investigated for transdermal delivery, although with variable success. Encapsulation of antigen in negatively charged poly(lactic acid) (PLA) nanoparticles did not enhance antigen delivery when applied on intact skin (Mattheolabakis et al. 2010). The nanoparticles were detected in the duct of the hair follicles indicating that the nanoparticles can penetrate the skin barrier through the hair follicles. However, when combining the microneedle approach (see below) with antigen-loaded PLGA nanoparticles, Zaric et al. observed efficient antitumour and antiviral immune responses upon transcutaneous vaccination (Zaric et al. 2013). In contrast, smaller VLPs (40 nm) adjuvanted with CpG were able to induce antigen-specific immune responses in mice characterized by high levels of IFN-γ and IgG1

(Young et al. 2006). Mittal et al. delivered ovalbumin-containing negatively charged poly(lactide-co-glycolide) (PLGA) or positively charged chitosan-coated PLGA nanoparticles to APCs in hair follicles, without any disruption of the skin (Mittal et al. 2013). Both formulations improved the delivery efficiency of ovalbumin into the hair follicles on excised pig ears by a factor of 2–3 compared to an ovalbumin solution, but it remains to be investigated if this improved delivery results in enhanced immune responses.

Slutter et al. compared different vaccine delivery systems for intradermal administration and found that N-trimethyl chitosan (TMC) nanoparticles were more effective carriers than PLGA nanoparticles (Slutter et al. 2010), positively charged liposomes (Slütter et al. 2011) and chitosan nanoparticles (Slütter et al. 2009). Bal et al. applied TMC nanoparticles loaded with diphtheria toxoid on skin pre-treated with microneedles to overcome the skin barrier (Bal et al. 2010a). After 1 hour of application of the nanoparticles, there was no enhancement of the immune response compared to a diphtheria toxoid solution. However, the authors suggest that TMC nanoparticle diffusion might be an important limiting factor for potency in TCI since the nanoparticles were more efficient in potentiating the immune response than a diphtheria toxoid solution when utilizing longer application times (Bal et al. 2010b).

Co-encapsulation of additional immunopotentiators with the ovalbumin antigen into TMC nanoparticles further improved the immunogenicity of the vaccine, since after intradermal vaccination, ovalbumin-loaded TMC nanoparticles modified with CpG and LPS provoked higher IgG titres than plain ovalbumin-loaded TMC nanoparticles (Bal et al. 2012). The potential of TMC as adjuvant was further increased by conjugating the antigen to the polymer, thereby creating a smaller unit (Slütter et al. 2010). Bal et al. found that TMC-ovalbumin conjugates were more immunogenic than physical mixtures of TMC and ovalbumin and ovalbumin-loaded nanoparticles after transcutaneous administration, likely because they penetrate the skin more easily than nanoparticles and consequently are better delivered to DCs (Bal et al. 2011). Size, choice of immunopotentiator and the use of combination approaches incorporating physical disruption of the SC thus play an important role for transcutaneous immunization.

18.5.4 Microneedle Arrays

Microneedle (MCN) arrays are novel drug delivery devices for percutaneous administration of bioactives developed in the 1970s by Gerstel and Place (1976). MCNs are breakthrough systems facilitating transdermal delivery by transiently and physically disrupting the SC and creating micron-sized pores. MCNs are attractive delivery devices because they allow painless drug delivery (Kaushik et al. 2001). Although, the length of the needles can be up to 1,000 μm and are likely to penetrate into the superficial dermis where pain receptors are located, the micron-sizes of needles reduce the chances of encountering and stimulating nerves (Prausnitz 2004).

MCNs have great market potential due to their low manufacturing and product distribution costs and the fact that they are easy to use do not require vaccine-administration expertize (Birchall et al. 2011).

18.5.4.1 Designs and Modes of Action

MCNs disrupt the SC and allow drug to pass through the skin. MCNs generally have a pyramidal shape with a sharp or dull tip and can be manufactured in different ways from a variety of materials. They are divided into four general categories depending on their mode of action (Fig. 18.2).

Solid MCNs: "Poke and Patch"

The "poke and patch" approach is to utilize MCNs to create micro-channels and then apply vaccine patches or formulations to the skin. Drugs penetrate into the skin via simple diffusion (McAllister et al. 2003). Solid MCNs were first used to enhance calcein permeation (Henry et al. 1999). Multiple studies have since reported the use of MCNs to enhance skin permeability, including studies using solid MCNs to transport recombinant virus (Carey et al. 2011; Hirschberg et al. 2012) and protein (Kumar et al. 2011; Ding et al. 2011) vaccines into the skin. Needles can be prepared using a variety of materials. Silicon has been commonly used to prepare MCN arrays. However, the fabrication of microneedles from silicon requires expensive microfabrication procedures and silicon needles may break off in the skin due to the brittle nature of silicon. Nowadays solid MCNs are usually made from polymers such as polyvinyl acetate (Donnelly et al. 2011) and polyetherimide (You et al. 2010). Their mechanical strength reduces the risk of needle breakage in the skin (Park et al. 2005).

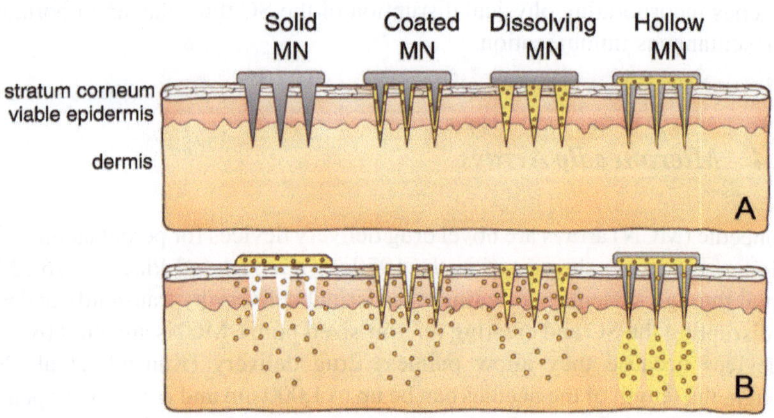

Fig. 18.2 Types of MCNs used for transdermal drug delivery. Adapted from Kim et al. (2012b)

Solid MCNs: "Coat and Poke"

The "coat and poke" approach is similar to the first approach except that the drug is not applied to the skin but is instead coated onto the needle surface. The solid-state vaccine on the surface of needle dissolves off in the skin following MCN insertion. Coated MCNs are an attractive approach as solid-state formulations are stable for longer periods of time as compared to liquid formulations (Kim et al. 2010). However, the amount of vaccine that can be coated onto the needles is limited. As a result, newer vaccine-coating processes have been developed in order to achieve increased vaccine coating. An example of this is an embossing process that fabricates groove-embedded MCNs (Han et al. 2009). An issue encountered with coated MCNs is loss of vaccine immunogenicity (Kim et al. 2011) and the use of stabilizers such as trehalose is essential to prevent this occurring (Kim et al. 2010).

Dissolving MCNs

Dissolving MCNs were developed due to environmental contamination issues arising upon improper disposal of used solid MCNs (Kim et al. 2012a). Dissolving MCNs are made from biodegradable materials such as polymers (Sullivan et al. 2008; Lee et al. 2008) and sugars (Lee et al. 2011; Martin et al. 2012) which dissolve upon exposure to intracellular fluids in the skin. Vaccines entrapped in the polymer matrix release into the skin after matrix degradation. Additionally, dissolving MCNs containing entrapped nanoparticles have been developed as complex controlled-release drug delivery devices (Kang et al. 2006).

Hollow MCNs

Hollow MCNs utilize the same mechanism of action as that used for traditional needle injection. Liquid vaccine formulations are transferred into the skin by active fluid flow or pressure-driven flow. Hollow MCNs are generally used with syringes and existing vaccine formulations but the injection rate through hollow MCNs is faster than subcutaneous injection (Burton et al. 2011). BD Soluvia™ and the MicronJet Needle (NanoPass) are examples of commercial hollow MCNs in the market.

18.5.5 Laserporation

Lasers have been used in medicine since the 1980s to remove or destroy tissue. Much work has focused on developing technologies that can be accurately targeted and have reduced heating and damage of surrounding tissue (Scheiblhofer et al. 2013). Ablative fractional laser (AFL) technologies are now available which can generate a predefined pattern of micropores. A proposed advantage of AFL over other penetration enhancing technologies includes the degree of precision possible with laser technologies in the creation of the size and depth of microchannels which heal quickly to maintain skin integrity. Re-epithelialization of channels 71 μm wide and 40 μm deep was reported to occur within 24 h (Chen et al. 2012). The P.L.E.A.S.E.® (Precise Laser Epidermal System) technology uses a diode-pumped Er:YAG laser to painlessly create several hundred micropores with a typical diameter of 100–150 μm at a targeted depth sequentially in only a few seconds in an area with a diameter of approximately 3 cm (Yu et al. 2011). Studies using this technology have demonstrated the induction of both T- and B-cell responses that appear to be dependent upon antigen presentation by langerin negative DCs (Weiss et al. 2012). Interestingly laserporation has been reported to bias responses towards a Th2 phenotype; however this appears to be at least partially dependent upon the layer of skin targeted and could be modified through the inclusion of adjuvants into the vaccine (Weiss et al. 2012). As well as being used for TCI, laserporation can also be used to improve immune responses to vaccines delivered intramuscularly (Zeira et al. 2003) and intradermally (Zeira et al. 2007).

18.5.6 Other Emerging Technologies

Many different technologies are being investigated for the delivery of drugs into and through the skin. This includes the use of technologies such as electroporation, iontophoresis, sonoporation and jet injection (reviewed in Gratieri et al. 2013). Less work has been done utilizing these systems for vaccine delivery. However electroporation has been utilized for the delivery of DNA vaccines in to a variety of animal species including non-human primates. Laddy et al. compared immune responses to vaccination with an avian influenza DNA vaccine delivered either intramuscularly (i.m.) or intradermally (i.d.) using electroporation to macaques (Laddy et al. 2009). They found that while i.m. immunization induced superior antibody responses i.d. immunization provided better protection, suggesting the importance of cellular immunity in protection against this infection. Electroporation has also been used in combination with intradermal jet injection (whereby a CO_2-propelled needle-free device injects vaccine as a liquid stream into skin) in mice to deliver high doses of plasmid DNA (Hallengard et al. 2012).

18.6 Conclusions

The importance of being able to deliver vaccines without needles in a simple manner that does not require medical personnel or expensive or technical equipment should not be underestimated. While much of the research here is still in the early stages it is easy to imagine such vaccines being available in the future.

However research still needs to be done to develop formulations that efficiently activate the most relevant populations of APCs and induce the appropriate immune response. Such research will require multidisciplinary research teams including immunologists and pharmaceutical scientists.

References

Ainbinder D, Touitou E (2005) Testosterone ethosomes for enhanced transdermal delivery. Drug Deliv 12(5):297–303. doi:10.1080/10717540500176910

Akhtar N (2011) Penetration enhancing effect of polysorbate 20 and 80 on the in vitro percutaneous absorption of L-ascorbic acid. Trop J Pharm Res 10(3):281–288. doi:10.4314/tjpr.v10i3.1

Arsic I, Vuleta G (1999) Influence of liposomes on the stability of vitamin a incorporated in polyacrylate hydrogel. Int J Cosmet Sci 21(4):219–225. doi:10.1046/j.1467-2494.1999.181682.x

Aylward B, Lloyd J, Zaffran M, Mcnairscott R, Evans P (1995) Reducing the risk of unsafe injections in immunization programs—financial and operational implications of various injection technologies. Bull World Health Organ 73(4):531–540

Bal S, Ding Z, Kersten GA, Jiskoot W, Bouwstra J (2010a) Microneedle-based transcutaneous immunisation in mice with N-trimethyl chitosan adjuvanted diphtheria toxoid formulations. Pharm Res 27(9):1837–1847. doi:10.1007/s11095-010-0182-y

Bal SM, Ding Z, van Riet E, Jiskoot W, Bouwstra JA (2010b) Advances in transcutaneous vaccine delivery: do all ways lead to Rome? J Control Release 148(3):266-282; doi:http://dx.doi.org/10.1016/j.jconrel.2010.09.018

Bal SM, Slütter B, Jiskoot W, Bouwstra JA (2011) Small is beautiful: N-trimethyl chitosan–ovalbumin conjugates for microneedle-based transcutaneous immunisation. Vaccine 29(23):4025–4032. doi:10.1016/j.vaccine.2011.03.039

Bal SM, Slütter B, Verheul R, Bouwstra JA, Jiskoot W (2012) Adjuvanted, antigen loaded N-trimethyl chitosan nanoparticles for nasal and intradermal vaccination: adjuvant- and site-dependent immunogenicity in mice. Eur J Pharm Sci 45(4):475–481; doi:http://dx.doi.org/10.1016/j.ejps.2011.10.003

Barry BW (2001) Novel mechanisms and devices to enable successful transdermal drug delivery. Eur J Pharm Sci 14(2):101–114

Bender J, Simonsson C, Smedh M, Engstrom S, Ericson MB (2008) Lipid cubic phases in topical drug delivery: visualization of skin distribution using two-photon microscopy. J Control Release 129(3):163–169. doi:10.1016/j.jconrel.2008.04.020

Benson HA (2006) Transfersomes for transdermal drug delivery. Expert Opin Drug Deliv 3(6):727–737. doi:10.1517/17425247.3.6.727

Betz G, Aeppli A, Menshutina N, Leuenberger H (2005) In vivo comparison of various liposome formulations for cosmetic application. Int J Pharm 296(1–2):44–54. doi:10.1016/j.ijpharm.2005.02.032

Birchall J, Clemo R, Anstey A, John D (2011) Microneedles in clinical practice–an exploratory study into the opinions of healthcare professionals and the public. Pharm Res 28(1):95–106. doi:10.1007/s11095-010-0101-2

Bobr A, Olvera-Gomez I, Igyarto BZ, Haley KM, Hogquist KA, Kaplan DH (2010) Acute ablation of Langerhans cells enhances skin immune responses. J Immunol 185(8):4724–4728. doi:10.4049/jimmunol.1001802

Boinpally RR, Zhou SL, Poondru S, Devraj G, Jasti BR (2003) Lecithin vesicles for topical delivery of diclofenac. Eur J Pharm Biopharm 56(3):389–392. doi:10.1016/S0939-6411(03)00143-7

Bolzinger M-A, Briançon S, Pelletier J, Chevalier Y (2012) Penetration of drugs through skin, a complex rate-controlling membrane. Curr Opin Coll Interface Sci 17(3):156–165. doi:10.1016/j.cocis.2012.02.001

Bos JD, Meinardi MMHM (2000) The 500 Dalton rule for the skin penetration of chemical compounds and drugs. Exp Dermatol 9(3):165–169

Bouwstra JA, de Vries MA, Gooris GS, Bras W, Brussee J, Ponec M (1991) Thermodynamic and structural aspects of the skin barrier. J Control Release 15(3):209–219. doi:10.1016/0168-3659(91)90112-q

Bouwstra JA, Honeywell-Nguyen PL, Gooris GS, Ponec M (2003) Structure of the skin barrier and its modulation by vesicular formulations. Prog Lipid Res 42(1):1–36

Boyd BJ (2003) Characterisation of drug release from cubosomes using the pressure ultrafiltration method. Int J Pharm 260(2):239–247. doi:10.1016/S0378-5173(03)00262-X

Brown MB, Martin GP, Jones SA, Akomeah FK (2006) Dermal and transdermal drug delivery systems: current and future prospects. Drug Deliv 13(3):175–187. doi:10.1080/10717540500455975

Brunel F, Darbouret A, Ronco J (1999) Cationic lipid DC-Chol induces an improved and balanced immunity able to overcome the unresponsiveness to the hepatitis B vaccine. Vaccine 17(17):2192–2203. doi:10.1016/s0264-410x(98)00492-7

Bursch LS, Wang L, Igyarto B, Kissenpfennig A, Malissen B, Kaplan DH, Hogquist KA (2007a) Identification of a novel population of Langerin(+) dendritic cells. J Exp Med 204(13):3147–3156

Bursch LS, Wang L, Igyarto B, Kissenpfennig A, Malissen B, Kaplan DH, Hogquist KA (2007b) Identification of a novel population of Langerin+ dendritic cells. J Exp Med 204(13):3147–3156. doi:10.1084/jem.20071966

Burton SA, Ng CY, Simmers R, Moeckly C, Brandwein D, Gilbert T, Johnson N, Brown K, Alston T, Prochnow G, Siebenaler K, Hansen K (2011) Rapid intradermal delivery of liquid formulations using a hollow microstructured array. Pharm Res 28(1):31–40. doi:10.1007/s11095-010-0177-8

Carey JB, Pearson FE, Vrdoljak A, McGrath MG, Crean AM, Walsh PT, Doody T, O'Mahony C, Hill AVS, Moore AC (2011) Microneedle array design determines the induction of protective memory CD8(+) T cell responses induced by a recombinant live malaria vaccine in mice. PLoS One 6(7):e22442. doi:10.1371/journal.pone.0022442

Castro GA, Ferreira LAM (2008) Novel vesicular and particulate drug delivery systems for topical treatment of acne. Expert Opin Drug Deliv 5(6):665–679. doi:10.1517/17425240802167092

Cevc G (1996) Transfersomes, liposomes and other lipid suspensions on the skin: permeation enhancement, vesicle penetration, and transdermal drug delivery. Crit Rev Ther Drug Carrier Syst 13(3–4):257–388

Cevc G, Blume G (1992) Lipid vesicles penetrate into intact skin owing to the transdermal osmotic gradients and hydration force. Biochim Biophys Acta 1104(1):226–232

Cevc G, Schatzlein A, Richardsen H (2002) Ultradeformable lipid vesicles can penetrate the skin and other semi-permeable barriers unfragmented. Evidence from double label CLSM experiments and direct size measurements. Biochim Biophy Acta 1564(1):21–30

Cevc G, Mazgareanu S, Rother M, Vierl U (2008) Occlusion effect on transcutaneous NSAID delivery from conventional and carrier-based formulations. Int J Pharm 359(1–2):190–197. doi:10.1016/j.ijpharm.2008.04.005

Chen D, Colditz IG, Glenn GM, Tsonis CG (2002) Effect of transcutaneous immunization with co-administered antigen and cholera toxin on systemic and mucosal antibody responses in sheep. Vet Immunol Immunopathol 86(3–4):177–182

Chen X, Shah D, Kositratna G, Manstein D, Anderson RR, Wu MX (2012) Facilitation of transcutaneous drug delivery and vaccine immunization by a safe laser technology. J Control Release 159(1):43–51; doi:http://dx.doi.org/10.1016/j.jconrel.2012.01.002

Combadiere B, Mahe B (2008) Particle-based vaccines for transcutaneous vaccination. Comp Immunol Microbiol Infect Dis 31(2–3):293–315. doi:10.1016/j.cimid.2007.07.015

Cosco D, Celia C, Cilurzo F, Trapasso E, Paolino D (2008) Colloidal carriers for the enhanced delivery through the skin. Expert Opin Drug Deliv 5(7):737–755. doi:10.1517/17425247.5.7.737

Coudeville L, Brunot A, Szucs TD, Dervaux B (2005) The economic value of childhood varicella vaccination in France and Germany. Value Health 8(3):209–222

Davidsen J, Rosenkrands I, Christensen D, Vangala A, Kirby D, Perrie Y, Agger EM, Andersen P (2005) Characterization of cationic liposomes based on dimethyldioctadecylammonium and synthetic cord factor from M. tuberculosis (trehalose 6,6′-dibehenate)—a novel adjuvant inducing both strong CMI and antibody responses. Biochim Biophy Acta 1718(1–2):22–31. doi:10.1016/j.bbamem.2005.10.011

Dayan N, Touitou E (2000) Carriers for skin delivery of trihexyphenidyl HCl: ethosomes vs. liposomes. Biomaterials 21(18):1879–1885. doi:10.1016/s0142-9612(00)00063-6

Dearman RJ, Bhushan M, Cumberbatch M, Kimber I, Griffiths CEM (2004) Measurement of cytokine expression and Langerhans cell migration in human skin following suction blister formation. Exp Dermatol 13(7):452–460

Deshmukh DD, Ravis WR, Betageri GV (2008) Improved delivery of cromolyn from oral proliposomal beads. Int J Pharm 358(1–2):128–136. doi:10.1016/j.ijpharm.2008.02.026

Díez-Sales O, Garrigues TM, Herráez JV, Belda R, Martín-Villodre A, Herráez M (2005) In vitro percutaneous penetration of acyclovir from solvent systems and Carbopol 971-P hydrogels: influence of propylene glycol. J Pharm Sci 94(5):1039–1047. doi:10.1002/jps.20317

Ding Z, Bal SM, Romeijn S, Kersten GFA, Jiskoot W, Bouwstra JA (2011) Transcutaneous immunization studies in mice using diphtheria toxoid-loaded vesicle formulations and a microneedle array. Pharm Res 28(1):145–158. doi:10.1007/s11095-010-0093-y

Donnelly RF, Majithiya R, Singh TRR, Morrow DIJ, Garland MJ, Demir YK, Migalska K, Ryan E, Gillen D, Scott CJ, Woolfson AD (2011) Design, optimization and characterisation of polymeric microneedle arrays prepared by a novel laser-based micromoulding technique. Pharm Res 28(1):41–57. doi:10.1007/s11095-010-0169-8

Dubey V, Mishra D, Nahar M, Jain NK (2007) Vesicles as tools for the modulation of skin permeability. Expert Opin Drug Deliv 4(6):579–593. doi:10.1517/17425247.4.6.579

El Maghraby GM, Barry BW, Williams AC (2008) Liposomes and skin: from drug delivery to model membranes. Eur J Pharm Sci 34(4–5):203–222. doi:10.1016/j.ejps.2008.05.002

Elnekave M, Furmanov K, Nudel I, Arizon M, Clausen BE, Hovav A-H (2010) Directly transfected Langerin+ dermal dendritic cells potentiate CD8+ T cell responses following intradermal plasmid DNA Immunization. J Immunol 185(6):3463–3471. doi:10.4049/jimmunol.1001825

Esposito E, Cortesi R, Drechsler M, Paccamiccio L, Mariani P, Contado C, Stellin E, Menegatti E, Bonina F, Puglia C (2005) Cubosome dispersions as delivery systems for percutaneous administration of indomethacin. Pharm Res 22(12):2163–2173. doi:10.1007/s11095-005-8176-x

Eyles JE, Elvin SJ, Westwood A, LeButt CS, Alpar HO, Somavarapu S, Williamson ED (2004) Immunisation against plague by transcutaneous and intradermal application of subunit antigens. Vaccine 22(31–32):4365–4373. doi:10.1016/j.vaccine.2004.02.049

Fang JY, Hung CF, Fang YP, Chan TF (2004) Transdermal iontophoresis of 5-fluorouracil combined with electroporation and laser treatment. Int J Pharm 270(1–2):241–249. doi:10.1016/j.ijpharm.2003.10.025

Fang JY, Fang CL, Liu CH, Su YH (2008a) Lipid nanoparticles as vehicles for topical psoralen delivery: solid lipid nanoparticles (SLN) versus nanostructured lipid carriers (NLC). Eur J Pharm Biopharm 70(2):633–640. doi:10.1016/j.ejpb.2008.05.008

Fang YP, Tsai YH, Wu PC, Huang YB (2008b) Comparison of 5-aminolevulinic acid-encapsulated liposome versus ethosome for skin delivery for photodynamic therapy. Int J Pharm 356(1–2):144–152. doi:10.1016/j.ijpharm.2008.01.020

Flacher V, Tripp CH, Stoitzner P, Haid B, Ebner S, Del Frari B, Koch F, Park CG, Steinman RM, Idoyaga J, Romani N (2010) Epidermal Langerhans cells rapidly capture and present antigens from C-type lectin-targeting antibodies deposited in the dermis. J Invest Dermatol 130(3):755–762. doi:10.1038/Jid.2009.343

Fuchs E, Raghavan S (2002) Getting under the skin of epidermal morphogenesis. Nat Rev Genet 3(3):199–209

Garçon N, Chomez P, Van Mechelen M (2007) GlaxoSmithKline adjuvant systems in vaccines: concepts, achievements and perspectives. Expert Rev Vaccines 6(5):723–739

Gerstel MSPA, Place, CA, Virgil A (1976) Drug delivery device. United States, Palo Alto, CA. Patent 3964482

Ginhoux F, Tacke F, Angeli V, Bogunovic M, Loubeau M, Dai X-M, Stanley ER, Randolph GJ, Merad M (2006) Langerhans cells arise from monocytes in vivo. Nat Immunol 7(3):265–273; doi:http://www.nature.com/ni/journal/v7/n3/suppinfo/ni1307_S1.html

Glenn GM, Scharton-Kersten T, Vassell R, Matyas GR, Alving CR (1999) Transcutaneous immunization with bacterial ADP-ribosylating exotoxins as antigens and adjuvants. Infect Immun 67(3):1100–1106

Glenn GM, Kenney RT, Ellingsworth LR, Frech SA, Hammond SA, Zoeteweij JP (2003) Transcutaneous immunization and immunostimulant strategies: capitalizing on the immunocompetence of the skin. Expert Rev Vaccines 2(2):253–267. doi:10.1586/14760584.2.2.253

Gordon S, Young K, Wilson R, Rizwan S, Kemp R, Rades T, Hook S (2012) Chitosan hydrogels containing liposomes and cubosomes as particulate sustained release vaccine delivery systems. J Liposome Res 22(3):193–204. doi:10.3109/08982104.2011.637502

Gratieri T, Alberti I, Lapteva M, Kalia YN (2013) Next generation intra- and transdermal therapeutic systems: using non- and minimally-invasive technologies to increase drug delivery into and across the skin. Eur J Pharm Sci 18;50(5):609–222; doi:http://dx.doi.org/10.1016/j.ejps.2013.03.019

Hadgraft J, Walters KA (1992) Skin penetration enhancers. In: Juninger HE (ed) Drug targeting and delivery. Ellis Horwood, Chichester, pp 169–177

Hallengard D, Brave A, Isaguliants M, Blomberg P, Enger J, Stout R, King A, Wahren B (2012) A combination of intradermal jet-injection and electroporation overcomes in vivo dose restriction of DNA vaccines. Genet Vaccines Ther 10(1):5–13. doi:10.1186/1479-0556-10-5

Han M, Kim DK, Kang SH, Yoon HR, Kim BY, Lee SS, Kim KD, Lee HG (2009) Improvement in antigen-delivery using fabrication of a grooves-embedded microneedle array. Sensors Actuators B Chem 137(1):274–280. doi:10.1016/j.snb.2008.11.017

Heard CM, Kung D, Thomas CP (2006) Skin penetration enhancement of mefenamic acid by ethanol and 1,8-cineole can be explained by the 'pull' effect. Int J Pharm 321(1–2):167–170. doi:10.1016/j.ijpharm.2006.05.018

Henri S, Poulin LF, Tamoutounour S, Ardouin L, Guilliams M, de Bovis B, Devilard E, Viret C, Azukizawa H, Kissenpfennig A, Malissen B (2010) CD207+ CD103+ dermal dendritic cells cross-present keratinocyte-derived antigens irrespective of the presence of Langerhans cells. J Exp Med 207(1):189–206. doi:10.1084/jem.20091964

Henry S, McAllister DV, Allen MG, Prausnitz MR (1999) Microfabricated microneedles: a novel approach to transdermal drug delivery. J Pharm Sci 88(9):948. doi:10.1021/js990783q

Hirschberg H, van Kuijk S, Loch J, Jiskoot W, Bouwstra J, Kersten G, Amorij J-P (2012) A combined approach of vesicle formulations and microneedle arrays for transcutaneous immunization against hepatitis B virus. Eur J Pharm Sci 46(1–2):1–7. doi:10.1016/j.ejps.2012.01.013

Holten-Andersen L, Doherty TM, Korsholm KS, Andersen P (2004) Combination of the cationic surfactant dimethyl dioctadecyl ammonium bromide and synthetic mycobacterial cord factor as an efficient adjuvant for tuberculosis subunit vaccines. Infect Immun 72(3):1608–1617. doi:10.1128/iai.72.3.1608-1617.2004

Honda T, Nakajima S, Egawa G, Ogasawara K, Malissen B, Miyachi Y, Kabashima K (2010) Compensatory role of Langerhans cells and langerin-positive dermal dendritic cells in the sensitization phase of murine contact hypersensitivity. J Allergy Clin Immunol 125(5):1154–1156. e1152. doi:10.1016/j.jaci.2009.12.005

Igyártó Botond Z, Haley K, Ortner D, Bobr A, Gerami-Nejad M, Edelson Brian T, Zurawski Sandra M, Malissen B, Zurawski G, Berman J, Kaplan Daniel H (2011) Skin-resident murine dendritic cell subsets promote distinct and opposing antigen-specific T helper cell responses. Immunity 35(2):260–272. doi:10.1016/j.immuni.2011.06.005

Kahlon R, Dutz JP (2003) Skin immune responses to peptide and protein antigen are TLR4 independent. Cell Immunol 226(2):116–123. doi:10.1016/j.cellimm.2003.11.007

Kai T, Mak VHW, Potts RO, Guy RH (1990) Mechanism of percutaneous penetration enhancement—effect of N-alkanols on the permeability barrier of hairless mouse skin. J Control Release 12(2):103–112

Kang EN, Wang HF, Kwon IK, Robinson J, Park K, Cheng JX (2006) In situ visualization of paclitaxel distribution and release by coherent anti-stokes Raman scattering microscopy. Anal Chem 78(23):8036–8043

Kaushik S, Hord AH, Denson DD, McAllister DV, Smitra S, Allen MG, Prausnitz MR (2001) Lack of pain associated with microfabricated microneedles. Anesth Analg 92(2):502–504

Kim YC, Quan FS, Compans RW, Kang SM, Prausnitz MR (2010) Formulation and coating of microneedles with inactivated influenza virus to improve vaccine stability and immunogenicity. J Control Release 142(2):187–195. doi:10.1016/j.jconrel.2009.10.013

Kim Y-C, Quan F-S, Compans R, Kang S-M, Prausnitz M (2011) Stability kinetics of influenza vaccine coated onto microneedles during drying and storage. Pharm Res 28(1):135–144. doi:10.1007/s11095-010-0134-6

Kim YC, Jarrahian C, Zehrung D, Mitragotri S, Prausnitz MR (2012a) Delivery systems for intradermal vaccination. Curr Top Microbiol Immunol 351:77–112. doi:10.1007/82_2011_123

Kim YC, Park JH, Prausnitz MR (2012b) Microneedles for drug and vaccine delivery. Adv Drug Deliv Rev. doi:10.1016/j.addr.2012.04.005

Kobayashi D, Matsuzawa T, Sugibayashi K, Morimoto Y, Kimura M (1994) Analysis of the combined effect of 1-menthol and ethanol as skin permeation enhancers based on a two-layer skin model. Pharm Res 11(1):96–103. doi:10.1023/a:1018953929457

Krueger GG, Stingl G (1989) Immunology inflammation of the skin—a 50-year perspective. J Invest Dermatol 92(4):S32–S51. doi:10.1111/1523-1747.Ep13074960

Kubo A, Nagao K, Yokouchi M, Sasaki H, Amagai M (2009) External antigen uptake by Langerhans cells with reorganization of epidermal tight junction barriers. J Exp Med 206(13): 2937–2946. doi:10.1084/jem.20091527

Kumar A, Li XR, Sandoval MA, Rodriguez BL, Sloat BR, Cui ZR (2011) Permeation of antigen protein-conjugated nanoparticles and live bacteria through microneedle-treated mouse skin. Int J Nanomedicine 6:1253–64. doi:10.2147/Ijn.S20413

Kwon TK, Kim J-C (2010) Preparation and in vitro skin permeation of cubosomes containing hinokitiol. J Dispers Sci Technol 31(7):1004–1009. doi:10.1080/01932690903224862

Laddy DJ, Yan J, Khan AS, Andersen H, Cohn A, Greenhouse J, Lewis M, Manischewitz J, King LR, Golding H, Draghia-Akli R, Weiner DB (2009) Electroporation of synthetic DNA antigens offers protection in nonhuman Primates challenged with highly pathogenic avian influenza virus. J Virol 83(9):4624–4630

Lambert PH, Laurent PE (2008) Intradermal vaccine delivery: will new delivery systems transform vaccine administration? Vaccine 26(26):3197–3208. doi:10.1016/j.vaccine.2008.03.095

Lampe MA, Burlingame AL, Whitney J, Williams ML, Brown BE, Roitman E, Elias PM (1983) Human stratum-corneum lipids—characterization and regional variations. J Lipid Res 24(2):120–130

Lee JW, Park J-H, Prausnitz MR (2008) Dissolving microneedles for transdermal drug delivery. Biomaterials 29(13):2113–2124. doi:10.1016/j.biomaterials.2007.12.048

Lee K, Lee CY, Jung H (2011) Dissolving microneedles for transdermal drug administration prepared by stepwise controlled drawing of maltose. Biomaterials 32(11):3134–3140. doi:10.1016/j.biomaterials.2011.01.014

Li N, Peng LH, Chen X, Nakagawa S, Gao JQ (2011) Effective transcutaneous immunization by antigen-loaded flexible liposome in vivo. Int J Nanomedicine 6:3241–3250

Liu X, Grice JE, Lademann J, Otberg N, Trauer S, Patzelt A, Roberts MS (2011) Hair follicles contribute significantly to penetration through human skin only at times soon after application as a solvent deposited solid in man. Br J Clin Pharmacol 72(5):768–774. doi:10.1111/j.1365-2125.2011.04022.x

Lopes LB, Speretta FFF, Vitoria M, Bentley LB (2007) Enhancement of skin penetration of vitamin K using monoolein-based liquid crystalline systems. Eur J Pharm Sci 32(3):209–215. doi:10.1016/j.ejps.2007.07.006

Madsen HB, Ifversen P, Madsen F, Brodin B, Hausser I, Nielsen HM (2009) In vitro cutaneous application of ISCOMs on human skin enhances delivery of hydrophobic model compounds through the stratum corneum. AAPS J 11(4):728–739. doi:10.1208/s12248-009-9149-5

Martin CJ, Allender CJ, Brain KR, Morrissey A, Birchall JC (2012) Low temperature fabrication of biodegradable sugar glass microneedles for transdermal drug delivery applications. J Control Release 158(1):93–101. doi:10.1016/j.jconrel.2011.10.024

Mattheolabakis G, Lagoumintzis G, Panagi Z, Papadimitriou E, Partidos CD, Avgoustakis K (2010) Transcutaneous delivery of a nanoencapsulated antigen: induction of immune responses. Int J Pharm 385(1–2):187–193; doi:http://dx.doi.org/10.1016/j.ijpharm.2009.10.033

McAllister DV, Wang PM, Davis SP, Park J-H, Canatella PJ, Allen MG, Prausnitz MR (2003) Microfabricated needles for transdermal delivery of macromolecules and nanoparticles: fabrication methods and transport studies. Proc Natl Acad Sci 100(24):13755–13760. doi:10.1073/pnas.2331316100

Menon GK (2002) New insights into skin structure: scratching the surface. Adv Drug Deliv Rev 54:S3–S17

Merad M, Ginhoux F, Collin M (2008) Origin, homeostasis and function of Langerhans cells and other langerin-expressing dendritic cells. Nat Rev Immunol 8(12):935–947. doi:10.1038/nri2455

Mezei M, Gulasekharam V (1980) Liposomes—a selective drug delivery system for the topical route of administration. Life Sci 26(18):1473–1477

Michaels AS, Chandrasekaran SK, Shaw JE (1975) Drug permeation through human skin: theory and in vitro experimental measurement. Am Inst Chem Eng J 21(5):985–996. doi:10.1002/aic.690210522

Miller MA, Pisani E (1999) The cost of unsafe injections. Bull World Health Organ 77(10):808–811

Mishra D (2010) Evaluation of solid lipid nanoparticles as carriers for delivery of Hepatitis B surface antigen for vaccination using subcutaneous route. J Pharm Pharm Sci 13(4):495

Mishra D, Dubey V, Asthana A, Saraf DK, Jain NK (2006) Elastic liposomes mediated transcutaneous immunization against Hepatitis B. Vaccine 24(22):4847–4855. doi:10.1016/j.vaccine.2006.03.011

Mishra D, Mishra PK, Dabadghao S, Dubey V, Nahar M, Jain NK (2010) Comparative evaluation of hepatitis B surface antigen-loaded elastic liposomes and ethosomes for human dendritic cell uptake and immune response. Nanomedicine 6(1):110–118. doi:10.1016/j.nano.2009.04.003

Mitsui H, Watanabe T, Saeki H, Mori K, Fujita H, Tada Y, Asahina A, Nakamura K, Tamaki K (2004) Differential expression and function of Toll-like receptors in Langerhans cells: comparison with splenic dendritic cells. J Invest Dermatol 122(1):95–102

Mittal A, Raber AS, Schaefer UF, Weissmann S, Ebensen T, Schulze K, Guzmán CA, Lehr C-M, Hansen S (2013) Non-invasive delivery of nanoparticles to hair follicles: A perspective for transcutaneous immunization. Vaccine 31(34):3442–3451; doi:http://dx.doi.org/10.1016/j.vaccine.2012.12.048

Morgen M, Lu GW, Du D, Stehle R, Lembke F, Cervantes J, Ciotti S, Haskell R, Smithey D, Haley K, Fan C (2011) Targeted delivery of a poorly water-soluble compound to hair follicles using polymeric nanoparticle suspensions. Int J Pharm 416(1):314–322. doi:10.1016/j.ijpharm.2011.06.019

Morimoto Y, Wada Y, Seki T, Sugibayashi K (2002) In vitro skin permeation of morphine hydrochloride during the finite application of penetration-enhancing system containing water, ethanol and l-menthol. Biol Pharm Bull 25(1):134–136

Nestle FO, Di Meglio P, Qin JZ, Nickoloff BJ (2009) Skin immune sentinels in health and disease. Nat Rev Immunol 9(10):679–691

Nokhodchi A, Shokri J, Dashbolaghi A, Hassan-Zadeh D, Ghafourian T, Barzegar-Jalali M (2003) The enhancement effect of surfactants on the penetration of lorazepam through rat skin. Int J Pharm 250(2):359–369. doi:10.1016/s0378-5173(02)00554-9

Norlen L, Al-Amoudi A (2004) stratum corneum keratin structure, function, and formation: the cubic rod-packing and membrane templating model. J Invest Dermatol 123(4):715–732; doi:http://www.nature.com/jid/journal/v123/n4/suppinfo/5602511s1.html

Obata Y, Takayama K, Maitani Y, Machida Y, Nagai T (1993) Effect of ethanol on skin permeation of nonionized and ionized diclofenac. Int J Pharm 89(3):191–198. doi:10.1016/0378-5173(93)90243-9

Ogiso T, Yamaguchi T, Iwaki M, Tanino T, Miyake Y (2001) Effect of positively and negatively charged liposomes on skin permeation of drugs. J Drug Target 9(1):49

O'Hagan DT, MacKichan ML, Singh M (2001) Recent developments in adjuvants for vaccines against infectious diseases. Biomol Eng 18(3):69–85

Otberg N, Richter H, Schaefer H, Blume-Peytavi U, Sterry W, Lademann J (2004) Variations of hair follicle size and distribution in different body sites. J Invest Dermatol 122(1):14–19. doi:10.1046/j.0022-202X.2003.22110.x

Paolino D, Lucania G, Mardente D, Alhaique F, Fresta M (2005) Ethosomes for skin delivery of ammonium glycyrrhizinate: In vitro percutaneous permeation through human skin and in vivo anti-inflammatory activity on human volunteers. J Control Release 106(1–2):99–110. doi:10.1016/j.jconrel.2005.04.007

Parhi R, Suresh P, Mondal S, Kumar PM (2012) Novel penetration enhancers for skin applications: a review. Curr Drug Deliv 9(2):219–230

Park JH, Allen MG, Prausnitz MR (2005) Biodegradable polymer microneedles: Fabrication, mechanics and transdermal drug delivery. J Control Release 104(1):51–66. doi:10.1016/j.jconrel.2005.02.002

Patzelt A, Richter H, Dähne L, Walden P, Wiesmüller K-H, Wank U, Sterry W, Lademann J (2011) Influence of the vehicle on the penetration of particles into hair follicles. Pharmaceutics 3(2):307–314

Paul A, Cevc G, Bachhawat BK (1998) Transdermal immunisation with an integral membrane component, gap junction protein, by means of ultradeformable drug carriers, transfersomes. Vaccine 16(2–3):188–195. doi:10.1016/s0264-410x(97)00185-0

Pearton M, Kang S-M, Song J-M, Anstey AV, Ivory M, Compans RW, Birchall JC (2010) Changes in human Langerhans cells following intradermal injection of influenza virus-like particle vaccines. PLoS One 5(8):e12410. doi:10.1371/journal.pone.0012410

Prausnitz MR (2004) Microneedles for transdermal drug delivery. Adv Drug Deliv Rev 56(5):581–587. doi:10.1016/j.addr.2003.10.023

Proksch E, Jensen J-M (2008) Skin as an organ of protection. In: Goldsmith LA, Katz SI, Gilchrest GA, Paller AS, Leffell DJ, Wolff K (eds) Fitzpatrick's dermatology in general medicine. McGraw-Hill, New York, pp 383–95

Rattanapak T, Birchall J, Young K, Ishii M, Meglinski I, Rades T, Hook S (2013) Transcutaneous immunization using microneedles and cubosomes: Mechanistic investigations using Optical Coherence Tomography and Two-Photon Microscopy. J Control Release 172(3):894–903; doi:http://dx.doi.org/10.1016/j.jconrel.2013.08.018

Rizwan SB, Hanley T, Boyd BJ, Rades T, Hook S (2009) Liquid crystalline systems of phytantriol and glyceryl monooleate containing a hydrophilic protein: characterisation, swelling and release kinetics. J Pharm Sci 98(11):4191–4204. doi:10.1002/jps.21724

Romani N, Clausen BE, Stoitzner P (2010) Langerhans cells and more: langerin-expressing dendritic cell subsets in the skin. Immunol Rev 234(1):120–141. doi:10.1111/j.0105-2896.2009.00886.x

Scharton-Kersten T, Glenn GM, Vassell R, Yu JM, Walwender D, Alving CR (1999) Principles of transcutaneous immunization using cholera toxin as an adjuvant. Vaccine 17:S37–S43

Scheiblhofer S, Thalhamer J, Weiss R (2013) Laser microporation of the skin: prospects for painless application of protective and therapeutic vaccines. Expert Opin Drug Deliv 10(6):761–773. doi:10.1517/17425247.2013.773970

Sharma BB, Jain SK, Vyas SP (1994) Topical liposome system bearing local-anesthetic lignocaine—preparation and evaluation. J Microencapsul 11(3):279–286

Shokri J, Nokhodchi A, Dashbolaghi A, Hassan-Zadeh D, Ghafourian T, Barzegar Jalali M (2001) The effect of surfactants on the skin penetration of diazepam. Int J Pharm 228(1–2):99–107. doi:10.1016/s0378-5173(01)00805-5

Singh M, O'Hagan DT (2003) Recent advances in veterinary vaccine adjuvants. Int J Parasitol 33(5–6):469–478. doi:10.1016/s0020-7519(03)00053-5

Skountzou I, Quan FS, Jacob J, Compans RW, Kang SM (2006) Transcutaneous immunization with inactivated influenza virus induces protective immune responses. Vaccine 24(35–36):6110–6119. doi:10.1016/j.vaccine.2006.05.014

Slütter B, Plapied L, Fievez V, Alonso Sande M, des Rieux A, Schneider Y-J, Van Riet E, Jiskoot W, Préat V (2009) Mechanistic study of the adjuvant effect of biodegradable nanoparticles in mucosal vaccination. J Control Release 138(2):113–121; doi:http://dx.doi.org/10.1016/j.jconrel.2009.05.011

Slutter B, Bal S, Keijzer C, Mallants R, Hagenaars N, Que I, Kaijzel E, van Eden W, Augustijns P, Lowik C, Bouwstra J, Broere F, Jiskoot W (2010) Nasal vaccination with N-trimethyl chitosan and PLGA based nanoparticles: Nanoparticle characteristics determine quality and strength of the antibody response in mice against the encapsulated antigen. Vaccine 28(38):6282–6291. doi:10.1016/j.vaccine.2010.06.121

Slütter B, Soema PC, Ding Z, Verheul R, Hennink W, Jiskoot W (2010) Conjugation of ovalbumin to trimethyl chitosan improves immunogenicity of the antigen. J Control Release 143(2):207–214; doi:http://dx.doi.org/10.1016/j.jconrel.2010.01.007

Slütter B, Bal SM, Ding Z, Jiskoot W, Bouwstra JA (2011) Adjuvant effect of cationic liposomes and CpG depends on administration route. J Control Release 154(2):123–130; doi:http://dx.doi.org/10.1016/j.jconrel.2011.02.007

Stoecklinger A, Eticha TD, Mesdaghi M, Kissenpfennig A, Malissen B, Thalhamer J, Hammerl P (2011) Langerin+ dermal dendritic cells are critical for CD8+ T cell activation and IgH γ-1 class switching in response to gene gun vaccines. J Immunol 186(3):1377–1383. doi:10.4049/jimmunol.1002557

Sullivan SP, Murthy N, Prausnitz MR (2008) Minimally invasive protein delivery with rapidly dissolving polymer microneedles. Adv Mater 20(5):933. doi:10.1002/adma.200701205

Sznitowska M, Janicki S, Williams AC (1998) Intracellular or intercellular localization of the polar pathway of penetration across stratum corneum. J Pharm Sci 87(9):1109–1114. doi:10.1021/js980018w

Takeuchi Y, Yasukawa H, Yamaoka Y, Kato Y, Morimoto Y, Fukumori Y, Fukuda T (1992) Effects of fatty acids, fatty amines and propylene glycol on rat stratum corneum lipids and proteins in vitro measured by Fourier transform infrared/attenuated total reflection (FT-IR/ATR) spectroscopy. Chem Pharm Bull (Tokyo) 40(7):1887–1892

Torchilin VP (1996) Liposomes as delivery agents for medical imaging. Mol Med Today 2(6):242–249. doi:10.1016/1357-4310(96)88805-8

Touitou E, Dayan N, Bergelson L, Godin B, Eliaz M (2000) Ethosomes—novel vesicular carriers for enhanced delivery: characterization and skin penetration properties. J Control Release 65(3):403–418

Touitou E, Godin B, Dayan N, Weiss C, Piliponsky A, Levi-Schaffer F (2001) Intracellular delivery mediated by an ethosomal carrier. Biomaterials 22(22):3053–3059

Valladeau J, Ravel O, Dezutter-Dambuyant C, Moore K, Kleijmeer M, Liu Y, Duvert-Frances V, Vincent C, Schmitt D, Davoust J, Caux C, Lebecque S, Saeland S (2000) Langerin, a novel C-type lectin specific to Langerhans cells, is an endocytic receptor that induces the formation of Birbeck granules. Immunity 12(1):71–81

Verma P, Pathak K (2010) Therapeutic and cosmeceutical potential of ethosomes: an overview. J Adv Pharm Technol Res 1(3):274–282. doi:10.4103/0110-5558.72415

Wang L, Bursch LS, Kissenpfennig A, Malissen B, Jameson SC, Hogquist KA (2008) Langerin expressing cells promote skin immune responses under defined conditions. J Immunol 180(7):4722–4727

Weiss R, Hessenberger M, Kitzmüller S, Bach D, Weinberger EE, Krautgartner WD, Hauser-Kronberger C, Malissen B, Boehler C, Kalia YN, Thalhamer J, Scheiblhofer S (2012) Transcutaneous vaccination via laser microporation. J Control Release 162(2):391–399; doi:http://dx.doi.org/10.1016/j.jconrel.2012.06.031

You S-K, Noh Y-W, Park H-H, Han M, Lee SS, Shin S-C, Cho C-W (2010) Effect of applying modes of the polymer microneedle-roller on the permeation of l-ascorbic acid in rats. J Drug Target 18(1):15–20. doi:10.3109/10611860903115274

Young SL, Wilson M, Wilson S, Beagley KW, Ward V, Baird MA (2006) Transcutaneous vaccination with virus-like particles. Vaccine 24(26):5406–5412; doi:http://dx.doi.org/10.1016/j.vaccine.2006.03.052

Yu J, Kalaria DR, Kalia YN (2011) Erbium:YAG fractional laser ablation for the percutaneous delivery of intact functional therapeutic antibodies. J Control Release 156(1):53–59; doi:http://dx.doi.org/10.1016/j.jconrel.2011.07.024

Zaric M, Lyubomska O, Touzelet O, Poux C, Al-Zahrani S, Fay F, Wallace L, Terhorst D, Malissen B, Henri S, Power UF, Scott CJ, Donnelly RF, Kissenpfennig A (2013) Skin dendritic cell targeting via microneedle arrays laden with antigen encapsulated poly-D-L-lactide-co-glycolide nanoparticles induces efficient anti-tumour and anti-viral immune responses. ACS Nano. doi:10.1021/nn304235j

Zeira E, Manevitch A, Khatchatouriants A, Pappo O, Hyam E, Darash-Yahana M, Tavor E, Honigman A, Lewis A, Galun E (2003) Femtosecond infrared laser—an efficient and safe in vivo gene delivery system for prolonged expression. Mol Ther 8(2):342–350

Zeira E, Manevitch A, Manevitch Z, Kedar E, Gropp M, Daudi N, Barsuk R, Harati M, Yotvat H, Troilo PJ, Griffiths TG, Pacchione SJ, Roden DF, Niu Z, Nussbaum O, Zamir G, Papo O, Hemo I, Lewis A, Galun E (2007) Femtosecond laser: a new intradermal DNA delivery method for efficient, long-term gene expression and genetic immunization. FASEB J 21(13):3522–3533. doi:10.1096/fj.06-7528com

Part IV
Pharmaceutical Analysis and Quality Control of Vaccines

Chapter 19
Pharmaceutical Analysis and Quality Control of Vaccines

Michele Pallaoro

19.1 Background

Subunit vaccines can be classified as special biological products in the sense that they share many common features with more classical Biologics like therapeutic proteins. In fact, subunit vaccines are composed of recombinant proteins (Zepp 2010), glycoconjugates (Avci et al. 2011; Rappuoli and De Gregorio 2011) or variations of these such that they only include the components (antigens) that best stimulate the immune system, instead of using the entire virus or bacterium (Rappuoli et al. 2011). Over the years, subunit vaccines have been evolving towards the goal of further refining antigens towards enrichment with the small antigen portions called epitopes, which are the very specific parts of the antigen that antibodies or T-cells recognize and bind to. Some recent examples describe epitope optimization, grafting, and epitope combination to improve even further this approach (Scarselli et al. 2011). Because subunit vaccines contain only the essential antigens and not all the other molecules that make up the microbe, the risk of adverse reactions to the vaccine is reduced, and this constitutes a big safety advantage (Ahmed et al. 2011); this is due to the intrinsic nature of the microbial components that the innate immune system has evolved to recognize on the spot to mount an immediate reaction against. This advantage has been regarded as one of those rare scientific quantum leaps that happen every now and then, but it has enabled the preparation of vaccines using highly pure and well-defined reagents, in a way very similar to Biologics.

Subunit vaccines can contain a variable number of antigens, generally several (3–5), but there is no intrinsic limit for the number of antigens, which can be above 20 or as little as one. Identifying which antigens stimulate the immune system in an optimal way and how to combine them with the adjuvant, develop and characterize

M. Pallaoro (✉)
Novartis Vaccines, Via Fiorentina 1, 53100 Siena, Italy
e-mail: Michele.Pallaoro@Novartis.com

C. Foged et al. (eds.), *Subunit Vaccine Delivery*, Advances in Delivery Science and Technology, DOI 10.1007/978-1-4939-1417-3_19

the components for a new vaccine is a complicated and long process. However, independently of how researchers approach this Holy Grail of vaccinology, subunit vaccines are made either by tearing apart bacteria or viruses and selecting the relevant portions by purification or, alternatively, are manufactured by producing the microbial antigenic molecules using recombinant DNA technology. Vaccines produced this way are referred to as subunit or recombinant subunit vaccines respectively.

The first recombinant subunit vaccine ever made was a vaccine against the hepatitis B virus (Andre and Safary 1987). Scientists inserted the relevant hepatitis B genes coding for the desired antigens into common baker's yeast. The yeast then produced the antigens, which were collected and purified for use in the vaccine. Since then, many additional vaccines have been produced on the basis of a similar approach (CDC 1999; Giuliani et al. 2006; McNeil 2006).

As far as what has been described to date, there is little or no difference between the production of subunit vaccines and classic biological products. However, what differentiate vaccines from classical biological products are the frequent addition of an adjuvant and the schedule of administration. The adjuvant, by definition from the Latin word *adjuvare*, is a component that is added to the antigens to help obtaining a stronger, more durable immune response. Classically, adjuvants are suspensions of aluminum salts (Chap. 3) or oil-in-water emulsions (Chap. 4); adjuvants can be classified as delivery systems and immunopotentiators at the same time, depending on their characteristics, and are not limited to the ones cited above, as many additional adjuvants are under development.

It is important also to spend a few words on the frequency of administration; vaccines are usually administered only a few times in a lifespan as their effect is long lasting, which is different for many Biologics that usually are administered repeatedly and regularly over the course of many years.

Given the similarities to classical Biologics described above, vaccines are a special case for several reasons; for example (1) the presence of the adjuvant complicates the characterization of the antigens, (2) it is more difficult to establish a clinical potency correlate, (3) vaccines usually target a prophylactic patient population, in contrast to a therapeutic population, and finally (4) it is more difficult to measure changes of the critical quality attributes (CQA), as per definition of a quality-by-design (QbD) approach, for vaccines than for other types of Biologics.

19.2 State of the Art

When determining the stability and quality of a vaccine it is important to understand and define the key parameters of that particular biopharmaceutical product (EMA 2005). It is also important to have a comprehensive knowledge of the stressors present in each step along the way from antigen preparation, to formulation and storage and the impact they have on the drug substance and drug product in relation to their CQAs. Subsequently, it is essential to develop key analytical assays for characterization of the biophysical and physicochemical characteristics of the biologic compounds under investigation.

As of today, there are six pillars defining the characteristics that a vaccine should meet according to regulatory agencies: (1) identity, (2) integrity, (3) quantity, (4) functionality, (5) adsorption (where applicable), and (6) desorption (where applicable) (EMA 2005).

In brief, the purpose of identity is to ensure that materials used for immunization are consistent with the drug substance, where drug substance/s refer to the active single components that once mixed together produce the drug product, more simply, antigens are the quintessential drug substances of vaccines. The purpose of integrity is to ensure that stability assays for formulated/desorbed antigens can be used to detect antigen changes. Quantity determination is important to make sure that the antigen dose is accurate, whereas the functionality aspect ensures via functional assays to evaluate the impact of the formulation on the antigen's stability/functionality and aims to establish a correlation between in vivo protection and resistance to different stresses in order to make sure that a clear relationship can be established between a stress, the induced biophysical change, and immunological protection. Adsorption, where applicable, ensures that antigens used for immunization are stably adsorbed on the surface of the adjuvant, and finally desorption ensures optimized desorption/separation methods are developed for proper analysis.

As mentioned above, it is important to have a detailed understanding of what could go wrong during manufacture and storage by performing a risk evaluation of all the critical steps leading to the final product. For example, it could be that some molecules do have weak points that could be due to (1) the intrinsic characteristics of the antigen molecule itself, (2) the sequence in the case of deamidation, (3) adsorption and slow changes on the surface of the adjuvant, and (4) the presence of residual detergents or particular excipients that can destabilize the drug substance or drug product in other cases. Stressors in processing and formulation could be hidden in each step and may include production and storage conditions, adsorption, excipient combination, time, temperature, freeze-thaw, mechanical stresses, pH, oxidizing agents, metals, light, leachables, and interactions with the container. All these factors may have an impact on the stability, aggregation behavior, and functionality of the antigen(s).

For these reasons, appropriate analysis methods for physical and chemical modifications that could happen during drug substance and drug product production and storage should be established to monitor reversible/irreversible aggregation, precipitation and partitioning, creaming, changes in structure, function, particle size, hydrophobicity, surface charge, deamidation, oxidation, hydrolysis, fragmentation, and many more.

Given the information above it appears key to propose a stepwise approach where it is first necessary to define the critical parameters having an impact on efficacy and safety of the product. Second, it is necessary to develop analytical assays for characterization of biophysical, physical, and chemical changes in the characteristics of the antigen(s) and of the adjuvant under evaluation. Finally, it is recommended to determine the root cause of any change, if possible, by deliberately applying predetermined stresses and observing the effect. This approach has the advantage of generating awareness on these factors and the level at which they induce a determined change thus becoming stressors, which will help correlating cause and effect. This concept is of paramount importance, as it is used to establish the boundaries within which a certain component is stable and that cannot be crossed without compromising its stability.

To develop the required analytical assays with the appropriate characteristics needed, it is necessary to take into consideration some general characteristics of many vaccines. In particular, it should be taken into account that antigens are usually formulated at rather low concentrations, and, differently from many classical Biologics that may be formulated at a concentration of tens of mg/ml, antigens are usually formulated in the range of 0.1 mg/mL. In the cases where there is the need to desorb antigens for characterization, the final concentration may be reduced even further depending on the desorption protocol. In addition, several antigens are usually present concomitantly, which poses further challenges in terms of optimal separation or reciprocal interference.

One of the major challenges is due to the presence of the adjuvant. In these cases, special measures to desorb and/or separate the antigens from the matrix of the adjuvant are required to eliminate most if not all of its interference. All these challenges live together with the need to provide the information required to answer to the questions raised by the characterization pillars mentioned above.

At this point, there is a clear distinction on how vaccines should be treated before characterization; the first difference lies in the presence or absence of an adjuvant whose matrix has the consequence of interfering with the vast majority of available analytical techniques. The second difference depends on the type of adjuvant. Each adjuvant has its own characteristics, and we will restrict the following discussion to the two major classes present in most of the marketed vaccines. The two classes of adjuvant are very different in nature and require the use of different approaches.

19.3 Characterization of Aluminum-Adjuvanted Vaccines

Aluminum salts based adjuvants (Chap. 3) have three major characteristics; a very large surface area, a highly charged surface, either positive or negative, and a well-defined particle size distribution (Gupta 1995; Hem and White 1995). The combination of these characteristics, and others, give them a very high adsorptive capacity, which is extended to a broad set of antigens. Adsorbed antigens are attracted to the surface of the adjuvant and adsorb to it with different strengths. The parameter describing this process is referred to as the adsorptive coefficient.

The characteristics described introduce the need to analyze and control additional parameters, as compared to an unadjuvanted formulation, namely particle size, antigen adsorption and adsorption stability, and the impact of adsorption on antigen identity, integrity, and immunogenicity. In the case of more complex adjuvant mixtures like AS04, which is a combination of a medium-sized molecule of natural origin, monophosphoryl lipid A (MPL) (Desombere et al. 2002), subjected to chemical detoxification reactions, and aluminum hydroxide, there are even more layers of complexity. These are given by the need to develop all the assays required for the characterization of MPL, its adsorption to alum, and the impact it has on the other components of the formulation. While some of these characteristics may simply be measured using appropriate techniques, others may require the development of very specific assays.

One of the most important parameters to follow is the antigen adsorption during the entire lifespan of the drug product, according to the guidelines set by EMA (EMA 2005). Antigen adsorption can be divided into (1) the analysis of adsorption over a specified time and temperature range and (2) the analysis of the antigen post adsorption. Although there are no guidelines specifying if the antigen needs to be adsorbed or not, there are clear guidelines specifying that whatever the adsorption extent is, it needs to remain constant for all the points of the analysis. This is pretty intuitive if we consider that the characteristics of the product and of the components need to remain constant, as mentioned earlier. Due to the broad nature of the adsorptive capacity and the different interactions occurring between the antigen(s) and the mineral salt adjuvant, it is common to observe complete or nearly complete antigen adsorption. However, the consequences of a lower desorption over time are yet unclear, as well as the pitfalls of basic adsorption characterization that should be avoided.

Formulations containing aluminum salts, for example, can be mildly centrifuged to easily separate the dispersed adjuvant from the liquid phase, and the amount of antigen in the liquid phase can then be quantified by different means with different degrees of accuracy by determining the level of un-adsorbed antigen(s) and indirectly the amount of adsorbed antigen(s). However, it is more important to quantify the amount of adsorbed antigen(s) than determining the amount of un-adsorbed antigen(s). In this case, there are currently two major possibilities; dissolving the adjuvant gel or treating the adjuvant in order to release the bound antigen(s). The first solution might be challenging as many mineral salts may not easily dissolve in the presence of citrate or phosphate buffer. The second solution might be easier and could be performed by treating the aluminum pellet with agents that interfere with antigen(s) binding, respectively. In our experience, this is the preferred and always the first choice, but often there is still reference to the other approach which is honestly more dated and much less efficacious at least in our experience. Detailed knowledge of the nature of the force(s) driving antigen adsorption onto the surface of the adjuvant is a critical step in this phase as it suggests how to attempt antigen desorption.

Antigen adsorption onto the surface of aluminum salts is mostly driven by three different mechanisms. The strongest is ligand exchange with potential to establish a dative bond between the antigen and the adjuvant. Second in order of strength is electrostatic interaction, which in most cases drives the adsorption process. Last but not least are the hydrophobic interactions, van der Waals forces, and hydrogen bonding, which act in a more proximal situation, but provide a very high number of low-energy bonds that eventually may become the most important factor involved. Thus, to interfere with adsorption, the most commonly used techniques employ combinations of (1) salts, citrate, or phosphate ions at concentrations and/or pH that do not support antigen adsorption interfering with all charge-related contributions, and (2) ethylene glycol or detergents to interfere with the hydrophobic interactions. Often, a combination of these agents is required to optimize desorption conditions. As an alternative, it is possible to dissolve the aluminum gel by using sodium citrate solutions (Hem and White 1995).

In all cases it is very relevant, but often neglected, to set appropriate controls to understand the impact of the formulation and desorption conditions used on the stability of the antigen(s) per se, as a non-successful antigen recovery from the formulation supernatant or desorption process may be misinterpreted as a strong adsorption of the antigen or antigen precipitation. Thus, before attempting to study antigen desorption, it is crucial to assess antigen stability under formulation conditions in the absence of adjuvant, followed by understanding the impact of the desorption conditions on the physicochemical stability of the antigen; once conditions maintaining appropriate antigen stability are established, desorption can be attempted. It is apparent that antigen desorption is not as straightforward as it seems because it depends on the combined nature of the antigen/adjuvant interaction and on the stability of the antigen under the formulation and desorption conditions. Therefore, starting from a restricted panel of conditions interfering with the basic nature of each of the factors influencing adsorption, each antigen may need development of a specific "desorption cocktail." It is also important to consider that adsorption and formulation aging do themselves induce changes in protein conformation impacting antigen's structural stability, as reported in a few recent reports (Watkinson et al. 2013). As a matter of fact, it has been reported in several papers and for several different antigens that desorption becomes more difficult as the formulation matures over time; this phenomenon has been suggested to be associated with antigenic structural changes interpreted as an indication of partial unfolding, although there are no definitive indications on this. Such unfolding might result in a greater degree of contact between the antigen and the adjuvant's surface and therefore in an increased resistance to desorption (Jones et al. 2005; Vessely et al. 2009).

Particular attention needs to be paid also to the characterization methods used to quantify the antigen(s) because of the differential nature of the interference that the specific excipient combination used for desorption may cause in different assays. A very common semiquantitative method largely employed because of the low impact of various interferences is sodium dodecyl sulfate polyacrylamide gel electrophoresis (SDS-PAGE). The method is simple and fast and allows for Western blotting to determine the identity of the antigen(s), but it has limitations in terms of accuracy and detection of small qualitative modifications, but also in terms of the total amount of phosphate that remains soluble under denaturing conditions and can be loaded in each well of the SDS-PAGE gel without compromising the solubility of the antigens, eventually resulting in technical artifacts. Other methods for quantification rely on UV or fluorescence, but these methods on their own are more prone to misinterpretation as the amount of information they provide is lower than SDS-PAGE. It would be ideal to combine them, especially in the cases where more than one antigen is present and recoveries are not complete. Reverse phase chromatography and immunoassays provide a much greater degree of accuracy and information allowing simultaneous quantification of several antigens but also providing additional information on stability, integrity, and functionality of the antigens, depending on how the assay is developed or the nature of the reagents used (monoclonal antibodies).

In conclusion, antigen desorption can be challenging, and several parameters need to be kept under consideration for an accurate evaluation. Furthermore, this needs to be considered in conjunction with the limitations posed by the desired analysis of the desorbed antigen(s). Depending on the level of accuracy desired in the analysis of the desorbed antigen(s), different scenarios can be envisaged; each scenario may have an increasing level of complexity and challenge, which is directly proportional to the increasing level of the desired characterization.

19.4 Characterization of Vaccines Adjuvanted with Oil-in-Water Emulsions

Only two oil-in-water emulsions are approved for human use, and both are a combination of squalene oil and Tween 80 surfactant with either a second surfactant like SPAN 85 (MF59) or α-tochopherol (AS03) in mildly acidic citrate buffer or neutral phosphate-buffered saline (PBS), respectively. Oil-in-water emulsions appear very different from the mineral salts discussed earlier and are characterized by a narrow oil droplet size distribution, and the presence of larger particles is considered a consequence of aging and is constantly monitored (McClements 2007). Several techniques are available to monitor interactions among emulsion components, other to determine component concentration, particle size, charge and other interfacial properties, and new techniques are available to study the antigen in the presence of the adjuvant and to monitor the quality and quantity of the excipients used for the emulsion.

As in the case of formulations containing mineral salts, also in the case of emulsions, the antigen(s) and the adjuvant need to maintain constant characteristics upon formulation. The presence of the emulsified oil droplets and of some spare detergent offers the antigen(s) a very different environment, as compared to other formulations, and these pose different challenges. The hydrophobic surface of the droplets and the residual free detergent could be viewed from a biochemical point of view as more challenging for the development of a long-term stable protein formulation. Thus, the impact on antigen stability needs to be carefully addressed, especially in light of long-term storage.

Also in this case the analysis of antigen post formulation needs attention. Usually, an ultracentrifugation step is required to separate the oil droplets from the remaining solution. If no strong interaction occurs between the antigen(s) and the droplets, the antigen can be recovered from the clear bulk solution, and SDS-PAGE could again be of great help for a basic semiquantitative evaluation. Also in this case there are various pitfalls that need to be considered, especially if the desired final accuracy is high, but again immune-reagents and reverse phase chromatography are excellent tools available to researchers for a more accurate quantification and stability analysis.

19.5 How to Study Antigen/Adjuvant Interaction

We have seen that the nature of the adjuvant and the antigen(s) influence one another in a sort of never-ending dialogue; this dialogue is mediated by the different nature of the interactions being developed and is subject to changes over time or as the nature of the components change. This to say, that it is difficult to establish a generalized way of looking at these interactions and how they evolve, although the central point is noting that, these interactions will evolve (Jones et al. 2005; Estey et al. 2009; Vessely et al. 2009).

For this reason it is of crucial importance to understand the tools available to study such evolution and the information they provide. Only in this way it would be possible to design the appropriate controls and develop new ideas to address the unanswered questions we can foresee at the moment.

Table 19.1 highlights some of the major techniques that have been employed recently to get a deeper understanding of these interactions and their evolution trying to avoid extensive sample manipulation.

As an example, for the characterization of aluminum-based formulations, only a limited subset of the characteristics can be followed with the different techniques described, but also that these can be generally referred to as energy exchange, particle scattering, charge, overall shape, antibody recognition, and spectroscopic characteristics. However, none of these, with the exception of the antibody-based techniques and NMR, has the level of detail required for an in-depth structural characterization of the changes occurring. More generally, all these techniques can be used to describe average changes like rearrangements of elements of secondary

Table 19.1 Recently described techniques used for direct vaccine characterization

Direct	Suspension	Emulsion
Isothermal titration calorimetry (ITC)	x	x[a]
Differential scanning nano calorimetry (nDSC)	x	x
Dynamic light scattering	x	x
Zeta potential	x	x
Static light scattering	x	x
Electron microscopy (EM)	x	x
Flow cytometry	x	–
Direct alhydrogel formulation immunoassay (DAFIA)	x	–
Fourier transformed infrared spectroscopy (FTIR)	x	–
solid state Circular Dichroism (ssCD)	x	x
Front face fluorescence spectroscopy	x	–
Raman spectroscopy	x[a]	–
Bio-layer interferometry	–	x
Single particle optical sensing	–	x
NMR	–	x
Surface tension	–	x

[a]Theoretically possible but not tested

structure, stability of thermal domains, or changes in exposure of aromatic or hydrophobic residues as a consequence of the measurement conditions employed (Jones et al. 2005; Ausar et al. 2011).

Currently, by taking advantage of these techniques we understand that both the adjuvant and the antigen(s) change as a matter of their reciprocal interaction. It is also clear that these changes depend not only on the adjuvant but also on the nature of the antigen(s), and as the nature of the antigen changes, the interactions will change with all the consequences.

The next question relates to the kinetics of these changes; once the antigen(s) and the adjuvant interact, they might change all at once so that these interactions remain constant over time, or the changes might occur more gradually with a different kinetics. By following the same samples over time it has been established that the situation is not static. Although data are not conclusive because only a few examples are reported in the literature, all data available seem to provide the same indication, which suggest that the antigen(s)–adjuvant interaction becomes stronger over time, resulting in lower desorption, changes in the thermal profile or changes in the exposure of secondary elements of structure or of hydrophobic residues. This being said, it is possible to prepare reference control samples where the adjuvant is missing but the treatment of the test sample is the same in order to discriminate between the changes induced by the process and those induced by adsorption on the adjuvant.

Finally, in light of the technical improvements in the past 30 years and a more favorable product complexity, a QbD approach extended to vaccine research could offer new avenues to determine with unprecedented precision drug product break points and offer the possibility to define the boundaries of formulation stability, efficacy, and safety. The consequence of that might be that understanding the nature and intensity of the parameters that affect stability, efficacy, and safety of the product will offer a way to approach them and improve the CQAs by pinpointing and correcting the exact residue(s) responsible for the undesired effect.

19.6 Perspectives for Future Development

Since subunit vaccines represent a significantly lower composition complexity, as compared to other classes of vaccines, they represent a better model to study more in detail some outstanding questions. As of today, for example, we still have no definitive answer on the relationship between immunogenicity and its relationship with the adsorption-induced structural and stability changes that the antigens undergo. It is clear that immunogenicity is impacted by antigen adsorption to the adjuvant, but we are still far away from understanding the factors that make an antigen more immunogenic as systematic studies that compare antigen adsorption capacity, adsorption coefficient, desorption potential, and stability in relationship to immune response are only a handful. An interesting study in this regard has recently been published where, for the first time, the authors attempt a systematic study of these different parameters and substantiate the hypothesis present in the literature that a strong interaction between the antigen and the adjuvant may hamper processing of the antigens by the immune system resulting in a lower immune response and suggest the importance of studying the antigen–adjuvant adsorption (Ausar et al. 2011). From this question, which is probably the central one, more questions arise on how to understand and control adsorption maturation and the impact on the immune response: This question could be very important to connect formulation maturation to immunogenicity and one day may have the potential to impact how we will determine the shelf-life of a product.

A question is how to study in more detail the structural changes that the antigens undergo upon adsorption. This last point will require very close collaboration between formulations scientists, immunologists, and structural biologists in complex experimental settings where a high degree of structural information and immune-tools are available. In my opinion, understanding the immunological determinants that make an antigen more immunogenic under varying conditions will be crucial to preserve and follow the subtended antigenic elements impacted by the different conditions explored. Until we understand this, we will not be in the position to make rational changes to improve the formulation's stability and efficacy.

From a characterization point of view, we can foresee that assays based on the direct alhydrogel formulation immunoassay (DAFIA) or flow cytometry (Zhu et al. 2009; Ugozzoli et al. 2011) developed to exploit monoclonal antibodies may be of more immediate use than other techniques to shed light on the antigenic elements that do best stimulate antibody-mediated protection.

In conclusion, the day we will be able to uncover the protective elements of each antigen we will have an idea of the antigenic determinants that should be preserved and presented to the immune and will be able to develop analytical tools necessary to follow and characterize them.

References

Ahmed SS, Plotkin SA et al (2011) Assessing the safety of adjuvanted vaccines. Sci Transl Med 3(93):93rv72

Andre FE, Safary A (1987) Summary of clinical findings on Engerix-B, a genetically engineered yeast derived hepatitis B vaccine. Postgrad Med J 63(Suppl 2):169–177

Ausar SF, Chan J et al (2011) Application of extrinsic fluorescence spectroscopy for the high throughput formulation screening of aluminum-adjuvanted vaccines. J Pharm Sci 100(2): 431–440

Avci FY, Li X et al (2011) A mechanism for glycoconjugate vaccine activation of the adaptive immune system and its implications for vaccine design. Nat Med 17(12):1602–1609

Centers for Disease Control and Prevention (CDC) (1999) Availability of lyme disease vaccine. MMWR Morb Mortal Wkly Rep 48:35–36

Costantino P, Viti S et al (1992) Development and phase-1 clinical-testing of a conjugate vaccine against meningococcus-a and meningococcus-C. Vaccine 10(10):691–698

Desombere I, Van der Wielen M et al (2002) Immune response of HLA DQ2 positive subjects, vaccinated with HBsAg/AS04, a hepatitis B vaccine with a novel adjuvant. Vaccine 20(19–20): 2597–2602

EMA (2005) Guideline on adjuvants in vaccines for human use. EMEA/CHMP/VEG/134716/2004. C. F. M. P. F. H. USE.

Estey T, Vessely C et al (2009) Evaluation of chemical degradation of a trivalent recombinant protein vaccine against botulinum neurotoxin by LysC peptide mapping and MALDI-TOF mass spectrometry. J Pharm Sci 98(9):2994–3012

Giuliani MM, Adu-Bobie J et al (2006) A universal vaccine for serogroup B meningococcus. Proc Natl Acad Sci U S A 103(29):10834–10839

Gupta (1995) Adjuvant properties of aluminum and calcium compounds. Pharm Biotechnol 6:229–248

Hem SL, White JL (1995) Structure and properties of aluminum-containing adjuvants. Pharm Biotechnol 6:249–276

Jones LS, Peek LJ et al (2005) Effects of adsorption to aluminum salt adjuvants on the structure and stability of model protein antigens. J Biol Chem 280(14):13406–13414

McClements DJ (2007) Critical review of techniques and methodologies for characterization of emulsion stability. Crit Rev Food Sci Nutr 47(7):611–649

McNeil C (2006) Who invented the VLP cervical cancer vaccines? J Natl Cancer Inst 98(7):433

Rappuoli R, De Gregorio E (2011) A sweet T cell response. Nat Med 17(12):1551–1552

Rappuoli R, Black S et al (2011) Vaccine discovery and translation of new vaccine technology. Lancet 378(9788):360–368

Scarselli M, Arico B et al (2011) Rational design of a meningococcal antigen inducing broad protective immunity. Sci Transl Med 3(91):91ra62

Ugozzoli M, Laera D et al (2011) Flow cytometry: an alternative method for direct quantification of antigens adsorbed to aluminum hydroxide adjuvant. Anal Biochem 418(2):224–230

Vessely C, Estey T et al (2009) Stability of a trivalent recombinant protein vaccine formulation against botulinum neurotoxin during storage in aqueous solution. J Pharm Sci 98(9): 2970–2993

Watkinson A et al (2013) Increasing the potency of an alhydrogel-formulated anthrax vaccine by minimizing antigen-adjuvant interactions. Clin Vaccine Immunol 20(11):1659–1668. doi:10.1128/CVI.00320-13

Zepp F (2010) Principles of vaccine design-Lessons from nature. Vaccine 28(Suppl 3):C14–C24

Zhu D, Huang S et al (2009) Development of a Direct Alhydrogel Formulation Immunoassay (DAFIA). J Immunol Methods 344(1):73–78

Chapter 20
The Physical Analysis of Vaccines

Yuan Cheng, Justin C. Thomas, Sangeeta B. Joshi, David B. Volkin, and C. Russell Middaugh

20.1 Methods for the Physical Analysis of Vaccines

20.1.1 Analysis of Primary Structure

The primary structure (i.e., chemical structure and post-translational modifications) of protein and virus-like particle-based (VLP) vaccines (see Chap. 9) can be easily probed by the use of peptide mapping. This is now routinely done with high-performance liquid chromatography mass spectrometry (HPLC-MS) methods in which proteins are first fragmented with enzymes of high specificity and the fragments are subsequently identified with MS. Peptide maps can be used to establish the integrity of the entire sequence of a protein as well as the presence of modifications such as glycosylation and/or chemically degraded amino acid side chains by reactions such as oxidation and deamidation. In the case of vaccines that contain more than one protein antigen, this method is often used to examine the proteins independently after their chromatographic separation. Once this is done, more routine methods such as reversed phase HPLC (RP-HPLC) and capillary isoelectric focusing (cIEF) can often be used for this purpose to increase throughput. In practice, such methods could also be used for viral antigens, but the presence of only

Y. Cheng
Department of Discovery Pharmaceutics, Pharmaceutical Candidate Optimization, Bristol-Myers Squibb, Princeton, NJ 08540, USA
e-mail: yuan.cheng@bms.com

J.C. Thomas
Eli Lilly and Company, Indianapolis, IN 46221, USA
e-mail: thomas_justin_c@lilly.com

S.B. Joshi • D.B. Volkin • C.R. Middaugh (✉)
Department of Pharmaceutical Chemistry, Macromolecule and Vaccine Stabilization Center, University of Kansas, Lawrence, KS 66047, USA
e-mail: joshi@ku.edu; volkin@ku.edu; middaugh@ku.edu

C. Foged et al. (eds.), *Subunit Vaccine Delivery*, Advances in Delivery Science and Technology, DOI 10.1007/978-1-4939-1417-3_20

small quantities of some viral proteins in vaccine formulations has made this difficult to implement on any routine basis.

Standard methods of nucleic acid sequencing and restriction mapping are also available for nucleic acid-based vaccines. With the increasing availability of highly specific glycosidases and certain spectrometric methods, chemical analysis of the carbohydrate components of vaccines is also now fairly routine (Alley et al. 2013; An et al. 2013). In addition, nuclear magnetic resonance (NMR) assays have become more routinely used to monitor the composition and chemical stability of sugar units comprising carbohydrate-based vaccines (Jones 2005). At much lower resolution, methods such as polyacrylamide gel electrophoresis [PAGE, with and without sodium dodecyl sulfate (SDS) and reducing agents] for proteins and agarose gel electrophoresis for nucleic acids can be used to detect chemical changes to a limited extent.

20.1.2 Analysis of Secondary Structure

The secondary structure types of most concern to a vaccinologist are the helices, sheets, and turns in proteins and the various types of helices found in nucleic acids. Three methods are commonly used to evaluate the secondary structure of vaccine antigens. Circular dichroism (CD) in the far ultraviolet (250–180 nm) region is probably the most widely employed. By measuring the difference in absorbance of left- and right-handed circularly polarized light, signals characteristic of optically active chromophores can be observed and directly related to both the type and amount of secondary structure in proteins and nucleic acids. By varying concentration and optical pathlength, measurements can be made over a wide array of conditions of pH, temperature, protein concentration, and in the presence of other solutes. Deconvolution of spectra and use of reference spectra permit estimates of the amount of secondary structure with accuracy on the order of a few percent. While this procedure works for purified macromolecules, it can only sometimes be used quantitatively for more complex systems such as viruses. This is because the spectra obtained reflect contributions from all components present (e.g., each protein and the nucleic acid present in a virus). In many cases, a single component may dominate the spectrum (such as a viral coat protein), greatly simplifying interpretation. The particulate nature of many vaccines does make CD susceptible to a number of artifacts including absorption flattening (shadowing of one particle by another) and differential scattering (scattering of one handedness of light more than another) which require careful consideration by the analyst.

Fourier transform infrared (FTIR) spectroscopy provides a second method that can be used to monitor secondary structure. For proteins, the amide bands (most commonly the Amide I or III) which contain distinct signals for α-helix, β-sheet, turns and random structure can be obtained by derivative and/or deconvolution analysis and used to obtain quantitative estimates of the relative amounts of secondary structure. Distinct peaks for intermolecular β-sheet structure are also often resolvable. Although at one time considered much less sensitive than CD (due to the need to use D_2O as a solvent to shift water bands so Amide signals could be measured), recent developments, especially those employing attenuated total reflectance (ATR) geometries, allow this technique to be used at much lower protein concentration in aqueous conditions. An advantage of FTIR spectroscopy is that it can also be used with solid (dried) samples. This ability has frequently been employed to look at the structure of lyophilized and spray-dried macromolecular systems. Of course, the same problem of interpretation with complex samples remains with this method. While lipids and carbohydrates are often CD transparent, FTIR spectroscopy may permit one to simultaneously detect signals from these moieties, making this method useful for carbohydrate- and lipid-containing vaccines. DNA molecules contain structurally sensitive spectral features as well. Signals from the nucleotide bases, carbohydrates and phosphate groups are all easily resolved, some of which are sensitive to the nature of any secondary structure present (e.g., A-form, B-form, Z-form DNA). FTIR can be especially useful for DNA-based vaccines which employ a carrier (e.g., cationic lipid, polyethylenimine) since the structure of both the DNA and cationic partner can be simultaneously examined (Middaugh and Ramsey 2007).

The third technique that can be used to examine the secondary structure of vaccine components is Raman spectroscopy. Like FTIR spectroscopy, Raman methods are used to examine vibrational transitions; but, rather than being based on the absorption of IR photons, small shifts in the frequency of scattered photons through interactions with vibrational states are detected. The selection rules for Raman spectroscopy are different than those for IR spectroscopy (i.e., a change in polarizability rather than an alteration in dipole moment is required). One consequence of this is that water causes less interference in the Raman technique. It is, however, significantly less sensitive than FTIR spectroscopy, usually requiring >5 mg/mL of protein for good signals. In the past several years, however, two versions of the Raman technique have been developed which at least partially alleviate this problem (Wen 2007). In the resonance Raman method, the sample is excited within an electronic absorption band. This produces a dramatic enhancement in the intensity of the vibrational transitions in the excited chromophores. Using an ultraviolet laser, both the peptide bonds and the aromatic side chains can be examined. This permits proteins to be characterized at very low concentrations. In surface-enhanced Raman, samples are adsorbed to a rough metal surface resulting in signals as much as 10^{10}–10^{11} times that of their usual magnitude. This method has not yet seen application to vaccines but may in the future prove of use. Similar comments are appropriate for surface-enhanced IR absorption (Brown et al. 2013). A major advantage of Raman spectroscopy is that it can simultaneously be used to examine secondary structure and aromatic- and sulfhydryl-containing side chains. Extensive studies of viruses have been performed with the Raman technique (Blanch et al. 2003), but it has not yet been extensively applied to vaccines. Finally, it is probable that at some point NMR will be used in some instances to characterize the secondary structure of vaccine components. Although natural isotope abundance has been successfully employed with small proteins, the necessity for isotope labeling and the size of vaccines currently prohibit its general application for examining the higher order structures of vaccines.

20.1.3 Analysis of Tertiary Structure

A wide variety of methods are available for tertiary structure analysis of vaccine antigens, most of which depend on detecting changes in the environment of aromatic side chains or nucleic acid bases or the use of environmentally sensitive extrinsic dyes. The three most commonly used techniques for this purpose are near-UV absorption, intrinsic tryptophan fluorescence, and near-UV CD spectroscopy. In near UV absorption, all three aromatic side chains (Phe, Tyr, Trp) contribute to the observed spectra between 250 and 300 nm. Their individual contribution can be resolved by derivative spectroscopy. Since Phe side chains are usually buried, while Tyr are interfacial due to their hydroxyl group and Trp residues more randomly dispersed, specific regions of structural change can occasionally be detected by using this method. Some additional information can sometimes be obtained by perturbing proteins with variable conditions of pH and temperature. Information about protein dynamic structure can also be obtained from the pre-transition slope of a second derivate peak (in a wavelength versus temperature plot) (Esfandiary et al. 2009) and shifts in peak position induced by added cations (Lucas et al. 2006). In nucleic acids, the spectra of the bases are very sensitive to their local intramolecular interactions and this can be used for detection of secondary structure changes. In the case of RNA, such changes may also be sensitive to alteration in tertiary structure. In both instances, the technique is most often used in a temperature-variation (melting) format. Additionally, the absorption spectra of a wide variety of external dyes can be altered by changes in the structure of proteins and nucleic acids (most often by binding at hydrophobic sites in proteins or between the bases/in the grooves of nucleic acids). In some situations, they can be used as probes for tertiary structure in RNA. Furthermore, a number of dyes can interact with membranes either at their surface or within a lipid bilayer. They can also be used to probe membrane integrity as well as fluidity. Most often, however, it is the fluorescent properties of such probes that are used as structural tools.

Fluorescence tends to be more sensitive than absorbance when changes in environment are monitored. The emission of fluorescence from Trp residues in proteins is often very sensitive to small structural changes. Interpretation of such spectral changes is often complicated by the presence of multiple Trp residues in proteins (Phe and Tyr residues usually produce quite limited fluorescence and this can be minimized by excitation at ≥295 nm). If, however, a component protein in a vaccine lacks Trp residues, this can allow localization of signal changes to other protein(s). There are a variety of ways in which intrinsic fluorescence can be used to characterize vaccines including quenching (using various probes such as O_2, acrylamide, and iodide to detect relative Trp exposure and molecular dynamics), resonance energy transfer (to measure the distance between fluorophores) and polarization (to characterize molecular mobility), but we will not consider these approaches further here. In general, nucleic acids lack fluorophores (the bases are not fluorescent), with the exception of the rarely encountered pseudouridine in RNA.

Although some macromolecular systems may lack intrinsic fluorophores, they may still be characterized using extrinsic fluorescent probes. In the case of proteins, many of the relevant probes are only weakly fluorescent in polar solutions, but strongly emit light when they bind to apolar sites. Probably the most commonly used dye is 8-anilino-napthalene sulfonic acid (ANS), but a wide variety of extrinsic fluorophores are available for varied uses. Molecules which bind either between the bases (intercalation) or in one of the grooves can be applied in a similar way to nucleic acids. For example, release of DNA or RNA from viruses during degradation can be measured in this manner (Volkin et al. 1997). Fluorescent probes can also be used to study membranes. Of particular note is the interaction of probes with lipid bilayers which can be used to probe the fluidity of the membrane.

As noted above, the near-UV region contains CD spectral peaks from the aromatic groups in proteins (as well as disulfides). Thus, it can also be used to detect tertiary structure changes. Very conveniently, under appropriate conditions of concentration and pathlength, the near and far UV region in a CD spectrum can be acquired in a single scan (Hu et al. 2011a). Tertiary structure changes in RNA can also, in principle, be studied by CD, but it is difficult to separate any such alteration from secondary structure changes. Although FTIR spectra are notoriously insensitive to tertiary structure changes, this is not the case with Raman spectroscopy where a variety of such signals are available. As in the case with secondary structure analysis, NMR can ultimately be expected to be a particularly powerful tertiary structure method. For the reasons mentioned above, however, it is currently of limited use.

20.1.4 Analysis of Macromolecular Size

Size is often a particularly important attribute of vaccine antigens. This is especially the case with particulate macromolecular systems such as viruses, VLPs and bacteria. We can differentiate between two different types of intermolecular association. In stoichiometric systems, complexes of well-defined monomer content (such as protein oligomeric association or VLPs) are formed. In the second, much more extensive association occurs, typically of ill-defined stoichiometry and more amorphous structure. This is generally referred to as aggregation. Similar analytical methods are often used to study both phenomena (Wang et al. 2013).

Size exclusion chromatography (SEC) is probably the most commonly used method to study macromolecular association. While calibration with size standards makes this a useful technique for monomeric and oligomeric species, larger aggregates and particles are frequently found in SEC column void volumes or irreversibly bound to the column matrix, making it less quantitative than desired. Furthermore, the dilution which takes place during chromatographic separation makes the method less useful for analysis of associating systems.

Light scattering methods are particularly useful for size analysis. In static light scattering experiments, the intensity of light scattered at one or more angles from a sample is monitored. In an ideal situation, the resultant data can be used to estimate size (the radius of gyration) and the molecular weight. Simple increases in scattering intensity can be used to follow aggregation. This can be performed using scattering monitored during intrinsic fluorescence measurements (by a dedicated detector or as part of the emission spectrum itself) or in the form of turbidity (optical density) using a conventional absorbance spectrometer. Light scattering detectors are also available for addition onto chromatography systems. In dynamic light scattering (DLS), the fluctuations in intensity of scattered light are monitored in the form of an auto-correlation function. This permits the diffusion coefficient of the scattering particles to be obtained and converted to a hydrodynamic radius by using the Stokes-Einstein equation. Either an average value and polydispersity of the target molecule can be obtained or the auto-correlation function can be deconvoluted into individual particle distributions, albeit with considerable uncertainty. It should be noted that size determination by DLS is generally limited to 1 μm or less.

The most accurate method of molecular size determination is probably analytical ultracentrifugation (AUC) in either the velocity or equilibrium mode (Philo 2009). In the sedimentation velocity method, the rate of migration of a particle during centrifugation is measured and quantified in the form of a sedimentation coefficient. This can be converted to a molecular weight if the diffusion coefficient is known (often obtained by DLS) and the partial specific volume of the particle is known (usually obtained by some type of density measurement) or can be estimated. Individual species quite close in size can often be resolved by this method.

In sedimentation equilibrium, the sample is spun to an equilibrium state in which its mass is distributed throughout the sample-containing centrifuge tube. This distribution can be converted to a molecular weight or analyzed to define the nature of any association present. Both AUC methods are somewhat specialized but often provide less ambiguous results than SEC or DLS. It is, of course, also possible to analyze the size of particles by methods such as transmission or scanning electron microscopy (TEM/SEM) or some version of scanning tunneling or atomic force microscopy. Although possessing some intuitive advantages, microscopy methods suffer from the artificial state of the sample during observation.

A wide variety of other methods are available for the characterization of large particles and aggregates; we will mention only a few here. These methods are usually used in combination because the coverage of size range varies with each method. They are often employed in conjugation with the methods described above. In nanoparticle tracking analysis, single particle DLS is performed with a laser through a microscope. Not only can the size of the scattering particles be determined (in the range 50–1,000 nm) but their number can be counted and size distributions obtained. Another method that can be employed to detect aggregation is the quartz crystal microbalance (QCM). In a recent version of this method ("the Archimedes instrument"), a mechanically resonant sampling device in the form of a hollow beam undergoes changes in its resonance frequency when it is modified by the presence of a sample's mass. This permits information about particle size, mass, volume, density, and concentration to be obtained over the range of approximately 50 nm–5 μm. Of particular importance, protein aggregates can be distinguished from air bubbles and silicone oil droplets by this method. For large particles, the coulter method is often used. In this technique particles are passed through a pore resulting in conductance changes which are proportional to a particle's volume. This allows samples in the range of 0.5–50 μm to be analyzed.

Another class of methods used to characterize large particles is based on light obscuration and digital imaging (Narhi et al. 2009). Samples are passed through a light illuminated orifice and the decrease in light intensity due the presence of the particles can be related to the size and volume of the particles. It is most effective in the range of 2–100 μm. In microflow digital imaging, samples are passed through a flow cell and magnified images are obtained by using a digital camera. Analysis of the images obtained permit information about particle size and morphology as well as the total number of particles. Like the recent QCM method, protein aggregates can be distinguished from air bubbles and silicone oil droplets by this method as well. Other techniques which are often useful for particle analysis include light and fluorescence microscopy and visual assessments, Zeta potential (surface charge) measurements, fluorescence activated cell sorting (FACS) and FTIR/Raman microscopy.

It is often necessary to examine antigens adsorbed to the surface of aluminum salt adjuvants (see Chaps. 3 and 19). The number of techniques available for such analyses is limited but includes fluorescence, FTIR and Raman spectroscopy as well as differential scanning calorimetry (DSC) (Jones et al. 2005; Peek et al. 2007; Katzenstein et al. 1986). This will be considered in greater detail in the examples described below.

It has recently been recognized that the array of data often obtained in vaccine characterization can be conveniently summarized in various graphical forms. This was initially done using an approach known as the empirical phase diagram (EPD). In this method, normalized values of experimental methods such as CD, fluorescence, and light scattering are presented in the form of multidimensional vectors in which the components of the vectors represent the experimental measurements (Maddux et al. 2011; Joshi et al. 2010). The resultant vectors are converted to colors using a red-green-blue (RGB) scheme. Recently, the colors have been assigned meaning in terms of secondary, tertiary, and associated structures. New versions of these diagrams have also been developed in the form of radar diagrams in which the length of the axes of polygons represent the measurements and Chernoff faces in which the experimental values are mapped to facial features. Details of these methods are described elsewhere (Kim et al. 2012). A method to compare macromolecular structure known as comparative signature diagrams has also been developed based on combining various types of measurements (Iyer et al. 2013). All of the above data visualization diagrams are usually generated as a function of variables such as temperature, pH, concentration, and ionic strength as well as pharmaceutical stresses such as agitation and freeze/thaw events.

We will now illustrate the use of the biophysical methods described above for vaccine characterization with examples. Due to familiarity, we will focus on work from our own laboratory but the approaches outlined in the following case studies are similar to those described by others.

20.2 Examples of the Physical Analysis of Vaccines

20.2.1 Protein-Based Vaccines

An example of the physical analysis of protein-based vaccines involves the development of novel powder formulations of a recombinant Protective Antigen (rPA)-based anthrax vaccine for nasal mucosal delivery (Jiang et al. 2006). In this case, the physical stability of rPA as a function of pH and temperature was first examined by CD, UV-Visible absorption, intrinsic and ANS fluorescence spectroscopies (Fig. 20.1a–e). To summarize and interpret the data from these biophysical techniques, an EPD was constructed using the results of CD, intrinsic and ANS fluorescence (Fig. 20.1f). The EPD clearly revealed changes in the conformational stability of the protein antigen as a function of pH and temperature and facilitated the selection of optimal conditions for subsequent excipient screening. Initial biophysical assessment identified aggregation as a major degradation pathway for rPA. As a result, a turbidity-based high throughput screening assay was employed to identify excipients that would inhibit the aggregation of rPA. The effect of the aggregation inhibitors identified on the conformational stability of rPA was then examined by CD and intrinsic Trp fluorescence. The aggregation inhibitors that also stabilized the secondary and tertiary structures of rPA were chosen for subsequent preparation and characterization of dry powder formulations. The effect of spray-freeze drying (SFD) on the stability of rPA was investigated using SDS-PAGE and SEC. These techniques were also used in assessing the storage stability of the SFD powder formulation by accelerated degradation studies at various temperatures. The results suggested that the SFD powder formulation developed in this study had much higher stability than liquid formulations. In combination with a noninvasive nasal delivery system, this SFD formulation can serve as a potential alternative to conventional parental delivery by injection.

A second example of a protein-based vaccine involves the development of a non-glycosylated (NG) protein antigen, EBA-175 RII-NG, for use as a candidate malaria vaccine (Peek et al. 2006). Since malaria largely exists in tropical regions where a cold chain is often not readily available, the stability of EBA-175 RII-NG becomes a particularly important issue for its use in a vaccine. This study established a systematic approach to identification of the optimal stabilizing formulations for EBA-175 RII-NG. Structural changes of EBA-175 RII-NG in response to thermal stress under various pH conditions were first examined using high-resolution second derivative absorbance spectroscopy in the absence or presence of 5 % sucrose. The tertiary structure of the protein was evaluated by analyzing shifts in the UV absorption spectrum's second derivative negative peak positions as a function of temperature and pH. The EPDs constructed from the secondary derivative absorbance data (Fig. 20.2a, b) reveal that the presence of 5 % sucrose has very little effect on the conformational stability of EBA-175 RII-NG. To obtain greater detail concerning the degradation pathway of EBA-175 RII-NG, structural changes in EBA-175 RII-NG as a function of temperature and pH were analyzed by a variety

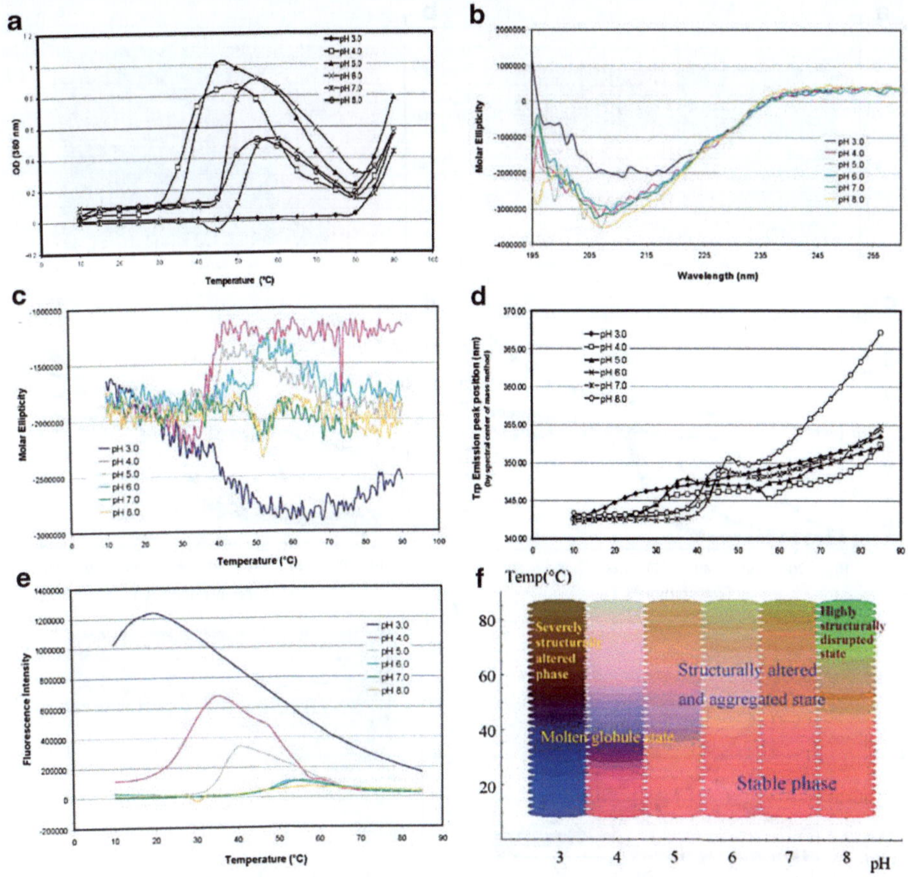

Fig. 20.1 Physical stability of recombinant Protective Antigen (rPA) as a function of pH and temperature as determined by optical density at 360 nm (**a**); CD spectra of rPA at 10 °C (**b**); CD signal at 222 nm (**c**); Trp fluorescence peak position (**d**); and ANS fluorescence intensity (**e**). Empirical phase diagram (EPD) of rPA generated using CD, Trp fluorescence, and ANS fluorescence measurements (**f**). Regions of similar color signify a similar physical state of rPA while changes in color represent transitions between such states. Five distinct phases were observed and defined as follows: (1) most stable phase [*red*-colored region in the *lower right*-hand corner]; (2) molten globule-like state [*blue/purple* area at pH 3, <45 °C]; (3) severely structurally altered phase [*dark brown* area at pH 3, >45 °C]; (4) structurally altered and aggregated state [*light purple–light brown* region at pH 4–7, >50 °C and pH 8, 50–65 °C]; (5) highly structurally disrupted form [*green* area at pH 8, >65 °C]. From: Jiang G, Joshi SB, Peek LJ, Brandau DT, Huang J, Ferriter MS, Woodley WD, Ford BM, Mar KD, Mikszta JA, Hwang CR, Ulrich R, Harvey NG, Middaugh CR, Sullivan VJ. Anthrax vaccine powder formulations for nasal mucosal delivery. Journal of Pharmaceutical Sciences 95 (1):80-96. Copyright 2006 by John Wiley Sons, Inc. Reprinted by permission of John Wiley & Sons, Inc.

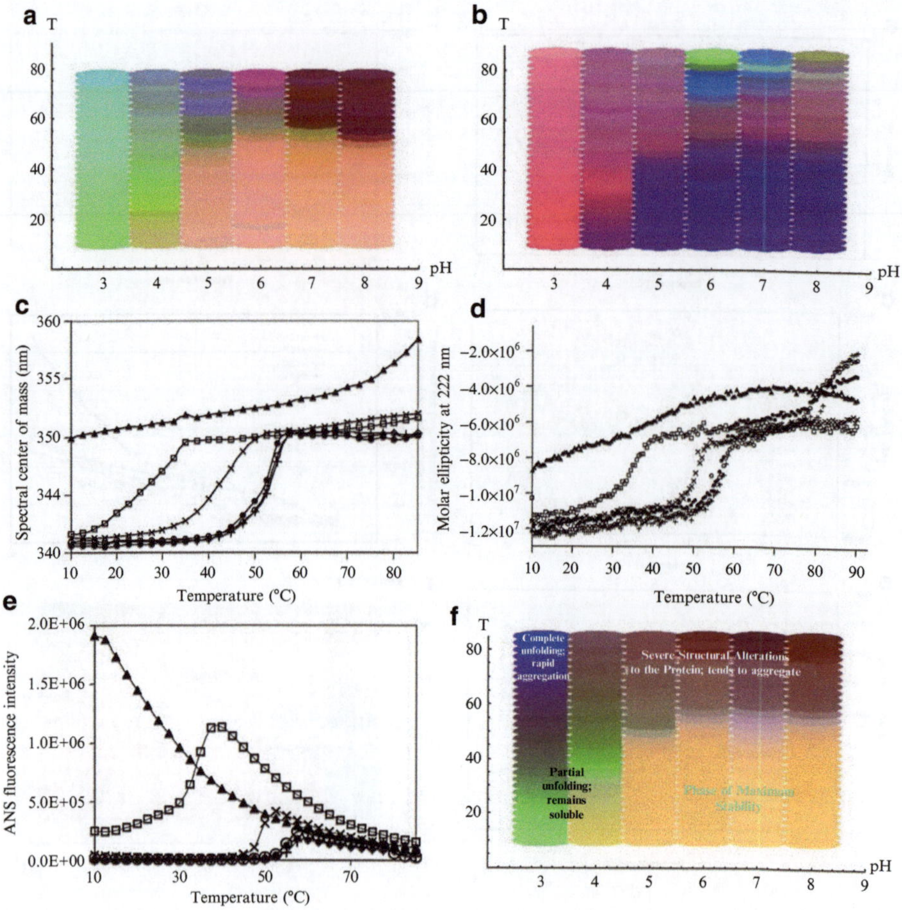

Fig. 20.2 EPDs created using high-resolution second derivative absorbance spectroscopy data for EBA-175 RII-RG protein antigen in the presence of 5 % sucrose (**a**) and in the absence of sucrose (**b**). Physical stability of EBA-175 RII-RG in the presence of 5 % sucrose as a function of temperature and pH as determined by changes in Trp fluorescence peak position (**c**); CD signal at 222 nm (**d**); and ANS fluorescence intensity (**e**). The six pH conditions evaluated were pH 3.0 (*filled triangle*), pH 4.0 (*open square*), pH 5.0 (*multiplication sign*), pH 6.0 (*open circle*), pH 7.0 (*plus sign*), and pH 8.0 (*filled circle*). EPD generated for EBA-175 RII-RG in the presence of 5 % sucrose using ANS fluorescence intensity, CD signal as 222 nm, and fluorescence peak position (**f**). The labels indicate the physical state of the protein within a particular color regime and are based on visual inspection of individual stability measurements. The region of greatest structural stability is within the pH range 5–8 at lower temperatures. Reprinted from Vaccine 24 (31–32), Peek LJ, Brandau DT, Jones LS, Joshi SB, Middaugh CR, A systematic approach to stabilizing EBA-175 RII-NG for use as a malaria vaccine, Pages 5839-5851, Copyright (2006), with permission from Elsevier

of techniques including CD, intrinsic and extrinsic fluorescence spectroscopies (Fig. 20.2c–e). An EPD was then developed to summarize the results from these techniques and select the condition used for subsequent screening of stabilizing excipients for the protein (Fig. 20.2f). Screening of stabilizers started from the identification of aggregation inhibitors using a turbidity assay. The effect of some selected aggregation inhibitors on the conformational stability of the protein was then examined using CD and fluorescence spectroscopies. Combinations of the best stabilizers were then examined to identify the optimal stabilizing conditions. As the last step in this study, the interaction of EBA-175 RII-NG with an aluminum salt adjuvant, including adsorption and desorption, was examined under various conditions. The systematic approach employed in this study can serve as the guide for future routine development of malaria protein therapeutics.

More examples examining the chemical and physical characterization of protein-based vaccines can be found from other previously published work, including studies of a 2-fluorohistidine-labeled analogue of recombinant anthrax protective antigen (Hu et al. 2012), *Clostridium difficile* toxins and toxoids (Salnikova et al. 2008a, b), gram-negative bacterial type III secretion system-based protein vaccines (Barrett et al. 2010; Markham et al. 2010), an aluminum salt-adjuvanted trivalent recombinant protein-based vaccine candidate against *Streptococcus pneumonia* (Iyer et al. 2012), and a recombinant pneumolysin protein antigen as a pneumococcal vaccine candidate (Hu et al. 2013).

20.2.2 *Virus-Like Particle-Based Vaccines*

VLPs are typically formed by the self-assembly of recombinantly expressed virus capsid proteins and may also contain other viral surface proteins or membrane proteins from the cell-based production source (see Chap. 9). VLPs mimic the surface protein structure of viruses, but are noninfectious due to the lack of viral genetic materials. As a result, VLPs usually serve as more effective vaccine antigens than monomeric subunit viral proteins. Two successful examples of VLP-based vaccines include hepatitis B virus (HBV) and, more recently, human papilloma virus (HPV) vaccines. The methods for the physical analysis of VLPs are largely similar to those described above for protein antigens. The major difference arises from the more complex structure and particulate nature of VLPs. Signals from VLPs are usually composite, reflecting contributions from all structural components. Accordingly, it is often difficult to identify the origin of structural perturbations induced by environmental stresses. Here we will discuss recent examples of the physical analysis of prospective VLP-based antigens.

The first example involves the conformational stability of Norwalk VLPs (Ausar et al. 2006). Norwalk virus (NV) is the major cause of acute nonbacterial gastroenteritis, accounting for 23 million reported cases annually in the United States. Consequently, NV vaccines are of vital importance. Previous studies found that NV-VLPs, formed by the self-assembly of 180 copies of the major NV capsid

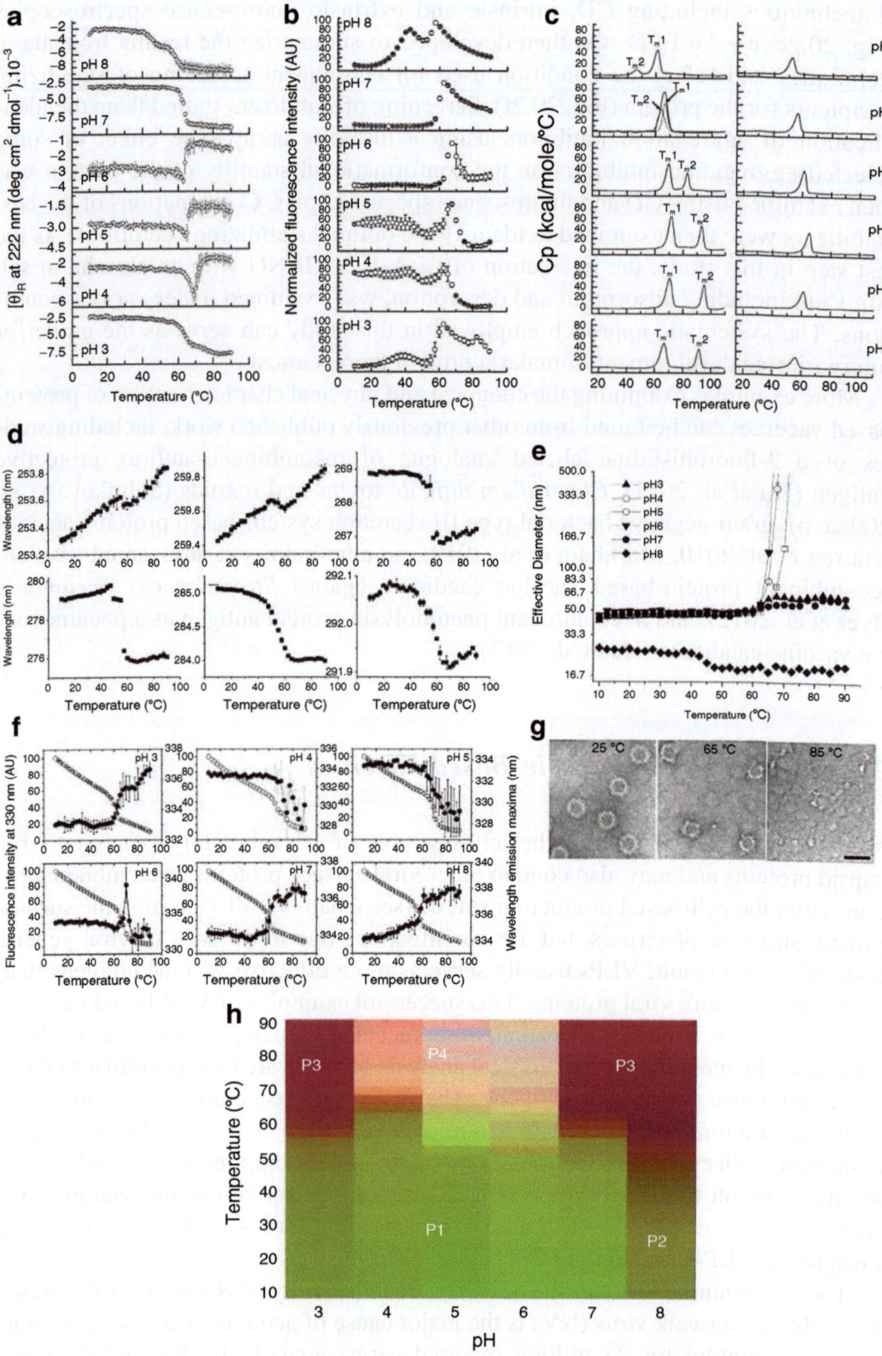

Fig. 20.3 Physical stability of Norwalk virus-like particles (NV-VLPs) as a function of temperature and pH as determined by changes in CD signal at 222 nm (**a**), ANS fluorescence intensity (**b**); DSC heat capacity (**c**); high-resolution second derivative absorbance spectroscopy at pH 7 (**d**);

protein (58 kDa), serve as an appropriate vaccine antigen. In this case, the inherent stability of NV-VLPs as a function of pH and temperature was examined by various biophysical techniques including high-resolution second derivative UV absorption spectroscopy, CD, DLS, DSC, TEM and both intrinsic and extrinsic fluorescence spectroscopies (Fig. 20.3a–g). An EPD was developed to summarize the results from these analyses (Fig. 20.3h). The EPD revealed that the thermal stability of NV-VLPs was pH dependent. The NV-VLPs displayed very good stability over the range of pH 3–7 up to 55 °C. At pH 8, however, reversible capsid dissociation was observed accompanied by perturbations in secondary and tertiary structure. Turbidity-based thermal melt analysis under a suboptimal condition identified from the initial biophysical characterization was employed to screen for stabilizers of NV-VLPs (Kissmann et al. 2008a). The stabilizers identified by turbidity analysis were then examined for their effects on the conformational stability of NV-VLPs using CD, DSC, and ANS fluorescence. The stabilizing conditions identified in this manner were used for the preparation of the NV-VLP formulation employed for the ensuing human clinical studies.

Another example of VLP-based vaccines subject to physical analysis involves the characterization and formulation of an engineered H1N1 influenza VLP, which represents an important step toward the development of a commercially viable vaccine (Kissmann et al. 2008a). Respiratory infection caused by the influenza virus is a leading cause of mortality and morbidity worldwide. Recombinant enveloped VLPs that consist of self-assembled murine leukemia virus (MLV) gag proteins and surrounding the membrane-bound primary influenza antigens hemagglutinin (HA) and neuraminidase (NA) have been developed as promising H1N1 influenza vaccine antigens. A combination of spectroscopic and light scattering techniques was employed in the initial assessment of the stability of the influenza VLP as a function of pH and temperature (Fig. 20.4a–d). The membrane-sensitive fluorescence dye Laurdan was used to study changes in membrane fluidity in response to changes in temperature and pH (Fig. 20.4e). The EPD constructed from the initial biophysical assessment revealed approximately ten distinct phases, reflecting the structural complexity of the influenza VLP and its distinctive response to pH and temperature. A turbidity-based analysis was employed as an initial screening assay for stabilizers, with the excipients identified being further screened by intrinsic and Laurdan fluorescence for conformational stabilizers. Trehalose, sorbitol, and glycine were found to best stabilize the influenza vaccine based on all of these criteria.

Fig. 20.3 (continued) DLS (**e**); and Trp fluorescence intensity and peak position (**f**). TEM images of NV-VLPs at various temperatures (**g**). EPD generated for NV-VLPs based on second derivate absorbance spectroscopy, Trp and ANS fluorescence, and CD (**h**). Four distinct phases of the NV-VLP were observed; *P1* native, intact form; *P2* disassembled; *P3* soluble VP1 oligomers; *P4* aggregated. This research was originally published in The Journal of Biological Chemistry. Ausar SF, Foubert TR, Hudson MH, Vedvick TS, Middaugh CR. Conformational stability and disassembly of Norwalk virus-like particles. Effect of pH and temperature. *The Journal of Biological Chemistry*. 2006; 281 (28):19478-19488. Copyright The American Society for Biochemistry and Molecular Biology

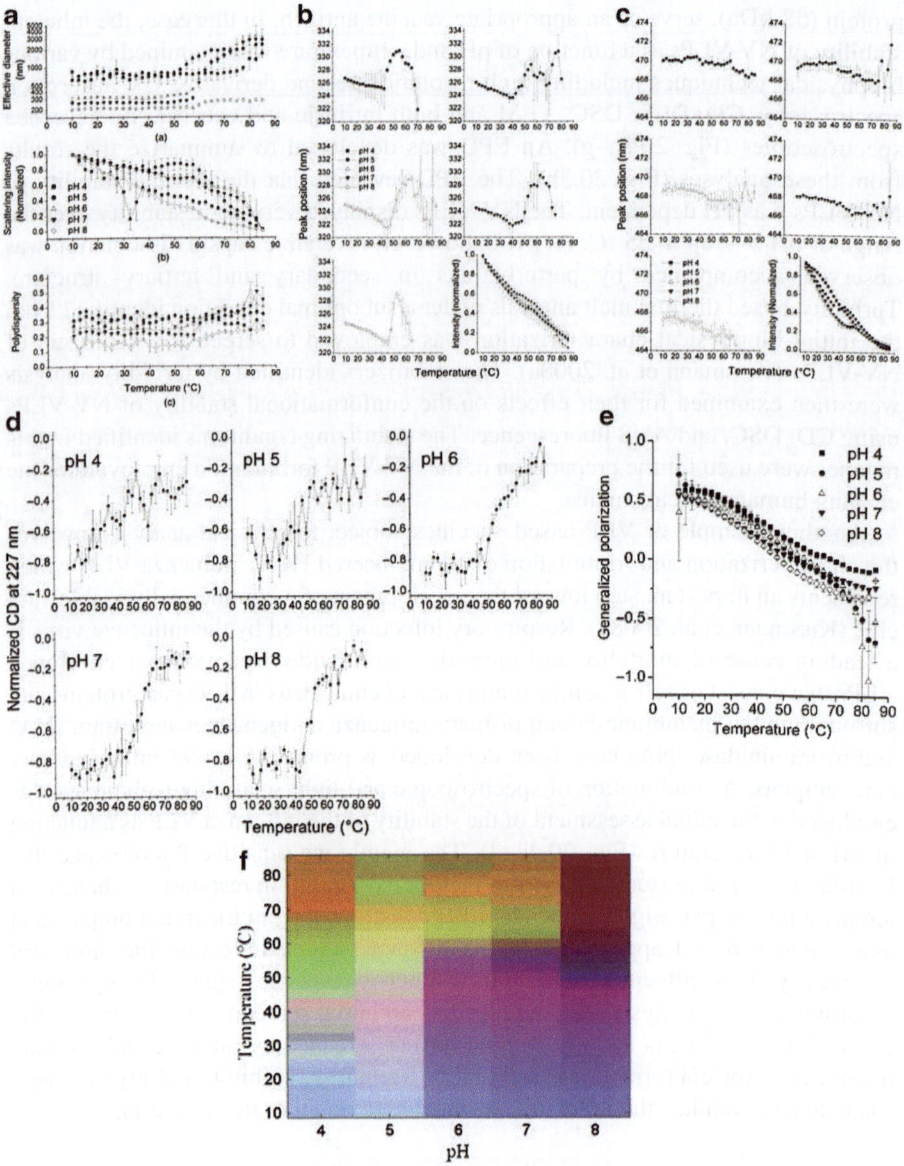

Fig. 20.4 Physical stability of H1N1 influenza VLPs as a function of temperature and pH as determined by DLS (**a**); fluorescence peak position and intensity (**b**); ANS fluorescence peak position and intensity (**c**); CD signal at 227 nm (**d**); Laurdan fluorescence (**e**). EPD of influenza VLPs (**f**). The EPD was prepared from temperature- and pH-dependent effective DLS diameters, static light scattering, polydispersity, CD at 227 nm, intrinsic fluorescence (peak position and relative intensity at 330 nm), ANS fluorescence (peak position and relative intensity at 485 nm), and generalized polarization of laurdan fluorescence data. The major phases observed are: the least structurally disrupted state of the VLPs (pH 6–8, low temperature, *blue*); a transition region that appears above the *blue* phase between 35 and 55 °C for pH 6 and 7, and from 35 to 50 °C at pH 8 (*purple*);

A third example of the physical analysis of VLP-based vaccines is the characterization of Ebola (eVLPs) and Marburg VLPs (mVLPs) (Hu et al. 2011b). As members of the Filoviridae family, the Ebola (EBOV) and Marburg viruses (MARV) cause severe hemorrhagic fevers in both humans and nonhuman primates. There are currently no licensed vaccines available for the prevention or treatment of these usually fatal viral infections. eVLPs and mVLPs, comprising assembled version of glycoprotein VP40 and NP, are considered promising vaccine antigens. Following an initial biophysical characterization by DLS, CD, and intrinsic fluorescence, an EPD was developed to display the conformational changes and aggregation behavior of the two enveloped VLPs in response to pH and temperature [due to the complexity of the figures, readers are referred to the original paper for details (Hu et al. 2011b)]. The EPDs showed that the stability of both eVLPs and mVLPs were highly pH-dependent, being more stable at higher pH values (Fig. 20.5) with maximal thermal stability observed within the pH range of 7–8. Both VLPs were least stable at pH 3 and 4 due to the susceptibility of the VLPs to aggregation. The knowledge gained from this study provided a basis for additional formulation development and long-term stability studies of these two vaccine candidates.

Two additional examples of the biophysical analysis of VLP-like systems are the needles of the type III secretion system of gram-negative bacteria (Barrett et al. 2010) and the particles formed by the E1 glycoprotein of the hepatitis C virus (He et al. 2009), both candidates for use as novel vaccine antigens.

20.2.3 *Virus-Based Vaccines*

As one of the most commonly used and effective approaches to vaccine development, the attenuated live virus approach has led to the development of a series of vaccines including measles, mumps, rubella, varicella, and rotavirus. In addition, inactivated viral vaccines have been successfully developed including polio and hepatitis A. Virus-based vaccines typically feature complex structures and significant lability. Physical analysis of virally based vaccines not surprisingly can resemble that of VLP-based vaccines. While biological assays have previously been the basis of viral vaccine structural characterization and stability studies, more physical approaches are beginning to make inroads into these activities.

Fig. 20.4 (continued) the variably colored area above 60 °C for pH 6 and 7 corresponding to particle aggregation; a conformationally altered state that lacks significant aggregation (pH 8, high temperature, *dark red*); a significantly structurally disrupted phase (pH 4 and 5, low temperature, *light blue*); and various conformationally altered states that give rise to multiple phases (above 35 °C in the low pH region, *green/orange*). From: Kissmann J, Joshi SB, Haynes JR, Dokken L, Richardson C, Middaugh CR. H1N1 influenza virus-like particles: physical degradation pathways and identification of stabilizers. Journal of Pharmaceutical Sciences 100 (2):634-645.

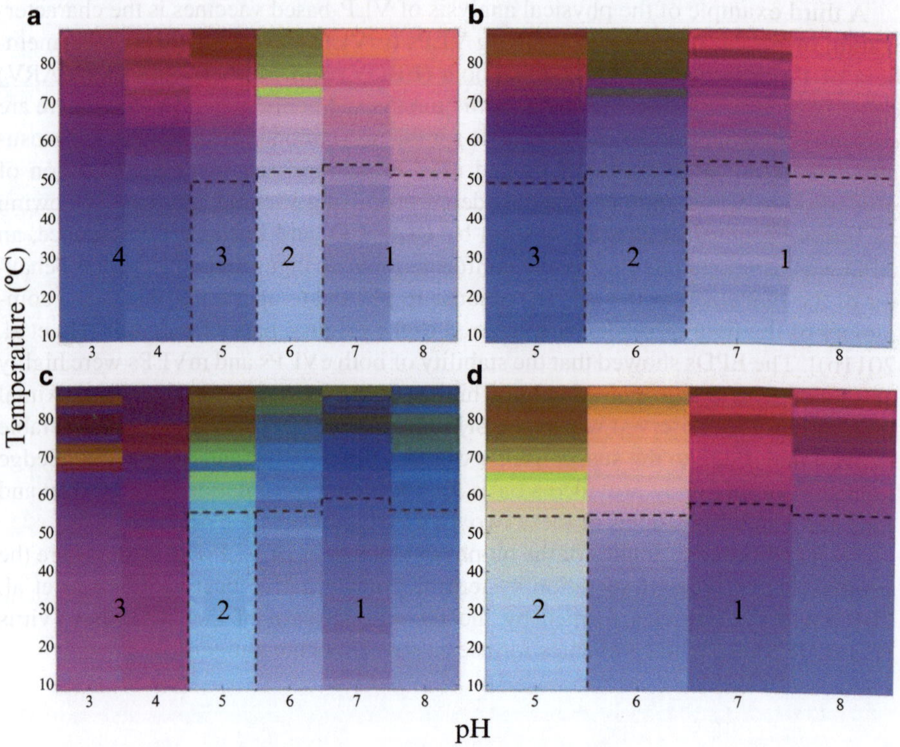

Fig. 20.5 EPDs of eVLPs (**a**, **b**) and mVLPs (**c**, **d**). The phases in the eVLP phase diagram (**a**) were assigned as follows: (1) native-like state and thermally stable; (2 and 3) intermediate phase with marginal stability; (4) structurally altered and prone to aggregation. In the mVLP phase diagram (**c**), the distinct phases present are: (1) native-like state and thermally stable (less stable in sample at pH 6); (2) intermediate phase with marginal stability; (3) structurally altered and prone to aggregation. These transitions were also evident in the data from the specific techniques used to generate the EPDs. The EPDs for (**b**) and (**d**) were prepared from dynamic light scattering (DLS), static light scattering, CD at 225 nm, tryptophan fluorescence (peak position and normalized intensity), and Laurden fluorescence data collected over the temperature range 10–87.5 °C. The EPDs for (**a**) and (**c**) were generated with the sample biophysical data, excluding DLS measurements. From Hu L, Trefethen JM, Zeng Y, Yee L, Ohtake S, Lechuga-Ballesteros D, Warfield KL, Aman MJ, Shulenin S, Unfer R, Enterlein SG, Truong-Le V, Volkin DB, Joshi SB, Middaugh CR. Biophysical characterization and conformational stability of Ebola and Marburg virus-like particles. Journal of Pharmaceutical Sciences 100 (12):5156-5173. Copyright 2011 by John Wiley Sons, Inc. Reprinted by permission of John Wiley & Sons, Inc.

In the first example, we consider human respiratory syncytial virus (RSV), which is the major cause of both lower respiratory tract infection among young children worldwide and of hospitalization in adults with community-acquired respiratory diseases (Ausar et al. 2005). Despite decades of effort, an effective vaccine for the prevention and treatment of RSV infection remains unavailable. Live attenuated RSV viruses have long been considered as promising vaccine antigens. The development of virus-based RSV vaccines, however, is very challenging due to their labile nature and complex structure. RSV is a single-stranded negative-sense RNA virus comprising ten different proteins, a genome of approximately 15,222 nucleotides and a lipid envelope. In this study, a crude human RSV was first purified by sucrose gradient centrifugation to remove cell impurities that might interfere with subsequent biophysical characterization. The thermal stability of RSV at different pH values was then examined using a variety of techniques, including high-resolution second derivative absorption spectroscopy, CD, DLS, turbidity, and both intrinsic and ANS fluorescence (Fig. 20.6a–f). The EPD constructed from the data (Fig. 20.6g) demonstrated that the secondary, tertiary, and quaternary structures of RSV are susceptible to both pH and temperature stresses, with maximal stability observed at near neutral pH. Turbidity and DLS data showed that RSV was prone to aggregation below pH 6. Based on these observations, turbidity analysis was selected as the initial screening approach to identify stabilizers that inhibit the aggregation of RSV (Ausar et al. 2007). The results revealed a variety of aggregation inhibitors of RSV including sugars, amino acids, polyols, and polyanions, with the latter especially effective. The effects of these aggregation inhibitors on the conformational stability of RSV was further examined using CD, DLS, UV absorption, and the generalized polarization of Laurdan fluorescence. Sugars, polyols, and polyanions all demonstrated an enhancement in the thermal stability of the secondary and tertiary structures of RSV proteins, as well as the lipid membrane of RSV. These results provided valuable information for future rational development of stabilizing formulations for live attenuated RSV vaccines.

Another example of a virally based vaccine that has been subjected to physical analysis is the measles virus, which also belongs to the family of negative-sense single-stranded RNA viruses (Kissmann et al. 2008b). Besides the RNA genome, the measles virion consists of six different structural proteins and an associated lipid membrane. The labile nature of existing attenuated live measles virus (MV) vaccines has hindered their use in some parts of the developing world. This study again started with a comprehensive characterization of the thermal stability of MV under different pH conditions using a variety of biophysical techniques including static light scattering, CD, intrinsic and ANS fluorescence (Fig. 20.7a–d). In addition, the fluidity of the lipid membrane was examined by Laurdan fluorescence (Fig. 20.7e). The results of these studies were synthesized into an EPD that allows the simultaneous visualization of the entire multi-dimensional data set (Fig. 20.7f). As seen with RSV, the EPD obtained is complex and contains at least six regions of distinct conformational states, reflecting the multifaceted structure of measles viruses. Guided by the EPD, screening assays were developed to identify stabilizers that inhibit the aggregation of viral particles, and enhance the structural integrity of the viral proteins and membrane. The stabilizers identified were demonstrated to significantly improve the infectivity of MV. This example provides additional evidence that accelerated thermal stability evaluation can be used to identify conditions that protect the biological activity of MV vaccines.

Similar physical characterization has been performed with different rotavirus serotypes (Esfandiary et al. 2010) and multiple types of adenoviruses (Rexroad et al. 2006a; Rexroad et al. 2006b; Rexroad et al. 2003; He et al. 2010).

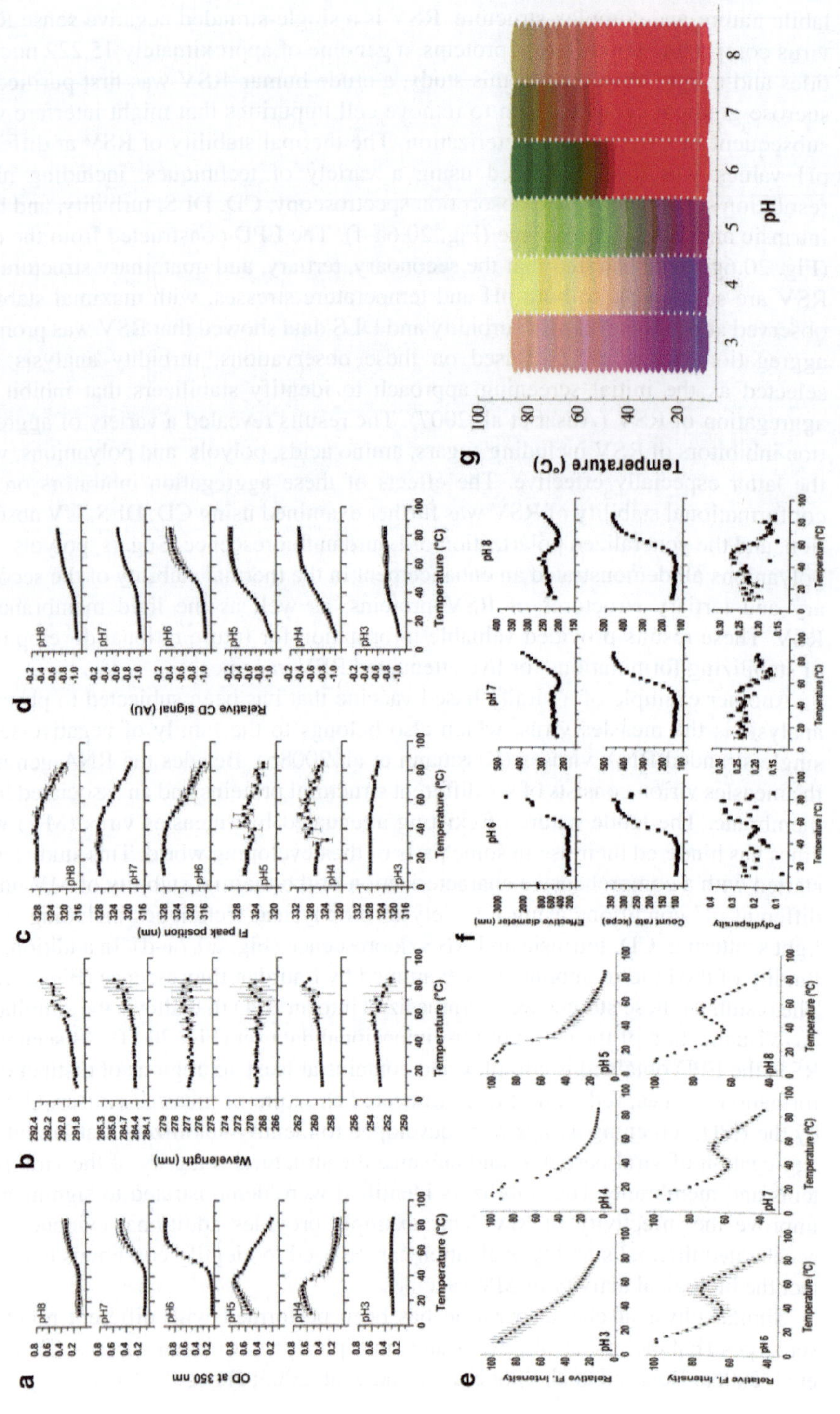
a
b
c
d
e
f
g
OD at 350 nm
Wavelength (nm)
Fl peak position (nm)
Relative CD signal (AU)
Relative Fl. Intensity
Effective diameter (nm)
Count rate (kcps)
Polydispersity
Temperature (°C)
pH
pH3
pH4
pH5
pH6
pH7
pH8

Fig. 20.6 Physical stability of live attenuated human respiratory syncytial virus (RSV) as a function of temperature and pH as determined by turbidity (OD at 350 nm) (**a**); high-resolution second derivative absorbance spectroscopy at pH 7 (**b**); Trp fluorescence peak position (**c**); CD signal at 222 nm (**d**); ANS fluorescence intensity (**e**); and DLS (**f**). EPD generated for live attenuated human RSV over the pH range of 3–8 (**g**). The major phases observed are: most stable phase (low temperatures, pH 6–8, *red*); conformationally altered and aggregated phase (pH 4–8, above ~40 °C, *yellow-green*); and thermally labile and self-associated phase (low pH, *purple-blue*). The data included in the generation of the EPD were second-derivative UV peaks, CD signal at 222 nm, optical density at 350 nm, Trp fluorescence peak position and intensity at 330 nm, and ANS fluorescence intensity at 485 nm. Reprinted with permission from Ausar SF, Rexroad J, Frolov VG, Look JL, Konar N, Middaugh CR. Analysis of the thermal and pH stability of human respiratory syncytial virus. Molecular Pharmaceutics 2 (6):491-499. Copyright 2005 American Chemical Society

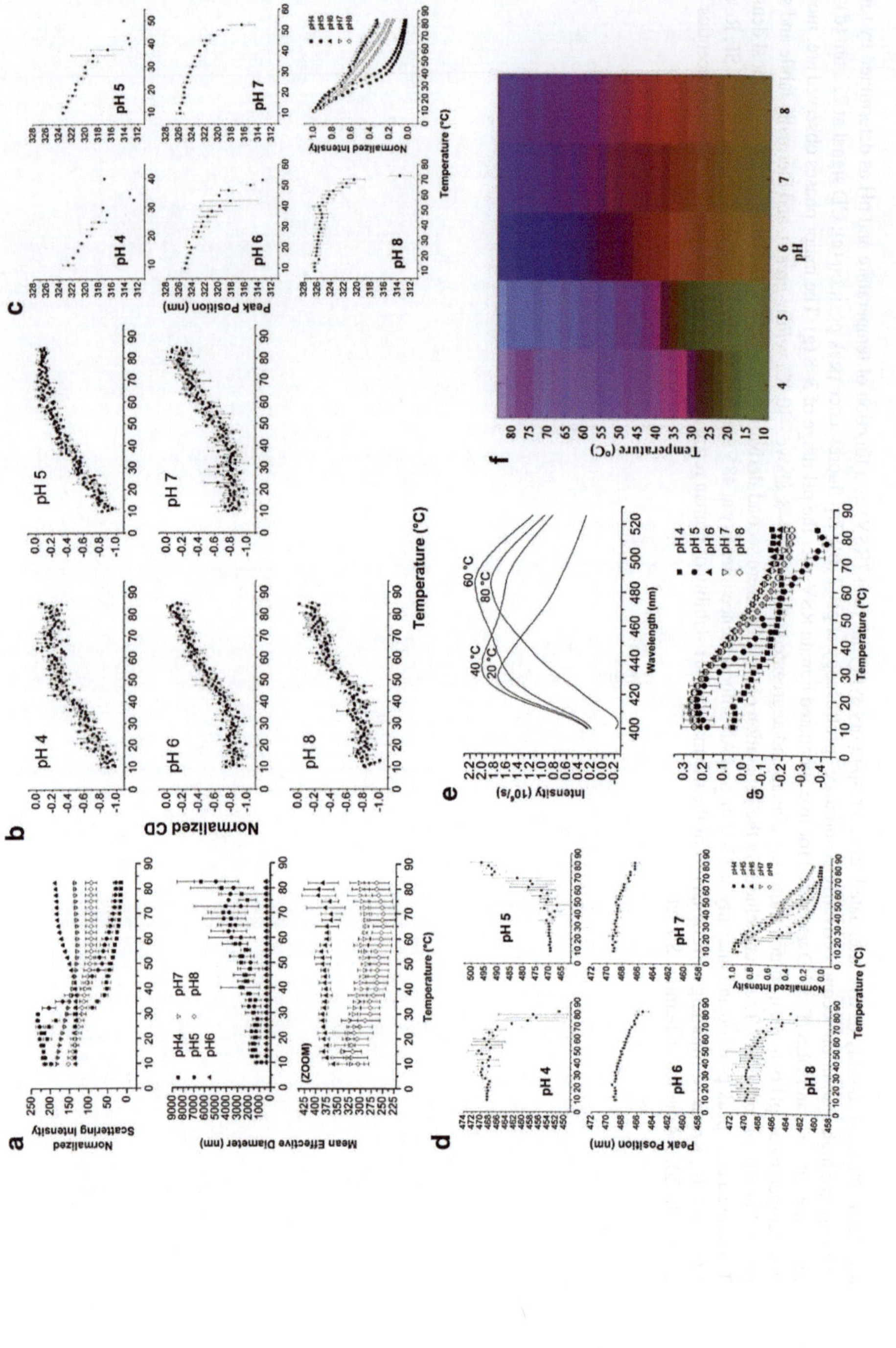
a
Normalized Scattering Intensity
pH4
pH5
pH6
pH7
pH8
Mean Effective Diameter (nm)
(ZOOM)
Temperature (°C)
b
Normalized CD
pH 4
pH 5
pH 6
pH 7
pH 8
Temperature (°C)
c
Peak Position (nm)
Normalized Intensity
d
e
Intensity (10⁶/s)
20 °C
40 °C
60 °C
80 °C
Wavelength (nm)
GP
f
Temperature (°C)
pH

Fig. 20.7 Physical stability of live attenuated measles viruses (MV) a function of temperature and pH as determined by static light scattering and DLS (**a**); CD signal at 222 nm (**b**); intrinsic fluorescence peak position and intensity (**c**); ANS fluorescence peak position and intensity (**d**); and Laurdan fluorescence (**e**). The EPD of MV is shown in (**f**). The EPD was generated using measurements of mean effective diameter, intensity of (562 nm) light scattering at 90°, CD at 222 nm, intrinsic fluorescence intensity at 322 nm, ANS peak position, ANS fluorescence intensity at 469 nm, and GP of Laurdan fluorescence. Reproduced with permission from Landes Bioscience, Inc. Kissmann J, Ausar S, Rudolph A, Braun C, Cape S, Sievers R, Federspiel M, Joshi S, Middaugh C. Stabilization of measles virus for vaccine formulation. Human Vaccines 2008; 4:350–359

20.2.4 Bacterial Vaccines

Bacterial vaccines are also characterized by a labile nature and complex structures. Physical analysis of bacterial vaccines resembles that of virally based vaccines, focusing on the assessment of the two major structural components of bacteria, their proteins and lipid membranes. An example of the physical analysis of bacterial vaccines involves the Ty21a typhoid vaccine developed from the wild type *Salmonella enterica* Serovar Typhi strain through chemical mutagenesis (Zeng et al. 2009). Despite their excellent potency, the liquid formulations of this vaccine exhibit low thermal stability. Thus, development of more stable liquid formulations is of significant importance. In this case, the integrity of the cell membrane was examined by *Bac*Light™ fluorescence (Fig. 20.8a), and the membrane fluidity was assessed by Laurdan fluorescence (Fig. 20.8b). In addition, the thermal stability of the protein components of Ty21a cells was examined by CD (Fig. 20.8c). The EPD constructed from these studies provided a global picture of the thermal stability of Ty21a cells over the pH and temperature ranges examined (Fig. 20.8d). On the basis of the EPD, a *Bac*Light™ fluorescence-based assay was employed to screen for excipients that enhance the thermal stability of the bacterial cell membranes. The stabilizing effect of selected excipients in a foam-dried formulation was also evaluated. The results suggest that the stabilizers identified for liquid formulations also enhance the stability of Ty21a cells in dried foam formulations, indicating that information gleaned from the physical characterization of a liquid formulation of something as complex as a live bacterial cell can facilitate the development of other types of formulations.

20.2.5 DNA Vaccines

DNA vaccines represent a new generation of vaccines that hold significant promise because of several advantages over conventional vaccines. There have been several successful developments of veterinary DNA vaccines, but human DNA vaccines are still in clinical trials and not yet available. The methods commonly used for protein analysis, such as CD, DLS, DSC, and FTIR, are also applicable to the physical analysis of DNA vaccines. Since DNA is not intrinsically fluorescent, extrinsic fluorescent dyes are often used for the analysis of DNA by fluorescence spectroscopy. DNA is usually characterized by very high melting temperatures. Consequently, temperature is not usually employed as an environmental stress in the physical analysis of the structural stability of DNA.

The example to be discussed here involves the characterization of polymeric and liposomal gene delivery systems (Ruponen et al. 2006). A plasmid DNA and its complexes with four cationic carriers at various ionic strengths and pH were examined by DLS, CD, and extrinsic fluorescence (Fig. 20.9a–c). The EPDs synthesized from the results of these techniques revealed two to three distinct regions for the

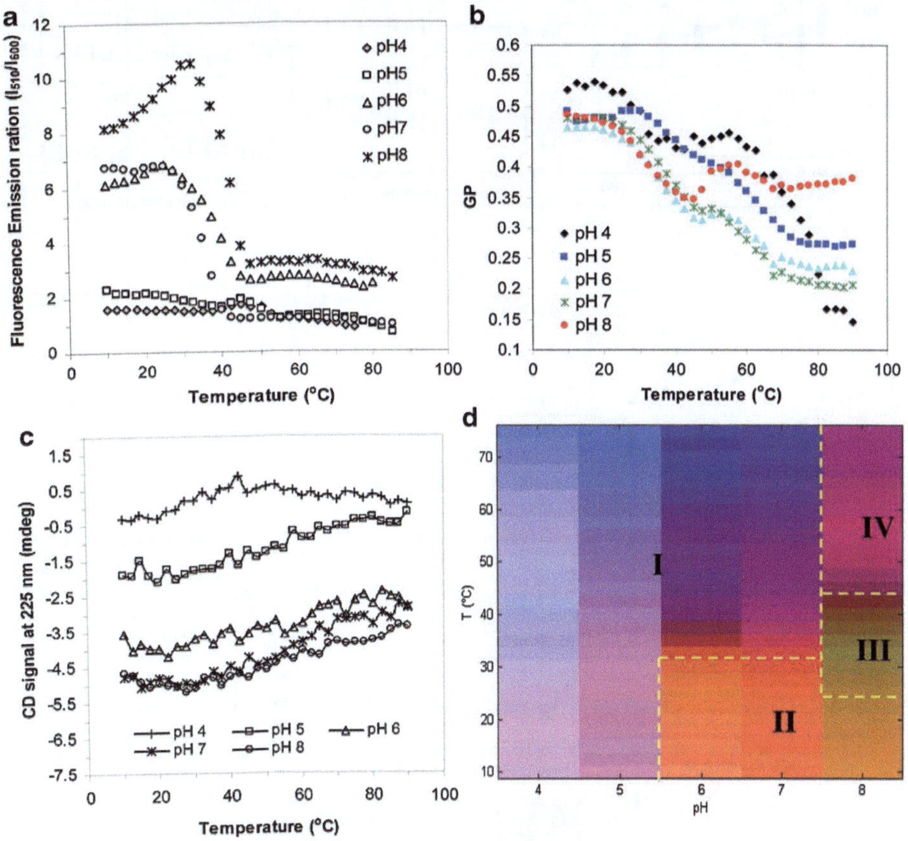

Fig. 20.8 Physical stability of Ty21a cells a function of temperature and pH as determined by *Bac*Light™ fluorescence intensity ratios (I_{510}/I_{600}) (**a**); Laurdan fluorescence (**b**); and CD signal at 225 nm (**c**). EPD of Ty21a cells constructed with fluorescence and CD data shows at least four apparent phases that correspond to Ty21a cells of different viabilities (**d**). The *orange* colored phase II is a relatively stable phase in which the cells are primarily viable. In the other three phases, cell integrity is disrupted to various extents. Reproduced with permission from Landes Bioscience, Inc. Zeng Y, Fan H, Chiueh G, Pham B, Martin R, Lechuga-Ballesteros D, Truong V, Joshi S, Middaugh C. Toward development of stable formulations of a live attenuated bacterial vaccine: A preformulation study facilitated by a biophysical approach. Human Vaccine 2009; 5:322–331

plasmid itself (Fig. 20.9d), but three to five such regions for the complexes of the plasmid DNA with cationic carriers (Fig. 20.10), implying that the presence of cationic carriers altered the nature of the plasmid DNA and its response to ionic strength and pH. A few structural features, including size, extent of collapse, and conformation of the DNA, were assigned to the regions present in the EPDs. The approach established here is also applicable to the physical analysis of other DNA vaccines, as well as those containing an RNA component.

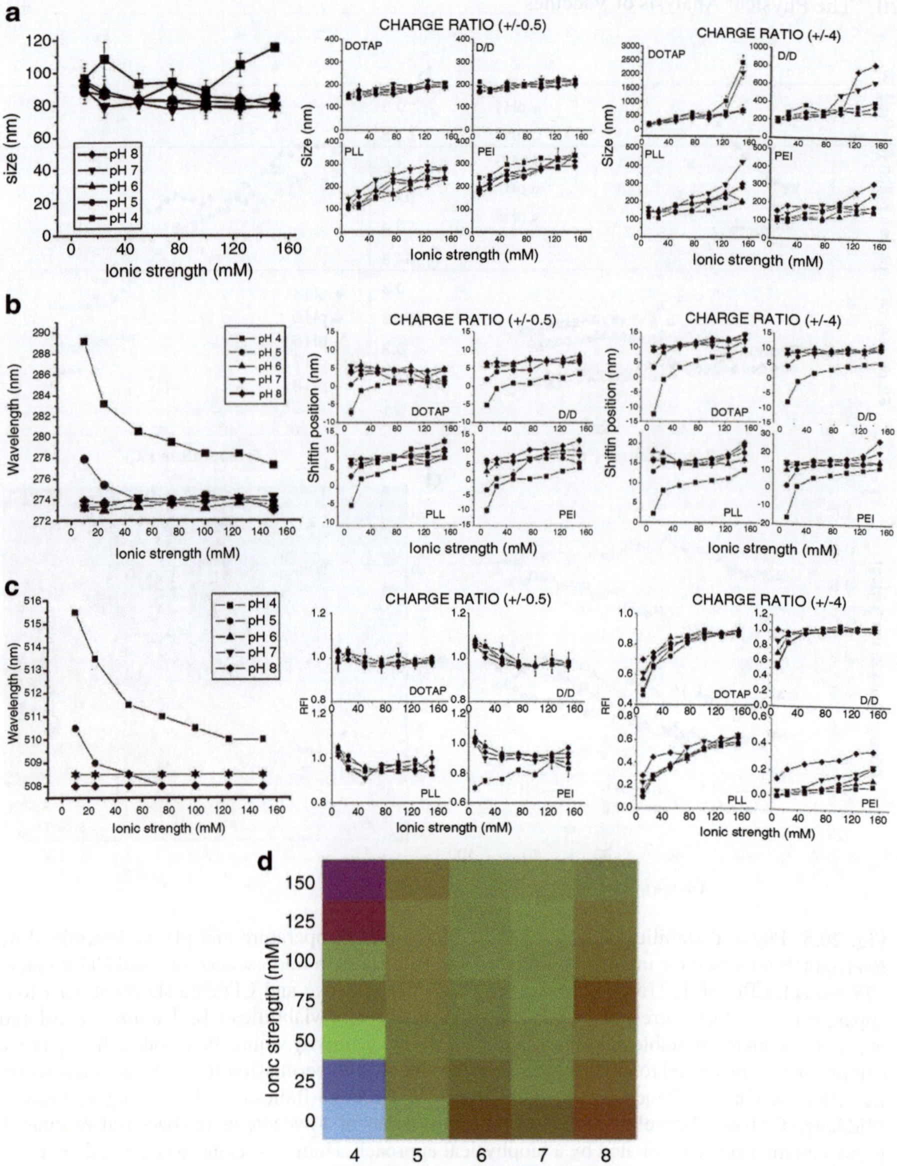

Fig. 20.9 Physical stability of uncomplexed plasmid DNA and its complexes with cationic carriers as a function of ionic strength and pH as determined by DLS (**a**, size of the plasmid), CD (**b**, changes in the secondary structure), and YOYO-1 fluorescence (**c**, condensation of the plasmid). In (**a–c**), the *left panel* is data acquired on uncomplexed DNA, the *middle panel* is data acquired on DNA complexed with cationic carriers at a charge ratio of 0.5, and the *third panel* is data acquired on DNA complexed with cationic carriers at a charge ratio of 4.0. The EPD for uncomplexed DNA (**d**) was constructed using the data shown in the *left panels* of (**a–c**). Regions of similar color signify a similar physical state of the plasmid while changes in color represent transitions between such states. From: Ruponen M, Braun CS, Middaugh CR. Biophysical characterization of polymeric and liposomal gene delivery systems using empirical phase diagrams. Journal of Pharmaceutical Sciences 95 (10):2101-2114. Copyright 2006 by John Wiley Sons, Inc. Reprinted by permission of John Wiley & Sons, Inc.

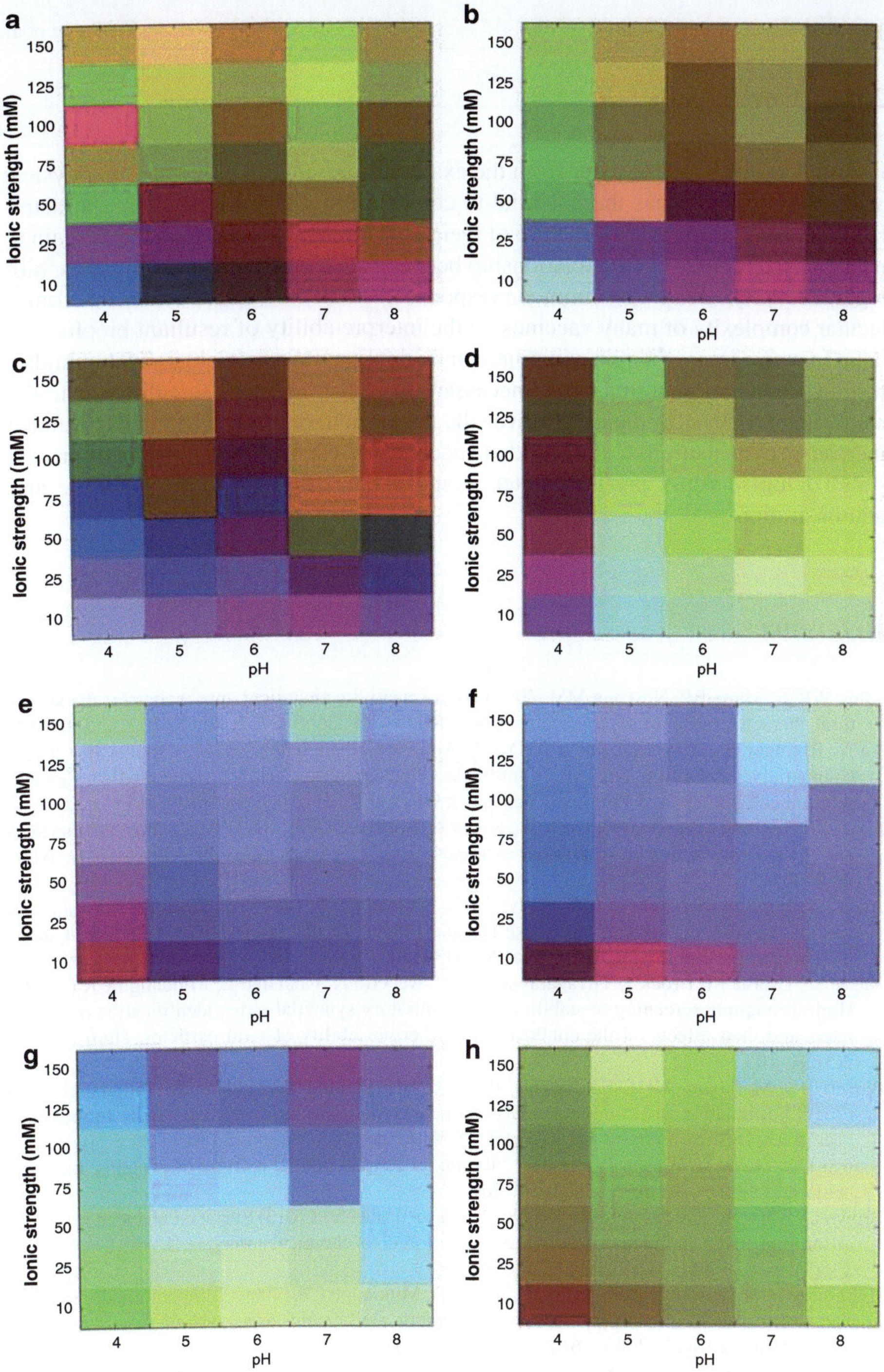

Fig. 20.10 EPDs of various nonviral gene delivery complexes at positive:negative charge ratios of 0.5 (**a**–**d**) and 4.0 (**e**–**h**) as a function of ionic strength and pH (**a**, **e**, DOTAP; **b**, **f**, DOTAP/DOPE [D/D]; **c**, **g**, PLL; **d**, **h**, PEI). The EPDs were generated using biophysical data collected from DLS (Fig. 20.9a), CD (Fig. 20.9b), and fluorescence (Fig. 20.9c). Regions of similar color signify a similar physical state of the gene delivery complex while changes in color represent transitions between such states. From: Ruponen M, Braun CS, Middaugh CR. Biophysical characterization of polymeric and liposomal gene delivery systems using empirical phase diagrams. Journal of Pharmaceutical Sciences 95 (10):2101-2114. Copyright 2006 by John Wiley Sons, Inc. Reprinted by permission of John Wiley & Sons, Inc.

20.3 Conclusion

In summary, it should be clear from the examples presented above that the physical analysis of vaccines has the potential to contribute significantly to their characterization, formulation, and evaluation of their stability. Major questions that remain to be answered include (1) the relationship between such physical measurements, biological activity, and in vivo immune responses, and (2) the effect of the macromolecular complexity of many vaccines on the interpretability of resultant biophysical data. In many cases, a single component may dominate the data obtained, but higher resolution data may ultimately be necessary to further increase the utility of lower resolution biophysical data. Whatever the case, such approaches promise to make increasing contributions to the world of vaccinology, especially in terms of improved stabilization to better preserve potency and extend shelf-life during storage and administration.

References

Alley WR Jr, Mann BF, Novotny MV (2013) High-sensitivity analytical approaches for the structural characterization of glycoproteins. Chem Rev 113(4):2668–2732. doi:10.1021/cr3003714

An Y, Rininger JA, Jarvis DL, Jing X, Ye Z, Aumiller JJ, Eichelberger M, Cipollo JF (2013) Comparative glycomics analysis of influenza Hemagglutinin (H5N1) produced in vaccine relevant cell platforms. J Proteome Res 12(8):3707–3720. doi:10.1021/pr400329k

Ausar SF, Rexroad J, Frolov VG, Look JL, Konar N, Middaugh CR (2005) Analysis of the thermal and pH stability of human respiratory syncytial virus. Mol Pharm 2(6):491–499. doi:10.1021/mp0500465

Ausar SF, Foubert TR, Hudson MH, Vedvick TS, Middaugh CR (2006) Conformational stability and disassembly of Norwalk virus-like particles. Effect of pH and temperature. J Biol Chem 281(28):19478–19488. doi:10.1074/jbc.M603313200

Ausar SF, Espina M, Brock J, Thyagarayapuran N, Repetto R, Khandke L, Middaugh CR (2007) High-throughput screening of stabilizers for respiratory syncytial virus: identification of stabilizers and their effects on the conformational thermostability of viral particles. Hum Vaccin 3(3):94–103

Barrett BS, Markham AP, Esfandiary R, Picking WL, Picking WD, Joshi SB, Middaugh CR (2010) Formulation and immunogenicity studies of type III secretion system needle antigens as vaccine candidates. J Pharm Sci 99(11):4488–4496. doi:10.1002/jps.22180

Blanch EW, Hecht L, Barron LD (2003) Vibrational Raman optical activity of proteins, nucleic acids, and viruses. Methods 29(2):196–209

Brown LV, Zhao K, King N, Sobhani H, Nordlander P, Halas NJ (2013) Surface-enhanced infrared absorption using individual cross antennas tailored to chemical moieties. J Am Chem Soc 135(9):3688–3695. doi:10.1021/ja312694g

Esfandiary R, Hunjan JS, Lushington GH, Joshi SB, Middaugh CR (2009) Temperature dependent 2nd derivative absorbance spectroscopy of aromatic amino acids as a probe of protein dynamics. Protein Sci 18(12):2603–2614. doi:10.1002/pro.264

Esfandiary R, Yee L, Ohtake S, Martin RA, Truong-Le VL, Lechuga-Ballesteros D, Moore DS, Joshi SB, Middaugh CR (2010) Biophysical characterization of rotavirus serotypes G1, G3 and G4. Hum Vaccin 6(5):390–398

He F, Joshi SB, Bosman F, Verhaeghe M, Middaugh CR (2009) Structural stability of hepatitis C virus envelope glycoprotein E1: effect of pH and dissociative detergents. J Pharm Sci 98(9):3340–3357. doi:10.1002/jps.21657

He F, Joshi SB, Moore DS, Shinogle HE, Ohtake S, Lechuga-Ballesteros D, Martin RA, Truong-Le VL, Middaugh CR (2010) Using spectroscopic and microscopic methods to probe the structural stability of human adenovirus type 4. Hum Vaccin 6(2):202–211

Hu L, Olsen C, Maddux N, Joshi SB, Volkin DB, Middaugh CR (2011a) Investigation of protein conformational stability employing a multimodal spectrometer. Anal Chem 83(24):9399–9405. doi:10.1021/ac201995c

Hu L, Trefethen JM, Zeng Y, Yee L, Ohtake S, Lechuga-Ballesteros D, Warfield KL, Aman MJ, Shulenin S, Unfer R, Enterlein SG, Truong-Le V, Volkin DB, Joshi SB, Middaugh CR (2011b) Biophysical characterization and conformational stability of Ebola and Marburg virus-like particles. J Pharm Sci 100(12):5156–5173. doi:10.1002/jps.22724

Hu L, Joshi SB, Andra KK, Thakkar SV, Volkin DB, Bann JG, Middaugh CR (2012) Comparison of the structural stability and dynamic properties of recombinant anthrax protective antigen and its 2-fluorohistidine-labeled analogue. J Pharm Sci 101(11):4118–4128. doi:10.1002/jps.23294

Hu L, Joshi SB, Liyanage MR, Pansalawatta M, Alderson MR, Tate A, Robertson G, Maisonneuve J, Volkin DB, Middaugh CR (2013) Physical characterization and formulation development of a recombinant pneumolysoid protein-based pneumococcal vaccine. J Pharm Sci 102(2):387–400. doi:10.1002/jps.23375

Iyer V, Hu L, Liyanage MR, Esfandiary R, Reinisch C, Meinke A, Maisonneuve J, Volkin DB, Joshi SB, Middaugh CR (2012) Preformulation characterization of an aluminum salt-adjuvanted trivalent recombinant protein-based vaccine candidate against Streptococcus pneumoniae. J Pharm Sci 101(9):3078–3090. doi:10.1002/jps.23175

Iyer V, Maddux N, Hu L, Cheng W, Youssef AK, Winter G, Joshi SB, Volkin DB, Middaugh CR (2013) Comparative signature diagrams to evaluate biophysical data for differences in protein structure across various formulations. J Pharm Sci 102(1):43–51. doi:10.1002/jps.23367

Jiang G, Joshi SB, Peek LJ, Brandau DT, Huang J, Ferriter MS, Woodley WD, Ford BM, Mar KD, Mikszta JA, Hwang CR, Ulrich R, Harvey NG, Middaugh CR, Sullivan VJ (2006) Anthrax vaccine powder formulations for nasal mucosal delivery. J Pharm Sci 95(1):80–96. doi:10.1002/jps.20484

Jones C (2005) NMR assays for carbohydrate-based vaccines. J Pharm Biomed Anal 38(5):840–850. doi:10.1016/j.jpba.2005.01.044

Jones LS, Peek LJ, Power J, Markham A, Yazzie B, Middaugh CR (2005) Effects of adsorption to aluminum salt adjuvants on the structure and stability of model protein antigens. J Biol Chem 280(14):13406–13414. doi:10.1074/jbc.M500687200, pii: M500687200

Joshi SB, Bhambhani A, Zeng Y, Middaugh CR (2010) An empirical phase diagram/high throughput screening approach to the characterization and formulation of biopharmaceuticals. In: Jameel F, Hershenson S (eds) Formulation and process development strategies for manufacturing biopharmaceuticals. Wiley, Hoboken

Katzenstein GE, Vrona SA, Wechsler RJ, Steadman BL, Lewis RV, Middaugh CR (1986) Role of conformational changes in the elution of proteins from reversed-phase HPLC columns. Proc Natl Acad Sci U S A 83(12):4268–4272

Kim JH, Iyer V, Joshi SB, Volkin DB, Middaugh CR (2012) Improved data visualization techniques for analyzing macromolecule structural changes. Protein Sci 21(10):1540–1553. doi:10.1002/pro.2144

Kissmann J, Ausar SF, Foubert TR, Brock J, Switzer MH, Detzi EJ, Vedvick TS, Middaugh CR (2008a) Physical stabilization of Norwalk virus-like particles. J Pharm Sci 97(10):4208–4218. doi:10.1002/jps.21315

Kissmann J, Ausar SF, Rudolph A, Braun C, Cape SP, Sievers RE, Federspiel MJ, Joshi SB, Middaugh CR (2008b) Stabilization of measles virus for vaccine formulation. Hum Vaccin 4(5):350–359

Lucas LH, Ersoy BA, Kueltzo LA, Joshi SB, Brandau DT, Thyagarajapuram N, Peek LJ, Middaugh CR (2006) Probing protein structure and dynamics by second-derivative ultraviolet absorption analysis of cation-{pi} interactions. Protein Sci 15(10):2228–2243. doi:10.1110/ps.062133706

Maddux NR, Joshi SB, Volkin DB, Ralston JP, Middaugh CR (2011) Multidimensional methods for the formulation of biopharmaceuticals and vaccines. J Pharm Sci. doi:10.1002/jps.22618

Markham AP, Barrett BS, Esfandiary R, Picking WL, Picking WD, Joshi SB, Middaugh CR (2010) Formulation and immunogenicity of a potential multivalent type III secretion system-based protein vaccine. J Pharm Sci 99(11):4497–4509. doi:10.1002/jps.22195

Middaugh CR, Ramsey JD (2007) Analysis of cationic-lipid-plasmid-DNA complexes. Anal Chem 79(19):7240–7248

Narhi LO, Jiang Y, Cao S, Benedek K, Shnek D (2009) A critical review of analytical methods for subvisible and visible particles. Curr Pharm Biotechnol 10(4):373–381

Peek LJ, Brandau DT, Jones LS, Joshi SB, Middaugh CR (2006) A systematic approach to stabilizing EBA-175 RII-NG for use as a malaria vaccine. Vaccine 24(31–32):5839–5851. doi:10.1016/j.vaccine.2006.04.067

Peek LJ, Martin TT, Elk Nation C, Pegram SA, Middaugh CR (2007) Effects of stabilizers on the destabilization of proteins upon adsorption to aluminum salt adjuvants. J Pharm Sci 96(3):547–557. doi:10.1002/jps.20762

Philo JS (2009) A critical review of methods for size characterization of non-particulate protein aggregates. Curr Pharm Biotechnol 10(4):359–372

Rexroad J, Wiethoff CM, Green AP, Kierstead TD, Scott MO, Middaugh CR (2003) Structural stability of adenovirus type 5. J Pharm Sci 92(3):665–678. doi:10.1002/jps.10340

Rexroad J, Evans RK, Middaugh CR (2006a) Effect of pH and ionic strength on the physical stability of adenovirus type 5. J Pharm Sci 95(2):237–247. doi:10.1002/jps.20496

Rexroad J, Martin TT, McNeilly D, Godwin S, Middaugh CR (2006b) Thermal stability of adenovirus type 2 as a function of pH. J Pharm Sci 95(7):1469–1479. doi:10.1002/jps.20617

Ruponen M, Braun CS, Middaugh CR (2006) Biophysical characterization of polymeric and liposomal gene delivery systems using empirical phase diagrams. J Pharm Sci 95(10):2101–2114. doi:10.1002/jps.20621

Salnikova MS, Joshi SB, Rytting JH, Warny M, Middaugh CR (2008a) Physical characterization of clostridium difficile toxins and toxoids: effect of the formaldehyde crosslinking on thermal stability. J Pharm Sci 97(9):3735–3752. doi:10.1002/jps.21261

Salnikova MS, Joshi SB, Rytting JH, Warny M, Middaugh CR (2008b) Preformulation studies of Clostridium difficile toxoids A and B. J Pharm Sci 97(10):4194–4207. doi:10.1002/jps.21300

Volkin DB, Burke CJ, Marfia KE, Oswald CB, Wolanski B, Middaugh CR (1997) Size and conformational stability of the hepatitis A virus used to prepare VAQTA, a highly purified inactivated vaccine. J Pharm Sci 86(6):666–673. doi:10.1021/js960475h

Wang T, Joshi SB, Kumru OS, Telikepalli S, Middaugh CR, Volkin DB (2013) Case studies applying biophysical techniques to better characterize protein aggregates and particulates of varying size. In: Narhi LO (ed) Biophysics for therapeutic protein development. Springer, New York

Wen ZQ (2007) Raman spectroscopy of protein pharmaceuticals. J Pharm Sci 96(11):2861–2878. doi:10.1002/jps.20895

Zeng Y, Fan H, Chiueh G, Pham B, Martin R, Lechuga-Ballesteros D, Truong VL, Joshi SB, Middaugh CR (2009) Towards development of stable formulations of a live attenuated bacterial vaccine: a preformulation study facilitated by a biophysical approach. Hum Vaccin 5(5): 322–331

Chapter 21
Characterizing the Association Between Antigens and Adjuvants

Mette Hamborg and Camilla Foged

21.1 Introduction

Subunit vaccines consist of two principal components: Antigen(s) and adjuvant. The selected antigen(s) should be the component(s) of the pathogen that best stimulate the immune system, and the choice of adjuvant should be based on the type of desired immune response. In addition, the adjuvant should be formulated with the antigen(s) in a way that ensures an optimal type of response and minimal side effects. Thus, during the development of new subunit vaccines it is crucial to characterize and possibly optimize the antigen–adjuvant interactions (association and de-association) to ensure optimal efficacy, safety, and quality of novel vaccine formulations.

Compared to the vast number of studies in the literature describing new antigens and/or adjuvants and their immunological profiles, studies regarding the physicochemical characterization of antigen–adjuvant interactions are sparse. Thus, the physicochemical characterization of the association and de-association between the two components is often neglected in the early stages of development mainly because vaccine formulations are usually characterized by a very high degree of complexity, and the combination of low antigen doses and colloidal systems poses analytical challenges (Volkin and Middaugh 2010; Dormitzer et al. 2008). In addition, demonstrating direct links between specific physicochemical formulation characteristics

M. Hamborg (✉)
Department of International Health, Microbiology and Immunology, Centre for Medical Parasitology, Faculty of Health and Medical Sciences, University of Copenhagen, 2100 Copenhagen Ø, Denmark
e-mail: mette.hamborg@sund.ku.dk

C. Foged
Department of Pharmacy, Faculty of Health and Medical Sciences, University of Copenhagen, Universitetsparken 2, 2100 Copenhagen Ø, Denmark
e-mail: camilla.foged@sund.ku.dk

C. Foged et al. (eds.), *Subunit Vaccine Delivery*, Advances in Delivery Science and Technology, DOI 10.1007/978-1-4939-1417-3_21

and vaccine efficacy may in practice be very challenging (Mortellaro and Ricciardi-Castagnoli 2011).

Nevertheless, a thorough understanding of the physicochemical properties of vaccines is an important and necessary step on the way to an optimal vaccine formulation, and throughout the vaccine development process, sufficient knowledge about a vaccine formulation should be gathered to directly or indirectly link changes in the physicochemical parameters to the performance of the vaccine. This includes identifying and optimizing critical parameters, such as the degree of antigen adsorption and the colloidal stability that may have an impact on the efficacy, safety, and stability of the vaccine product and to optimize these. Also parameters which describe the interactions that cannot be linked to the efficacy and safety may still be monitored to guarantee the quality and consistency of the vaccine production process (Clapp et al. 2011).

Stanley Hem and coworkers pioneered the field emphasizing the importance of characterizing antigen–adjuvant interactions by combining thorough physicochemical characterization with immunological studies, and today the various vaccine guidelines from regulatory agencies require that aspects that are critical for the biological properties of the adjuvant–antigen combination, such as antigen association/de-association and binding characteristics, should be identified and monitored (EMA, FDA, WHO). More specifically, the EMA guideline states that it is crucial to define and describe the mechanism of association (and de-association) and the association efficiency between antigen(s) and adjuvant. Quality evaluation of a vaccine–adjuvant formulation must therefore include a proof of adequate and consistent association of the antigen(s) with the adjuvant, also throughout the shelf life of the vaccine, and demonstrate that no significant de-association takes place in the course of the shelf life (Table 21.1) (EMEA 2005).

The understanding of aluminum adjuvant–antigen association/de-association and of the important factors to address when preparing vaccine formulations containing aluminum-based adjuvants has been greatly improved during the past decades (Clapp et al. 2011; Hem and Hogenesch 2007). The experiences can serve as inspiration in the characterization of other particulate adjuvant–antigen interactions, although each new type of adjuvant system has its own challenges. Therefore, the important characterization parameters for each type of antigen–adjuvant formulation must be selected on a case-by-case basis, depending on the physicochemical properties of the adjuvant and the vaccine's mechanism of action. The key is to maintain both the antigen's and the adjuvant's integrity.

Proteins are delicate three-dimensional molecules with distinct hydrophobic regions and localized charges that govern the protein folding. Any small changes in

Table 21.1 Aspects of adjuvant–antigen interactions which should be appropriately characterized

Physical characteristics (e.g., particle size, size distribution, surface charge, sedimentation)
Level and consistency of antigen association to the adjuvant
Physical and chemical integrity of the antigen
Extent of antigen release (de-association)

the primary, secondary, or tertiary structure of a protein antigen upon interaction with the adjuvant can potentially impact its immunogenicity, either by altering the three-dimensional structure which can potentially affect the structural epitopes or by chemical alterations which can potentially alter the linear epitopes. Additionally, changes in the antigen conformation and flexibility may also potentially affect processing and presentation by antigen-presenting cells (APCs) to T-cells (Dai et al. 2001; Tynan et al. 2005; Thai et al. 2004).

In the following the interactions between adjuvants and antigens are presented and discussed. Emphasis is put on aluminum and emulsions adjuvants, which are components of marketed vaccines, as well as on lipid vesicles for which we present brief examples from our own research. The discussion highlights the important physicochemical parameters that describe the interactions and the analytical techniques currently used to address these.

21.2 Aluminum Salts

21.2.1 Physical Characteristics

The most commonly used aluminum salts are aluminum phosphate and aluminum hydroxide (Chap. 3). They have different physical characteristics: Aluminum hydroxide has a point of zero charge (PZC) of 4–5.5, whereas aluminum phosphate has a PZC of 11 (Lindblad 2004). Thus, at neutral pH aluminum hydroxide is net positively charged while aluminum phosphate is net negatively charged. Furthermore, aluminum hydroxide is a crystalline salt and has only hydroxyl groups on its surface, whereas aluminum phosphate, which is amorphous, has both hydroxyl and phosphate groups on its surface (Hem and Hogenesch 2007). Aluminum salts are dispersions that sediment over time and thus the speed of sedimentation is used as a measure of the colloidal stability of the formulation. Sedimentation speed may in fact be affected by antigen adsorption providing a tool to detect differences in the degree of antigen association and batch-to-batch variations (Fox et al. 2013a). Changes in sedimentation speed upon formulation have been addressed by standard spectrophotometric methods using a 500 nm laser beam (Capelle et al. 2005) or by laser scattering optical profiling (Fox et al. 2013a). Here the light transmission from a 870 nm laser is measured in a setup, which is suited for obtaining real-time kinetic analysis throughout the vertical profile of the sample (Fox et al. 2013a).

21.2.2 Antigen Association

The two predominant mechanisms of antigen adsorption to aluminum-containing adjuvants are electrostatic interactions and ligand exchange, although hydrophobic interactions and van der Waal forces may also be involved (Seeber et al. 1991;

Alshakhshir et al. 1994; Rinella et al. 1998a; Rinella et al. 1998b; Iyer et al. 2004; Iyer et al. 2003a). The weak non-covalent interaction forces are highly dependent on the buffer species, pH, ionic strength, and the isoelectric point (pI) of the antigen. Thus, the degree of antigen adsorption can be adjusted via selection of buffer conditions (a high ionic strength can reduce the electrostatic forces). Ligand exchange is an adsorption mechanism that involves the exchange of a phosphate group of the antigen with a surface hydroxyl group of the adjuvant (Hem et al. 2010). It means that the antigen must be phosphorylated to covalently interact with alum via ligand exchange (Lu et al. 2013). Of notice is that phosphate buffer salts can interfere with alum and modify its surface by exchanging hydroxyl groups on the surface of alum with phosphate. Consequently, phosphate buffer is usually avoided in aluminum-based formulations because antigen adsorption could be affected during storage.

In general, antigens with a pI above 7.4 will adsorb stronger to aluminum phosphate, while antigens with a pI below 7.4 will adsorb better onto aluminum hydroxide (Lindblad 2004; Gupta 1998). Antigens adsorbed via electrostatic attractions are less strongly adsorbed and do generally desorb from the adjuvant under physiologically relevant conditions. On the other hand, antigens adsorbed by ligand exchange are strongly adsorbed and show a more irreversible nature of adsorption (Hem and Hogenesch 2007; Iyer et al. 2003b). The optimal strength of adsorption seems to be a delicate balance between on the one hand having the antigen adsorbed but on the other hand not having it too strongly adsorbed. A strong adsorption might impede the immune response (Lu et al. 2013), and the strength of antigen adsorption to aluminum-based adjuvants has in fact been shown to be inversely related to the immune response (Hem et al. 2010; Chang et al. 2001; Iyer et al. 2003b).

The most common way to quantify antigen adsorption is an indirect method involving centrifugation of the aluminum salt particles and assaying the supernatant for unbound protein to extrapolate the amount of bound material (Lindblad and Schonberg 2010). This approach can be used to construct adsorption isotherms from which the strength of adsorption and adsorption capacity may be estimated (Clapp et al. 2011). The mechanism(s) of antigen adsorption and the degree of desorption can be investigated by testing the degree of elution in different solvents (Iyer et al. 2004), i.e., if the antigen elutes when increasing the ionic strength, electrostatic attraction is likely to be the main force driving adsorption.

21.2.3 Antigen Integrity

The physical and chemical stability of proteins upon adsorption onto aluminum adjuvants have been studied to some extent, but due to analytical challenges such as analytical incompatibilities with the aluminum particles as well as low antigen concentration, biophysical studies in particular are not easily performed. Jones et al. presented one of the first studies, which demonstrated that the secondary structure of a number of model proteins was perturbed upon adsorption to alum (Jones et al. 2005). Since then, several studies have been carried out with vaccine-relevant antigens. However, a number of discrepancies are evident and no general statements regarding the structural perturbation and its significance for efficacy can be deduced. Some proteins unfold upon immobilization (Jones et al. 2005; Zheng et al. 2007a; Zheng et al. 2007b; Regnier et al. 2012) while others maintain their conformational integrity and/or are even stabilized (Dong et al. 2006; Agopian et al. 2007). The discrepancies between the studies can been explained by (1) the highly unique characteristics of each protein, (2) the difference in the analytical approach taken, and (3) protein concentration may influence and change protein adsorption patterns (Bai and Dong 2009; Clapp et al. 2011). Direct characterization of aluminum-bound antigen has been achieved by using techniques such as fluorescence, Fourier transform infrared spectroscopy (FTIR), microcalorimetry techniques, and circular dichroism (CD) (Regnier et al. 2012). Since the particulate nature of alum is a potential source of artifacts for the above-mentioned techniques, some antigen integrity studies are performed upon desorption/separation from the adjuvant (Hutcheon et al. 2006). Desorption generally requires high salt and/or surfactant concentrations, depending on the mechanism of adsorption, which has the obvious disadvantage that the separation process itself may compromise antigen integrity. For the biophysical methods, additional complications are the low dose of antigen and the choice of desorption-medium, which should be compatible with the specific assay (Clapp et al. 2011). SDS-polyacrylamide gel electrophoresis (SDS-PAGE) and immunodetection techniques such as enzyme-linked immunosorbent assay (ELISA) and Western blotting (WB) offer good alternatives to the biophysical methods based on spectrophotometry. These methods are not compromised by the low antigen concentrations and recently more nondestructive characterization methods (performed in the presence of alum) have been developed, e.g., binding assays and intrinsic fluorescence (Mulder et al. 2012; Wagner et al. 2012). Changes in the chemical stability of antigens upon adsorption to alum are often neglected for vaccine formulations, but reactions such as oxidation and deamidation could disrupt linear (and structural) epitopes. D´Souza et al. observed rapid deamidation of an anthrax antigen when adsorbed on aluminum hydroxide gel which correlated with reduced potency of the vaccine. This and other studies highlight the importance of also characterizing the chemical stability upon adsorption (Table 21.2) (D'Souza et al. 2013; Estey et al. 2009).

Table 21.2 Analytical tools used to characterize alum adjuvant–antigen interactions

Interaction parameter	Techniques	Challenges	Reference
Physical characteristics	Visual appearance		Lindblad and Schonberg (2010) Capelle et al. (2005)
	Particle size and distribution (DLS, SLS)		
	Zeta potential		
	Sedimentation		
Adsorption	Adsorption isotherms		Clapp et al. (2011)
Structural integrity of the antigen	DSC	Vaccine concentration is usually too low compared to the sensitivity of the method	Jones et al. (2005), Bai and Dong (2009)
	FTIR		
	Intrinsic fluorescence	Light scattering artifacts	
	CD		
Structural integrity of specific epitopes (binding assays)	ELISA	Monoclonal antibodies that bind to structural epitopes difficult to produce/expensive. Extraction of the antigen from the alum is sometimes needed.	Mulder et al. (2012), Hutcheon et al. (2006)
	Surface plasmon resonance		
	Western blotting		
Chemical stability	MS	Separation necessary	D'Souza et al. (2013)

21.3 Emulsions

21.3.1 Physical Characteristics

Adjuvants based on emulsions are dispersions, which consist of a disperse phase in a continuous phase, e.g., water droplets dispersed in a continuous oil phase (w/o emulsion) or oil droplets dispersed in a continuous aqueous phase (o/w emulsion) (Chap. 4). Emulsions are thermodynamically unstable systems and the dispersed droplets may eventually coalesce and separate out into the component phases. The droplet size is essential for the stability of emulsions since small droplets are generally better stabilized compared to larger droplets due to Brownian motion (Wilde 2000).

Because the single components of an emulsion adjuvant rarely are very immunogenic on their own the characteristics of the emulsion (droplet size and antigen association) are likely to play an important role for evoking the intended immune response (Calabro et al. 2013; Brito et al. 2013). For example, the droplet size of MF59 is believed to be critical for the adjuvant effect (Ott et al. 2000).

21.3.2 Antigen Association

Antigen association to oil droplets does not seem to be a prerequisite for emulsions to be effective adjuvants (Brito et al. 2013). For MF59 and AS03 association between antigen and the emulsion droplets has not shown to be critical for the adjuvant activity (Ott et al. 2000; Lai et al. 2012; Brito et al. 2013). Antigen distribution and association in the emulsion is not easy to address and it is frequently based on an indirect estimate which requires a phase separation of the emulsion followed by a determination of the amount of antigen in the aqueous phase (Zhu et al. 2011). Alternatively, the importance of association has been studied by separate injections of the antigen and the emulsion in animals. MF59 that was injected 24 h prior to the antigen or 1 h after the antigen still elicited the same response as the co-administration of the antigen and MF59 (Dupuis et al. 1999). Similar findings were shown for AS03 which could also be administered 1 h prior to the antigen and still elicit the same result as upon co-administration, however co-localization appeared to be important (Morel et al. 2011). This suggests that direct association is not crucial for the vaccine response.

21.3.3 Antigen Integrity

Emulsions possess oil–water interfaces that protein antigens are prone to adsorb to and the adsorption can influence the stability of the emulsion as well as the protein stability (Wilde 2000; Jorgensen et al. 2004; Jorgensen et al. 2003). The particulate and turbid nature of emulsions is an analytical challenge as it interferes with many of the conventional methods including reverse phase high pressure liquid chromatography (HPLC), size exclusion chromatography, dynamic light scattering (DLS), and CD. The difference in the refractive index between the aqueous phase and the oil phase causes light scattering artifacts. Husband et al. showed how refractive index matching between the two phases by addition of glycerol can circumvent the artifacts caused by light scattering (Husband et al. 2001). Nevertheless, most emulsions are characterized by different refractive indexes and thus the challenges caused by light scattering have to be overcome in other ways. One possibility is to separate the oil phase and the water phase by destabilizing the emulsion, e.g., by benzyl alcohol as done for vaccines containing Montanide ISA 720 (Miles et al. 2005). The protein integrity may be analyzed in the water phase but a prerequisite is that the protein affinity for the water phase is favorable and that phase separation of the emulsion is feasible. If the protein can be recovered in the aqueous phase, the antigen stability can be tested by SDS-PAGE, WB, ELISA, and N-terminal sequencing (Miles et al. 2005). For MF59, it has been suggested that a phase separation via removal of the oil droplets by high speed centrifugation may be undertaken to allow for the use of some of the traditional methods described above. This procedure will minimize the number of particles/lipid droplets present, but not all of the particles are removed (Brito et al. 2013). Another possibility is to analyze the antigen structure by techniques commonly used for therapeutic proteins which are less sensitive to light scattering such as FTIR, front-face fluorescence, and differential scanning calorimetry (Jorgensen et al. 2004; Jorgensen et al. 2003). As compared to vaccine formulations, therapeutic protein formulations contain higher protein concentrations, and the methods mentioned above are currently not sufficiently sensitive to allow for measuring structural changes at the low antigen concentrations existing in an emulsion matrix. Alternatively, protein gel electrophoresis and ELISA have been applied to study the integrity of the antigen gp140 upon interaction with MF59 (Lai et al. 2012). Minor changes in protein conformation were observed after extracting the antigen from the adjuvant (Lai et al. 2012).

The plethora of potential interactions and conformational changes between protein antigens and emulsion formulations emphasize the importance of structural characterization, and several studies have shown that the secondary and tertiary structure of proteins is altered in emulsion formulations as compared to aqueous solutions. In some cases these interactions, though detectable, did not appear to affect the immunogenicity. Fox et al. tested a panel of emulsions and found that although structural changes of the H5N1 recombinant antigen were apparent from intrinsic fluorescence measurements, the differences did not appear to correlate to immunological activity (Fox et al. 2013a). Many different antigens have successfully been formulated with the MF59 formulation (Chap. 4). A few antigens were not stable post formulation with MF59. HIV gp120 underwent conformational changes over time in MF59 whereas CMV Gb existed in an equilibrium of different multimeric stages, which was shifted in the presence of MF59. Therefore these vaccines are formulated as dual vials, which are mixed immediately prior to immunization (Ott et al. 2000).

As in the case of alum, it is not straightforward to analyze the antigen integrity in the presence of emulsion-based adjuvants and it is hardly possible without altering the formulation either by extracting the antigens from the emulsion or increasing the antigen concentration. Both approaches leave the open question of whether the antigen in the altered formulation still represents the antigen in the actual vaccine formulation (Table 21.3).

Table 21.3 Analytical tools used to characterize emulsion adjuvant–antigen interactions

Interaction parameter	Techniques	Challenges	Reference
Physical characteristics	Visual appearance	–	Xue et al. (2010), Ott et al. (2000)
	Zeta potential		
	Particle size (DLS/SLS)		
	Viscosity		
Antigen association/ integrity	Protein gel electrophoresis	Requires separation of the bound and unbound protein. Usually estimated from the antigen content in the aqueous phase after phase separation caused by centrifugation	Ott et al. (2000), Lai et al. (2012), Xue et al. (2010)
Integrity of the antigen	Intrinsic fluorescence	The antigen concentration is usually too low compared to the sensitivity of DSC and FTIR	Fox et al (2013a), Jorgensen et al. (2004), Jorgensen et al. (2003), Miles et al. (2005)
	DSC		
	FTIR		
	N-terminal sequencing		
	SDS-PAGE		
Structural integrity of specific epitopes (binding assay)	ELISA	Monoclonal antibodies difficult to access/ expensive. Extraction of the antigen from the emulsion is needed	Lai et al. (2012), Xue et al. (2010), Miles et al. (2005)
	Western blot		

21.4 Lipid Vesicles

21.4.1 Physical Characteristics

Adjuvants based on lipid vesicles are self-assembling systems consisting of lipid bilayer(s) with an aqueous core. Classical liposomes are composed of phospholipids and cholesterol but many other surfactants can also form lipid vesicles, and by modifying the surfactant composition their surface charge and fluidity can be altered. In the following all lipid vesicles are referred to as liposomes. The effect of protein interaction with lipid vesicles can be characterized by measuring the size distribution, the zeta potential, and the thermotropic phase behavior of the liposomes. Additionally, microscopy-based techniques (such as cryo-TEM) are also useful to study morphological changes of the liposomes.

21.4.2 Antigen Adsorption

Protein antigens may be associated with liposomes in a number of ways: Encapsulated in the aqueous core, intercalated into the bilayer, covalently bound to modified lipids, or adsorbed to the surface of the liposomes. In most liposome adjuvant formulations, the antigen is externally associated, either by simple mixing or by surface adsorption. In this case, the main driving force is electrostatic interaction, although hydrophobic interactions may also come into play (Henriksen-Lacey et al. 2011). One exception are virosomes in which virus antigens are intercalated into the bilayer mainly through hydrophobic interactions (Herzog et al. 2009). Encapsulation may also be approached if the antigen needs to be protected against enzymatic degradation. However, loading efficiencies can sometimes be low due to the surface active nature of the proteins which will challenge their encapsulation (Gregoriadis et al. 1999).

Charged liposomes are ideal for surface adsorption of antigens but the formulation pH, the pI of the protein, the excipients, and the ionic strength will influence adsorption efficiency. In general, most antigens with a pI below the pH (typically 7.4) will adsorb to cationic liposomes while antigens with a pI above the pH will adsorb to anionic liposomes (Henriksen-Lacey et al. 2010b; Henriksen-Lacey et al. 2011) The importance of antigen association to the liposomes tends to vary for different systems. In a study by Yanasarn et al. no correlation could be found between the degree of adsorption and the immunological response: The anionic liposomes showed a similar immunogenicity as compared to cationic liposomes demonstrating higher levels of association (Yanasarn et al. 2011). However, for the CAF01 adjuvant, which besides the cationic surfactant dimethyldioctadecylammonium (DDA) bromide also contains the immunopotentiator trehalose dibehenate, a certain degree of antigen adsorption (>50 %) and co-localization of the antigen and the adjuvant has been shown to be important to achieve an immune response (Henriksen-Lacey et al. 2010b; Kamath et al. 2012). Yet, for many other liposomal formulations little is known about the degree of adsorption or its importance, e.g., for a malaria vaccine formulation containing the liposomal system AS01 and the malaria antigen RTS,S for which no information is available regarding the degree or mechanism(s) of absorption (Fox et al. 2013b).

The degree of adsorption can in principle be determined in the same manner as for alum. However, in contrast to alum, which easily sediments upon centrifugation liposomes often have a lower density than water and therefore require a larger gravitational force to sediment (Davidsen et al. 2005; Henriksen-Lacey et al. 2010a). This practical challenge can be overcome by using ultracentrifugation or by adding additional oppositely charged protein which causes liposomes to aggregate, facilitating sedimentation (Henriksen-Lacey et al. 2010a).

21.4.3 Antigen Integrity

As described for emulsions, there are several challenges involved in characterizing the antigen in an emulsion and the same is the case when replacing the emulsion with liposomes. The stabilized particulate structures induce light scattering artifacts when

Table 21.4 Analytical tools used to characterize liposome adjuvant–antigen interactions

Interaction parameter	Techniques	Challenges	Reference
Physical characteristics	Visual appearance	The resolution limit of cryo-TEM is 4–5 nm	Davidsen et al. (2005)
	Particle size and distribution (DLS, NTA)		
	Zeta potential		
	Thermotropic phase behavior (DSC)		
	Morphology (cryo-TEM)		
Adsorption	Sedimentation of the liposomes with the bound protein followed by a quantification of the antigen content in the supernatant	Requires separation of the bound and the unbound protein	Davidsen et al. (2005), Hamborg et al. (2013a), Henriksen-Lacey et al. (2010a)
Structural integrity of the antigen	Intrinsic fluorescence	Vaccine concentration is typically too low compared to the sensitivity of the method	Hamborg et al. (2013a), Hamborg et al. (2013b)
	DSC		
	FTIR		

using most spectroscopy-based techniques and depending on the antigen-to-lipid ratio, the colloidal stability of the liposomes may be more or less affected (Hamborg et al. 2013a). Very few studies have addressed antigen integrity in vaccine formulations adjuvanted with liposomes probably due to the analytical challenges described above. However by choosing suitable analytical techniques such as DSC, intrinsic fluorescence, and FTIR, antigen integrity can be measured. For tuberculosis antigens and model antigens adsorbed to cationic liposomes, no structural changes were observed (Hamborg et al. 2013a; Hamborg et al. 2013b). To make DSC and FTIR studies feasible, the system was modified to contain higher protein concentrations than the relevant concentration in vaccines (Table 21.4) (<0.1 mg/mL).

21.5 Conclusion

In the past decade the approaches taken to describe and characterize subunit vaccines have progressed from labelling it an unknown mixture to a well-defined formulation. This change has been driven by research activities combining the fields of immunology and pharmaceutical sciences, along with the development of methods to characterize aluminum-based (or subunit) vaccine formulations. Some of these methods are today generally accepted, e.g., "determining the degree of antigen adsorption." Yet, several characterization aspects remain to be developed into standard tools to characterize new vaccine formulations, e.g., antigen integrity which is only sometimes feasible to characterize depending on the availability of a suitable method. However, the most suitable analytical methods for a given formulation must be identified on a case-by-case basis. Also for emulsion-based vaccines which have been on the market for several years, there is no general consensus for how to characterize the formulation and which parameters to characterize. The reason for this is probably found in the high degree of complexity of this type of formulation and the lack of optimal analytical methods. There are no easy choices and proper characterization of subunit vaccines is an evolving field which requires expertise and experience in all aspects of its assessment.

References

Agopian A, Ronzon F, Sauzeat E et al (2007) Secondary structure analysis of HIV-1-gp41 in solution and adsorbed to aluminum hydroxide by Fourier transform infrared spectroscopy. Biochim Biophys Acta 1774:351–358

Alshakhshir R, Regnier F, White JL et al (1994) Effect of protein adsorption on the surface-charge characteristics of aluminum-containing adjuvants. Vaccine 12:472–474

Bai SF, Dong AC (2009) Effects of immobilization onto aluminum hydroxide particles on the thermally induced conformational behavior of three model proteins. Int J Biol Macromol 45:80–85

Brito LA, Malyala P, O´Hagan DT (2013) Vaccine adjuvant formulations: a pharmaceutical perspective. Semin Immunol 25:130–145

Calabro S, Tritto E, Pezzotti A et al (2013) The adjuvant effect of MF59 is due to the oil-in-water emulsion formulation, none of the individual components induce a comparable adjuvant effect. Vaccine 31:3363–3369

Capelle MAH, Brugger P, Arvinte T (2005) Spectroscopic characterization of antibodies adsorbed to aluminium adjuvants: correlation with antibody vaccine immunogenicity. Vaccine 23:1686–1694

Chang M, Shi Y, Nail SL et al (2001) Degree of antigen adsorption in the vaccine or interstitial fluid and its effect on the antibody response in rabbits. Vaccine 19:2884–2889

Clapp T, Siebert P, Chen DX et al (2011) Vaccines with aluminum-containing adjuvants: optimizing vaccine efficacy and thermal stability. J Pharm Sci 100:388–401

D'Souza AJM, Mar KD, Huang J et al (2013) Rapid deamidation of recombinant protective antigen when adsorbed on aluminum hydroxide gel correlates with reduced potency of vaccine. J Pharm Sci 102:454–461

Dai GX, Steede NK, Landry SJ (2001) Allocation of helper T-cell epitope immunodominance according to three-dimensional structure in the human immunodeficiency virus type I envelope glycoprotein gp120. J Biol Chem 276:41913–41920

Davidsen J, Rosenkrands I, Christensen D et al (2005) Characterization of cationic liposomes based on dimethyldioctadecylammonium and synthetic cord factor from *M. tuberculosis* (trehalose 6,6'-dibehenate)—novel adjuvant inducing both strong CMI and antibody responses. Biochim Biophys Acta 1718:22–31

Dong AC, Jones LS, Kerwin BA et al (2006) Secondary structures of proteins adsorbed onto aluminum hydroxide: infrared spectroscopic analysis of proteins from low solution concentrations. Anal Biochem 351:282–289

Dormitzer PR, Ulmer JB, Rappuoli R (2008) Structure-based antigen design: a strategy for next generation vaccines. Trends Biotechnol 26:659–667

Dupuis M, McDonald DM, Ott G (1999) Distribution of adjuvant MF59 and antigen gD2 after intramuscular injection in mice. Vaccine 18:434–439

EMEA (2005) Guideline on adjuvants in vaccines for human use. http://www.emea.europa.eu/docs/en_GB/document_library/Scientific_guideline/2009/09/WC500003809.pdf. Accessed 2 May 2014

Estey T, Vessely C, Randolph TW et al (2009) Evaluation of chemical degradation of a trivalent recombinant protein vaccine against botulinum neurotoxin by LysC peptide mapping and MALDI-TOF mass spectrometry. J Pharm Sci 98:2994–3012
Fox CB, Barnes VL, Evers T et al (2013a) Adjuvanted pandemic influenza vaccine: variation of emulsion components affects stability, antigen structure, and vaccine efficacy. Influenza Other Respir Viruses 7:815–826
Fox CB, Kramer RM, Barnes V et al (2013b) Working together: interactions between vaccine antigens and adjuvants. Ther Adv Vaccines 1:7–20
Gregoriadis G, McCormack B, Obrenovic M et al (1999) Vaccine entrapment in liposomes. Methods 19:156–162
Gupta RK (1998) Aluminum compounds as vaccine adjuvants. Adv Drug Deliv Rev 32:155–172
Hamborg M, Jorgensen L, Bojsen AR et al (2013a) Protein antigen adsorption to the DDA/TDB liposomal adjuvant: effect on protein structure, stability, and liposome physicochemical characteristics. Pharm Res 30:140–155
Hamborg M, Kramer R, Schante CE et al (2013b) The physical stability of the recombinant tuberculosis fusion antigens H1 and H56. J Pharm Sci 102:3567–3578
Hem SL, Hogenesch H (2007) Relationship between physical and chemical properties of aluminum-containing adjuvants and immunopotentiation. Expert Rev Vaccines 6:685–698
Hem SL, Hogenesch H, Middaugh CR et al (2010) Preformulation studies—the next advance in aluminum adjuvant-containing vaccines. Vaccine 28:4868–4870
Henriksen-Lacey M, Bramwell VW, Christensen D et al (2010a) Liposomes based on dimethyldioctadecylammonium promote a depot effect and enhance immunogenicity of soluble antigen. J Control Release 142:180–186
Henriksen-Lacey M, Christensen D, Bramwell VW et al (2010b) Liposomal cationic charge and antigen adsorption are important properties for the efficient deposition of antigen at the injection site and ability of the vaccine to induce a CMI response. J Control Release 145:102–108
Henriksen-Lacey M, Korsholm KS, Andersen P et al (2011) Liposomal vaccine delivery systems. Expert Opin Drug Deliv 8:505–519
Herzog C, Hartmann K, Kunzi V et al (2009) Eleven years of Inflexal (R) V-a virosomal adjuvanted influenza vaccine. Vaccine 27:4381–4387
Husband FA, Garrood MJ, Mackie AR et al (2001) Adsorbed protein secondary and tertiary structures by circular dichroism and infrared spectroscopy with refractive index matched emulsions. J Agric Food Chem 49:859–866
Hutcheon CJ, Becker JO, Russell BA et al (2006) Physiochemical and functional characterization of antigen proteins eluted from aluminum hydroxide adjuvant. Vaccine 24:7214–7225
Iyer S, Hogenesch H, Hem SL (2003a) Effect of the degree of phosphate substitution in aluminum hydroxide adjuvant on the adsorption of phosphorylated proteins. Pharm Dev Technol 8:81–86
Iyer S, Hogenesch H, Hem SL (2003b) Relationship between the degree of antigen adsorption to aluminum hydroxide adjuvant in interstitial fluid and antibody production. Vaccine 21: 1219–1223
Iyer S, Robinett RSR, Hogenesch H et al (2004) Mechanism of adsorption of hepatitis B surface antigen by aluminum hydroxide adjuvant. Vaccine 22:1475–1479
Jones LS, Peek LJ, Power J et al (2005) Effects of adsorption to aluminum salt adjuvants on the structure and stability of model protein antigens. J Biol Chem 280:13406–13414
Jorgensen L, Vermehren C, Bjerregaard S et al (2003) Secondary structure alterations in insulin and growth hormone water-in-oil emulsions. Int J Pharm 254:7–10
Jorgensen L, van De Weert M, Vermehren C et al (2004) Probing structural changes of proteins incorporated into water-in-oil emulsions. J Pharm Sci 93:1847–1859
Kamath AT, Mastelic B, Christensen D et al (2012) Synchronization of dendritic cell activation and antigen exposure is required for the induction of Th1/Th17 responses. J Immunol 188: 4828–4837
Lai RPJ, Seaman MS, Tonks P et al (2012) Mixed adjuvant formulations reveal a new combination that elicit antibody response comparable to Freund's adjuvants. PLoS ONE 7:e35083

Lindblad EB (2004) Aluminium compounds for use in vaccines. Immunol Cell Biol 82:497–505
Lindblad EB, Schonberg NE (2010) Aluminum adjuvants: preparation, application, dosage, and formulation with antigen. Methods Mol Biol 626:41–58
Lu FJ, Boutselis I, Borch RF et al (2013) Control of antigen-binding to aluminum adjuvants and the immune response with a novel phosphonate linker. Vaccine 31:4362–4367
Miles AP, McClellan HA, Rausch KM et al (2005) Montanide((R)) ISA 720 vaccines: quality control of emulsions, stability of formulated antigens, and comparative immunogenicity of vaccine formulations. Vaccine 23:2530–2539
Morel S, Didierlaurent A, Bourguignon P et al (2011) Adjuvant System AS03 containing alpha-tocopherol modulates innate immune response and leads to improved adaptive immunity. Vaccine 29:2461–2473
Mortellaro A, Ricciardi-Castagnoli P (2011) From vaccine practice to vaccine science: the contribution of human immunology to the prevention of infectious disease. Immunol Cell Biol 89:332–339
Mulder AM, Carragher B, Towne V et al (2012) Toolbox for non-intrusive structural and functional analysis of recombinant VLP based vaccines: a case study with hepatitis B vaccine. PLoS ONE 7:e33235
Ott G, Radhakrishnan R, Fang JH et al (2000) The adjuvant MF59: a 10-year perspective. In: O'Hagan (ed) Vaccine adjuvants, vol 42, Methods in molecular medicine. Humana Press Inc., Tayowa, pp 211–228
Regnier M, Metz B, Tilstra W et al (2012) Structural perturbation of diphtheria toxoid upon adsorption to aluminium hydroxide adjuvant. Vaccine 30:6783–6788
Rinella JV, White JL, Hem SL (1998a) Effect of pH on the elution of model antigens from aluminum-containing adjuvants. J Colloid Interface Sci 205:161–165
Rinella JV, Workman RF, Hermodson MA et al (1998b) Elutability of proteins from aluminum-containing vaccine adjuvants by treatment with surfactants. J Colloid Interface Sci 197:48–56
Seeber SJ, White JL, Hem SL (1991) Solubilization of aluminum-containing adjuvants by constituents of interstitial fluid. J Parenter Sci Technol 45:156–159
Thai R, Moine G, Desmadril M et al (2004) Antigen stability controls antigen presentation. J Biol Chem 279:50257–50266
Tynan FE, Elhassen D, Purcell AW et al (2005) The immunogenicity of a viral cytotoxic T cell epitope is controlled by its MHC-bound conformation. J Exp Med 202:1249–1260
Volkin DB, Middaugh CR (2010) Vaccines as physically and chemically well-defined pharmaceutical dosage forms. Expert Rev Vaccines 9:689–691
Wagner L, Verma A, Meade BD et al (2012) Structural and immunological analysis of anthrax recombinant protective antigen adsorbed to aluminum hydroxide adjuvant. Clin Vaccine Immunol 19:1465–1473
Wilde PJ (2000) Interfaces: their role in foam and emulsion behaviour. Curr Opin Colloid Int 5:176–181
Xue XY, Ding F, Zhang QF et al (2010) Stability and potency of the Plasmodium falciparum MSP1-19/AMA-1 (III) chimeric vaccine candidate with Montanide ISA720 adjuvant. Vaccine 28:3152–3158
Yanasarn N, Sloat BR, Cui Z (2011) Negatively charged liposomes show potent adjuvant activity when simply admixed with protein antigens. Mol Pharm 8:1174–1185
Zheng YW, Lai XX, Ipsen H et al (2007a) Structural changes of protein antigens due to adsorption onto and release from aluminium hydroxide using FTIR-ATR. Spectroscopy 21:211–226
Zheng YW, Lai XX, Ipsen H et al (2007b) The structural stability of protein antigens adsorbed by aluminium hydroxide in comparison to the antigens in solutions. Spectroscopy 21:257–268
Zhu DM, McClellan H, Dai WL et al (2011) Long term stability of a recombinant Plasmodium falciparum AMA1 malaria vaccine adjuvanted with Montanide (R) ISA 720 and stabilized with glycine. Vaccine 29:3640–3645

Index

C. Foged et al. (eds.), *Subunit Vaccine Delivery*, Advances in Delivery Science and Technology, DOI 10.1007/978-1-4939-1417-3

MIX
Papier aus verantwortungsvollen Quellen
Paper from responsible sources
FSC® C105338

If you have any concerns about our products,
you can contact us on
ProductSafety@springernature.com

In case Publisher is established outside the EU,
the EU authorized representative is:
Springer Nature Customer Service Center GmbH
Europaplatz 3, 69115 Heidelberg, Germany

Printed by Libri Plureos GmbH
in Hamburg, Germany